WITHDRAWN BY THE
UNIVERSITY OF MICHIGAN

Applied Mathematical Sciences
Volume 157

Editors
S.S. Antman J.E. Marsden L. Sirovich

Advisors
J.K. Hale P. Holmes J. Keener
J. Keller B.J. Matkowsky A. Mielke
C.S. Peskin K.R. Sreenivasan

Springer
New York
Berlin
Heidelberg
Hong Kong
London
Milan
Paris
Tokyo

Applied Mathematical Sciences

1. *John:* Partial Differential Equations, 4th ed.
2. *Sirovich:* Techniques of Asymptotic Analysis.
3. *Hale:* Theory of Functional Differential Equations, 2nd ed.
4. *Percus:* Combinatorial Methods.
5. *von Mises/Friedrichs:* Fluid Dynamics.
6. *Freiberger/Grenander:* A Short Course in Computational Probability and Statistics.
7. *Pipkin:* Lectures on Viscoelasticity Theory.
8. *Giacaglia:* Perturbation Methods in Non-linear Systems.
9. *Friedrichs:* Spectral Theory of Operators in Hilbert Space.
10. *Stroud:* Numerical Quadrature and Solution of Ordinary Differential Equations.
11. *Wolovich:* Linear Multivariable Systems.
12. *Berkovitz:* Optimal Control Theory.
13. *Bluman/Cole:* Similarity Methods for Differential Equations.
14. *Yoshizawa:* Stability Theory and the Existence of Periodic Solution and Almost Periodic Solutions.
15. *Braun:* Differential Equations and Their Applications, 3rd ed.
16. *Lefschetz:* Applications of Algebraic Topology.
17. *Collatz/Wetterling:* Optimization Problems.
18. *Grenander:* Pattern Synthesis: Lectures in Pattern Theory, Vol. I.
19. *Marsden/McCracken:* Hopf Bifurcation and Its Applications.
20. *Driver:* Ordinary and Delay Differential Equations.
21. *Courant/Friedrichs:* Supersonic Flow and Shock Waves.
22. *Rouche/Habets/Laloy:* Stability Theory by Liapunov's Direct Method.
23. *Lamperti:* Stochastic Processes: A Survey of the Mathematical Theory.
24. *Grenander:* Pattern Analysis: Lectures in Pattern Theory, Vol. II.
25. *Davies:* Integral Transforms and Their Applications, 2nd ed.
26. *Kushner/Clark:* Stochastic Approximation Methods for Constrained and Unconstrained Systems.
27. *de Boor:* A Practical Guide to Splines: Revised Edition.
28. *Keilson:* Markov Chain Models—Rarity and Exponentiality.
29. *de Veubeke:* A Course in Elasticity.
30. *Sniatycki:* Geometric Quantization and Quantum Mechanics.
31. *Reid:* Sturmian Theory for Ordinary Differential Equations.
32. *Meis/Markowitz:* Numerical Solution of Partial Differential Equations.
33. *Grenander:* Regular Structures: Lectures in Pattern Theory, Vol. III.
34. *Kevorkian/Cole:* Perturbation Methods in Applied Mathematics.
35. *Carr:* Applications of Centre Manifold Theory.
36. *Bengtsson/Ghil/Källén:* Dynamic Meteorology: Data Assimilation Methods.
37. *Saperstone:* Semidynamical Systems in Infinite Dimensional Spaces.
38. *Lichtenberg/Lieberman:* Regular and Chaotic Dynamics, 2nd ed.
39. *Piccini/Stampacchia/Vidossich:* Ordinary Differential Equations in $\mathbf{R}^n$.
40. *Naylor/Sell:* Linear Operator Theory in Engineering and Science.
41. *Sparrow:* The Lorenz Equations: Bifurcations, Chaos, and Strange Attractors.
42. *Guckenheimer/Holmes:* Nonlinear Oscillations, Dynamical Systems, and Bifurcations of Vector Fields.
43. *Ockendon/Taylor:* Inviscid Fluid Flows.
44. *Pazy:* Semigroups of Linear Operators and Applications to Partial Differential Equations.
45. *Glashoff/Gustafson:* Linear Operations and Approximation: An Introduction to the Theoretical Analysis and Numerical Treatment of Semi-Infinite Programs.
46. *Wilcox:* Scattering Theory for Diffraction Gratings.
47. *Hale/Magalhães/Oliva:* Dynamics in Infinite Dimensions, 2nd ed.
48. *Murray:* Asymptotic Analysis.
49. *Ladyzhenskaya:* The Boundary-Value Problems of Mathematical Physics.
50. *Wilcox:* Sound Propagation in Stratified Fluids.
51. *Golubitsky/Schaeffer:* Bifurcation and Groups in Bifurcation Theory, Vol. I.
52. *Chipot:* Variational Inequalities and Flow in Porous Media.
53. *Majda:* Compressible Fluid Flow and Systems of Conservation Laws in Several Space Variables.
54. *Wasow:* Linear Turning Point Theory.
55. *Yosida:* Operational Calculus: A Theory of Hyperfunctions.
56. *Chang/Howes:* Nonlinear Singular Perturbation Phenomena: Theory and Applications.
57. *Reinhardt:* Analysis of Approximation Methods for Differential and Integral Equations.
58. *Dwoyer/Hussaini/Voigt (eds):* Theoretical Approaches to Turbulence.
59. *Sanders/Verhulst:* Averaging Methods in Nonlinear Dynamical Systems.

(continued following index)

Tomasz Kaczynski   Konstantin Mischaikow
Marian Mrozek

Computational Homology

With 78 Figures

Tomasz Kaczynski
Department of Mathematics
and Computer Science
University of Sherbrooke
Quebec J1K 2R1
Canada
kaczyn@dmi.usherb.ca

Konstantin Mischaikow
School of Mathematics
Georgia Institute of
Technology
Atlanta, GA 30332-0160
USA
mischaik@math.gatech.edu

Marian Mrozek
Institute of Computer
Science
Jagiellonian University
ul. Nawojki 11
31-072 Kraków
Poland
mrozek@ii.uj.edu.pl

Editors:
S.S. Antman
Department of Mathematics
and
Institute for Physical Science
and Technology
University of Maryland
College Park, MD 20742-4015
USA
ssa@math.umd.edu

J.E. Marsden
Control and Dynamical
Systems, 107-81
California Institute of
Technology
Pasadena, CA 91125
USA
marsden@cds.caltech.edu

L. Sirovich
Division of Applied
Mathematics
Brown University
Providence, RI 02912
USA
chico@camelot.mssm.edu

Mathematics Subject Classification (2000): 55-02, 55M20, 55M25, 37B10, 37B30, 37M99, 68T10, 68U10

Library of Congress Cataloging-in-Publication Data
Kaczynski, Tomasz.
   Computational homology / Tomasz Kaczynski, Konstantin Mischaikow, Marian Mrozek.
     p. cm.
   Includes bibliographical references and index.
   ISBN 0-387-40853-3 (acid-free paper)
    1. Homology theory.   I. Mischaikow, Konstantin Michael. II. Mrozek, Marian. III. Title.
   QA612.3.K33 2003
   514'.23—dc22   2003061109

ISBN 0-387-40853-3   Printed on acid-free paper.

© 2004 Springer-Verlag New York, Inc.
All rights reserved. This work may not be translated or copied in whole or in part without the written permission of the publisher (Springer-Verlag New York, Inc., 175 Fifth Avenue, New York, NY 10010, USA), except for brief excerpts in connection with reviews or scholarly analysis. Use in connection with any form of information storage and retrieval, electronic adaptation, computer software, or by similar or dissimilar methodology now known or hereafter developed is forbidden.
The use in this publication of trade names, trademarks, service marks, and similar terms, even if they are not identified as such, is not to be taken as an expression of opinion as to whether or not they are subject to proprietary rights.

Printed in the United States of America.   (MVY)

9 8 7 6 5 4 3 2 1   SPIN 10951408

Springer-Verlag is a part of *Springer Science+Business Media*

springeronline.com

To Claude,
        Françoise,
                and Joanna

Preface

At the turn of the 20th century, mathematicians had a monopoly on well-defined complicated global problems, such as celestial mechanics, fixed points of high-dimensional nonlinear functions, the geometry of level sets of differentiable functions, algebraic varieties, distinguishing topological spaces, etc. Algebraic topology, of which homology is a fundamental part, was developed in response to such challenges and represents one of the great achievements of 20th-century mathematics. While its roots can be traced to the middle of the 18th-century with Euler's famous formula that for the surface of a convex polyhedron

$$\text{faces} - \text{edges} + \text{vertices} = 2,$$

it is fair to say that it began as a subject in its own right in the seminal works of Henri Poincaré on "Analysis Situs." Though Poincaré was motivated by analytic problems, as his techniques developed they took on a combinatorial form similar in spirit to Euler's formula.

The power of algebraic topology lies in its coarseness. To understand this statement, consider Euler's formula. Observe, for instance, that the size of the polyhedron is of no importance. In particular, therefore, small changes in the shape of the polyhedron do not alter the formula. On the other hand, if one begins with a polyhedron that is punctured by k holes as opposed to a convex polyhedron, then the formula becomes

$$\text{faces} - \text{edges} + \text{vertices} = 2 - 2k.$$

As a result, counting local objects—the faces, edges, and vertices—of a polyhedron allows us to determine a global property, how many holes it has. Furthermore, it shows that formulas of this form can be used to distinguish objects with important different geometric properties.

The potential of Poincaré's revolutionary ideas for dealing with global problems was quickly recognized, and this led to a broad development of the subject. However, as is to be expected, the form of development matched the problems of interest. As indicated above, a typical question might be concerned with the structure of the level set of a differentiable function. Solving such problems using purely combinatorial arguments suggested by Euler's formula, that is, cutting the set into a multitude of small pieces and then counting, is in general impractical. This led to the very formidable and powerful algebraic machinery that is now referred to as algebraic topology. In its simplest form this tool takes objects defined in terms of traditional mathematical formulas and produces algebraic invariants that provide fundamental information about geometric properties of the objects.

As we begin the 21st century, complexity has spread beyond the realm of mathematics. With the advent of computers and sophisticated sensing devices, scientists, engineers, doctors, social scientists, and business people all have access to, or through numerical simulation can create, huge data files. Furthermore, for some of these data sets the crucial information is geometric in nature, but it makes little sense to think of these geometric objects as being presented in terms of or derived from traditional mathematical formulas. As an example, think of medical imaging. Notice that even though the input is different, the problems remain the same: identifying and classifying geometric properties or abnormalities. Furthermore, inherent in numerical or experimental data is error. What is needed is a framework in which geometrical objects can be recognized even in the presence of small perturbations.

Hopefully these arguments suggest that the extraordinary success of algebraic topology in the traditional domains of mathematics can be carried over to this new set of challenges. However, to do so requires the ability to efficiently compute the algebraic topological quantities starting with experimental or numerical data—information that is purely combinatorial in nature.

The purpose of this book is to present a computational approach to homology with the hope that such a theory will prove beneficial to the analysis and understanding of today's complex geometric challenges. Naturally this means that our intended audience includes computer scientists, engineers, experimentalists, and theoreticians in nonlinear science. As such we have tried to keep the mathematical prerequisites to an absolute minimum. At the same time we, the authors, are mathematicians and proud of our trade. We believe that the most significant applications of the theory will be realized by those who understand the fundamental concepts of the theory. Therefore, we have insisted on a rigorous development of the subject. Thus this book can also be used as an introductory text in homology for mathematics students. It differs from the traditional introductory topology books in that a great deal of effort is spent discussing the computational aspects of the subject.

The broad range of background and interests of the intended readers leads to organizational challenges. With this in mind we have tried on a variety of levels to make the book as modular as possible. On the largest scale the book is divided into three parts: Part I, which contains the core material on computational homology; Part II, which describes applications and extensions; and Part III, which contains a variety of preliminary material. These parts are described in greater detail below. There is, however, another natural division—the homology of spaces and the homology of maps—with different potential applications. Cognizant of the fact that the applications are of primary interest to some readers, we have attempted to organize the chapters in Parts I and II along these lines. Thus, as is indicated in Figure 1, a variety of options concern the order in which the material can be read.

We have already argued that algebraic topology holds tremendous potential for applications, but homology is also a beautiful subject in that topology is transformed into algebra and from the algebra one can recover aspects of the topology. However, this process involves some deep ideas and it is easy to lose track of the big picture in the midst of the mathematical technicalities. With this in mind we begin Part I with a preview both to the applications and to the homology of spaces. The ideas are sketched through very simple examples and without too much concern for rigor. Since the homology of maps depends deeply on the homology of spaces, we postpone a preview of this subject to Chapter 5.

As mentioned above, Part I contains the core material. It provides a rigorous introduction to the homology of spaces and continuous functions and is meant to be read sequentially. In Chapter 2 we define homology and investigate its most elementary properties. In particular, we explain how to each topological space we can assign a sequence of abelian groups called the homology groups of the space. There is a caveat: that the topological spaces we consider must be built out of d-dimensional unit cubes with vertices on the integer lattice. This is in contrast to the standard combinatorial approach, which is based on simplices. There are two reasons for this. The first comes from applications. Consider digital images. The basic building blocks are pixels, which are easily identified with squares. Similarly, an experimental or numerically generated data point comes with errors. Thus a d-dimensional data point can be thought of as lying inside a d-dimensional cube whose width is determined by the error bounds. The second reason—which will be made clear in the text—has to do with the simplicity of algorithms.

In Chapter 3 we show that homology is computable by presenting in detail an algorithm based on linear algebra over the integers. This is essential because it demonstrates that the homology group of a topological space made up of cubes is computable. However, for spaces made up of many cubes, the algorithm is of little immediate practical value. Therefore, in Chapter 4 we introduce combinatorial techniques for reducing the number of elements involved in the computation.

The contents of Chapter 4 also naturally foreshadow questions concerning maps between topological spaces and maps between the associated homology groups. The construction of homology maps is done in Chapter 6. We approach this problem from the viewpoint of multivalued maps. This has an extremely important consequence. One can efficiently compute homology maps—an absolutely essential feature given the purposes of this endeavor. We know of no practical algorithms for producing simplicial maps that approximate continuous maps. Algorithms for the computation of homology maps are discussed in Chapter 7.

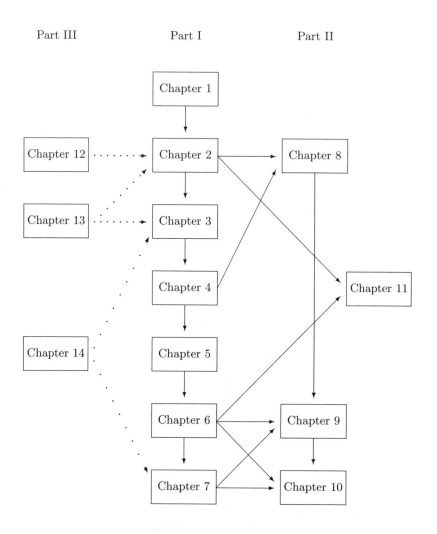

Fig. 1. Chapter dependence chart.

As mentioned earlier, we have attempted to keep the necessary prerequisites for this book to a minimum. Ideally the reader would have had an introductory course in point-set topology, an abstract algebra course, and familiarity with computer algorithms. On the other hand, such a reader would know far more topology, algebra, and computer science than is really necessary. Furthermore, the union of these subjects in the typical curriculum of our desired readers is probably fairly rare. For this reason we have included in Part III brief introductions to these subjects. Perhaps a more traditional organization would have placed Chapters 12, 13, and 14 at the beginning of the book. However, it is our opinion that nothing kills the excitement of studying a new subject as rapidly as the idea of multiple chapters of preliminary material. We suggest that the reader begins with Chapter 1 and consults the last three chapters on a need-to-know basis.

We argue at the beginning of this preface that algebraic topology has an important role to play in the analysis of numerical and experimental data. In Part II we elaborate on these ideas. It should be mentioned that applications of homology to these types of problems is a fairly new idea and still in a fairly primitive state. Thus the focus of this part is on conveying the potential rather than elaborate applications. We begin in Chapter 8 with a discussion that relates cubical complexes to image data and numerically generated data. Even the simple examples presented here suggest the need for more sophisticated ideas from homology. Therefore, Chapter 9 introduces more sophisticated algebraic concepts and computational techniques that are at the heart of homological algebra. In Chapter 10 we indicate how homology can be used to study nonlinear dynamics. In some sense this is one of the most traditional applications of the subject—the focus of much of Poincaré's work was related to differential equations and, in fact, he set the foundations for what is known today as dynamical systems. The modern twist is that these ideas can be tied to numerical methods to obtain computer-assisted proofs in this subject. Finally, though the cubical theory presented in this book has concrete advantages, it is also in many ways too rigid. In Chapter 11 we extend the cubical theory to polyhedra, thereby tying the results and techniques of this book to the enormous body of work known as algebraic topology.

Preliminary versions of this book have been used by the authors to teach one-semester or full-year courses at both the undergraduate and graduate levels and for a mixed audience of mathematics, physics, computer science, and engineering students. As mentioned several times, Part I contains the essential material. It can be covered in a single quarter, but at the expense of any applications or extensions presented in Part II. On the other hand, as is indicated in Figure 1, the material in Chapter 8 is conceptually accessible immediately following Chapter 2. However, the description of the computational techniques used to analyze the images in Chapter 8 is presented in Chapters 3 and 4. The first two sections of Chapter 11 concerning simplicial complexes can be read right after (or even parallel with) Chapter 2, but the last section on the homology functor can only be attained after reading Chapter 6. Similarly, Sections 10.4 and 10.5 of Chapter 10 can be read following Chapter 6. However, the material of Sections 10.6 and 10.7 depends heavily on topics in Chapter 9.

This book provides the conceptual background for computational homology. However, interesting applications also require efficient code, which as one might expect, is evolving rapidly. Software and examples are available at the *Computational Homology Program* (CHomP) web site, accessible via *www.springeronline.com*. This site also contains errata and additional links.

Preparing this book has been an exciting and challenging project. The material presented here is a unique combination of current research and classical mathematics that fundamentally depends on an interplay among mathematical rigor, computation, and application. We could not have completed this project without the support and assistance of numerous individuals.

Results presented here for the first time have been shaped by our collaboration with students and colleagues: Madjid Allili, Sarah Day, Marcio Gameiro, Bill Kalies, Howard Karloff, Paweł Pilarczyk, Andrzej Szymczak, and Thomas Wanner. We would like also to thank Stanisław Sędziwy for his continuous encouragement to write this book.

As mentioned above, earlier versions of this text have been used in courses and seminars. The feedback from participants too numerous to mention has led to its greatly improved current form. In particular, we thank Philippe Barbe, Bogdan Batko, Sylvain Bérubé, David Corriveau, Anna Danielewska, Rob Ghrist, Michał Jaworski, Janusz Mazur, Todd Moeller, Stephan Siegmund, David Smith, Krzysztof Szyszkiewicz-Warzecha, and Anik Trahan for detailed comments on significant portions of the text.

This project has been supported in part by the NSF USA, NSERC Canada, FCAR Québec, KBN of Poland,[1] CRM Montreal, Faculty of Science of Université de Sherbrooke, the Center for Dynamical Systems and Nonlinear Studies at Georgia Tech, and the Department of Computer Science, Jagiellonian University. Furthermore, we would also like to express thanks for the opportunity to teach this material when it was still at a fairly experimental stage. For this we thank Fred Andrew and Richard Duke. We would never have finished the necessary corrections to the galley proofs in the allotted time were it not for the efficient and thorough assistance of Annette Rohrs.

Finally, we are most deeply indebted to our families (who for some inexplicable reason don't share our enthusiasm for the intricacies of computational homology) for putting up with shortened vacations and our absence during late nights, long weekends, and extended travel.

Sherbrooke, Quebec, Canada *Tomasz Kaczynski*
Atlanta, Georgia, USA *Konstantin Mischaikow*
Kraków, Poland *Marian Mrozek*
June 2003

---

[1] KBN, Grant No. 2 P03A 011 18 and 2 P03A 041 24.

Contents

Preface VII

Part I Homology

1 Preview 3
   1.1 Analyzing Images 3
   1.2 Nonlinear Dynamics 13
   1.3 Graphs 17
   1.4 Topological and Algebraic Boundaries 19
   1.5 Keeping Track of Directions 24
   1.6 Mod 2 Homology of Graphs 26

2 Cubical Homology 39
   2.1 Cubical Sets 39
      2.1.1 Elementary Cubes 40
      2.1.2 Cubical Sets 42
      2.1.3 Elementary Cells 44
   2.2 The Algebra of Cubical Sets 47
      2.2.1 Cubical Chains 47
      2.2.2 Cubical Chains in a Cubical Set 53
      2.2.3 The Boundary Operator 54
      2.2.4 Homology of Cubical Sets 60
   2.3 Connected Components and $H_0(X)$ 66
   2.4 Elementary Collapses 70
   2.5 Acyclic Cubical Spaces 79
   2.6 Homology of Abstract Chain Complexes 85
   2.7 Reduced Homology 88
   2.8 Bibliographical Remarks 91

3 Computing Homology Groups 93
- 3.1 Matrix Algebra over **Z** 94
- 3.2 Row Echelon Form 107
- 3.3 Smith Normal Form 117
- 3.4 Structure of Abelian Groups 125
- 3.5 Computing Homology Groups 132
- 3.6 Computing Homology of Cubical Sets 134
- 3.7 Preboundary of a Cycle—Algebraic Approach 139
- 3.8 Bibliographical Remarks 141

4 Chain Maps and Reduction Algorithms 143
- 4.1 Chain Maps 143
- 4.2 Chain Homotopy 149
- 4.3 Internal Elementary Reductions 155
  - 4.3.1 Elementary Collapses Revisited 155
  - 4.3.2 Generalization of Elementary Collapses 157
- 4.4 CCR Algorithm 165
- 4.5 Bibliographical Remarks 171

5 Preview of Maps 173
- 5.1 Rational Functions and Interval Arithmetic 174
- 5.2 Maps on an Interval 176
- 5.3 Constructing Chain Selectors 185
- 5.4 Maps of Γ^1 189

6 Homology of Maps 199
- 6.1 Representable Sets 199
- 6.2 Cubical Multivalued Maps 206
- 6.3 Chain Selectors 210
- 6.4 Homology of Continuous Maps 215
  - 6.4.1 Cubical Representations 216
  - 6.4.2 Rescaling 222
- 6.5 Homotopy Invariance 231
- 6.6 Bibliographical Remarks 234

7 Computing Homology of Maps 235
- 7.1 Producing Multivalued Representation 236
- 7.2 Chain Selector Algorithm 240
- 7.3 Computing Homology of Maps 242
- 7.4 Geometric Preboundary Algorithm (optional section) 244
- 7.5 Bibliographical Remarks 253

Part II Extensions

8 Prospects in Digital Image Processing 257
    8.1 Images and Cubical Sets 257
    8.2 Patterns from Cahn–Hilliard 259
    8.3 Complicated Time-Dependent Patterns..................... 266
    8.4 Size Function 269
    8.5 Bibliographical Remarks 277

9 Homological Algebra 279
    9.1 Relative Homology 279
        9.1.1 Relative Homology Groups 279
        9.1.2 Maps in Relative Homology......................... 286
    9.2 Exact Sequences... 289
    9.3 The Connecting Homomorphism 292
    9.4 Mayer–Vietoris Sequence 299
    9.5 Weak Boundaries.. 303
    9.6 Bibliographical Remarks 306

10 Nonlinear Dynamics 307
    10.1 Maps and Symbolic Dynamics............................ 308
    10.2 Differential Equations and Flows 318
    10.3 Ważewski Principle 320
    10.4 Fixed-Point Theorems 324
        10.4.1 Fixed Points in the Unit Ball 324
        10.4.2 The Lefschetz Fixed-Point Theorem 326
    10.5 Degree Theory 332
        10.5.1 Degree on Spheres 333
        10.5.2 Topological Degree 336
    10.6 Complicated Dynamics.................................. 342
        10.6.1 Index Pairs and Index Map 343
        10.6.2 Topological Conjugacy 357
    10.7 Computing Chaotic Dynamics............................ 361
    10.8 Bibliographical Remarks 375

11 Homology of Topological Polyhedra 377
    11.1 Simplicial Homology 378
    11.2 Comparison of Cubical and Simplicial Complexes 385
    11.3 Homology Functor...................................... 388
        11.3.1 Category of Cubical Sets 389
        11.3.2 Connected Simple Systems 390
    11.4 Bibliographical Remarks 393

Part III Tools from Topology and Algebra

12 Topology 397
    12.1 Norms and Metrics in $\mathbf{R}^d$ 397
    12.2 Topology 402
    12.3 Continuous Maps 407
    12.4 Connectedness 411
    12.5 Limits and Compactness 415

13 Algebra 419
    13.1 Abelian Groups 419
        13.1.1 Algebraic Operations 419
        13.1.2 Groups 420
        13.1.3 Cyclic Groups and Torsion Subgroup 422
        13.1.4 Quotient Groups 424
        13.1.5 Direct Sums 426
    13.2 Fields and Vector Spaces 427
        13.2.1 Fields 427
        13.2.2 Vector Spaces 429
        13.2.3 Linear Combinations and Bases 430
    13.3 Homomorphisms 433
        13.3.1 Homomorphisms of Groups 433
        13.3.2 Linear Maps 437
        13.3.3 Matrix Algebra 438
    13.4 Free Abelian Groups 441
        13.4.1 Bases in Groups 441
        13.4.2 Subgroups of Free Groups 446
        13.4.3 Homomorphisms of Free Groups 447

14 Syntax of Algorithms 451
    14.1 Overview 451
    14.2 Data Structures 453
        14.2.1 Elementary Data Types 453
        14.2.2 Lists 454
        14.2.3 Arrays 455
        14.2.4 Vectors and Matrices 456
        14.2.5 Sets 457
        14.2.6 Hashes 458
    14.3 Compound Statements 459
        14.3.1 Conditional Statements 459
        14.3.2 Loop Statements 459
        14.3.3 Keywords **break** and **next** 460
    14.4 Function and Operator Overloading 461
    14.5 Analysis of Algorithms 462

References 465

Symbol Index 471

Subject Index 475

Part I

Homology

1
Preview

Homology is a very powerful tool in that it allows one to draw conclusions about *global* properties of spaces and maps from *local* computations. It also involves a wonderful mixture of algebra, combinatorics, computation, and topology. Each of these subjects is, of course, interesting in its own right and appears as the subject of multiple sections in this book. But our primary objective is to see how they can be combined to produce homology, how homology can be computed efficiently, and how homology provides us with information about the geometry and topology of nonlinear objects and functions. Given the amount of theory that needs to be developed, it is easy to lose sight of these objectives along the way.

Therefore, we begin with a preview. We start by considering applications of homology and then provide a heuristic introduction to the underlying mathematical ideas of the subject. The point of this chapter is to provide intuition for the big picture. As such we will, without explanation, introduce some fundamental terminology with the expectation that the reader will remember the words. Precise definitions and mathematical details are provided in the rest of the book.

1.1 Analyzing Images

It is hard to think of a scientific or engineering discipline that does not generate computational simulations or make use of recording devices or sensors to produce or collect image data. It is trivial to record simple color videos of events, but it should be noted that such a recording can easily require about 25 megabytes of data per second. While obviously more difficult, it is possible, using X-ray computed tomography, to visualize cardiovascular tissue with a resolution on the order of 10 μm. Because this can be done at a high speed, timed sequences of three-dimensional images can be constructed. This technique can be used to obtain detailed information about the geometry and function of the heart, but it entails large amounts of data. Obviously, the size and complexity of this data will grow as the sophistication of the sensors or simulations increases.

These large amounts of data are a mixed blessing; while we can be more confident that the desired information is captured, extracting the relevant information in a sea of data can become more difficult. One solution is to develop automated methods for image processing. These techniques are often, if somewhat artificially, separated into two categories: *low-level vision* and *high-level vision*. Typical examples of the latter include object recognition, optical character and handwriting recognizers, and robotic control by visual feedback. Low-level vision, on the other hand, focuses on the geometric structures of the objects being imaged. As such it often is a first step toward higher-level tasks. It is our belief that computational homology has the potential to play an important role in low-level vision. Notice the phrasing of this last sentence: The use of algebraic topology in imaging is, at the time that this book is being written, a very new subject. We hope that this work will open doors to exciting endeavors.

Let us begin by using some extremely simple figures to get a feel for what homology measures. Consider the line segments in Figure 1.1(a) and (b). For a topologist the most important distinguishing properties of these figures is not that they are made up of straight line segments, nor that the lengths of the line segments are different, but rather that in Figure 1.1(a) we have an object that consists of one *connected component* and in Figure 1.1(b) the object consists of two distinct pieces. Homology provides us with a means of measuring this. In particular, if we call the object in Figure 1.1(a) X and the object in Figure 1.1(b) Y, then the *zeroth homology groups* of these figures are

$$H_0(X) \cong \mathbf{Z} \quad \text{and} \quad H_0(Y) \cong \mathbf{Z}^2.$$

We will explain later what this means or how it is computed. For the moment it is sufficient to know that homology allows us to assign to a topological space (e.g., X or Y) an *abelian group* and that the dimension of this group counts the number of distinct pieces that make up the space.

(a)   (b)

Fig. 1.1. (a) A line segment consisting of one connected component. (b) A divided line segment consisting of two connected components.

Given that there is a zeroth homology group, the reader might suspect that there is also a *first homology group*. This is correct and it measures different topological information. Again, let us consider some simple figures. In Figure 1.2(a) we see a line segment in the plane. If we think of this as a piece of fencing, it is clear that it does not enclose any region in the plane. On the other hand, Figure 1.2(b) clearly encloses the square $(0,1) \times (0,1)$. If we add more segments we can enclose more squares as in Figure 1.2(c), though, of course, some segments need not enclose any region. Finally, as indicated by the shaded square in Figure 1.2(d), by filling in some regions we can eliminate enclosed areas.

The first homology group measures the number of these enclosed regions and for each of these figures, denoted respectively by X_a, X_b, X_c, and X_d, it is as follows:

$$H_1(X_a) = 0, \quad H_1(X_b) \cong \mathbf{Z}, \quad H_1(X_c) \cong \mathbf{Z}^2, \quad \text{and} \quad H_1(X_d) \cong \mathbf{Z}.$$

Notice that even though the drawings in Figure 1.2(b)–(d) are more elaborate than those of Figure 1.1, the number of connected components is always one, thus

$$H_0(X_i) \cong \mathbf{Z}, \quad i = a, b, c, d.$$

One final, but extremely important, comment is that the homology of an object does not depend on the ambient space. Hence if we were to lift the objects of Figure 1.2 from the plane and think of them as existing in $\mathbf{R}^3$ or in any abstract higher-dimensional space, their homology groups would not change. After reading this book the reader will realize that the homologies of such different things as a garden hose and a coffee mug are the same as the homology of X_b. Actually, homology is not measuring size or even a specific shape; rather it measures very fundamental properties like the number of holes and pieces. We will return to this point shortly.

If it were not for the fact that we introduce the words "homology group" in our discussion of Figures 1.1 and 1.2, the previous comments would be completely trivial. So let us try to rephrase the statements in the form of a nontrivial question.

Can we develop a computationally efficient algebraic tool that tells us how many connected components and enclosed regions a geometric object contains?

For example, it would be nice to be able to enter the mazelike object of Figure 1.3 into a computer and have the computer tell us whether the figure consisting of all black line segments is connected and whether there are one or more enclosed white regions in the figure. By the end of Chapter 4 the reader will be able to do this and much more.

This last suggestion raises a fundamental question: What does it mean to enter a figure into the computer? Consider, for example, the left-hand image of Figure 1.4, which shows two circles that intersect. There are a variety of ways in which we could attempt to represent the circles. In a mathematics class they would typically be understood to be smooth curves, composed of uncountably many points. However, this picture was produced by a computer, and thus only a finite amount of information is presented. The right-hand image of

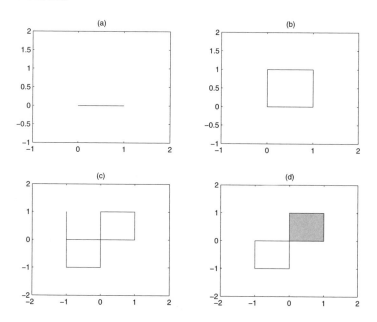

Fig. 1.2. (a) A simple line segment in the plane does not enclose any region. (b) These four line segments enclose the region $(0,1) \times (0,1)$. (c) It is easy to bound more than one region. (d) By filling in a region we can eliminate enclosed regions.

Figure 1.4 is obtained by zooming in on the upper point of the intersection of the circles. Observe that the smooth curves have become chains of small squares. These squares represent the smallest geometric units of information presented in the image on the left-hand side.

Given our goal of an automated method for image processing, it seems that these squares are a natural input for the computer. Hence we could enter the circles in Figure 1.4 into the computer by listing the black squares.

More interesting images are, of course, more complicated. Consider the photo in Figure 1.5 of the moon's surface in the Sea of Tranquillity taken from the Apollo 10 spacecraft. This is a black-and-white photo that, if rendered on a computer screen, would be presented as a rectangular set of elements each one of which is called a picture element, or a *pixel* for short. Each pixel has a specific light intensity determined by an integer gray-scale value between 0 and 255. This rendering captures the essential elements of a digital image: The image must be defined on a discretized domain (the array of pixels), and similarly the observed values of the image must lie in a discrete set (gray-scale values).

On the other hand, the Sea of Tranquillity is an analog rather than a digital object. Its physical presence is continuous in space and time. Furthermore, the visual intensity of the image that we would see if we were observing the moon directly also takes a continuum of values. Clearly, information is being lost

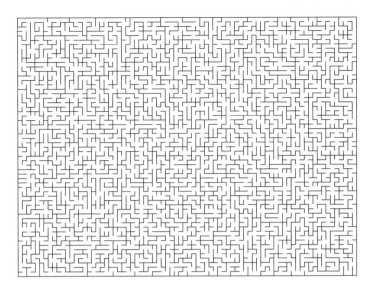

Fig. 1.3. How many distinct bounded regions are in this complicated maze?

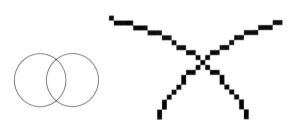

Fig. 1.4. On the left we have a simple image of two intersecting circles produced by a computer. On the right is a magnification of the upper point at which the circles intersect. Observe that the smooth curve is now a chain of squares that intersect either at a vertex or on a face.

in the process of describing an analog object in terms of digital information. This is an important point and is the focus of considerable research in the image processing community. However, these problems lie outside the scope of this book and, with the exception of Section 8.1 and occasional comments, will not be discussed further.

On a more positive note, we can use this simple example to indicate the value of being able to count "holes." One of the more striking features of

8   1 Preview

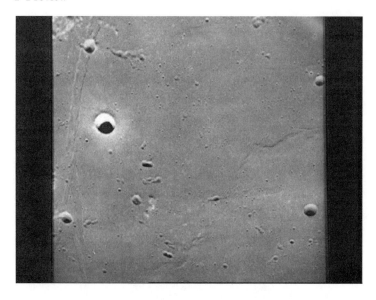

Fig. 1.5. Near-vertical photograph taken from the Apollo 10 Command and Service Modules shows features typical of the Sea of Tranquillity near Apollo Landing Site 2. The original is a 70-mm black-and-white photo (see the NASA web page http://images.jsc.nasa.gov/iams/images/pao/AS10/10075149.jpg).

Figure 1.5 is the craters and the natural question is: How many are there? To answer this we first need to decide which pixels represent the smooth surface and which represent the cratered surface. Thus we want to reduce this picture to a binary image that distinguishes between smooth and cratered surface areas.

The simplest approach for reducing a gray-scale image to a binary image is *image thresholding*. One chooses a range of gray-scale values $[T_0, T_1]$; all pixels whose values lie in $[T_0, T_1]$ are colored black while the others are left white. Of course, the resulting binary image depends greatly on the values of T_0 and T_1 that are chosen. Again, the question of the optimal choice of $[T_0, T_1]$ and the development of more sophisticated reduction techniques are serious problems that are not dealt with in this book.

Having acknowledged the simplistic approach we are adopting here, consider Figure 1.6. These binary images were obtained from Figure 1.5 as follows. Recall that in a gray-scale image each pixel has a value between 0 and 255, where 0 is black and 255 is white. The craters are darker in color (which corresponds to a lower gray-scale value). Since there is no absolute definition of which gray scales correspond to which craters, we choose two threshold intervals: $[95, 255]$ and $[105, 255]$. Pixels in theses ranges are taken to represent the smooth surface: the black pixels in Figure 1.6 correspond to the lightest pixels in Figure 1.5 and the white pixels to the darkest. In other words, the black region is indicative of the smooth surface and the white regions are craters. It should not come as a surprise that different thresholding intervals result in different binary images.

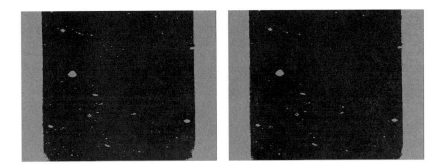

Fig. 1.6. The left figure was obtained from Figure 1.5 by choosing the threshold interval [95, 255] while the right figure corresponds to the threshold interval [105, 225].

Counting the number of holes (white regions bounded by black pixels) in the binary pictures provides an approximation of the number of craters in the picture. Observe that we are back to the problem motivated by Figures 1.2 and 1.3. Thus we want to be able to compute the first homology group for the object defined by the black pixels.

The examples presented so far, namely Figures 1.2 and 1.3 and the lunar photograph, were included—under the assumption that a picture saves a thousand words—to help develop intuition. The full mathematical machinery of homology is not necessary to analyze such simple images in the plane, and we do not recommend the material in this book for a reader whose only interest is in counting craters on the moon. For example, we could have identified the craters by choosing to threshold with the intervals [0, 95] and [0, 105] and then counting the number of connected pieces. This latter task requires no knowledge of homology. On the other hand, many physical problems are higher-dimensional, where our visual intuition fails and topological reasoning becomes crucial.

To choose a specific problem where computational topology has been employed, we turn to the subject of metallurgy and in particular to the work of [53, 39]. Consider a binary alloy consisting of iron (Fe) and chromium (Cr) created by heating the metals to a high temperature. The initial configuration of the iron and chromium atoms is essentially spatially homogeneous, up to small, random variations. However, upon cooling, the iron and chromium atoms separate, leading to a two-phase microstructure; that is, the material divides into regions that consist primarily of iron atoms or chromium atoms, but not both. These regions, which we will denote by $F(t)$ for iron and $C(t)$ for chromium, are obviously three-dimensional structures, can be extremely complicated, and furthermore, change their form with time t. Current technology allows for accurate three-dimensional measurements to be performed on the atomic level (essentially the material is serially sectioned and then examined with an atomic probe; see [53] for details). Thus these regions can be experimentally determined. Of course, for each sample the actual geometric structure of $F(t)$ and $C(t)$ will be different since the initial configurations of iron and chromium atoms are distinct. On the theoretical side there are mathematical models meant to describe this process of decomposition. The easiest to state mathematically takes the form of the Cahn–Hilliard equation

$$\frac{\partial u}{\partial t} = -\Delta(\epsilon^2 \Delta u + u - u^3), \quad x \in \Omega, \tag{1.1}$$

$$n \cdot \nabla u = n \cdot \nabla \Delta u = 0, \quad x \in \partial\Omega, \tag{1.2}$$

where n is the outward normal to $\partial\Omega$. Of particular interest is the case where $\epsilon > 0$ but small, since under this condition solutions to this equation produce complicated patterns. For example, Figure 1.7 contains a plot of the level set S defined by $u(x, y, z, \tau) = 0$ on the domain $\Omega = [0, 1]^3$ where $\epsilon = 0.1$. This was obtained by starting with a small but random initial condition $u_0(x, y, z)$ satisfying

$$\int\int\int_\Omega u_0(x,y,z)\,dx\,dy\,dz = 0.$$

Since (1.1) is a nonlinear partial differential equation, there is no hope of obtaining an analytic formula for the solution. Thus $u(x, y, z, \tau)$ was computed numerically on a grid consisting of $128 \times 128 \times 128$ cubical elements until time $t = \tau$. This means that $u(x, y, z, \tau)$ is approximated by a set of numbers $\{u(i, j, k, \tau) \mid 1 \leq i, j, k \leq 128\}$.

Returning to the issue of modelling alloys, the assumption is that positive values of u indicate higher density of one element and that negative values of u indicate higher density of the other element. More precisely, for some $\delta > 0$ but small, the region of one phase is given by $R_1(t) := \{x \in \Omega \mid u(x,t) > 1 - \delta\}$ and the other by $R_2(t) := \{x \in \Omega \mid u(x,t) < -1 + \delta\}$ (see [71] for a broad introduction to models of pattern formation). With appropriate modifications to (1.1) (see [53] for the details), one can then numerically simulate the iron–chromium alloy; in particular, one can try to compare $F(t)$ with $R_1(t)$ and $C(t)$ with $R_2(t)$.

At this point this problem's need for an algebraic measure of the topological structure can be made clear. Recall that in the material one starts with an essentially random configuration of iron and chromium atoms. To model this numerically, one begins with a random initial condition $u(x, 0)$ where $|u(x, 0)| < \mu$ for all $x \in \Omega$ and μ is small. Equation (1.1) is then solved until $t = t_0$. We now wish to compare $F(t_0)$ against $R_1(t_0)$ and $C(t_0)$ against $R_2(t_0)$. However, since the initial conditions are random, it makes no sense to demand that $F(t_0) = R_1(t_0)$ or even that they be close to one another. Instead we ask if they are similar in the sense of their topology. More precisely,

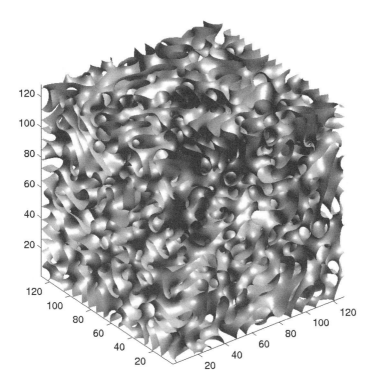

Fig. 1.7. A graphical rendering of the level set S defined by $u(x, y, z, \tau) = 0$. This was obtained by starting with a small but random initial condition on a grid consisting of $128 \times 128 \times 128$ elements and using a finite-element method numerically solving (1.1) until time $t = \tau$. It should be noted that the object is constructed by means of a triangulated surface using the data points $\{u(i, j, k, \tau) \mid 1 \leq i, j, k \leq 128\}$. This is a standard procedure in computer graphics as it produces an object that is much more pleasant to view. The homology groups for the triangulated surface S are $H_0(S) = \mathbf{Z}$, $H_1(S) = \mathbf{Z}^{1701}$, and $H_2(S) = 0$.

each of the regions $F(t_0)$, $C(t_0)$, or $R_i(t_0)$ can be made up of a multitude of components. There can be tunnels that pass through the regions and even hollow cavities.

We have suggested that the zeroth and first homology groups can be used to identify the number of components and tunnels. Thus we could try to compare

$$H_0(F(t)) \text{ versus } H_0(R_1(t)) \quad \text{and} \quad H_0(C(t)) \text{ versus } H_0(R_2(t))$$

and

$$H_1(F(t)) \text{ versus } H_1(R_1(t)) \quad \text{and} \quad H_1(C(t)) \text{ versus } H_1(R_2(t)).$$

As we saw with the examples of Figure 1.2, even if the objects are not identical in shape they can have the same homology groups.

Of course, these comparisons ignore the question of cavities. For this we need the *second homology group*; that is, we should compare

$$H_2(F(t)) \text{ versus } H_2(R_1(t)) \quad \text{and} \quad H_2(C(t)) \text{ versus } H_2(R_2(t)).$$

These types of comparisons are performed in [39]. However, the reader who consults [39] will not find the words "homology group" in the text. There are probably two reasons for this. First, this vocabulary is not common knowledge in the metallurgy community. Second, and more importantly, because the regions are three-dimensional, computational tricks can be employed to circumvent the need to explicitly compute the homology groups.

The intention of the last sentence by no means is to suggest that the reader who is interested in these kinds of problems can avoid learning homology theory. The tricks need to be justified, and the simplest justification makes essential use of algebraic topology. This dichotomy between the algebraic theory and the computational methods will be made clear in this book. After all, our goal is not only to compute homology groups, but to do so in an efficient manner. Thus in Chapter 3 we provide a purely algebraic algorithm for computing homology. This guarantees that homology groups are always computable. However, this method is extremely inefficient. Thus in Chapter 4 we introduce reduction algorithms that are combinatorial in nature and reasonably fast. Nevertheless, the justification of these latter algorithms depends crucially on a solid understanding of the algebraic theory.

In the next example, which is essentially four-dimensional, the tricks employed by [39] no longer work, and even on the level of language we can begin to appreciate the advantage of an abstract algebraic approach to the topic.

Figure 1.8 is a tomographic image of a horizontal slice of a human heart. Depending on the machine, several such images representing different cross sections can be taken simultaneously. Combining these two-dimensional images results in a three-dimensional image made up of three-dimensional cubes or *voxels*. As in the case of the lunar photo, a gray scale is assigned to each voxel. Assume that by using appropriate thresholding techniques we can identify those voxels that correspond to heart tissue. Then, using the language of the previous example, cavities could be identified with chambers and tunnels might indicate blood vessels, valves, or even defects such as holes in the heart.

However, medical technology allows us to go further. Multiple images can be obtained within the time span of a single heartbeat. Thus the full data set results in a four-dimensional object—three space dimensions and one time dimension—where the individual data elements are four-dimensional cubes or *tetrapus*. At this stage standard English begins to fail. As a simple example, consider a chamber of the heart at an instant of time when the valves are closed. In this case, the chamber is a cavity in a three-dimensional object. However, if we include time, then the valve will open and close so the three-dimensional cavity does not lead to a four-dimensional cavity. Which of the homology groups characterizes such a region? By the end of Chapter 2 the reader will know the answer.

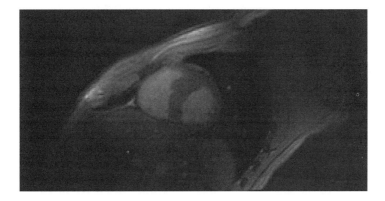

Fig. 1.8. A tomographic section of a human heart produced by the Surgical Planning Lab, Department of Radiology, Brigham and Women's Hospital.

1.2 Nonlinear Dynamics

In the previous section we present examples that suggest the need for algorithms to analyze the geometric structure of objects. We now turn to problems where it is important to have algorithms for studying nonlinear functions.

As motivation we turn to a simple model for population dynamics. Let y_n represent a population at time n. The simplest possible assumption is that y_{n+1}, the population one time unit later, is proportional to y_n. In this case $y_{n+1} = r y_n$, where r is the growth rate. An obvious problem with this model is that when $r > 1$, the population can grow to arbitrary sizes, which cannot happen, because the resources are limited. This issue can be avoided by assuming that the rate of growth is a decreasing function of the population. For simplicity let $r(y) = K - y$ for some constant K. Then the population at time $n + 1$ is given by

$$y_{n+1} = r(y_n) y_n = (K - y_n) y_n. \tag{1.3}$$

If we use the change of variables $x = y/K$, (1.3) takes the form

$$x_{n+1} = K x_n (1 - x_n).$$

Notice that the new variable x_n represents the scaled population level at time n.

To simplify the discussion, let us choose $K = 4$ and write

$$x_{n+1} = f(x_n) = 4x_n(1 - x_n).$$

A natural question is: Given an initial population level x_0, what are the future levels of the population? Notice that producing the sequence of population levels

$$x_0, x_1, x_2, \ldots, x_n, \ldots$$

is equivalent to iterating the function f. Clearly, one can write a simple program that performs such an operation. However, before doing so we wish to make two observations. First, because we are talking about populations, we are only interested in $x \geq 0$. Thus the model obviously has some flaws since if $x_0 > 1$, then $x_1 = f(x_0) < 0$. For this reason we will assume that $0 \leq x_0 \leq 1$. Second, $f([0,1]) = [0,1]$. So if we begin with the restricted initial conditions, then our population always remains nonnegative.

Figure 1.9 shows a sequence of populations $\{x_i \mid i = 0, \ldots, 100\}$ where $x_0 = 0.1$. One of the most striking features of this plot is the lack of a specific pattern. This raises a series of ever more difficult questions.

1. Do initial conditions exist for which the population is fixed, that is, $x_0 = x_1 = x_2 = \ldots$? Observe that this is equivalent to asking if there is a solution to the equation $f(x) = x$.
2. Do initial conditions exist that lead to *periodic orbits* of a given period? More specifically, given a positive integer k, does an initial condition x_0 exist such that $f^k(x_0) = x_0$ but $f^j(x_0) \neq x_0$ for all $j = 1, 2, \ldots, k-1$? In this case we would say that we have found a periodic orbit with *minimal period k*.
3. What is the set of all k for which there exists a periodic orbit with minimal period k? How many such orbits are there?
4. Are there many orbits that, like that of Figure 1.9, seem to have no predictable pattern? Are there any simple rules that such orbits must satisfy?

For this map the answer to the first question is obvious:

$$f(x) = x \quad \text{if and only if} \quad x = 0 \text{ or } x = \frac{3}{4}.$$

The astute reader will realize that, for any fixed k, finding a periodic orbit of that period is the same as solving $f^k(x) - x = 0$. But finding explicit solutions to $f^k(x) - x$ for $k > 2$ is a difficult problem. Moreover, in many applications one is forced to deal with a map $f : X \to X$, where either X is a potentially high-dimensional and/or complicated space, or f is not known explicitly; for example, f is the time one map of a differential equation and can be found only by numerical integration.

For this particular map, answers to even the third and fourth questions are reasonably well understood [73]. However, given a particular function $f : \mathbf{R}^n \to \mathbf{R}^n$, for $n \geq 2$, our knowledge of the dynamics of f most likely comes from numerical simulations. With this in mind, let us return to the numerically computed orbit of Figure 1.9 and ask ourselves if we can trust

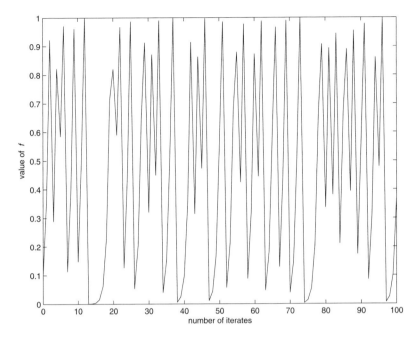

Fig. 1.9. A computed orbit for the logistic map $f(x) = 4x(1-x)$ with initial condition $x_0 = 0.1$.

the computation. Figure 1.10 shows two sequences of population levels. The circles indicate the sequence $\{x_i \mid i = 0, \ldots, 15\}$, where $x_0 = 0.1000$ and the stars represent $\{y_i \mid i = 0, \ldots, 15\}$, where $y_0 = 0.1001$. Observe that after 10 steps there is little correlation between the two sequences. In fact,

$$|x_{13} - y_{13}| \geq 0.9515.$$

Since the trajectories are forced to remain in the interval $[0,1]$, this effectively states that a mistake on the order of 10^{-4} leads to an error that is essentially the same size as the entire range of possible values. This kind of phenomenon is often referred to as chaotic dynamics. Since numerical computations induce errors merely by the fact that the computer is incapable of representing numbers to infinite precision, in a chaotic system any single computed trajectory is suspect.

Hopefully this discussion has demonstrated that numerical simulations of dynamical systems need to be treated with respect. What is probably not clear is how computational homology can be used. A precise answer is the subject of Chapter 10. For the moment we will have to settle for some suggestive comments.

Let $f : \mathbf{R} \to \mathbf{R}$ be continuous and let us try to determine if f has a fixed point, that is, if there is a solution to the problem $f(x) = x$. Observe that

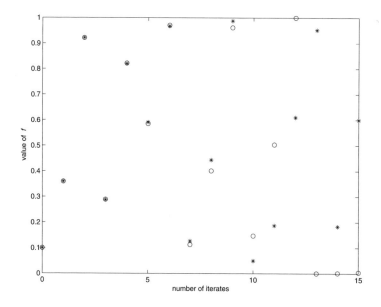

Fig. 1.10. Two computed orbits for the logistic map $f(x) = 4x(1-x)$. The $\circ$ and $*$ correspond to initial conditions $x_0 = 0.1$ and $y_0 = 0.1001$, respectively.

this is equivalent to showing that $f(x) - x = 0$. Let $g(x) = f(x) - x$, and assume that we determine that $g(a) < 0$ and $g(b) > 0$ for some $a < b$. By the intermediate value theorem, there exists $c \in (a, b)$ such that $g(c) = 0$ and hence $f(c) = c$.

Observe that a key element in this approach is the assumption that f and hence g are continuous. This is a topological assumption. Also notice that we do not need to know $g(a)$ nor $g(b)$ precisely; it is sufficient to show that the inequalities are satisfied. We know that the typical numerical computation will result in errors, thus a result of this form is encouraging. Therefore, let us try to be a little more precise.

Given an input x, we would like to have the output $g(x)$. However, the computer produces a potentially different value, which we will denote by $g_{num}(x)$. Assume that we can obtain an error bound for g_{num}, that is a number μ such that $|g(x) - g_{num}(x)| < \mu$. Returning to the fixed-point problem; if the computer determines that $g_{num}(a) < -\mu$ and $g_{num}(b) > \mu$, then we know that $g(a) < 0$ and $g(b) > 0$ and hence we can conclude the existence of a fixed point.

Generalizing this result to higher dimensions is not trivial. In the previous section we promise that Chapter 2 will show how homology groups are generated by topological spaces. In Chapter 6 we will go a step further and show that given a continuous map $f : X \to Y$ between topological spaces, there are unique linear maps $f_{*k} : H_k(X) \to H_k(Y)$, called the *homology maps*, from the homology groups of one space to the homology groups of the other. Furthermore, we will show that to correctly compute the homology maps we do not need to know the function f explicitly, but rather it is sufficient to know a numerical approximation, f_{num}, and an appropriate error bound μ.

One of the most remarkable theorems involving homology is the Lefschetz fixed-point theorem (see Theorem 10.46), which guarantees the existence of a fixed point if the traces of the homology maps satisfy a simple condition. Since even with numerical error we can compute these homology maps correctly, this allows us to use the computer to give mathematically rigorous arguments for the existence of fixed points for high-dimensional nonlinear functions. In fact, at the end of Chapter 10 we show that generalizations of these types of arguments can be used to rigorously answer any of the four questions posed at the beginning of this section.

1.3 Graphs

The previous two sections are meant to be an enticement, suggesting the power and potential for homology in a variety of applications. But no attempt is made to explain how or why there should be an algebraic theory that can perform the needed measurements. We hope to rectify this, at least on a heuristic level, in the next few sections. Since we are still trying to motivate the subject, we will keep things as simple as possible. In particular, let us stick to objects such as those of Figures 1.2 and 1.3.[1]

To do mathematics we need to make sure that these simple objects are well defined. Graphs provide a nice starting point.

Definition 1.1 A *graph* G is a subset of $\mathbf{R}^3$ made up of a finite collection of points $\{v_1, \ldots, v_n\}$, called *vertices*, together with straight-line segments $\{e_1, \ldots, e_m\}$, joining vertices, called *edges*, which satisfy the following intersection conditions:

1. The intersection of distinct edges either is empty or consists of exactly one vertex, and
2. if an edge and a vertex intersect, then the vertex is an endpoint of the edge.

More explicitly, an edge joining vertices v_0 and v_1 is the set of points

$$\{x \in \mathbf{R}^3 \mid x = tv_0 + (1-t)v_1,\ 0 \leq t \leq 1\},$$

which is denoted by $[v_0, v_1]$.

A *path* in G is an ordered sequence of edges of the form

---
[1] We temporarily ignore the previously discussed two-dimensional pixel structure visible when zooming in on computer-generated figures and we view them here as one-dimensional objects composed of line segments.

$$\{[v_0, v_1], [v_1, v_2], \ldots, [v_{l-1}, v_l]\}.$$

This path *begins* at v_0 and *ends* at v_l. Its *length* is l, the number of its edges. A graph G is *connected* if, for every pair of vertices $u, w \in G$, there is a path in G that begins at u and ends at w. A *loop* is a path that consists of distinct edges and begins and ends at the same vertex. A connected graph that contains no loops is a *tree*.

Observe that Figures 1.2(a), (b), c) and Figure 1.3 are graphs. Admittedly they are drawn as subsets of $\mathbf{R}^2$, but we can think of $\mathbf{R}^2 \subset \mathbf{R}^3$. Figure 1.2(d) is not a graph. It is also easy to see that Figures 1.2(b) and (c) contain loops, and that Figure 1.2(a) is a tree. What about Figure 1.3?

Graphs, and more generally topological spaces, *cannot* be entered into a computer. As defined above a graph is a subset of $\mathbf{R}^3$ and so consists of infinitely many points, but a computer can only store a finite amount of data. Of course, there is a very natural way to *represent* a graph, which only involves a finite amount of information.

Definition 1.2 A *combinatorial graph* is a pair $(\mathcal{V}, \mathcal{E})$, where $\mathcal{V}$ is a finite set whose elements are called *vertices* and $\mathcal{E}$ is a collection of pairs of distinct elements of $\mathcal{V}$ called *edges*. If an edge e of a combinatorial graph consists of the pair of vertices v_0 and v_1, we will write $e = [v_0, v_1]$.

It may seem at this point that we are making a big deal out of a minor issue. Consider, however, the sphere, $\{(x, y, z) \in \mathbf{R}^3 \mid x^2 + y^2 + z^2 = 1\}$. It is not so clear how to represent this in terms of a finite amount of data, but as we shall eventually see, we can compute its homology via a combinatorial representation. In fact, homology is a combinatorial object; this is what makes it such a powerful tool.

Even in the setting of graphs the issue of combinatorial representations is far from clear. Consider, for example, the graph $G = [0, 1] \subset \mathbf{R}$. How should we represent it as a combinatorial graph? Since we have not said what the vertices and edges are, the most obvious answer is to let $\mathcal{V} = \{0, 1\}$ and $\mathcal{E} = \{[0, 1]\}$. However, G could also be thought of as the graph containing the vertices 0, $1/2$, 1 and the edges $[0, 1/2]$, $[1/2, 1]$, in which case the natural combinatorial representation is given by $\mathcal{V}_2 = \{0, 1/2, 1\}$ and $\mathcal{E}_2 = \{[0, 1/2], [1/2, 1]\}$. More generally, we can think of G as being made up of many short segments leading to the combinatorial representation

$$\mathcal{V}_n := \{j/n \mid j = 0, \ldots, n\}, \quad \mathcal{E}_n := \{[j/n, (j+1)/n] \mid j = 0, \ldots, n-1\}.$$

We are motivating homology in terms of graphs (i.e., subsets of $\mathbf{R}^3$), however, the input data to the computer is in the form of a combinatorial graph, namely a finite list. Thus, to prove that homology is an invariant of a graph, we have to show that given any two combinatorial graphs that represent the same set, the corresponding homology is the same. This is not trivial! In fact, it will not be proven until Chapter 6.

On the other hand, in this chapter we are not supposed to be worrying about details, so we won't. However, we should not forget that there is this potential problem:

Can we make sure that two different combinatorial objects that give rise to the same set also give rise to the same homology?

Before turning to the algebra, we want to prove a simple property about trees.

A vertex that only intersects a single edge is called a *free vertex*.

Proposition 1.3 *Every tree contains at least one free vertex.*

Proof. Assume not. Then there exists a tree T with 0 free vertices. Let n be the number of edges in T. Let e_1 be an edge in T. Label its vertices by v_1^- and v_1^+. Since T has no free vertices, there is an edge e_2 with vertices $v_2^{\pm}$ such that $v_1^+ = v_2^-$. Continuing in this manner we can label the edges by e_i and the vertices by $v_i^{\pm}$, where $v_i^- = v_{i-1}^+$. Since there are only a finite number of vertices, at some point in this procedure we get $v_i^+ = v_j^-$ for some $i > j \geq 1$. Then $\{e_j, e_{j+1}, \ldots, e_i\}$ forms a loop. This is a contradiction. □

Exercises

1.1 Associate combinatorial graphs to Figures 1.2(a), (b), and (c).

1.2 Among all paths joining two vertices v and w at least one has minimal length. Such a path is called minimal. Show that any two edges and vertices on a minimal path are different.

1.3 Let T be a tree with n edges. Prove that T has $n + 1$ vertices.
Hint: Argue by induction on the number of edges.

1.4 Topological and Algebraic Boundaries

As stated earlier, our goal is to develop an algebraic means of detecting whether a set bounds a region or not. So we begin with the two simple sets of Figure 1.11, an interval I and the perimeter Γ^1 of a square. We want to think of these sets as graphs. As is indicated in Figure 1.11, we represent both sets by graphs consisting of four edges. The difference lies in the set of vertices.

We mentioned earlier that homology has the remarkable property that local calculations lead to knowledge about global properties. As already observed, the difference between the combinatorial graphs of I and Γ^1 is found in the vertices, which are clearly local objects. So let us focus on vertices and observe that they represent the endpoints or, as we shall call them from now on, the boundary points of edges.

Consider both the graph and the combinatorial graph of I. The left-hand column of Table 1.1 indicates the boundary points of each of the edges. The

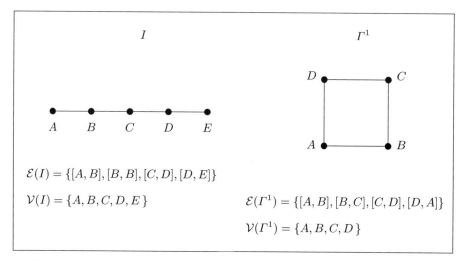

Fig. 1.11. Graphs and corresponding combinatorial graphs for $[0,1]$ and Γ^1.

right-hand column is derived from the combinatorial graph. To explain its meaning first recall that our goal is to produce an algebraic tool for understanding graphs. Instead of starting with formal definitions we are going to look for patterns. So for the moment the elements of the right-hand column can be considered to be algebraic quantities that correspond to elements of the combinatorial graph.

Table 1.1. Topological and Algebraic Boundaries in $[0,1]$

Topology	Algebra
$\operatorname{bd}[A,B] = \{A\} \cup \{B\}$	$\partial \widehat{[A,B]} = \widehat{A} + \widehat{B}$
$\operatorname{bd}[B,C] = \{B\} \cup \{C\}$	$\partial \widehat{[B,C]} = \widehat{B} + \widehat{C}$
$\operatorname{bd}[C,D] = \{C\} \cup \{D\}$	$\partial \widehat{[C,D]} = \widehat{C} + \widehat{D}$
$\operatorname{bd}[D,E] = \{D\} \cup \{E\}$	$\partial \widehat{[D,E]} = \widehat{D} + \widehat{E}$

On the topological level such basic algebraic notions as addition and subtraction of edges and points are not obvious concepts. The point of moving to an algebraic level is to allow ourselves this luxury. To distinguish between sets and algebra, we write the algebraic objects with a hat on top and allow ourselves to formally add them. For example, the topological objects such as a point $\{A\}$ and an edge $[A,B]$ become an algebraic object $\widehat{A}$ and $\widehat{[A,B]}$. Furthermore, we allow ourselves the luxury of writing expressions like $\widehat{A} + \widehat{B}$, $\widehat{[A,B]} + \widehat{[B,C]}$, or even $\widehat{A} + \widehat{A} = 2\widehat{A}$.

1.4 Topological and Algebraic Boundaries

How should we interpret the symbol ∂, called the *boundary operator*, which we have written in the table? We are doing algebra, so it should be some type of map that takes the algebraic object $[\widehat{A,B}]$ to the sum of $\widehat{A}$ and $\widehat{B}$. The nicest maps are *linear maps*. This would mean that

$$\partial([\widehat{A,B}] + [\widehat{B,C}]) = \partial([\widehat{A,B}]) + \partial([\widehat{B,C}])$$
$$\stackrel{(1)}{=} \widehat{A} + \widehat{B} + \widehat{B} + \widehat{C}$$
$$= \widehat{A} + 2\widehat{B} + \widehat{C},$$

where $\stackrel{(1)}{=}$ follows from Table 1.1.

The counterpart of this calculation on the topology level would be

$$\operatorname{bd}([A,B] \cup [B,C]) = \operatorname{bd}[A,C] = \{A\} \cup \{C\}.$$

If we think that $+$ on the algebraic side somehow matches $\cup$ on the topology side, then this suggests that we would like

$$\partial([\widehat{A,B}] + [\widehat{B,C}]) = \widehat{A} + \widehat{C} = \partial[\widehat{A,C}].$$

The only way that this can happen is for $2\widehat{B} = 0$. This may seem like a pretty strange relation and suggests that at this point there are three things we can do:

1. Give up;
2. start over and try to find a different definition for ∂; or
3. be stubborn and press on.

The fact that this book has been written suggests that we are not about to give up. We shall discuss option 2 in Section 1.5. For now we shall just press on and adopt the trick of counting *modulo 2*. This means that we will just check whether an element appears an odd or even number of times; if it is odd we keep the element, if it is even we discard the element, that is, we declare

$$0 = 2\widehat{A} = 2\widehat{B} = 2\widehat{C} = 2\widehat{D} = 2\widehat{E}.$$

Continuing to use the presumed linearity of ∂ and counting modulo 2, we have that

$$\partial\left([\widehat{A,B}] + [\widehat{B,C}] + [\widehat{C,D}] + [\widehat{D,E}]\right) = \widehat{A} + \widehat{B} + \widehat{B} + \widehat{C} + \widehat{C} + \widehat{D} + \widehat{D} + \widehat{E}$$
$$= \widehat{A} + \widehat{E}.$$

As an indication that we are not too far off track, observe that if we had begun with a representation of I in terms of the combinatorial graph

$$\mathcal{E}'(I) = \{[A,E]\} \quad \mathcal{V}'(I) = \{A,E\},$$

then $\operatorname{bd}[A,E] = \{A\} \cup \{E\}$.

Doing the same for the graph and combinatorial graph representing Γ^1 we get Table 1.2. Adding up the algebraic boundaries, we have

$$\partial\left([\widehat{A,B}] + [\widehat{B,C}] + [\widehat{C,D}] + [\widehat{D,A}]\right) = \widehat{A} + \widehat{B} + \widehat{B} + \widehat{C} + \widehat{C} + \widehat{D} + \widehat{D} + \widehat{A}$$
$$= \widehat{A} + \widehat{A} \qquad (1.4)$$
$$= 0.$$

Table 1.2. Topology and Algebra of Boundaries in Γ^1

Topology	Algebra
bd $[A, B] = \{A\} \cup \{B\}$	$\partial[\widehat{A,B}] = \widehat{A} + \widehat{B}$
bd $[B, C] = \{B\} \cup \{C\}$	$\partial[\widehat{B,C}] = \widehat{B} + \widehat{C}$
bd $[C, D] = \{C\} \cup \{D\}$	$\partial[\widehat{C,D}] = \widehat{C} + \widehat{D}$
bd $[D, A] = \{D\} \cup \{A\}$	$\partial[\widehat{D,A}] = \widehat{D} + \widehat{A}$

Based on these two examples one might make the extravagant claim that spaces with *cycles*—algebraic objects whose boundaries add up to zero—enclose regions. This is almost true.

To see how this fails, observe that we could "fill in" Γ^1 [think of Figure 1.2(d)]. How does this affect the algebra? To make sense of this we need to go beyond graphs into *cubical complexes*, which will be defined later. For the moment consider the picture and collection of sets in Figure 1.12. The new aspect is the square Q that has filled in the region bounded by Γ^1. This is coded in the combinatorial information as the element $\{Q\}$.

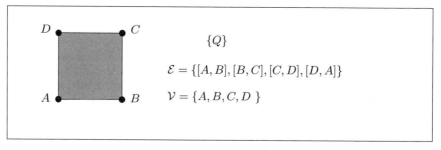

Fig. 1.12. The set Q and corresponding combinatorial data.

Observe that the perimeter or boundary of Q is Γ^1. Table 1.3 contains the topological boundary information and the associated algebra.

1.4 Topological and Algebraic Boundaries

Table 1.3. Topology and Algebra of Boundaries in Q

Topology	Algebra
$\operatorname{bd} Q = [A,B] \cup [B,C] \cup [C,D] \cup [D,A]$	$\partial \widehat{Q} = [\widehat{A,B}] + [\widehat{B,C}] + [\widehat{C,D}] + [\widehat{D,A}]$
$\operatorname{bd}[A,B] = \{A\} \cup \{B\}$	$\partial[\widehat{A,B}] = \widehat{A} + \widehat{B}$
$\operatorname{bd}[B,C] = \{B\} \cup \{C\}$	$\partial[\widehat{B,C}] = \widehat{B} + \widehat{C}$
$\operatorname{bd}[C,D] = \{C\} \cup \{D\}$	$\partial[\widehat{C,D}] = \widehat{C} + \widehat{D}$
$\operatorname{bd}[D,A] = \{D\} \cup \{A\}$	$\partial[\widehat{D,A}] = \widehat{D} + \widehat{A}$

Since $\Gamma^1 \subset Q$, it is not surprising to see the contents of Table 1.2 contained in Table 1.3. Now observe that

$$\partial \widehat{Q} = [\widehat{A,B}] + [\widehat{B,C}] + [\widehat{C,D}] + [\widehat{D,A}].$$

Equation (1.4) indicates that the cycle $[\widehat{A,B}] + [\widehat{B,C}] + [\widehat{C,D}] + [\widehat{D,A}]$ was the interesting algebraic aspect of Γ^1. In Q it appears as the boundary of an object. Our observation is that *cycles that are boundaries are uninteresting and should be ignored.*

Restating this purely algebraically, we are looking for cycles, that is elements of the *kernel* of the boundary operator. Furthermore, if a cycle is a *boundary*, namely it belongs to the *image* of the boundary operator, then we wish to ignore it. From an algebraic point of view this means we want, somehow, to set boundaries equal to zero.

The reader may have wondered why, after introducing the notion of a loop, we suddenly switched to the language of cycles. Loops are combinatorial objects—lists that our computer can store. Cycles, on the other hand, are algebraic objects. The comments of the previous paragraph have no simple conceptual correspondence on the combinatorial level.

We have by now introduced many vague and complicated notions. If you feel things are spinning out of control, don't worry. Admittedly, there are a lot of loose ends that we need to tie up, and we will begin to do so in the next chapter. The process of developing new mathematics typically involves developing new intuitions and finding new patterns—in this case we have the advantage of knowing that it will all work out in the end. For now let's just enjoy trying to match topology and algebra.

Exercises

1.4 Repeat the discussion of this section using Figures 1.2(c) and (d). In particular, make up a topology and algebra table and identify the cycles.

1.5 Repeat the above computations for a graph that represents a triangle in the plane.

24   1 Preview

1.5 Keeping Track of Directions

We want to repeat the discussion of the last section, but this time we will try to avoid counting the algebraic objects modulo 2. To do this we will consider I and Γ^1 as the explicit subsets of $\mathbf{R}^2$ indicated in Figure 1.13. Of course, this figure looks a lot like Figure 1.11. However, we have added arrows, which we can think of as the standard directions of the x- and y-axes.

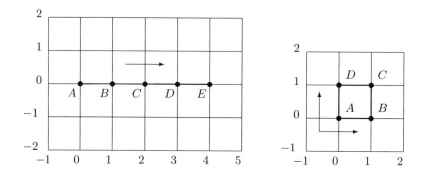

Fig. 1.13. The sets I and Γ^1 as explicit subsets of $\mathbf{R}^2$.

We will use these arrows to give a sense of direction to the edges and use the corresponding algebra to keep track of this as follows. Consider the edge $[A, B]$ in either I or Γ^1. On the combinatorial level it is defined by its endpoints A and B. Thus, in principle, we could denote the edge by $[A, B]$ or $[B, A]$. For the computations performed in the previous subsection, this would make no difference (check it). Now, however, we want to distinguish these two cases. In particular, since the x-coordinate value of the vertex A is less than that of B, we insist on writing the edge as $[A, B]$. Consider the edge with vertices C and D in I and Γ^1. In the first case we write $[C, D]$, but in the latter case we write $[D, C]$. In a similar vein, the edge with vertices A and D in Γ^1 is written as $[A, D]$ since the y-coordinate of A is less than the y-coordinate of D.

Having insisted on a specific order to denote the edges, we should not lose track of this when we apply the boundary operator. Let us declare that

$$\partial \widehat{[A, B]} := \widehat{B} - \widehat{A}$$

as a way to keep track of the fact that vertex A comes before vertex B. Of course, we will still insist that ∂ be linear.

Using this linearity on the algebra generated by the edges of I, we obtain

1.5 Keeping Track of Directions

$$\partial\left([\widehat{A,B}] + [\widehat{B,C}] + [\widehat{C,D}] + [\widehat{D,E}]\right) = \widehat{B} - \widehat{A} + \widehat{C} - \widehat{B} + \widehat{D} - \widehat{C} + \widehat{E} - \widehat{D}$$
$$= \widehat{E} - \widehat{A}.$$

Again, we see that there is consistency between the algebra and the topology, since if we think of I as a single edge, then $\operatorname{bd} I = \{E\} \cup \{A\}$ and the minus sign suggests traversing from A to E.

Applying the boundary operator to the algebraic objects generated by the edges of Γ^1 gives rise to Table 1.4.

Table 1.4. Topology and Algebra of Boundaries in Γ^1 Remembering the Directions of the x- and y-axes.

Topology	Algebra
$\operatorname{bd}[A,B] = \{A\} \cup \{B\}$	$\partial[\widehat{A,B}] = \widehat{B} - \widehat{A}$
$\operatorname{bd}[B,C] = \{B\} \cup \{C\}$	$\partial[\widehat{B,C}] = \widehat{C} - \widehat{B}$
$\operatorname{bd}[D,C] = \{C\} \cup \{D\}$	$\partial[\widehat{D,C}] = \widehat{C} - \widehat{D}$
$\operatorname{bd}[A,D] = \{D\} \cup \{A\}$	$\partial[\widehat{A,D}] = \widehat{D} - \widehat{A}$

When we go around Γ^1 counterclockwise and keep track of the directions through which we pass the edges, we see that $[D,C]$ and $[A,D]$ are traversed in the opposite order from the orientations of the x- and y-axes. To keep track of this on the algebraic level we will write

$$[\widehat{A,B}] + [\widehat{B,C}] - [\widehat{D,C}] - [\widehat{A,D}].$$

Notice that we see a significant difference with the purely topological representation

$$\Gamma^1 = [A,B] \cup [B,C] \cup [D,C] \cup [A,D].$$

However,

$$\partial\left([\widehat{A,B}] + [\widehat{B,C}] - [\widehat{D,C}] - [\widehat{A,D}]\right) = \partial\left([\widehat{A,B}]\right) + \partial\left([\widehat{B,C}]\right)$$
$$- \partial\left([\widehat{D,C}]\right) - \partial\left([\widehat{A,D}]\right)$$
$$= \widehat{B} - \widehat{A} + \widehat{C} - \widehat{B} + \widehat{D} - \widehat{C} + \widehat{A} - \widehat{D}$$
$$= 0.$$

So again, we see that the algebra that corresponds to the interesting topology is a cycle—a sum of algebraic objects whose boundaries add up to zero.

We still need to understand what happens to this algebra when we fill in Γ^1 by Q. Consider Figure 1.14. Table 1.5 contains the topological boundary information and algebra that we are associating to it.

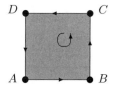

Fig. 1.14. The set Q and associated directions.

Since $\Gamma^1 \subset Q$, we again see the contents of Table 1.4 contained in Table 1.5. Keeping track of directions and walking around the edge of Q in a counterclockwise direction, it seems reasonable to define

$$\partial \widehat{Q} = [\widehat{A,B}] + [\widehat{B,C}] - [\widehat{D,C}] - [\widehat{A,D}].$$

Equation (1.4) indicates that the cycle $[\widehat{A,B}] + [\widehat{B,C}] - [\widehat{D,C}] - [\widehat{A,D}]$ is the interesting algebraic aspect of Γ^1. In Q it appears as the boundary of an object. Again, the observation that we will make is: Cycles that are boundaries should be considered uninteresting.

Table 1.5. Topology and Algebra of Boundaries in Figure 1.14

Topology	Algebra
$\operatorname{bd} Q = \Gamma^1 = [A,B] \cup [B,C] \cup [C,D] \cup [D,A]$	$\partial \widehat{Q} = [\widehat{A,B}] + [\widehat{B,C}] - [\widehat{D,C}] - [\widehat{A,D}]$
$\operatorname{bd}[A,B] = \{A\} \cup \{B\}$	$\partial[\widehat{A,B}] = \widehat{B} - \widehat{A}$
$\operatorname{bd}[B,C] = \{B\} \cup \{C\}$	$\partial[\widehat{B,C}] = \widehat{C} - \widehat{B}$
$\operatorname{bd}[D,C] = \{D\} \cup \{C\}$	$\partial[\widehat{D,C}] = \widehat{C} - \widehat{D}$
$\operatorname{bd}[A,D] = \{A\} \cup \{D\}$	$\partial[\widehat{A,D}] = \widehat{D} - \widehat{A}$

1.6 Mod 2 Homology of Graphs

We have done the same example twice using different types of arithmetic, but the conclusion is the same. We should look for a linear operator that somehow algebraically mimics what is done by taking the edge or boundary of a set. Then, having found this operator, we should look for cycles (elements of the kernel) but ignore boundaries (elements of the image). This is still pretty fuzzy so let's do it again; a little slower and more formally, but in the general setting of graphs and using linear algebra. We are not yet aiming for rigor, but we do want to suggest that there is a way to deal with all the ideas that are being thrown about in a systematic manner.

The linear algebra we are going to do, *mod 2 linear algebra*, may seem a little strange at first, but hopefully you will recognize that all the important ideas, such as vector addition, basis, matrices, etc., still hold. The difference is that unlike the traditional linear algebra, where the set of scalars by which vectors may be multiplied consists of all real numbers, in the mod 2 linear algebra we restrict the scalars to the set $\mathbf{Z}_2 := \{0, 1\}$. A nice consequence is that the number of algebraic elements is finite, so in many cases we can explicitly write them out.

Let G be a graph with a fixed representation as a combinatorial graph with vertices $\mathcal{V}$ and edges $\mathcal{E}$. We will construct two vector spaces $C_0(G; \mathbf{Z}_2)$ and $C_1(G; \mathbf{Z}_2)$ as follows. Declare the set of vertices $\mathcal{V}$ to be the set of basis elements of $C_0(G; \mathbf{Z}_2)$. Thus, if $\mathcal{V} = \{v_1, \ldots, v_n\}$, then the collection

$$\{\widehat{v}_i \mid i = 1, \ldots, n\}$$

is a basis for $C_0(G; \mathbf{Z}_2)$. Notice that we are continuing to use the hat notation to distinguish between an algebraic object (a basis element) and a combinatorial object (an element in a list).[2] Since $\{\widehat{v}_i \mid i = 1, \ldots, n\}$ is a basis, elements of $C_0(G; \mathbf{Z}_2)$ take the form

$$c = \alpha_1 \widehat{v}_1 + \alpha_2 \widehat{v}_2 + \cdots + \alpha_n \widehat{v}_n, \tag{1.5}$$

where $\alpha_i \in \mathbf{Z}_2$. An expression of this form will be referred to as a *chain*. Note that the symbol $\mathbf{Z}_2$ in the notation $C_0(G; \mathbf{Z}_2)$ and $C_1(G; \mathbf{Z}_2)$ is used to remind us that we are adding mod 2.

Example 1.4 Let us assume that $\mathcal{V} = \{v_1, v_2, v_3, v_4\}$. Then the basis elements for $C_0(G; \mathbf{Z}_2)$ are $\{\widehat{v}_1, \widehat{v}_2, \widehat{v}_3, \widehat{v}_4\}$. Therefore, every element of $C_0(G; \mathbf{Z}_2)$ can be written as

$$c = \alpha_1 \widehat{v}_1 + \alpha_2 \widehat{v}_2 + \alpha_3 \widehat{v}_3 + \alpha_4 \widehat{v}_4.$$

However, since $\alpha_i \in \{0, 1\}$, we can write out all the elements of $C_0(G; \mathbf{Z}_2)$. In particular,

---

[2] Since we are trying to be a little more formal in this section, perhaps this is a good place to emphasize the fact that elements of $\mathcal{V}$ are really combinatorial objects. These are the objects that we want the computer to manipulate; therefore, they are just elements of a list stored in the computer. Of course, since we are really interested in understanding the topology of the graph G, we make the identification of $v \in \mathcal{V}$ with $v \in G \subset \mathbf{R}^3$. But it is we who make this identification, not the computer. Furthermore, as will become clear especially in Part II where we discuss the applications of homology, the ability to make this identification, and thereby pass from combinatorics to topology, is a very powerful technique.

$$C_0(G; \mathbf{Z}_2) = \left\{ \begin{array}{c} 0,\ \widehat{v_1},\ \widehat{v_2},\ \widehat{v_3},\ \widehat{v_4}, \\ \widehat{v_1} + \widehat{v_2},\ \widehat{v_1} + \widehat{v_3},\ \widehat{v_1} + \widehat{v_4},\ \widehat{v_2} + \widehat{v_3},\ \widehat{v_2} + \widehat{v_4},\ \widehat{v_3} + \widehat{v_4}, \\ \widehat{v_1} + \widehat{v_2} + \widehat{v_3},\ \widehat{v_1} + \widehat{v_2} + \widehat{v_4},\ \widehat{v_1} + \widehat{v_3} + \widehat{v_4},\ \widehat{v_2} + \widehat{v_3} + \widehat{v_4}, \\ \widehat{v_1} + \widehat{v_2} + \widehat{v_3} + \widehat{v_4} \end{array} \right\},$$

Each element of $C_0(G; \mathbf{Z}_2)$ is a vector and, of course, we are allowed to add vectors, but mod 2, for example,

$$(\widehat{v_1} + \widehat{v_2} + \widehat{v_3}) + (\widehat{v_1} + \widehat{v_2} + \widehat{v_4}) = 2\widehat{v_1} + 2\widehat{v_2} + \widehat{v_3} + \widehat{v_4} = 0 + 0 + \widehat{v_3} + \widehat{v_4} = \widehat{v_3} + \widehat{v_4}.$$

Returning to the general discussion, let the set of edges $\mathcal{E}$ be the set of basis elements of $C_1(G; \mathbf{Z}_2)$. If $\mathcal{E} = \{e_1, \ldots, e_k\}$, then the collection $\{\widehat{e_i} \mid i = 1, \ldots, k\}$ is a basis for $C_1(G; \mathbf{Z}_2)$ and an element of $C_1(G; \mathbf{Z}_2)$ takes the form

$$c = \alpha_1 \widehat{e_1} + \alpha_2 \widehat{e_2} + \cdots + \alpha_k \widehat{e_k},$$

where again α_i is 0 or 1. The vector spaces $C_k(G; \mathbf{Z}_2)$ are called the *k-chains* for G. A word of warning is in order here. The vector space is an algebraic concept based on very geometric ideas. However, the reader should not seek any relation between the geometry of the vector spaces $C_k(G; \mathbf{Z}_2)$ and the geometry of the graph G. As we will see soon, even the dimensions of $C_k(G; \mathbf{Z}_2)$ have nothing to do with the dimension 3 of the space in which G is located.

It is convenient to introduce two more vector spaces $C_2(G; \mathbf{Z}_2)$ and $C_{-1}(G; \mathbf{Z}_2)$. We will always take $C_{-1}(G; \mathbf{Z}_2)$ to be the trivial vector space $\mathbf{0}$, that is, the vector space consisting of exactly one element 0. For graphs we will also set $C_2(G; \mathbf{Z}_2)$ to be the trivial vector space. As we will see in Chapter 2, for higher-dimensional spaces this is not the case.

We now need to formally define the boundary operators that were alluded to earlier. Let

$$\partial_0 : C_0(G; \mathbf{Z}_2) \to C_{-1}(G; \mathbf{Z}_2),$$
$$\partial_1 : C_1(G; \mathbf{Z}_2) \to C_0(G; \mathbf{Z}_2),$$
$$\partial_2 : C_2(G; \mathbf{Z}_2) \to C_1(G; \mathbf{Z}_2)$$

be *linear maps*. Since $C_{-1}(G; \mathbf{Z}_2) = \mathbf{0}$, it is clear that the image of ∂_0 must be zero. For similar reasons, the same must be true for ∂_2. Since we have chosen bases for the vector spaces $C_1(G; \mathbf{Z}_2)$ and $C_0(G; \mathbf{Z}_2)$, we can express ∂_1 as a matrix. The entries of this matrix are determined by how ∂_1 acts on the basis elements (i.e., the edges $\mathbf{e}_i$). In line with the previous discussion we make the following definition. Let the edge e_i have vertices v_j and v_k. Define

$$\partial_1 \widehat{e_i} := \widehat{v_j} + \widehat{v_k}.$$

In our earlier example we were interested in cycles, that is, objects that get mapped to 0 by ∂. In general, given a linear map A, the set of elements that get sent to 0 is called the *kernel* of A and is denoted by $\ker A$. Thus the

1.6 Mod 2 Homology of Graphs

set of cycles forms the kernel of ∂. Because the set of cycles plays such an important role, it has its own notation:

$$Z_0(G; \mathbf{Z}_2) := \ker \partial_0 = \{c \in C_0(G; \mathbf{Z}_2) \mid \partial_0 c = 0\},$$
$$Z_1(G; \mathbf{Z}_2) := \ker \partial_1 = \{c \in C_1(G; \mathbf{Z}_2) \mid \partial_1 c = 0\}.$$

Since $C_{-1}(G; \mathbf{Z}_2) = 0$, it is obvious that $Z_0(G; \mathbf{Z}_2) = C_0(G; \mathbf{Z}_2)$; namely everything in $C_0(G; \mathbf{Z}_2)$ gets sent to 0.

We also observed that cycles that are boundaries are not interesting. To formally state this, define the set of boundaries to be the image of the boundary operator. More precisely,

$$B_0(G; \mathbf{Z}_2) := \operatorname{im} \partial_1 = \{b \in C_0(G; \mathbf{Z}_2) \mid \exists c \in C_1(G; \mathbf{Z}_2) \text{ such that } \partial_1 c = b\},$$
$$B_1(G; \mathbf{Z}_2) := \operatorname{im} \partial_2 = \{b \in C_1(G; \mathbf{Z}_2) \mid \exists c \in C_2(G; \mathbf{Z}_2) \text{ such that } \partial_2 c = b\}.$$

Recall that we set $C_2(G; \mathbf{Z}_2) = 0$; thus $B_1(G; \mathbf{Z}_2) = \mathbf{0}$.

Observe that $B_0(G; \mathbf{Z}_2) \subset C_0(G; \mathbf{Z}_2) = Z_0(G; \mathbf{Z}_2)$, that is, every 0-boundary is a 0-cycle. We shall show later that every boundary is a cycle, that is,

$$B_k(G; \mathbf{Z}_2) \subset Z_k(G; \mathbf{Z}_2).$$

This is a very important fact—but not at all obvious at this point.

We can finally define homology in this rather special setting. For $k = 0, 1$, the kth homology with $\mathbf{Z}_2$ coefficients is defined to be the quotient space

$$H_k(G; \mathbf{Z}_2) := Z_k(G; \mathbf{Z}_2) / B_k(G; \mathbf{Z}_2).$$

If you have not worked with quotient spaces, then the obvious question is: *What does this notation mean?*

Let us step back for a moment and remember what we are trying to do. Recall that the interesting objects are cycles, but if a cycle is a boundary, then it is no longer interesting. Thus we begin with a cycle $z \in Z_k(G; \mathbf{Z}_2)$. It is possible that $z \in B_k(G; \mathbf{Z}_2)$. In this case we want z to be uninteresting. From an algebraic point of view we can take this to mean that we want to set z equal to 0.

Now consider two cycles $z_1, z_2 \in Z_i(G; \mathbf{Z}_2)$. What if there exists a boundary $b \in B_k(G; \mathbf{Z}_2)$ such that

$$z_1 + b = z_2?$$

Since boundaries are supposed to be 0, this suggests that b should be zero and hence that we want z_1 and z_2 to be the same.

Mathematically, when we want different objects to be the same, we form *equivalence classes*. With this in mind we define an equivalence class on the set of cycles by

$$z_1 \sim z_2 \quad \text{if and only if} \quad z_1 + b = z_2$$

for some $b \in B_k(G; \mathbf{Z}_2)$. The notation $\sim$ means equivalent. The equivalence class of the cycle $z \in Z_k(G; \mathbf{Z}_2)$ is the set of all cycles that are equivalent to z. It is denoted by $[z]$. Thus

$$[z] := \{z' \in Z_k(G; \mathbf{Z}_2) \mid z \sim z'\}.$$

Therefore, $z_1 \sim z_2$ is the same as saying $[z_1] = [z_2]$. The set of equivalence classes makes up the elements of $H_k(G; \mathbf{Z}_2)$.

This brief discussion contains several fundamental but nontrivial mathematical concepts, so to help make some of these ideas clearer consider the following examples.

Example 1.5 Let us start with the trivial graph consisting of a single point, $G = \{v\}$. Then

$$\mathcal{V} = \{v\}, \quad \mathcal{E} = \emptyset.$$

These are used to generate the bases for the chains. In particular, the basis for $C_0(G; \mathbf{Z}_2)$ is $\{\widehat{v}\}$. This means that any element $v \in C_0(G; \mathbf{Z}_2)$ has the form

$$c = \alpha \widehat{v}.$$

If $\alpha = 0$, then $c = 0$. If $\alpha = 1$, then $c = \widehat{v}$. Thus

$$C_0(G; \mathbf{Z}_2) = \{0, \widehat{v}\}.$$

Since, by definition, $\partial_0 = 0$, it follows that $Z_0(G; \mathbf{Z}_2) = C_0(G; \mathbf{Z}_2)$. Thus

$$Z_0(G; \mathbf{Z}_2) = \{0, \widehat{v}\}.$$

The basis for $C_1(G; \mathbf{Z}_2)$ is the empty set; therefore,

$$C_1(G; \mathbf{Z}_2) = \mathbf{0}.$$

By definition, $Z_1(G; \mathbf{Z}_2) \subset C_1(G; \mathbf{Z}_2)$, so $Z_1(G; \mathbf{Z}_2) = \mathbf{0}$.

Since $C_1(G; \mathbf{Z}_2) = \mathbf{0}$, the image of $C_1(G; \mathbf{Z}_2)$ under ∂_1 must also be trivial. Thus $B_0(G; \mathbf{Z}_2) = \mathbf{0}$. Now consider two cycles $z_1, z_2 \in Z_0(G; \mathbf{Z}_2)$. $B_0(G; \mathbf{Z}_2) = \mathbf{0}$ implies that $z_1 \sim z_2$ if and only if $z_1 = z_2$. Therefore,

$$H_0(G; \mathbf{Z}_2) = Z_0(G; \mathbf{Z}_2) = \{[0], [\widehat{v}]\}.$$

Of course, since $Z_1(G; \mathbf{Z}_2) = \mathbf{0}$, it follows that

$$H_1(G; \mathbf{Z}_2) = \mathbf{0}.$$

Example 1.6 Let G be the graph of Figure 1.11 representing I. Then,

$$\mathcal{V} = \{A, B, C, D, E\},$$
$$\mathcal{E} = \{[A, B], [B, C][C, D], [D, E]\}.$$

1.6 Mod 2 Homology of Graphs

Thus the basis for the 0-chains, $C_0(G; \mathbf{Z}_2)$, is

$$\{\widehat{A}, \widehat{B}, \widehat{C}, \widehat{D}, \widehat{E}\},$$

while the basis for the 1-chains, $C_1(G; \mathbf{Z}_2)$, is

$$\{\widehat{[A,B]}, \widehat{[B,C]}, \widehat{[C,D]}, \widehat{[D,E]}\}.$$

Unlike the previous example, we really need to write down the boundary operator $\partial_1 : C_1(G; \mathbf{Z}_2) \to C_0(G; \mathbf{Z}_2)$. Given the above-mentioned bases, since ∂_1 is a linear map it can be written as a matrix. To do this it is convenient to use the notation of column vectors. So let

$$\widehat{A} = \begin{bmatrix} 1 \\ 0 \\ 0 \\ 0 \\ 0 \end{bmatrix}, \ \widehat{B} = \begin{bmatrix} 0 \\ 1 \\ 0 \\ 0 \\ 0 \end{bmatrix}, \ \widehat{C} = \begin{bmatrix} 0 \\ 0 \\ 1 \\ 0 \\ 0 \end{bmatrix}, \ \widehat{D} = \begin{bmatrix} 0 \\ 0 \\ 0 \\ 1 \\ 0 \end{bmatrix}, \ \widehat{E} = \begin{bmatrix} 0 \\ 0 \\ 0 \\ 0 \\ 1 \end{bmatrix}$$

and

$$\widehat{[A,B]} = \begin{bmatrix} 1 \\ 0 \\ 0 \\ 0 \end{bmatrix}, \ \widehat{[B,C]} = \begin{bmatrix} 0 \\ 1 \\ 0 \\ 0 \end{bmatrix}, \ \widehat{[C,D]} = \begin{bmatrix} 0 \\ 0 \\ 1 \\ 0 \end{bmatrix}, \ \widehat{[D,E]} = \begin{bmatrix} 0 \\ 0 \\ 0 \\ 1 \end{bmatrix}.$$

With this convention, ∂_1 becomes the 5×4 matrix

$$\partial_1 = \begin{bmatrix} 1 & 0 & 0 & 0 \\ 1 & 1 & 0 & 0 \\ 0 & 1 & 1 & 0 \\ 0 & 0 & 1 & 1 \\ 0 & 0 & 0 & 1 \end{bmatrix}.$$

Let's do a quick check. For example,

$$\partial_1 \widehat{[B,C]} = \begin{bmatrix} 1 & 0 & 0 & 0 \\ 1 & 1 & 0 & 0 \\ 0 & 1 & 1 & 0 \\ 0 & 0 & 1 & 1 \\ 0 & 0 & 0 & 1 \end{bmatrix} \begin{bmatrix} 0 \\ 1 \\ 0 \\ 0 \end{bmatrix} = \begin{bmatrix} 0 \\ 1 \\ 1 \\ 0 \\ 0 \end{bmatrix} = \widehat{B} + \widehat{C}.$$

The next step is to compute the cycles—to find $Z_1(G; \mathbf{Z}_2) := \ker \partial_1$. Observe that by definition the chain $c \in C_1(G; \mathbf{Z}_2)$ is in $Z_1(G; \mathbf{Z}_2)$ if and only if $\partial_1 c = \widehat{0}$. Again, since $c \in C_1(G; \mathbf{Z}_2)$, it can be written as a sum of basis elements, that is,

$$c = \alpha_1 \widehat{[A,B]} + \alpha_2 \widehat{[B,C]} + \alpha_3 \widehat{[C,D]} + \alpha_4 \widehat{[D,E]}.$$

32   1 Preview

Writing this in the form of a column vector, we have

$$c = \alpha_1 \begin{bmatrix} 1 \\ 0 \\ 0 \\ 0 \end{bmatrix} + \alpha_2 \begin{bmatrix} 0 \\ 1 \\ 0 \\ 0 \end{bmatrix} + \alpha_3 \begin{bmatrix} 0 \\ 0 \\ 1 \\ 0 \end{bmatrix} + \alpha_4 \begin{bmatrix} 0 \\ 0 \\ 0 \\ 1 \end{bmatrix} = \begin{bmatrix} \alpha_1 \\ \alpha_2 \\ \alpha_3 \\ \alpha_4 \end{bmatrix}.$$

In this form, finding $c \in \ker \partial_1$ is equivalent to solving the equation

$$\partial_1 \begin{bmatrix} \alpha_1 \\ \alpha_2 \\ \alpha_3 \\ \alpha_4 \end{bmatrix} = \begin{bmatrix} 1 & 0 & 0 & 0 \\ 1 & 1 & 0 & 0 \\ 0 & 1 & 1 & 0 \\ 0 & 0 & 1 & 1 \\ 0 & 0 & 0 & 1 \end{bmatrix} \begin{bmatrix} \alpha_1 \\ \alpha_2 \\ \alpha_3 \\ \alpha_4 \end{bmatrix} = \begin{bmatrix} \alpha_1 \\ \alpha_1 + \alpha_2 \\ \alpha_2 + \alpha_3 \\ \alpha_3 + \alpha_4 \\ \alpha_4 \end{bmatrix} = \begin{bmatrix} 0 \\ 0 \\ 0 \\ 0 \\ 0 \end{bmatrix},$$

which implies that $\alpha_i = 0$ for $i = 1, \ldots, 4$. Thus the only element in $Z_1(G; \mathbf{Z}_2)$ is the zero vector and hence $Z_1(G; \mathbf{Z}_2) = \mathbf{0}$. Since $\partial_2 = 0$, we have $B_1(G; \mathbf{Z}_2) = \mathbf{0}$. So

$$H_1(G; \mathbf{Z}_2) := Z_1(G; \mathbf{Z}_2)/B_1(G; \mathbf{Z}_2) = \mathbf{0}.$$

Computing $H_0(G; \mathbf{Z}_2)$ is more interesting. Since $Z_0(G; \mathbf{Z}_2) = C_0(G; \mathbf{Z}_2)$, a basis for $Z_0(G; \mathbf{Z}_2)$ consists of

$$\widehat{A} = \begin{bmatrix} 1 \\ 0 \\ 0 \\ 0 \\ 0 \end{bmatrix}, \widehat{B} = \begin{bmatrix} 0 \\ 1 \\ 0 \\ 0 \\ 0 \end{bmatrix}, \widehat{C} = \begin{bmatrix} 0 \\ 0 \\ 1 \\ 0 \\ 0 \end{bmatrix}, \widehat{D} = \begin{bmatrix} 0 \\ 0 \\ 0 \\ 1 \\ 0 \end{bmatrix}, \widehat{E} = \begin{bmatrix} 0 \\ 0 \\ 0 \\ 0 \\ 1 \end{bmatrix}.$$

What about $B_0(G; \mathbf{Z}_2)$? This is the image of ∂_1 spanned by the images of each of the basis elements. This we can compute. In particular,

$$\partial_1 [\widehat{A, B}] = \begin{bmatrix} 1 & 0 & 0 & 0 \\ 1 & 1 & 0 & 0 \\ 0 & 1 & 1 & 0 \\ 0 & 0 & 1 & 1 \\ 0 & 0 & 0 & 1 \end{bmatrix} \begin{bmatrix} 1 \\ 0 \\ 0 \\ 0 \end{bmatrix} = \begin{bmatrix} 1 \\ 1 \\ 0 \\ 0 \\ 0 \end{bmatrix}.$$

Similarly,

$$\partial_1[\widehat{B,C}] = \begin{bmatrix} 0 \\ 1 \\ 1 \\ 0 \\ 0 \end{bmatrix}, \quad \partial_1[\widehat{C,D}] = \begin{bmatrix} 0 \\ 0 \\ 1 \\ 1 \\ 0 \end{bmatrix}, \quad \partial_1[\widehat{D,E}] = \begin{bmatrix} 0 \\ 0 \\ 0 \\ 1 \\ 1 \end{bmatrix}.$$

From this we can conclude that if $b \in B_0(G; \mathbf{Z}_2)$, then

1.6 Mod 2 Homology of Graphs

$$b = \alpha_1 \begin{bmatrix} 1 \\ 1 \\ 0 \\ 0 \\ 0 \end{bmatrix} + \alpha_2 \begin{bmatrix} 0 \\ 1 \\ 1 \\ 0 \\ 0 \end{bmatrix} + \alpha_3 \begin{bmatrix} 0 \\ 0 \\ 1 \\ 1 \\ 0 \end{bmatrix} + \alpha_4 \begin{bmatrix} 0 \\ 0 \\ 0 \\ 1 \\ 1 \end{bmatrix},$$

where $\alpha_i \in \{0, 1\}$.

It is easy to check that no basis element of $Z_0(G; \mathbf{Z}_2)$ is in $B_0(G; \mathbf{Z}_2)$. For example, if for some $\alpha_1, \alpha_2, \alpha_3, , \alpha_4, \alpha_5 \in \{0, 1\}$ we have

$$\widehat{A} = \begin{bmatrix} 1 \\ 0 \\ 0 \\ 0 \\ 0 \end{bmatrix} = \alpha_1 \begin{bmatrix} 1 \\ 1 \\ 0 \\ 0 \\ 0 \end{bmatrix} + \alpha_2 \begin{bmatrix} 0 \\ 1 \\ 1 \\ 0 \\ 0 \end{bmatrix} + \alpha_3 \begin{bmatrix} 0 \\ 0 \\ 1 \\ 1 \\ 0 \end{bmatrix} + \alpha_4 \begin{bmatrix} 0 \\ 0 \\ 0 \\ 1 \\ 1 \end{bmatrix},$$

then

$$\alpha_1 = 1,$$
$$\alpha_1 + \alpha_2 = 0,$$
$$\alpha_2 + \alpha_3 = 0,$$
$$\alpha_3 + \alpha_4 = 0,$$
$$\alpha_4 = 0.$$

Since we count modulo 2, we conclude from the second equation that $\alpha_2 = 1$. But then similarly also $\alpha_3 = \alpha_4 = 1$, which contradicts the last equation. This leads to the conclusion $[\widehat{A}] \neq [0]$ for the equivalence classes in $H_0(G; \mathbf{Z}_2)$. On the other hand, again counting mod 2,

$$\begin{bmatrix} 1 \\ 0 \\ 0 \\ 0 \\ 0 \end{bmatrix} + \begin{bmatrix} 1 \\ 1 \\ 0 \\ 0 \\ 0 \end{bmatrix} = \begin{bmatrix} 2 \\ 1 \\ 0 \\ 0 \\ 0 \end{bmatrix} = \begin{bmatrix} 0 \\ 1 \\ 0 \\ 0 \\ 0 \end{bmatrix}.$$

We can rewrite this equation as

$$\widehat{A} + \begin{bmatrix} 1 \\ 1 \\ 0 \\ 0 \\ 0 \end{bmatrix} = \widehat{B}.$$

Since

$$\begin{bmatrix} 1 \\ 1 \\ 0 \\ 0 \\ 0 \end{bmatrix} \in B_0(G; \mathbf{Z}_2),$$

this implies that
$$\widehat{A} \sim \widehat{B}.$$

Therefore, on the level of homology
$$[\widehat{A}] = [\widehat{B}] \in H_0(G; \mathbf{Z}_2).$$

We leave it to the reader to check that, in fact,
$$[\widehat{A}] = [\widehat{B}] = [\widehat{C}] = [\widehat{D}] = [\widehat{E}] \in H_0(G; \mathbf{Z}_2). \tag{1.6}$$

Thus, if we write out the elements of $H_0(G; \mathbf{Z}_2)$ without repetition, we get
$$\{[0], [\widehat{A}]\} \subset H_0(G; \mathbf{Z}_2). \tag{1.7}$$

The question is if we have found all homology classes in $H_0(G; \mathbf{Z}_2)$. To answer it we need to understand that the vector space $Z_0(G; \mathbf{Z}_2)$ *induces* the structure of a vector space on $H_0(G; \mathbf{Z}_2)$. For this end consider the two cycles $\widehat{A}, \widehat{B}$. They both induce the same element in $H_0(G; \mathbf{Z}_2)$ because $[\widehat{A}] = [\widehat{B}]$. Since $Z_0(G; \mathbf{Z}_2)$ is a vector space, we can add $\widehat{A}$ and $\widehat{B}$ to get $\widehat{A} + \widehat{B}$. Observe that in column notation

$$\widehat{A} + \widehat{B} = \begin{bmatrix} 1 \\ 0 \\ 0 \\ 0 \\ 0 \end{bmatrix} + \begin{bmatrix} 0 \\ 1 \\ 0 \\ 0 \\ 0 \end{bmatrix} = \begin{bmatrix} 1 \\ 1 \\ 0 \\ 0 \\ 0 \end{bmatrix} \in B_0(G; \mathbf{Z}_2).$$

Let us collect all this information in the context of the equivalence classes.
$$[0] \stackrel{1}{=} 2[\widehat{A}] = [\widehat{A}] + [\widehat{A}] \stackrel{2}{=} [\widehat{A}] + [\widehat{B}] \stackrel{3}{=} [\widehat{A} + \widehat{B}] \stackrel{4}{=} [0].$$

Equality 1 follows from the fact that we are still doing mod 2 arithmetic. Equality 2 holds because $\widehat{A} \sim \widehat{B}$. Equality 3 is what we mean by saying that $Z_0(G; \mathbf{Z}_2)$ induces the structure of a vector space on $H_0(G; \mathbf{Z}_2)$. More precisely, to add equivalence classes we just add representatives of the equivalence classes.[3] The last equality follows from the fact that $\widehat{A} + \widehat{B} \in B_0(G; \mathbf{Z}_2)$.

Equipped with the vector space structure of $H_0(G; \mathbf{Z}_2)$, we easily see that the homology class of an arbitrary cycle is

$$[\alpha_1 \widehat{A} + \alpha_2 \widehat{B} + \alpha_3 \widehat{C} + \alpha_4 \widehat{D} + \alpha_5 \widehat{E}] = (\alpha_1 + \alpha_2 + \alpha_3 + \alpha_4 + \alpha_5)[\widehat{A}] \in \{[0], [\widehat{A}]\}.$$

Therefore,
$$H_0(G; \mathbf{Z}_2) = \{[0], [\widehat{A}]\}. \tag{1.8}$$

---

[3] Of course, it needs to be checked that this addition is well defined. This is done in Chapter 13.

1.6 Mod 2 Homology of Graphs   35

A final comment is needed before ending this example. We motivated computing homology by arguing that we wanted an algebraic method for determining whether pictures have bounded regions or not. We introduced the combinatorial graphs as a combinatorial representation that was then turned into algebra. We do not want the final answer to depend on the graph. We also mentioned that proving this is difficult. But to emphasize this we will write the homology in terms of the set rather than the graph. In other words, what we claim (without proof!) is that

$$H_1(I; \mathbf{Z}_2) = \mathbf{0} \quad \text{and} \quad H_0(I; \mathbf{Z}_2) = \{[0], [\widehat{A}]\}.$$

Notice, however, that the chains depend explicitly on the combinatorial graph; therefore, it makes no sense to write $C_k(I; \mathbf{Z}_2)$.

Example 1.7 Let G be the graph of Figure 1.11 representing Γ^1. Then,

$$\mathcal{V} = \{A, B, C, D\},$$
$$\mathcal{E} = \{[A, B], [B, C], [D, C], [A, D]\}.$$

The computation of the homology begins just as in the previous example. The basis for the 0-chains, $C_0(G; \mathbf{Z}_2)$, is

$$\{\widehat{A}, \widehat{B}, \widehat{C}, \widehat{D}\},$$

while the basis for the 1-chains, $C_1(G; \mathbf{Z}_2)$, is

$$\{[\widehat{A, B}], [\widehat{B, C}], [\widehat{D, C}], [\widehat{A, D}]\}.$$

Again, using column vectors let

$$\widehat{A} = \begin{bmatrix} 1 \\ 0 \\ 0 \\ 0 \end{bmatrix}, \quad \widehat{B} = \begin{bmatrix} 0 \\ 1 \\ 0 \\ 0 \end{bmatrix}, \quad \widehat{C} = \begin{bmatrix} 0 \\ 0 \\ 1 \\ 0 \end{bmatrix}, \quad \widehat{D} = \begin{bmatrix} 0 \\ 0 \\ 0 \\ 1 \end{bmatrix}$$

and

$$[\widehat{A, B}] = \begin{bmatrix} 1 \\ 0 \\ 0 \\ 0 \end{bmatrix}, \quad [\widehat{B, C}] = \begin{bmatrix} 0 \\ 1 \\ 0 \\ 0 \end{bmatrix}, \quad [\widehat{D, C}] = \begin{bmatrix} 0 \\ 0 \\ 1 \\ 0 \end{bmatrix}, \quad [\widehat{A, D}] = \begin{bmatrix} 0 \\ 0 \\ 0 \\ 1 \end{bmatrix}.$$

With this convention, ∂_1 becomes the 4×4 matrix

$$\partial_1 = \begin{bmatrix} 1 & 0 & 0 & 1 \\ 1 & 1 & 0 & 0 \\ 0 & 1 & 1 & 0 \\ 0 & 0 & 1 & 1 \end{bmatrix}.$$

To compute $Z_1(G; \mathbf{Z}_2) := \ker \partial_1$, we need to solve the equation

$$\partial_1 \begin{bmatrix} \alpha_1 \\ \alpha_2 \\ \alpha_3 \\ \alpha_4 \end{bmatrix} = \begin{bmatrix} \alpha_1 + \alpha_4 \\ \alpha_1 + \alpha_2 \\ \alpha_2 + \alpha_3 \\ \alpha_3 + \alpha_4 \end{bmatrix} = \begin{bmatrix} 0 \\ 0 \\ 0 \\ 0 \end{bmatrix}.$$

Observe that, since we are using mod 2 arithmetic, $-1 = 1$ and any solution must satisfy

$$\alpha_1 = \alpha_2 = \alpha_3 = \alpha_4.$$

In particular, $\alpha_1 = \alpha_2 = \alpha_3 = \alpha_4 = 1$ is the only nonzero solution. Thus

$$Z_1(G; \mathbf{Z}_2) = \left\{ \begin{bmatrix} 0 \\ 0 \\ 0 \\ 0 \end{bmatrix}, \begin{bmatrix} 1 \\ 1 \\ 1 \\ 1 \end{bmatrix} \right\}.$$

As in the previous example, $B_1(G; \mathbf{Z}_2) = \mathbf{0}$. So

$$H_1(\varGamma^1; \mathbf{Z}_2) := Z_1(G; \mathbf{Z}_2) = \left\{ \begin{bmatrix} 0 \\ 0 \\ 0 \\ 0 \end{bmatrix}, \begin{bmatrix} 1 \\ 1 \\ 1 \\ 1 \end{bmatrix} \right\}.$$

Notice that this is different from the previous two examples, which has to do with the fact that $\varGamma^1$ encloses a region in the plane.

We still need to compute $H_0(\varGamma^1; \mathbf{Z}_2)$. We know that $Z_0(G; \mathbf{Z}_2) = C_0(G; \mathbf{Z}_2)$. The next step is to understand $B_0(G; \mathbf{Z}_2)$. Again, we look at how ∂_1 acts on the basis elements of $C_1(G; \mathbf{Z}_2)$ and conclude that

$$\partial_1[\widehat{A,B}] = \begin{bmatrix} 1 \\ 1 \\ 0 \\ 0 \end{bmatrix}, \quad \partial_1[\widehat{B,C}] = \begin{bmatrix} 0 \\ 1 \\ 1 \\ 0 \end{bmatrix}, \quad \partial_1[\widehat{D,C}] = \begin{bmatrix} 0 \\ 0 \\ 1 \\ 1 \end{bmatrix}, \quad \partial_1[\widehat{A,D}] = \begin{bmatrix} 1 \\ 0 \\ 0 \\ 1 \end{bmatrix}.$$

As before, $\widehat{A} \notin B_0(G; \mathbf{Z}_2)$, hence $[\widehat{A}] \neq [0]$ for elements of $H_0(\varGamma^1; \mathbf{Z}_2)$, but

$$\widehat{A} + \begin{bmatrix} 1 \\ 1 \\ 0 \\ 0 \end{bmatrix} = \widehat{B},$$

and so

$$\widehat{A} \sim \widehat{B}.$$

It is left to the reader to check that, as in the previous example,

$$H_0(\varGamma^1; \mathbf{Z}_2) = \{[0], [\widehat{A}]\}.$$

Remark 1.8 Several times in this section we have ended up with a vector space containing two elements $\{[0], [\widehat{A}]\}$. Observe that we have the following relations under vector addition with mod 2 arithmetic:

$$[0] + [0] = [0], \quad [0] + [\widehat{A}] = [\widehat{A}], \quad [\widehat{A}] + [\widehat{A}] = [0].$$

We can identify this with mod 2 arithmetic, that is, we can consider the set $\mathbf{Z}_2 = \{0, 1\}$ where

$$0 + 0 = 0, \quad 0 + 1 = 1, \quad 1 + 1 = 0.$$

From now on we will do this. In particular, this allows us to write the homology groups from the previous example as

$$H_0(\Gamma^1; \mathbf{Z}_2) \cong \mathbf{Z}_2, \quad H_1(\Gamma^1; \mathbf{Z}_2) \cong \mathbf{Z}_2.$$

Exercises

1.6 Verify Eq. (1.6).

1.7 Verify Eq. (1.8) by explicitly writing out all the elements of $Z_0(G; \mathbf{Z}_2)$ and for each of them find out whether its homology class is 0 or $[\widehat{A}]$.

1.8 Let G be the graph with edges $\mathcal{E} = \{[v_1, v_2]\}$ and vertices $\mathcal{V} = \{v_1, v_2\}$. Compute $H_k(G; \mathbf{Z}_2)$ for $k = 0, 1$.

1.9 Compute $H_k(G; \mathbf{Z}_2)$ for $k = 0, 1$, where G is the graph

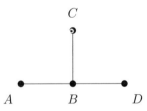

1.10 Compute $H_k(G; \mathbf{Z}_2)$ for $k = 0, 1$, where G is the graph with vertices $\{v_1, v_2, v_3\}$ and edges $\{[v_1, v_2], [v_2, v_3], [v_3, v_1]\}$.

1.11 Let G^n be a graph with n vertices where every pair of distinct vertices is connected by one edge. Create input files for the Homology program for several values of n starting from $n = 4$. Use the Homology program to compute the dimensions r_k of $H_k(G^n; \mathbf{Z}_2)$. Make a conjecture about the formula for every $n \geq 3$.

2
Cubical Homology

In Chapter 1 we have provided a heuristic introduction to homology using graphs that are made up of vertices (zero-dimensional cubes) and edges (one-dimensional cubes). Before that we motivated the need for a computational theory of homology using examples from image processing where it was natural to think of the images as being presented in terms of pixels (two-dimensional cubes), voxels (three-dimensional cubes), and even tetrapus (four-dimensional cubes). In Section 2.1 we formalize and generalize these examples to cubical complexes.

These complexes are based on cubes with vertices in an integer lattice. This may, at first glance, appear to be too restrictive. For example, in numerical analysis one often needs approximations that involve very fine grids. However, from a theoretical point of view, the size of a grid is just a question of choice of units. Thus, in principle one may assume that each cube is unitary, that it has sides of length 1 and vertices with integer coordinates.

Having established the geometric building blocks, in Section 2.2 we turn to defining homology groups. Sections 2.3 through 2.5 provide examples, theorems, and discussions of some of the more elementary relationships between homology groups and topological spaces. In Section 2.6 we provide a purely algebraic generalization of the concept of homology groups. This abstract approach is used in Chapter 4 to justify algorithms that are computationally more efficient. We conclude in Section 2.7 with a slightly different definition of homology that results in the reduced homology groups. As will become clear later, there is a variety of theorems in which the statements and proofs are clearer using this alternative approach.

2.1 Cubical Sets

Given a graph, we obtained a combinatorial object by defining it in terms of vertices and edges. Of course, these will not suffice to combinatorialize higher-dimensional topological spaces. Thus in this section we formally introduce the notion of cubes, which form the building blocks for the homology theory presented in this book.

2.1.1 Elementary Cubes

Definition 2.1 An *elementary interval* is a closed interval $I \subset \mathbf{R}$ of the form

$$I = [l, l+1] \quad \text{or} \quad I = [l, l]$$

for some $l \in \mathbf{Z}$. To simplify the notation, we write

$$[l] = [l, l]$$

for an interval that contains only one point. Elementary intervals that consist of a single point are *degenerate*, while those of length 1 are *nondegenerate*.

Example 2.2 The intervals $[2,3]$, $[-15,-14]$, and $[7]$ are all examples of elementary intervals. On the other hand, $[\frac{1}{2}, \frac{3}{2}]$ is not an elementary interval since the boundary points are not integers. Similarly, $[1,3]$ is not an elementary interval since the length of the interval is greater than 1.

Definition 2.3 An *elementary cube* Q is a finite product of elementary intervals, that is,

$$Q = I_1 \times I_2 \times \cdots \times I_d \subset \mathbf{R}^d,$$

where each I_i is an elementary interval. The set of all elementary cubes in $\mathbf{R}^d$ is denoted by $\mathcal{K}^d$. The set of all elementary cubes is denoted by $\mathcal{K}$, namely

$$\mathcal{K} := \bigcup_{d=1}^{\infty} \mathcal{K}^d.$$

Figure 2.1 indicates a variety of elementary cubes. Observe that the cube $[1,2] \subset \mathbf{R}$ is different from the cube $[1,2] \times [0] \subset \mathbf{R}^2$ since they are subsets of different spaces. Of course, using the inclusion map $i : \mathbf{R} \to \mathbf{R}^2$ given by $i(x) = (x,0)$, we can identify these two elementary cubes. However, we take great care in this book to explicitly state this identification if we make it. Thus, if the identification is not clearly stated, then they should be treated as distinct sets.

Another natural identification one should be aware of is the correspondence between a point in $\mathbf{R}^d$ and an elementary cube all of whose elementary intervals are degenerate. For example, $(1,0,2) \in \mathbf{R}^3$ is a point in a topological space, whereas $[1] \times [0] \times [2]$ is an elementary cube. The former is a topological object, while the latter will be used as a combinatorial element.

Here are some other examples of elementary cubes:

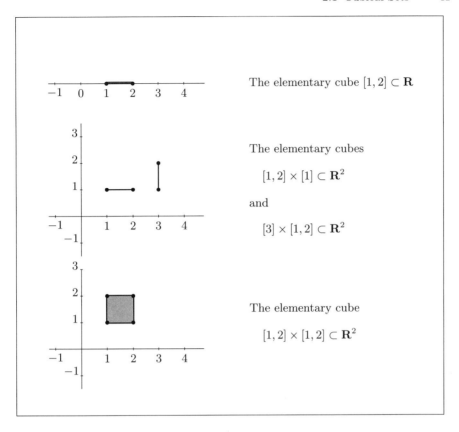

Fig. 2.1. Elementary cubes in $\mathbf{R}$ and $\mathbf{R}^2$.

$Q_1 := [1,2] \times [0,1] \times [-2,-1] \subset \mathbf{R}^3;$
$Q_2 := [1] \times [1,2] \times [0,1] = \{1\} \times [1,2] \times [0,1] \subset \mathbf{R}^3;$
$Q_3 := [1,2] \times [0] \times [-1] = [1,2] \times \{0\} \times \{-1\} \subset \mathbf{R}^3;$
$Q_4 := [0] \times [0] \times [0] = \{(0,0,0)\} \subset \mathbf{R}^3;$
$Q_5 := [-1,0] \times [3,4] \times [6] \times [1,2] = [-1,0] \times [3,4] \times \{6\} \times [1,2] \subset \mathbf{R}^4,$

which we shall not attempt to draw.

Definition 2.4 Let $Q = I_1 \times I_2 \times \cdots \times I_d \subset \mathbf{R}^d$ be an elementary cube. The *embedding number* of Q is denoted by $\operatorname{emb} Q$ and is defined to be d since $Q \subset \mathbf{R}^d$. The interval I_i is referred to as the ith *component* of Q and is written as $I_i(Q)$. The *dimension* of Q is defined to be the number of nondegenerate components in Q and is denoted by $\dim Q$.

Observe that if $\operatorname{emb} Q = d$, then $Q \in \mathcal{K}^d$. We also let

$$\mathcal{K}_k := \{Q \in \mathcal{K} \mid \dim Q = k\}$$

and
$$\mathcal{K}_k^d := \mathcal{K}_k \cap \mathcal{K}^d.$$

Example 2.5 Referring to the elementary cubes defined above, we have that $I_1(Q_3) = [1,2]$, $I_2(Q_3) = [0]$, and $I_3(Q_3) = [-1]$. Furthermore,

$$\operatorname{emb} Q_1 = 3 \text{ and } \dim Q_1 = 3,$$
$$\operatorname{emb} Q_2 = 3 \text{ and } \dim Q_2 = 2,$$
$$\operatorname{emb} Q_3 = 3 \text{ and } \dim Q_3 = 1,$$
$$\operatorname{emb} Q_4 = 3 \text{ and } \dim Q_4 = 0,$$
$$\operatorname{emb} Q_5 = 4 \text{ and } \dim Q_5 = 3.$$

In particular, the reader should observe that the only general relation between the embedding number and the dimension of an elementary cube Q is that

$$0 \leq \dim Q \leq \operatorname{emb} Q. \tag{2.1}$$

Proposition 2.6 *Let $Q \in \mathcal{K}_k^d$ and $P \in \mathcal{K}_{k'}^{d'}$. Then*

$$Q \times P \in \mathcal{K}_{k+k'}^{d+d'}.$$

Proof. Because $Q \in \mathcal{K}^d$, it can be written as the product of d elementary intervals:
$$Q = I_1 \times I_2 \times \ldots \times I_d.$$
Similarly,
$$P = J_1 \times J_2 \times \ldots \times J_{d'},$$
where each J_i is an elementary interval. Hence,
$$Q \times P = I_1 \times I_2 \times \ldots \times I_d \times J_1 \times J_2 \times \ldots \times J_{d'},$$
which is a product of $d + d'$ elementary intervals. It is left to the reader to check that $\dim(Q \times P) = \dim Q + \dim P$. □

It should be clear from the proof of Proposition 2.6 that though they lie in the same space $Q \times P \neq P \times Q$.

The following definition will allow us to decompose elementary cubes into lower-dimensional objects.

Definition 2.7 Let $Q, P \in \mathcal{K}$. If $Q \subset P$, then Q is a *face* of P. This is denoted by $Q \preceq P$. If $Q \preceq P$ and $Q \neq P$, then Q is a *proper face* of P, which is written as $Q \prec P$. Q is a *primary* face of P if Q is a face of P and $\dim Q = \dim P - 1$.

Example 2.8 Let $Q = [1,2] \times [1]$ and $P = [1,2] \times [1,2]$. Then $Q \prec P$ and Q is a primary face of P.

2.1.2 Cubical Sets

As mentioned before, elementary cubes make up the basic building blocks for our homology theory. Thus we begin by restricting our attention to the following class of topological spaces.[1]

---
[1] We consider more general spaces in Chapter 11.

Definition 2.9 A set $X \subset \mathbf{R}^d$ is *cubical* if X can be written as a finite union of elementary cubes.

If $X \subset \mathbf{R}^d$ is a cubical set, then we adopt the following notation:
$$\mathcal{K}(X) := \{Q \in \mathcal{K} \mid Q \subset X\}$$
and
$$\mathcal{K}_k(X) := \{Q \in \mathcal{K}(X) \mid \dim Q = k\}.$$
Observe that if $Q \subset X$ and $Q \in \mathcal{K}$, then $\operatorname{emb} Q = d$, since $X \subset \mathbf{R}^d$. This in turn implies that $Q \in \mathcal{K}^d$, so to use the notation $\mathcal{K}^d(X)$ is somewhat redundant, but it serves to remind us that $X \subset \mathbf{R}^d$. Therefore, when it is convenient we will write $\mathcal{K}^d(X)$ and also $\mathcal{K}_k^d(X) := \mathcal{K}^d(X) \cap \mathcal{K}_k(X)$. In analogy with graphs, the elements of $\mathcal{K}_0(X)$ are the *vertices* of X and the elements of $\mathcal{K}_1(X)$ are the *edges* of X. More generally, the elements of $\mathcal{K}_k(X)$ are the *k-cubes* of X.

Example 2.10 Consider the set $X = [0,1] \times [0,1] \times [0,1] \subset \mathbf{R}^3$. This is an elementary cube and, hence, is a cubical set. It is easy to check that

$$\mathcal{K}_3(X) = \{[0,1] \times [0,1] \times [0,1]\},$$
$$\mathcal{K}_2(X) = \{[0] \times [0,1] \times [0,1], [1] \times [0,1] \times [0,1],$$
$$[0,1] \times [0] \times [0,1], [0,1] \times [1] \times [0,1],$$
$$[0,1] \times [0,1] \times [0], [0,1] \times [0,1] \times [1]\},$$
$$\mathcal{K}_1(X) = \{[0] \times [0] \times [0,1], [0] \times [1] \times [0,1],$$
$$[0] \times [0,1] \times [0], [0] \times [0,1] \times [1],$$
$$[1] \times [0] \times [0,1], [1] \times [1] \times [0,1],$$
$$[1] \times [0,1] \times [0], [1] \times [0,1] \times [1],$$
$$[0,1] \times [0] \times [0], [0,1] \times [0] \times [1],$$
$$[0,1] \times [1] \times [0], [0,1] \times [1] \times [1]\},$$
$$\mathcal{K}_0(X) = \{[0] \times [0] \times [0], [0] \times [0] \times [1],$$
$$[0] \times [1] \times [0], [0] \times [1] \times [1],$$
$$[1] \times [0] \times [0], [1] \times [0] \times [1],$$
$$[1] \times [1] \times [0], [1] \times [1] \times [1]\}.$$

Example 2.11 It should be noted that the definition of a cubical set is extremely restrictive. For example, the unit circle $x^2 + y^2 = 1$ is not a cubical set. In fact, even a simple set such as a point may or may not be a cubical set. In particular, consider the set consisting of one point $P = \{(x,y,z)\} \subset \mathbf{R}^3$. P is a cubical set if and only if x, y, and z are all integers.

The following result uses some simple point-set topology (see Chapter 12).

Proposition 2.12 *If $X \subset \mathbf{R}^d$ is cubical, then X is closed and bounded.*

Proof. By definition a cubical set is the finite union of elementary cubes. By Exercise 12.13 an elementary cube is closed and by Theorem 12.18 the finite union of closed sets is closed.

To show that X is bounded, it is sufficient to prove that for some $R > 0$

$$X \subset B_0(0, R), \tag{2.2}$$

where $B_0(0, R)$ denotes the ball around the origin in the supremum norm (see Section 12.1). For this end let $Q \in \mathcal{K}(X)$. Then $Q = I_1 \times I_2 \times \cdots \times I_d$, where $I_i = [l_i]$ or $I_i = [l_i, l_i + 1]$. Let

$$\rho(Q) = \max_{i=1,\ldots d}\{|l_i| + 1\}.$$

Taking $R := \max_{Q \in \mathcal{K}(X)} \rho(Q)$ we easily verify (2.2). □

2.1.3 Elementary Cells

Elementary cubes are the building blocks for the homology theory that we are developing. However, for technical reasons it is useful to have additional sets to work with. For this reason we introduce the notion of elementary cells.

Definition 2.13 Let I be an elementary interval. The associated *elementary cell* is

$$\mathring{I} := \begin{cases} (l, l+1) & \text{if } I = [l, l+1], \\ [l] & \text{if } I = [l, l]. \end{cases}$$

We extend this definition to a general elementary cube $Q = I_1 \times I_2 \times \ldots \times I_d \subset \mathbf{R}^d$ by defining the associated *elementary cell* as

$$\mathring{Q} := \mathring{I}_1 \times \mathring{I}_2 \times \ldots \times \mathring{I}_d.$$

Example 2.14 Consider the elementary cube $Q = [1, 2] \times [3] \in \mathcal{K}_1^2$. The associated elementary cell is $\mathring{Q} = (1, 2) \times [3] \subset \mathbf{R}^2$.

Given a point in $\mathbf{R}^d$, we need to be able to describe the elementary cell or cube that contains it. For this, the following two functions are useful. Let $x \in \mathbf{R}$,

$$\text{floor}(x) := \max\{n \in \mathbf{Z} \mid n \leq x\},$$
$$\text{ceil}(x) := \min\{n \in \mathbf{Z} \mid x \leq n\}.$$

Some attributes of elementary cells are summarized in the following proposition.

Proposition 2.15 *Elementary cells have the following properties:*

(i) $\mathbf{R}^d = \bigcup\{\mathring{Q} \mid Q \in \mathcal{K}^d\}$.

(ii) $A \subset \mathbf{R}^d$ bounded implies that $\operatorname{card}\{Q \in \mathcal{K}^d \mid \mathring{Q} \cap A \neq \emptyset\} < \infty$ where, given a set S, $\operatorname{card}(S)$ stands for its cardinality, that is, the number of its elements.

(iii) If $P, Q \in \mathcal{K}^d$, then $\mathring{P} \cap \mathring{Q} = \emptyset$ or $P = Q$.

(iv) For every $Q \in \mathcal{K}$, $\operatorname{cl} \mathring{Q} = Q$.

(v) $Q \in \mathcal{K}^d$ implies that $Q = \bigcup \{\mathring{P} \mid P \in \mathcal{K}^d \text{ such that } \mathring{P} \subset Q\}$.

(vi) If X is a cubical set and $\mathring{Q} \cap X \neq \emptyset$ for some elementary cube Q, then $Q \subset X$.

Proof. (i) Obviously $\bigcup\{\mathring{Q} \mid Q \in \mathcal{K}^d\} \subset \mathbf{R}^d$. To prove the opposite inclusion, take a point $x = (x_1, x_2, \ldots, x_d) \in \mathbf{R}^d$ and set

$$I_i := [\operatorname{floor}(x_i), \operatorname{ceil}(x_i)].$$

Then $\mathring{Q} := \mathring{I}_1 \times \mathring{I}_2 \times \ldots \times \mathring{I}_d$ is an elementary cell and $x \in \mathring{Q}$. This proves (i).

(ii) The proof is straightforward.

(iii) The result is obvious for elementary cubes of dimension zero and one, that is, for elementary intervals. It extends immediately to all elementary cubes, because the intersection of products of intervals is the product of the intersections of the corresponding intervals.

(iv) See Definition 12.19, Exercise 12.29, and Exercise 12.30.

(v) Let $x = (x_1, x_2, \ldots, x_d) \in Q$. Define

$$J_i := \begin{cases} [x_i, x_i] & \text{if } x_i \text{ is an endpoint of } I_i(Q), \\ I_i(Q) & \text{otherwise}, \end{cases}$$

and set $P := J_1 \times J_2 \times \ldots \times J_d$. Then obviously $x \in \mathring{P}$ and $\mathring{P} \subset Q$. Hence x belongs to the right-hand side of (v).

(vi) Let $X = P_1 \cup P_2 \cup \ldots \cup P_m$, where the P_i are elementary cubes. Then $\mathring{Q} \cap P_i \neq \emptyset$ for some $i = 1, 2, \ldots, m$. It follows from (v) and (iii) that $\mathring{Q} \subset P_i$ and from (iv) that $Q \subset \operatorname{cl} P_i = P_i \subset X$. □

Exercises

2.1 The *unitary d-cube* is $[0,1]^d \subset \mathbf{R}^d$. This obviously is an elementary cube.

(a) List the faces contained in $[0,1]^2$.
(b) How many primary faces are in $[0,1]^d$?
(c) How many faces of all dimensions does $[0,1]^d$ contain?

2.2 Show that any $(k-1)$-dimensional face of a $(k+1)$-dimensional elementary cube Q is a common face of exactly two k-dimensional faces of Q.

The next two exercises are practice using the CubTop program. The reader should first consult the readme file in CubTop.

46   2 Cubical Homology

2.3 Open the file exC2d.txt in the folder CubTop/Examples of a set X presented there to see how is it constructed. You will see a list of elementary cubes in $\mathbf{R}^2$ there. Then perform the following tasks.

(a) Run CubTop with the option -g to create a bitmap file of X.
(b) Run CubTop properface with the options -c -o -g to see the proper faces of X. Present the output files.

2.4 Create input files for each of the three elementary cubes displayed in Figure 2.1. The list of elementary cubes in each file will contain one element only. Then perform the following tasks.

(a) Repeat step (b) of the previous exercise for each of your files.
(b) Run CubTop union with the options -c -o -g for the two elementary cubes in $\mathbf{R}^2$ to create their union Y.
(c) Repeat step (b) of the previous exercise for the file of Y to display the free faces of Y.

2.5 Two combinatorial graphs $(\mathcal{V}_1, \mathcal{E}_1)$ and $(\mathcal{V}_2, \mathcal{E}_2)$ are *equivalent* if there exists a bijection $f : \mathcal{V}_1 \to \mathcal{V}_2$ such that, for any $u, v \in \mathcal{V}_1$, $(u, v) \in \mathcal{E}_1$ if and only if $(f(u), f(v)) \in \mathcal{E}_2$.

Prove that any combinatorial graph that is a tree can be represented as a one-dimensional cubical set X in the sense that it is equivalent to the combinatorial graph $(\mathcal{K}_0(X), \mathcal{E}(X))$, where $\mathcal{E}(X)$ are pairs of vertices of edges in $\mathcal{K}_0(X)$.

2.6 Give an example of a combinatorial graph that cannot be represented by a one-dimensional cubical set.

2.7 Let Q^d be the unitary cube in Exercise 2.1 and let X be its one-dimensional skeleton, that is, the union of all edges of Q^d.

(a) For $d = 2, 3, 4, 5, 6$, determine the number of vertices of Q^d and the number of edges of Q^d (note that X has the same vertices and edges as Q^d).
(b) For the same values of d, prepare the input files and run the Homology program to compute the ranks r_0 and r_1 of $H_0(X)$ and $H_1(X)$, respectively.
(c) Make a guess, based on your empirical experience, on how the number of vertices, the number of edges, and the ranks r_0, r_1 are related together. Use more values of d if what you computed in (a) and (b) is not sufficient for making a good guess.

2.8 Let $X_i := \{x \in \mathbf{R}^d \mid \|x\|_i \leq r\}$. For which values of r and which norms (see Section 12.1) $i = 0, 1$, or 2 is X_i a cubical set?

2.9 What are the elementary cells of the unit cube in $\mathbf{R}^3$?

2.10 The file exR2d.txt in the folder CubTop/Examples contains a list of elementary cells. Run the CubTop program on the file exR2d.txt with the option -g to see what the union U of those cells looks like. Run CubTop closedhull with appropriate options to see the cubical set that is the closure of U.

2.2 The Algebra of Cubical Sets

In this section we finally present the formal definitions that we use to pass from the topology of a cubical set to the algebra of homology.

2.2.1 Cubical Chains

In Chapter 1 we have associated algebraic structures with graphs. Given the set of vertices $\{v_1, v_2, \ldots, v_n\}$ and edges $\{e_1, e_2, \ldots, e_m\}$, we have considered the 0-chains of the form

$$c = \alpha_1 \widehat{v_1} + \alpha_2 \widehat{v_2} + \ldots + \alpha_n \widehat{v_n}$$

and 1-chains of the form

$$d = \beta_1 \widehat{e_1} + \beta_2 \widehat{e_2} + \ldots + \beta_m \widehat{e_m},$$

respectively. The hat notation is used to distinguish between a vertex and an edge viewed as geometric objects and those viewed as algebraic objects. We shall now introduce an analogous algebraic structure for cubical sets of arbitrary dimension.

Mimicking what is done in Chapter 1, with each elementary k-cube $Q \in \mathcal{K}_k^d$ we associate an algebraic object $\widehat{Q}$ called an *elementary k-chain* of $\mathbf{R}^d$. The set of all elementary k-chains of $\mathbf{R}^d$ is denoted by

$$\widehat{\mathcal{K}}_k^d := \left\{ \widehat{Q} \mid Q \in \mathcal{K}_k^d \right\},$$

and the set of all *elementary chains* of $\mathbf{R}^d$ is given by

$$\widehat{\mathcal{K}}^d := \bigcup_{k=0}^{\infty} \widehat{\mathcal{K}}_k^d.$$

Given any *finite* collection $\{\widehat{Q}_1, \widehat{Q}_2, \ldots, \widehat{Q}_m\} \subset \widehat{\mathcal{K}}_k^d$ of k-dimensional elementary chains, we are allowed to consider sums of the form

$$c = \alpha_1 \widehat{Q}_1 + \alpha_2 \widehat{Q}_2 + \ldots + \alpha_m \widehat{Q}_m,$$

where α_i are arbitrary integers. If all the $\alpha_i = 0$, then we let $c = 0$. These can be thought of as our k-chains, the set of which is denoted by C_k^d. The addition of k-chains is naturally defined by

$$\sum \alpha_i \widehat{Q}_i + \sum \beta_i \widehat{Q}_i := \sum (\alpha_i + \beta_i) \widehat{Q}_i.$$

Observe that given an arbitrary k-chain $c = \sum_{i=0}^m \alpha_i \widehat{Q}_i$, there is an inverse element $-c = \sum_{i=0}^m (-\alpha_i) \widehat{Q}_i$ with the property that $c + (-c) = 0$. Therefore, C_k^d is an abelian group and, in fact, it is a free abelian group with basis $\widehat{\mathcal{K}}_k^d$.

The idea behind the presentation we just gave of the derivation of the abelian group of k-chains should be intuitively clear: Each elementary cube is used to generate a basis element that we called an elementary chain, and thus a chain is just defined in terms of a finite sum of elementary chains. There is, however, another prescription for using a set to generate a free abelian group. This involves viewing the chains as functions from $\mathcal{K}_k^d$ to $\mathbf{Z}$ (see Definition 13.67).

In particular, for each $Q \in \mathcal{K}_k^d$, define $\widehat{Q} : \mathcal{K}_k^d \to \mathbf{Z}$ by

$$\widehat{Q}(P) := \begin{cases} 1 & \text{if } P = Q, \\ 0 & \text{otherwise,} \end{cases} \qquad (2.3)$$

and in a slight abuse of notation let $0 : \mathcal{K}_k^d \to \mathbf{Z}$ be the zero function, namely $0(Q) = 0$ for all $Q \in \mathcal{K}_k^d$. Refining the definition presented above, $\widehat{Q}$ is the *elementary chain dual* to the elementary cube Q. Because the elementary chains take values in the integers, we are allowed to take finite sums of them.

Definition 2.16 The group C_k^d of k-*dimensional chains* of $\mathbf{R}^d$ (k-chains for short) is the free abelian group generated by the elementary chains of $\mathcal{K}_k^d$. Thus the elements of C_k^d are functions $c : \mathcal{K}_k^d \to \mathbf{Z}$ such that $c(Q) = 0$ for all but a finite number of $Q \in \mathcal{K}_k^d$. In particular, $\widehat{\mathcal{K}}_k^d$ is the basis for C_k^d. Using the notation of Definition 13.67,

$$C_k^d := \mathbf{Z}(\mathcal{K}_k^d).$$

If $c \in C_k^d$, then $\dim c := k$.

Obviously, since the elementary cubes are contained in $\mathbf{R}^d$, for $k < 0$ and $k > d$, the set $\mathcal{K}_k = \emptyset$ and the corresponding group of k-chains is $C_k^d = 0$.

Observe that since $\mathcal{K}_k^d$ is infinite, C_k^d is an infinitely generated free abelian group. In practice we are interested in the chains localized to cubical sets, and soon we will give an appropriate definition. However, as long as the localization is irrelevant, it is convenient not to bind ourselves to a particular cubical set.

Recall that, given an elementary cube Q, its dual elementary chain is $\widehat{Q}$. Similarly, given an elementary chain $\widehat{Q}$, we refer to Q as its dual elementary cube. This is justified by the following proposition.

Proposition 2.17 *The map $\phi : \mathcal{K}_k^d \to \widehat{\mathcal{K}}_k^d$ given by $\phi(Q) = \widehat{Q}$ is a bijection.*

Proof. Because $\widehat{\mathcal{K}}_k^d$ is defined to be the image of ϕ, it is obvious that ϕ is surjective. To prove injectivity, assume that $P, Q \in \mathcal{K}_k^d$ and $\widehat{P} = \widehat{Q}$. This implies that

$$1 = \widehat{P}(P) = \widehat{Q}(P)$$

and hence that $P = Q$. □

Observe that the inverse of ϕ allows us to pass from an algebraic object, the elementary chain $\widehat{Q}$, to a well-defined topological set, Q. We would like to be able to do this for general chains.

2.2 The Algebra of Cubical Sets

Definition 2.18 Let $c \in C_k^d$. The *support* of the chain c is the cubical set

$$|c| := \bigcup \{Q \in \mathcal{K}_k^d \mid c(Q) \neq 0\}.$$

Support has several nice geometric features.

Proposition 2.19 *Support satisfies the following properties:*

(i) $|c| = \emptyset$ *if and only if* $c = 0$.
(ii) Let $\alpha \in \mathbf{Z}$ *and* $c \in C_k^d$; *then*

$$|\alpha c| = \begin{cases} \emptyset & \text{if } \alpha = 0, \\ |c| & \text{if } \alpha \neq 0. \end{cases}$$

(iii) If $Q \in \mathcal{K}$, *then* $\left|\widehat{Q}\right| = Q$.
(iv) If $c_1, c_2 \in C_k^d$, *then* $|c_1 + c_2| \subset |c_1| \cup |c_2|$.

Proof. (i) In the case of the zero chain for every $Q \in \mathcal{K}_k^d$ the value $0(Q) = 0$; therefore, $|c| = \emptyset$. On the other hand, if $|c| = \emptyset$, then there is no Q such that $c(Q) \neq 0$; therefore, $c = 0$.

(ii) This follows directly from the definition of support and (i).

(iii) This follows from the definition of elementary chains.

(iv) Let $x \in |c_1 + c_2|$. Then $x \in Q$ for some $Q \in \mathcal{K}_k^d$ such that $(c_1+c_2)(Q) = c_1(Q)+c_2(Q) \neq 0$. It follows that either $c_1(Q) \neq 0$ or $c_2(Q) \neq 0$, hence $x \in |c_1|$ or $x \in |c_2|$. □

Example 2.20 It is not true in general that $|c_1 + c_2| = |c_1| \cup |c_2|$. Consider any chain c such that $|c| \neq \emptyset$. Observe that

$$\emptyset = |c - c| \neq |c| \cup |c| = |c| \neq \emptyset.$$

Similarly, notice that support does not define a bijection between the set of chains and cubical sets. In particular, for any chain c,

$$|2c| = |c|.$$

Consider an arbitrary chain $c \in C_k^d$. As indicated above, the set of elementary chains forms a basis for C_k^d. Thus it would be nice to have a simple formula that describes c in terms of the elements of $\widehat{\mathcal{K}}_k^d$. This is the motivation for the following definition, which is analogous to the dot product in a vector space.

Definition 2.21 Consider $c_1, c_2 \in C_k^d$, where $c_1 = \sum_{i=1}^m \alpha_i \widehat{Q}_i$ and $c_2 = \sum_{i=1}^m \beta_i \widehat{Q}_i$. The *scalar product* of the chains c_1 and c_2 is defined as

$$\langle c_1, c_2 \rangle := \sum_{i=1}^m \alpha_i \beta_i.$$

Proposition 2.22 *The scalar product defines a mapping*

$$\langle \cdot, \cdot \rangle : C_k^d \times C_k^d \to \mathbf{Z}$$
$$(c_1, c_2) \mapsto \langle c_1, c_2 \rangle,$$

which is bilinear.

Proof. We need to show that $\langle \alpha c_1 + \beta c_2, c_3 \rangle = \alpha \langle c_1, c_3 \rangle + \beta \langle c_2, c_3 \rangle$ and $\langle c_1, \alpha c_2 + \beta c_3 \rangle = \alpha \langle c_1, c_2 \rangle + \beta \langle c_1, c_3 \rangle$. For $j = 1, 2, 3$, let

$$c_j = \sum_{i=1}^{m} \gamma_{j,i} \widehat{Q}_i.$$

Then

$$\langle \alpha c_1 + \beta c_2, c_3 \rangle = \left\langle \alpha \sum_{i=1}^{m} \gamma_{1,i} \widehat{Q}_i + \beta \sum_{i=1}^{m} \gamma_{2,i} \widehat{Q}_i, \sum_{i=1}^{m} \gamma_{3,i} \widehat{Q}_i \right\rangle$$
$$= \left\langle \sum_{i=1}^{m} (\alpha \gamma_{1,i} + \beta \gamma_{2,i}) \widehat{Q}_i, \sum_{i=1}^{m} \gamma_{3,i} \widehat{Q}_i \right\rangle$$
$$= \sum_{i=1}^{m} (\alpha \gamma_{1,i} + \beta \gamma_{2,i}) \gamma_{3,i}$$
$$= \alpha \sum_{i=1}^{m} \gamma_{1,i} \gamma_{3,i} + \beta \sum_{i=1}^{m} \gamma_{2,i} \gamma_{3,i}$$
$$= \alpha \left\langle \sum_{i=1}^{m} \gamma_{1,i} \widehat{Q}_i, \sum_{i=1}^{m} \gamma_{3,i} \widehat{Q}_i \right\rangle + \beta \left\langle \sum_{i=1}^{m} \gamma_{2,i} \widehat{Q}_i, \sum_{i=1}^{m} \gamma_{3,i} \widehat{Q}_i \right\rangle$$
$$= \alpha \langle c_1, c_3 \rangle + \beta \langle c_2, c_3 \rangle.$$

The proof of the other equality is nearly the same. □

Using the scalar product notation, the formula for the support of a chain can be rewritten as

$$|c| = \bigcup_{\langle c, \widehat{Q}_i \rangle \neq 0} Q_i. \tag{2.4}$$

While (2.4) is a nice formula, it does have its restrictions. The most obvious is that the support of the k-chain is given only in terms of k-dimensional cubes. On the other hand, we know that cubes can be decomposed into lower-dimensional faces. Furthermore, the construction of the boundary operator in Chapter 1 is based on this decomposition. As will be made precise in the next section, the cubical boundary operator takes k-chains to $(k-1)$-chains, and for this reason we want some method to express the support of a k-chain in terms of lower-dimensional cubes. This is the motivation for defining the following product.

2.2 The Algebra of Cubical Sets

Definition 2.23 Given two elementary cubes $P \in \mathcal{K}_k^d$ and $Q \in \mathcal{K}_{k'}^{d'}$ set
$$\widehat{P} \diamond \widehat{Q} := \widehat{P \times Q}.$$
This definition extends to arbitrary chains $c_1 \in C_k^d$ and $c_2 \in C_{k'}^{d'}$ by
$$c_1 \diamond c_2 := \sum_{P \in \mathcal{K}_k, Q \in \mathcal{K}_{k'}} \langle c_1, \widehat{P} \rangle \langle c_2, \widehat{Q} \rangle \widehat{P \times Q}.$$
The chain $c_1 \diamond c_2 \in C_{k+k'}^{d+d'}$ is called the *cubical product* of c_1 and c_2.

Example 2.24 Let
$$P_1 = [0] \times [0,1], \; P_2 = [1] \times [0,1], \; P_3 = [0,1] \times [0], \; P_4 = [0,1] \times [1]$$
and
$$Q_1 = [-1, 0], \; Q_2 = [0, 1].$$
Observe that $\widehat{P}_i \in \widehat{\mathcal{K}}_1^2$ and $\widehat{Q}_i \in \widehat{\mathcal{K}}_1^1$. Given the chains $c_1 = \widehat{P}_1 + \widehat{P}_2 + \widehat{P}_3 + \widehat{P}_4$ and $c_2 = \widehat{Q}_1 + \widehat{Q}_2$, by definition
$$c_1 \diamond c_2 = \widehat{P_1 \times Q_1} + \widehat{P_2 \times Q_1} + \widehat{P_3 \times Q_1} + \widehat{P_4 \times Q_1}$$
$$+ \widehat{P_1 \times Q_2} + \widehat{P_2 \times Q_2} + \widehat{P_3 \times Q_2} + \widehat{P_4 \times Q_2}$$
while
$$c_2 \diamond c_1 = \widehat{Q_1 \times P_1} + \widehat{Q_1 \times P_2} + \widehat{Q_1 \times P_3} + \widehat{Q_1 \times P_4}$$
$$+ \widehat{Q_2 \times P_1} + \widehat{Q_2 \times P_2} + \widehat{Q_2 \times P_3} + \widehat{Q_2 \times P_4}.$$
Figure 2.2 indicates the support of the chains $c_1, c_2, c_1 \diamond c_2$, and $c_2 \diamond c_1$.

The cubical product has the following properties.

Proposition 2.25 *Let c_1, c_2, c_3 be any chains.*

(i) $c_1 \diamond 0 = 0 \diamond c_1 = 0.$
(ii) $c_1 \diamond (c_2 + c_3) = c_1 \diamond c_2 + c_1 \diamond c_3$, provided $c_2, c_3 \in C_k^d$.
(iii) $(c_1 \diamond c_2) \diamond c_3 = c_1 \diamond (c_2 \diamond c_3).$
(iv) If $c_1 \diamond c_2 = 0$, then $c_1 = 0$ or $c_2 = 0$.
(v) $|c_1 \diamond c_2| = |c_1| \times |c_2|.$

Proof. (i) and (ii) follow immediately from the definition.
(iii) The proof is straightforward.
(iv) Assume that $c_1 = \sum_{i=1}^m \alpha_i \widehat{P}_i$ and $c_2 = \sum_{j=1}^n \beta_j \widehat{Q}_j$. Then
$$\sum_{i=1}^m \sum_{j=1}^n \alpha_i \beta_j \widehat{P}_i \diamond \widehat{Q}_j = 0,$$

52   2 Cubical Homology

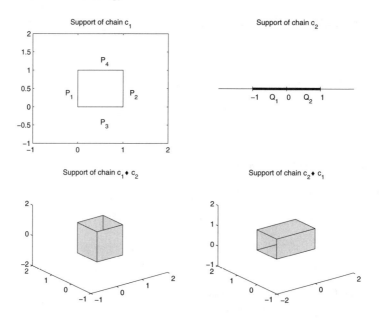

Fig. 2.2. The support of the chains c_1, c_2, $c_1 \diamond c_2$, and $c_2 \diamond c_1$.

that is, $\alpha_i \beta_j = 0$ for any $i = 1, 2, \ldots, m$, and $j = 1, 2, \ldots, n$. It follows that

$$0 = \sum_{i=1}^{m} \sum_{j=1}^{n} (\alpha_i \beta_j)^2 = \Big(\sum_{i=1}^{m} \alpha_i^2\Big)\Big(\sum_{j=1}^{n} \beta_j^2\Big),$$

hence $\sum_{i=1}^{m} \alpha_i^2 = 0$ or $\sum_{j=1}^{n} \beta_j^2 = 0$. Consequently, $c_1 = 0$ or $c_2 = 0$.
(v) We leave this as an exercise. □

Proposition 2.26 *Let $\widehat{Q}$ be an elementary cubical chain of $\mathbf{R}^d$ with $d > 1$. Then there exist unique elementary cubical chains $\widehat{I}$ and $\widehat{P}$ with $\operatorname{emb} I = 1$ and $\operatorname{emb} P = d - 1$ such that*

$$\widehat{Q} = \widehat{I} \diamond \widehat{P}.$$

Proof. Since $\widehat{Q}$ is an elementary cubical chain, Q is an elementary cube, namely
$$Q = I_1 \times I_2 \times \cdots \times I_d.$$
Set $I := I_1$ and $P := I_2 \times I_3 \times \cdots \times I_d$; then $\widehat{Q} = \widehat{I} \diamond \widehat{P}$.

We still need to prove that this is the unique decomposition. If $\widehat{Q} = \widehat{J} \diamond \widehat{P'}$ for some $J \in \mathcal{K}^1$ and $P' \in \mathcal{K}^{d-1}$, then $\widehat{I_1 \times P} = \widehat{J \times P'}$ and from Proposition 2.17 we obtain $I_1 \times P = J \times P'$. Since $I_1, J \subset \mathbf{R}$, it follows that $I_1 = J$ and $P = P'$. □

2.2.2 Cubical Chains in a Cubical Set

Having discussed chains in general, we now move to studying them in the context of a cubical set.

Definition 2.27 Let $X \subset \mathbf{R}^d$ be a cubical set. Let $\widehat{\mathcal{K}}_k(X) := \{\widehat{Q} \mid Q \in \mathcal{K}_k(X)\}$. $C_k(X)$ is the subgroup of C_k^d generated by the elements of $\widehat{\mathcal{K}}_k(X)$ and is referred to as the set of *k-chains of X*.

The reader can easily check that

$$C_k(X) = \{c \in C_k^d \mid |c| \subset X\}. \tag{2.5}$$

Since we know that $X \subset \mathbf{R}^d$, it is not necessary to write a superscript d in $\widehat{\mathcal{K}}_k(X)$ and $C_k(X)$.

It is left as an exercise to check that $\widehat{\mathcal{K}}_k(X)$ is a basis of $C_k(X)$. Moreover, since for any cubical set X the family $\mathcal{K}_k(X)$ is finite, $C_k(X)$ is a finite-dimensional free abelian group. Finally, given any $c \in C_k(X)$, we have the decomposition

$$c = \sum_{Q_i \in \mathcal{K}_k(X)} \alpha_i \widehat{Q}_i,$$

where $\alpha_i := c(Q_i)$. Rewriting this in terms of the scalar product, we obtain the following formula.

Proposition 2.28 *For any $c \in C_k(X)$,*

$$c = \sum_{Q \in \mathcal{K}_k(X)} \langle c, \widehat{Q} \rangle \widehat{Q}.$$

While the notation we are using for chains is consistent, some care must be taken when discussing certain 0-chains. For example, $0 \in C_k(X)$, the identity element of the group, satisfies $|0| = \emptyset$, while $\widehat{[0]}$ is the dual of the vertex located at the origin, that is, $|\widehat{[0]}| = \{0\} \subset \mathbf{R}$.

Example 2.29 Let $c = \widehat{A}_2 + \widehat{B}_1 - \widehat{B}_2 - \widehat{A}_1$, where

$$A_1 = [0] \times [0,1], \quad B_1 = [1] \times [0,1], \quad A_2 = [0,1] \times [0], \quad B_2 = [0,1] \times [1].$$

Then $|c|$ is the contour of the square $Q := [0,1]^2$ shown in Figure 2.3. One can give a geometric interpretation of the signs appearing in the expression for c by drawing appropriate arrows along the intervals marking the one-dimensional elementary chains. Taking the decomposition of the corresponding edge to the product of elementary intervals we find exactly one nondegenerate interval. Replacing this interval by its lower or upper end we obtain two 0-faces of the edge, the *lower vertex* and the *upper vertex*. The direction of the arrows is from the lower vertex to the upper vertex for positive one-dimensional elementary chains and reverse for negative elementary chains. In Figure 2.3 the arrows are marked accordingly to the signs of elementary chains in c. Thus positive or negative elementary chains represent the direction in which an edge is traversed. For example, we think of $\widehat{A}_1$ as indicating moving along the edge from $(0,0)$ to $(0,1)$ while $-\widehat{A}_1$ means traversing the edge in the opposite direction. With this in mind, c represents a counterclockwise closed path around the square.

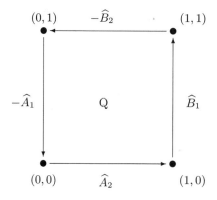

Fig. 2.3. Boundary of the unit square.

Example 2.30 Using the notation of the previous example, consider the chain $2c$. It is clear that $|2c| = |c|$, so both chains represent the same geometric object. The chain $2c$ can be interpreted as a path winding twice around the square in the counterclockwise direction. Similarly, the chain

$$(\widehat{A}_1 + \widehat{B}_2) + (\widehat{A}_2 + \widehat{B}_1)$$

could be interpreted as a "sum" of two different paths along the boundary of the square connecting $(0,0)$ to $(1,1)$.

2.2.3 The Boundary Operator

Our aim in this section is to generalize the definition of the boundary operator given in Section 1.4 and to prove its basic properties.

Definition 2.31 Given $k \in \mathbf{Z}$, the *cubical boundary operator* or *cubical boundary map*

$$\partial_k : C_k^d \to C_{k-1}^d$$

is a homomorphism of free abelian groups, which is defined for an elementary chain $\widehat{Q} \in \widehat{\mathcal{K}}_k^d$ by induction on the embedding number d as follows.

2.2 The Algebra of Cubical Sets

Consider first the case $d = 1$. Then Q is an elementary interval and hence $Q = [l] \in \mathcal{K}_0^1$ or $Q = [l, l+1] \in \mathcal{K}_1^1$ for some $l \in \mathbb{Z}$. Define

$$\partial_k \widehat{Q} := \begin{cases} 0 & \text{if } Q = [l], \\ \widehat{[l+1]} - \widehat{[l]} & \text{if } Q = [l, l+1]. \end{cases}$$

Now assume that $d > 1$. Let $I = I_1(Q)$ and $P = I_2(Q) \times \cdots \times I_d(Q)$. Then by Proposition 2.26,

$$\widehat{Q} = \widehat{I} \diamond \widehat{P}.$$

Define

$$\partial_k \widehat{Q} := \partial_{k_1} \widehat{I} \diamond \widehat{P} + (-1)^{\dim I} \widehat{I} \diamond \partial_{k_2} \widehat{P}, \tag{2.6}$$

where $k_1 = \dim I$ and $k_2 = \dim P$. Finally, we extend the definition to all chains by linearity; that is, if $c = \alpha_1 \widehat{Q}_1 + \alpha_2 \widehat{Q}_2 + \cdots + \alpha_m \widehat{Q}_m$, then

$$\partial_k c := \alpha_1 \partial_k \widehat{Q}_1 + \alpha_2 \partial_k \widehat{Q}_2 + \cdots + \alpha_m \partial_k \widehat{Q}_m.$$

Example 2.32 Let $Q = [l] \times [k]$. Then

$$\partial_0 \widehat{Q} = \partial_0 \widehat{[l]} \diamond \widehat{[k]} + (-1)^{\dim [l]} \widehat{[l]} \diamond \partial_0 \widehat{[k]}$$
$$= 0 \diamond \widehat{[k]} + \widehat{[l]} \diamond 0$$
$$= 0 + 0.$$

Example 2.33 Let $Q = [l, l+1] \times [k, k+1]$ (see Figure 2.4). Then

$$\partial_2 \widehat{Q} = \partial_1 \widehat{[l, l+1]} \diamond \widehat{[k, k+1]} + (-1)^{\dim [l, l+1]} \widehat{[l, l+1]} \diamond \partial_1 \widehat{[k, k+1]}$$
$$= (\widehat{[l+1]} - \widehat{[l]}) \diamond \widehat{[k, k+1]} - \widehat{[l, l+1]} \diamond (\widehat{[k+1]} - \widehat{[k]})$$
$$= \widehat{[l+1]} \diamond \widehat{[k, k+1]} - \widehat{[l]} \diamond \widehat{[k, k+1]} - \widehat{[l, l+1]} \diamond \widehat{[k+1]} + \widehat{[l, l+1]} \diamond \widehat{[k]}$$
$$= \widehat{[l+1] \times [k, k+1]} - \widehat{[l] \times [k, k+1]} + \widehat{[l, l+1] \times [k]} - \widehat{[l, l+1] \times [k+1]}$$
$$= \widehat{B}_1 - \widehat{A}_1 + \widehat{A}_2 - \widehat{B}_2,$$

where

$$A_1 = [l] \times [k, k+1],$$
$$B_1 = [l+1] \times [k, k+1],$$
$$A_2 = [l, l+1] \times [k],$$
$$B_2 = [l, l+1] \times [k+1].$$

By definition the domain of ∂_k consists of the k-chains. Thus, if we know that $c \in C_k^d$, then it is redundant to write $\partial_k(c)$. Even worse, it is also inconvenient. Consider the defining equation (2.6) where we are forced to include the subscripts k_1 and k_2, which obviously are determined by the dimensions of the appropriate elementary cubical chains. Therefore, to simplify the presentation we shall simplify the notation ∂_k to ∂. The following proposition demonstrates the wisdom of this (try writing the proposition using subscripts for the boundary operator).

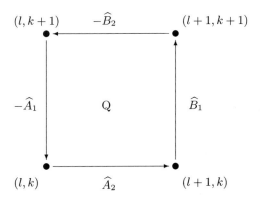

Fig. 2.4. Boundary of $[l, l+1] \times [k, k+1]$.

Proposition 2.34 *Let c and c' be cubical chains; then*

$$\partial(c \diamond c') = \partial c \diamond c' + (-1)^{\dim c} c \diamond \partial c'. \qquad (2.7)$$

Proof. Let us begin by showing that it is sufficient to prove the result for elementary cubical chains. More precisely, assume that for any $Q, Q' \in \mathcal{K}$,

$$\partial(\widehat{Q} \diamond \widehat{Q'}) = \partial\widehat{Q} \diamond \widehat{Q'} + (-1)^{\dim Q} \widehat{Q} \diamond \partial\widehat{Q'}. \qquad (2.8)$$

Let $c = \sum_{i=1}^{m} \alpha_i \widehat{Q_i}$ and let $c' = \sum_{j=1}^{m'} \alpha'_j \widehat{Q'_j}$. Observe that $\dim c = \dim Q_i$. Now,

$$\partial(c \diamond c') = \partial \left(\sum_{i=1}^{m} \alpha_i \widehat{Q_i} \diamond \sum_{j=1}^{m'} \alpha'_j \widehat{Q'_j} \right)$$

$$\stackrel{1}{=} \partial \left(\sum_{i=1}^{m} \sum_{j=1}^{m'} \alpha_i \alpha'_j \widehat{Q_i} \diamond \widehat{Q'_j} \right)$$

$$\stackrel{2}{=} \sum_{i=1}^{m} \sum_{j=1}^{m'} \alpha_i \alpha'_j \partial \left(\widehat{Q_i} \diamond \widehat{Q'_j} \right)$$

$$\stackrel{3}{=} \sum_{i=1}^{m} \sum_{j=1}^{m'} \alpha_i \alpha'_j \left(\partial\widehat{Q_i} \diamond \widehat{Q'_j} + (-1)^{\dim Q_i} \widehat{Q_i} \diamond \partial\widehat{Q'_j} \right)$$

$$= \sum_{i=1}^{m} \sum_{j=1}^{m'} \alpha_i \alpha'_j \left(\partial\widehat{Q_i} \diamond \widehat{Q'_j} + (-1)^{\dim c} \widehat{Q_i} \diamond \partial\widehat{Q'_j} \right)$$

$$= \sum_{i=1}^{m} \sum_{j=1}^{m'} \alpha_i \partial\widehat{Q_i} \diamond \alpha'_j \widehat{Q'_j} + (-1)^{\dim c} \sum_{i=1}^{m} \sum_{j=1}^{m'} \alpha_i \widehat{Q_i} \diamond \alpha'_j \partial\widehat{Q'_j}$$

$$= \partial c \diamond c' + (-1)^{\dim c} c \diamond \partial c'.$$

Equality 1 follows from the bilinearity of the cubical product. Equality 2 is due to the linearity of the boundary operator. Equality 3 follows from the assumption that (2.8) is satisfied.

Thus, to prove the proposition, it is sufficient to verify (2.8). The proof will be done by induction on $d := \operatorname{emb} Q$.

If $d = 1$, then the result follows immediately from Definition 2.31.

If $d > 1$, then we can decompose Q as in Proposition 2.26, namely $Q = I \times P$, where $\operatorname{emb} I = 1$ and $\operatorname{emb} P = d - 1$. Then, by the definition of the boundary operator,

$$\partial(\widehat{Q} \diamond \widehat{Q'}) = \partial(\widehat{I} \diamond \widehat{P} \diamond \widehat{Q'})$$
$$= \partial \widehat{I} \diamond \widehat{P} \diamond \widehat{Q'} + (-1)^{\dim I} \widehat{I} \diamond \partial(\widehat{P} \diamond \widehat{Q'}).$$

Since $\widehat{P} \diamond \widehat{Q'}$ satisfies the induction assumption, we see that

$$\partial(\widehat{Q} \diamond \widehat{Q'}) = \partial \widehat{I} \diamond \widehat{P} \diamond \widehat{Q'} + (-1)^{\dim I} \widehat{I} \diamond \left(\partial \widehat{P} \diamond \widehat{Q'} + (-1)^{\dim P} \widehat{P} \diamond \partial \widehat{Q'} \right)$$
$$= \partial \widehat{I} \diamond \widehat{P} \diamond \widehat{Q'} + (-1)^{\dim I} \widehat{I} \diamond \partial \widehat{P} \diamond \widehat{Q'} + (-1)^{\dim I + \dim P} \widehat{I} \diamond \widehat{P} \diamond \partial \widehat{Q'}$$
$$= \left(\partial \widehat{I} \diamond \widehat{P} + (-1)^{\dim I} \widehat{I} \diamond \partial \widehat{P} \right) \diamond \widehat{Q'} + (-1)^{\dim Q} \widehat{Q} \diamond \partial \widehat{Q'}$$
$$= \partial \widehat{Q} \diamond \widehat{Q'} + (-1)^{\dim Q} \widehat{Q} \diamond \partial \widehat{Q'},$$

where the last equality follows again from the definition of the boundary operator. □

By a straightforward induction argument we obtain the following corollary.

Corollary 2.35 *If $Q_1, Q_2, \ldots, Q_m$ are elementary cubes, then*

$$\partial(\widehat{Q}_1 \diamond \widehat{Q}_2 \diamond \cdots \diamond \widehat{Q}_m) = \sum_{j=1}^{m} (-1)^{\sum_{i=1}^{j-1} \dim Q_i} \widehat{Q}_1 \diamond \cdots \diamond \widehat{Q}_{j-1} \diamond \partial \widehat{Q}_j \diamond \widehat{Q}_{j+1} \diamond \cdots \diamond \widehat{Q}_m.$$

From this corollary one can immediately obtain the following proposition.

Proposition 2.36 *Let $Q \in \mathbf{R}^d$ be an n-dimensional elementary cube with decomposition into elementary intervals given by $Q = I_1 \times I_2 \times \cdots \times I_d \in \mathbf{R}^d$ and let the one-dimensional intervals in this decomposition be $I_{i_1}, I_{i_2}, \ldots, I_{i_n}$, with $I_{i_j} = [k_j, k_j + 1]$. For $j = 1, 2, \ldots, n$ let*

$$Q_j^- := I_1 \times \cdots \times I_{i_j - 1} \times [k_j] \times I_{i_j + 1} \times \cdots \times I_d,$$
$$Q_j^+ := I_1 \times \cdots \times I_{i_j - 1} \times [k_j + 1] \times I_{i_j + 1} \times \cdots \times I_d$$

denote the primary faces of Q. Then

$$\partial \widehat{Q} = \sum_{j=1}^{n} (-1)^{j-1} \left(\widehat{Q}_j^+ - \widehat{Q}_j^- \right).$$

The following proposition demonstrates a crucial property of the boundary operator.

Proposition 2.37
$$\partial \circ \partial = 0.$$

Proof. Because ∂ is a linear operator, it is enough to verify this property for elementary cubical chains. Again, the proof is by induction on the embedding number.

Let Q be an elementary interval. If $Q = [l]$, then by definition $\partial \widehat{Q} = 0$ so $\partial(\partial \widehat{Q}) = 0$. If $Q = [l, l+1]$, then

$$\begin{aligned}\partial(\partial \widehat{Q}) &= \partial(\partial \widehat{[l, l+1]}) \\ &= \partial(\widehat{[l+1]} - \widehat{[l]}) \\ &= \partial \widehat{[l+1]} - \partial \widehat{[l]} \\ &= 0 - 0 \\ &= 0.\end{aligned}$$

Now assume that $Q \in \mathcal{K}^d$ for $d > 1$. Then $Q = I \times P$, where $I = I_1(Q)$ and $P = I_2(Q) \times \cdots \times I_d(Q)$. So, by Proposition 2.26,

$$\begin{aligned}\partial(\partial \widehat{Q}) &= \partial(\partial(\widehat{I \times P})) \\ &= \partial(\partial(\widehat{I} \diamond \widehat{P})) \\ &= \partial\left(\partial\widehat{I} \diamond \widehat{P} + (-1)^{\dim \widehat{I}} \widehat{I} \diamond \partial \widehat{P}\right) \\ &= \partial\left(\partial\widehat{I} \diamond \widehat{P}\right) + (-1)^{\dim \widehat{I}} \partial\left(\widehat{I} \diamond \partial \widehat{P}\right) \\ &= \partial\partial\widehat{I} \diamond \widehat{P} + (-1)^{\dim \partial\widehat{I}} \partial\widehat{I} \diamond \partial\widehat{P} + (-1)^{\dim \widehat{I}} \partial\left(\widehat{I} \diamond \partial \widehat{P}\right) \\ &= (-1)^{\dim \partial\widehat{I}} \partial\widehat{I} \diamond \partial\widehat{P} + (-1)^{\dim \widehat{I}}\left(\partial\widehat{I} \diamond \partial\widehat{P} + (-1)^{\dim \widehat{I}} \widehat{I} \diamond \partial\partial\widehat{P}\right) \\ &= (-1)^{\dim \partial\widehat{I}} \partial\widehat{I} \diamond \partial\widehat{P} + (-1)^{\dim \widehat{I}} \partial\widehat{I} \diamond \partial\widehat{P}.\end{aligned}$$

The last step uses the induction hypothesis that the proposition is true if the embedding number is less than d.

Observe that if $\dim \widehat{I} = 0$, then $\partial \widehat{I} = 0$, in which case we have that each term in the sum is 0 and hence $\partial\partial \widehat{Q} = 0$. On the other hand, if $\dim \widehat{I} = 1$, then $\dim \partial\widehat{I} = 0$ and hence the two terms cancel each other, giving the desired result. □

Example 2.33 and Figure 2.4 might suggest that the algebraic and topological boundaries are closely related. We would like to know whether or not

$$|\partial c| = \mathrm{bd}\, |c|.$$

Exercise 2.15 shows that this is false. What is true in general is the following proposition.

2.2 The Algebra of Cubical Sets

Proposition 2.38 *For any chain $c \in C_k^d$,*

$$|\partial c| \subset |c|.$$

Moreover, $|\partial c|$ is contained in the $(k-1)$-dimensional skeleton of $|c|$, that is, the union of $(k-1)$-dimensional faces of $|c|$.

Proof. First consider the case when $c = \widehat{Q}$, where $Q \in \mathcal{K}_k$. It follows from Exercise 2.18 that $\left|\partial \widehat{Q}\right| \subset \bigcup \mathcal{K}_{k-1}(Q) \subset Q = \left|\widehat{Q}\right|$. If c is arbitrary, then $c = \sum_i \alpha_i \widehat{Q}_i$ for some $\alpha_i \neq 0$ and

$$|\partial c| = \left|\sum_i \alpha_i \partial \widehat{Q}_i\right| \subset \bigcup_i \left|\partial \widehat{Q}_i\right| \subset \bigcup_i \left|\widehat{Q}_i\right| = |c|. \qquad \square$$

We now go back to cubical chains in the setting of a fixed cubical set X. The first observation is that the boundary operator maps chains in X into chains in X. More precisely, we have the following proposition.

Proposition 2.39 *Let $X \subset \mathbf{R}^d$ be a cubical set. Then*

$$\partial_k(C_k(X)) \subset C_{k-1}(X).$$

Proof. Let $c \in C_k(X)$. Then by Eq. (2.5), $|c| \subset X$, and by Proposition 2.38, $|\partial_k(c)| \subset |c| \subset X$. Therefore, $\partial_k(c) \in C_{k-1}(X)$. $\square$

An immediate consequence of this proposition is that the restriction of the operator ∂ to chains in X, $\partial_k^X : C_k(X) \to C_{k-1}(X)$, given by

$$\partial_k^X(c) := \partial_k(c)$$

makes sense. This justifies the following definition.

Definition 2.40 The *boundary operator* for the cubical set X is defined to be

$$\partial_k^X : C_k(X) \to C_{k-1}(X)$$

obtained by restricting $\partial_k : C_k^d \to C_{k-1}^d$ to $C_k(X)$.

In the sequel we typically omit the superscript X in ∂_k^X whenever X is clear from the context.

Definition 2.41 The *cubical chain complex* for the cubical set $X \subset \mathbf{R}^d$ is

$$\mathcal{C}(X) := \{C_k(X), \partial_k^X\}_{k \in \mathbf{Z}},$$

where $C_k(X)$ are the groups of cubical k-chains generated by $\mathcal{K}_k(X)$ and ∂_k^X is the cubical boundary operator restricted to X.

2.2.4 Homology of Cubical Sets

We are now ready to give the most important definitions of this book. Let $X \subset \mathbf{R}^d$ be a cubical set. A k-chain $z \in C_k(X)$ is called a *cycle* in X if $\partial z = 0$. As Definition 13.36 shows, the kernel of a linear map is the set of elements are sent to zero and is a subgroup of the domain. Thus the set of all k-cycles in X, which is denoted by $Z_k(X)$, is $\ker \partial_k^X$ and forms a subgroup of $C_k(X)$. Because of its importance, we explicitly summarize these comments via the following set of relations:

$$Z_k(X) := \ker \partial_k^X = C_k(X) \cap \ker \partial_k \subset C_k(X). \tag{2.9}$$

A k-chain $z \in C_k(X)$ is called a *boundary* in X if there exists $c \in C_{k+1}(X)$ such that $\partial c = z$. Thus the set of boundary elements in $C_k(X)$, which is denoted by $B_k(X)$, consists of the image of ∂_{k+1}^X. Since ∂_{k+1}^X is a homomorphism, $B_k(X)$ is a subgroup of $C_k(X)$. Again, these comments can be summarized by

$$B_k(X) := \operatorname{im} \partial_{k+1}^X = \partial_{k+1}(C_{k+1}(X)) \subset C_k(X). \tag{2.10}$$

By Proposition 2.37, $\partial c = z$ implies $\partial z = \partial^2 c = 0$. Hence every boundary is a cycle and thus $B_k(X)$ is a subgroup of $Z_k(X)$. We are interested in cycles that are not boundaries. We want to treat cycles that are boundaries as trivial. In order to give the nontrivial cycles an algebraic structure, we introduce an equivalence relation. We say that two cycles $z_1, z_2 \in Z_k(X)$ are *homologous* and we write $z_1 \sim z_2$ if $z_1 - z_2$ is a boundary in X, that is, $z_1 - z_2 \in B_k(X)$. The equivalence classes are elements of the quotient group $Z_k(X)/B_k(X)$ (see Chapter 13 for the definition of the quotient group).

Definition 2.42 The kth *cubical homology group*, or briefly the kth homology group of X, is the quotient group

$$H_k(X) := Z_k(X)/B_k(X).$$

The homology of X is the collection of all homology groups of X. The shorthand notation for this is

$$H_*(X) := \{H_k(X)\}_{k \in \mathbf{Z}}.$$

As emphasized in the introduction, we will use the homology groups of the cubical set X to gain information about the topological structure of X. Thus the elements of $H_*(X)$ play an important role. However, we used chains to pass from the topology of X to the homology $H_*(X)$ and, therefore, we often find ourselves discussing the elements of $H_*(X)$ in terms of representative chains. For this reason we introduce the following notation.

Definition 2.43 Given $z \in Z_k(X)$, $[z]_X \in H_k(X)$ is the homology class of z in X. To simplify the notation, if the cubical set X is clear from the context of the discussion, then we let $[z] := [z]_X$.

2.2 The Algebra of Cubical Sets   61

Example 2.44 Let $X = \emptyset$. Then $C_k(X) = 0$ for all k and hence
$$H_k(X) = 0 \quad k = 0, 1, 2, \ldots.$$

Example 2.45 Let $X = \{x_0\} \subset \mathbf{R}^d$ be a cubical set consisting of a single point. Then $x_0 = [l_1] \times [l_2] \times \cdots \times [l_d]$. Thus
$$C_k(X) \cong \begin{cases} \mathbf{Z} & \text{if } k = 0, \\ 0 & \text{otherwise}. \end{cases}$$

Furthermore, $Z_0(X) \cong C_0(X) = \mathbf{Z}$. Since $C_1(X) = 0$, $B_0(X) = 0$ and therefore, $H_0(X) \cong \mathbf{Z}$. Since $C_k(X) = 0$ for all $k \geq 1$, $H_k(X) = 0$ for all $k \geq 1$. Therefore,
$$H_k(X) \cong \begin{cases} \mathbf{Z} & \text{if } k = 0, \\ 0 & \text{otherwise}. \end{cases}$$

Example 2.46 Recall the cubical set
$$\Gamma^1 = [0] \times [0,1] \cup [1] \times [0,1] \cup [0,1] \times [0] \cup [0,1] \times [1].$$

The sets of elementary cubes are
$$\mathcal{K}_0(\Gamma^1) = \{[0] \times [0], [0] \times [1], [1] \times [0], [1] \times [1]\},$$
$$\mathcal{K}_1(\Gamma^1) = \{[0] \times [0,1], [1] \times [0,1], [0,1] \times [0], [0,1] \times [1]\}.$$

Thus the bases for the sets of chains are
$$\widehat{\mathcal{K}}_0(\Gamma^1) = \{\widehat{[0] \times [0]}, \widehat{[0] \times [1]}, \widehat{[1] \times [0]}, \widehat{[1] \times [1]}\}$$
$$= \{\widehat{[0]} \diamond \widehat{[0]}, \widehat{[0]} \diamond \widehat{[1]}, \widehat{[1]} \diamond \widehat{[0]}, \widehat{[1]} \diamond \widehat{[1]}\},$$
$$\widehat{\mathcal{K}}_1(\Gamma^1) = \{\widehat{[0] \times [0,1]}, \widehat{[1] \times [0,1]}, \widehat{[0,1] \times [0]}, \widehat{[0,1] \times [1]}\}$$
$$= \{\widehat{[0]} \diamond \widehat{[0,1]}, \widehat{[1]} \diamond \widehat{[0,1]}, \widehat{[0,1]} \diamond \widehat{[0]}, \widehat{[0,1]} \diamond \widehat{[1]}\}.$$

To compute the boundary operator we need to compute the boundary of the basis elements.
$$\partial(\widehat{[0]} \diamond \widehat{[0,1]}) = -\widehat{[0]} \diamond \widehat{[0]} + \widehat{[0]} \diamond \widehat{[1]}.$$
$$\partial(\widehat{[1]} \diamond \widehat{[0,1]}) = -\widehat{[1]} \diamond \widehat{[0]} + \widehat{[1]} \diamond \widehat{[1]}.$$
$$\partial(\widehat{[0,1]} \diamond \widehat{[0]}) = -\widehat{[0]} \diamond \widehat{[0]} + \widehat{[1]} \diamond \widehat{[0]}.$$
$$\partial(\widehat{[0,1]} \diamond \widehat{[1]}) = -\widehat{[0]} \diamond \widehat{[1]} + \widehat{[1]} \diamond \widehat{[1]}.$$

We can put this into the form of a matrix
$$\partial_1 = \begin{bmatrix} -1 & 0 & -1 & 0 \\ 1 & 0 & 0 & -1 \\ 0 & -1 & 1 & 0 \\ 0 & 1 & 0 & 1 \end{bmatrix}.$$

62   2 Cubical Homology

To understand $Z_1(\Gamma^1)$, we need to know $\ker \partial_1$, that is, we need to solve the equation

$$\begin{bmatrix} -1 & 0 & -1 & 0 \\ 1 & 0 & 0 & -1 \\ 0 & -1 & 1 & 0 \\ 0 & 1 & 0 & 1 \end{bmatrix} \begin{bmatrix} \alpha_1 \\ \alpha_2 \\ \alpha_3 \\ \alpha_4 \end{bmatrix} = \begin{bmatrix} 0 \\ 0 \\ 0 \\ 0 \end{bmatrix}.$$

This in turn means solving

$$\begin{bmatrix} -\alpha_1 - \alpha_3 \\ \alpha_1 - \alpha_4 \\ -\alpha_2 + \alpha_3 \\ \alpha_2 + \alpha_4 \end{bmatrix} = \begin{bmatrix} 0 \\ 0 \\ 0 \\ 0 \end{bmatrix},$$

which gives

$$\alpha_1 = -\alpha_2 = -\alpha_3 = \alpha_4.$$

Hence

$$Z_1(\Gamma^1) = \left\{ \alpha[1, -1, -1, 1]^T \mid \alpha \in \mathbf{Z} \right\},$$

that is, $Z_1(\Gamma^1)$ is generated by

$$\widehat{[0]} \diamond \widehat{[0,1]} - \widehat{[1]} \diamond \widehat{[0,1]} - \widehat{[0,1]} \diamond \widehat{[0]} + \widehat{[0,1]} \diamond \widehat{[1]}.$$

Since $C_2(\Gamma^1) = 0$, $B_1(\Gamma^1) = 0$ and hence

$$H_1(\Gamma^1) = Z_1(\Gamma^1) \cong \mathbf{Z}.$$

We turn to computing $H_0(\Gamma^1)$. First observe that there is no solution to the equation

$$\begin{bmatrix} -1 & 0 & -1 & 0 \\ 1 & 0 & 0 & -1 \\ 0 & -1 & 1 & 0 \\ 0 & 1 & 0 & 1 \end{bmatrix} \begin{bmatrix} \alpha_1 \\ \alpha_2 \\ \alpha_3 \\ \alpha_4 \end{bmatrix} = \begin{bmatrix} 1 \\ 0 \\ 0 \\ 0 \end{bmatrix}.$$

This implies that $\widehat{[0]} \diamond \widehat{[0]} \notin B_0(\Gamma^1)$. On the other hand,

$$\partial \left(\widehat{[0]} \diamond \widehat{[0,1]} \right) = -\widehat{[0]} \diamond \widehat{[0]} + \widehat{[0]} \diamond \widehat{[1]},$$

$$\partial \left(\widehat{[0]} \diamond \widehat{[0,1]} + \widehat{[0,1]} \diamond \widehat{[1]} \right) = -\widehat{[0]} \diamond \widehat{[0]} + \widehat{[1]} \diamond \widehat{[1]},$$

$$\partial \left(\widehat{[0]} \diamond \widehat{[0,1]} + \widehat{[0,1]} \diamond \widehat{[1]} - \widehat{[1]} \diamond \widehat{[0,1]} \right) = -\widehat{[0]} \diamond \widehat{[0]} + \widehat{[1]} \diamond \widehat{[0]}.$$

Thus,

$$\left\{ \widehat{[0]} \diamond \widehat{[0]} - \widehat{[0]} \diamond \widehat{[1]},\ \widehat{[0]} \diamond \widehat{[0]} - \widehat{[1]} \diamond \widehat{[0]},\ \widehat{[0]} \diamond \widehat{[0]} - \widehat{[1]} \diamond \widehat{[1]} \right\} \subset B_0(\Gamma^1).$$

In particular, all the elementary chains are homologous, that is, $\widehat{[0]} \diamond \widehat{[0]} \sim \widehat{[0]} \diamond \widehat{[1]} \sim \widehat{[1]} \diamond \widehat{[0]} \sim \widehat{[1]} \diamond \widehat{[1]}$.

Now consider an arbitrary chain $z \in C_0(\Gamma^1)$. Then
$$z = \alpha_1 \widehat{[0]} \diamond \widehat{[0]} + \alpha_2 \widehat{[0]} \diamond \widehat{[1]} + \alpha_3 \widehat{[1]} \diamond \widehat{[0]} + \alpha_4 \widehat{[1]} \diamond \widehat{[1]}.$$

So on the level of homology

$$\begin{aligned}
[z]_{\Gamma^1} &= \left[\alpha_1 \widehat{[0]} \diamond \widehat{[0]} + \alpha_2 \widehat{[0]} \diamond \widehat{[1]} + \alpha_3 \widehat{[1]} \diamond \widehat{[0]} + \alpha_4 \widehat{[1]} \diamond \widehat{[1]}\right]_{\Gamma^1} \\
&= \alpha_1 \left[\widehat{[0]} \diamond \widehat{[0]}\right]_{\Gamma^1} + \alpha_2 \left[\widehat{[0]} \diamond \widehat{[1]}\right]_{\Gamma^1} + \alpha_3 \left[\widehat{[1]} \diamond \widehat{[0]}\right]_{\Gamma^1} + \alpha_4 \left[\widehat{[1]} \diamond \widehat{[1]}\right]_{\Gamma^1} \\
&= (\alpha_1 + \alpha_2 + \alpha_3 + \alpha_4) \left[\widehat{[0]} \diamond \widehat{[0]}\right]_{\Gamma^1},
\end{aligned}$$

where the last equality comes from that fact that all the elementary chains are homologous. Therefore, we can think of every element of $H_0(\Gamma^1) = Z_0(\Gamma^1)/B_0(\Gamma^1)$ as being generated by $\widehat{[0]} \diamond \widehat{[0]}$ and thus $\dim H_0(\Gamma^1) = 1$.

In particular, we have proven that

$$H_k(\Gamma^1) \cong \begin{cases} \mathbf{Z} & \text{if } k = 0, 1, \\ 0 & \text{otherwise.} \end{cases}$$

We could continue in this fashion for a long time computing homology groups, but as the reader hopefully has already seen this is a rather time-consuming process. Furthermore, even if one takes a simple set such as

$$X = [0,1] \times [0,1] \times [0,1] \times [0,1],$$

the number of elementary cubes is quite large and the direct computation of its homology is quite tedious. Obviously, we need to develop more efficient methods.

Exercises

2.11 Let $c \in C_k^d$.

1. Show that for any $Q \in \mathcal{K}_k^d$,
$$c(Q) = \langle c, \widehat{Q} \rangle.$$

2. Justify the following equalities:
$$c = \sum_{Q \in \mathcal{K}_k^d} c(Q) \widehat{Q} = \sum_{Q \in \mathcal{K}_k^d} \langle c, \widehat{Q} \rangle \widehat{Q}.$$

2.12 Prove Proposition 2.25(v).

2.13 For each Q given below determine the elementary cubes I and P of Proposition 2.26 that satisfy $\widehat{Q} = \widehat{I} \diamond \widehat{P}$. What are the dimensions of I and P?

(a) $Q = [2,3] \times [1,2] \times [0,1]$.

(b) $Q = [-1] \times [0,1] \times [2]$.

2.14 Let $c \in C_k^d$. Show that
$$|c| := \bigcup \left\{ Q \in \mathcal{K}^d \mid \langle c, \widehat{Q} \rangle \neq 0 \right\}.$$

2.15 Let
$$c_1 = \widehat{[-1,0]} \times \widehat{[0]} + \widehat{[0,1]} \times \widehat{[0]} + \widehat{[0]} \times \widehat{[-1,0]} + \widehat{[0]} \times \widehat{[0,1]}$$
and
$$c_2 = \widehat{[-1,0]} \times \widehat{[0]} + \widehat{[0,1]} \times \widehat{[0]} + \widehat{[0]} \times \widehat{[0,1]}.$$
Consider the sets $X = |c_1|$ and $Y = |c_2|$ in $\mathbf{R}^2$.

(a) Make a sketch of X and Y. Try to define the *topological boundary* of those sets by analogy to the discussion in Chapter 1. Note that this concept is not explicitly defined in that chapter and it is only used in very simple cases. According to your definition, is the vertex $(0,0)$ a part of the boundary of X or Y?
(b) Find ∂c_1 and ∂c_2. Determine the sets $|\partial c_1|$ and $|\partial c_2|$. Note that $(0,0) \in |\partial c_2|$, but $(0,0) \notin |\partial c_1|$. This shows that one has to be careful about the relation between the concept of algebraic boundary and topological boundary, whatever this second one could mean.

2.16 Create the input files for the CubTop program of the sets X and Y given in the previous exercise.

(a) Call CubTop for each file with the functions freefaces and topobound and the option -g. View the bitmap files. Does any of them fit your answer to the previous question?
(b) Repeat the same exercise for the file exC2d.txt in the folder CubTop/Examples.

2.17 Let $Q = [0,1]^d \subset \mathbf{R}^d$ and put
$$A_j := \{x \in Q \mid x_j = 0\},$$
$$B_j := \{x \in Q \mid x_j = 1\}.$$
Show that
$$\partial \widehat{Q} = \sum_{j=1}^d (-1)^j (\widehat{A_j} - \widehat{B_j}).$$

2.18 Let Q be an elementary cube of dimension k. Show that under the assumption that $Q_i = I_i(Q)$, the support of each nonzero term of the sum in Corollary 2.35 is the union of two parallel $(k-1)$-dimensional faces of Q.

2.2 The Algebra of Cubical Sets

2.19 Let P be an elementary k-dimensional cube. Let Q be a $(k-1)$-dimensional face of P. Show that

$$\langle \widehat{\partial P}, \widehat{Q} \rangle = \pm 1.$$

2.20 Let $X = \{0\} \times [-1,1] \cup [-1,1] \times \{0\} \subset \mathbf{R}^2$. Determine the cubical chain complex $\mathcal{C}(X)$ and compute $H_*(X)$.

2.21 Let X consist of the one-dimensional faces of $[0,2] \times [0,1]$. Determine the cubical chain complex $\mathcal{C}(X)$ and compute $H_*(X)$.

2.22 Let Γ^2 be the cubical set consisting of the two-dimensional faces of $[0,1]^3$. Determine the cubical chain complex $\mathcal{C}(\Gamma^2)$, and compute $H_*(\Gamma^2)$.

2.23 Let X be a cubical set obtained by removing the set $(1,2) \times (1,2) \times [0,1]$ from the solid rectangle $[0,3] \times [0,3] \times [0,1]$. Let T be the union of the free faces of X (compare this set with a torus discussed in Example 11.10).

(a) Prepare the data file for computing the chain complex $\mathcal{C}(X)$ of X by the Homology program. Run the program to find $\mathcal{C}(X)$, and $H_*(X, \mathbf{Z})$.
(b) Determine $\mathcal{C}(T)$, and compute $H_*(T)$.

2.24 Let L be the cubical set presented in Figure 1.3 (L is defined in the file maze.bmp). Run the Homology program to find the homology of L. Open two gates (i.e., remove two pixels) in opposite walls of the labyrinth and again run the program to find the homology of what is left. Make a guess about the solvability of the labyrinth, namely, a possibility of passing inside from one gate to another without crossing a wall.

2.25 Let P be the cubical set defined in the file `qprojpln.cub` in the folder Examples of Homology program. The name for this set and the origin of the file are explained in Chapter 11. Run the program to find the homology of P with respect to

(a) coefficients in Z;
(b) coefficients in $\mathbf{Z}_p$ for $p = 2, 3, 5, 7$.

2.26 Repeat the steps of Exercise 2.25 for the files:

(a) `qklein.cub`;
(b) `kleinbot.cub`.

Again, the names and origins of these files become more transparent in Chapter 11. Note that the presented cubical sets, although very distinct, have isomorphic homology groups.

2.3 Connected Components and $H_0(X)$

This chapter begins with a discussion of cubical sets, a very special class of topological spaces. We then move on to the combinatorics and algebra associated with these spaces, and finally, we define the homology of a cubical set. Though there is much more to be said, we have defined the essential steps in moving from topology to homology. It is worth noting how little topology is involved in the process; in fact, most of our discussion has revolved around the algebra induced by combinatorial data. The opposite sequence of relations has not been addressed; what do the homology groups imply about the topology of the set? In this section we begin to move in this direction. We show that the zero-dimensional homology group measures the number of connected components of the cubical set. However, to do so, we need to show first that in the case of cubical sets there is a purely combinatorial way to describe the topological concepts of connectedness and connected components.

Recall that for any topological space X and any point $x \in X$ the union of all connected subsets of X containing x is a connected subset of X (see Theorem 12.53). It is called the *connected component of x in X*. We denote it by $\mathrm{cc}_X(x)$.

Theorem 2.47 *For any $x, y \in X$, either $\mathrm{cc}_X(x) = \mathrm{cc}_X(y)$ or $\mathrm{cc}_X(x) \cap \mathrm{cc}_X(y) = \emptyset$.*

Proof. Assume that $\mathrm{cc}_X(x) \cap \mathrm{cc}_X(y) \neq \emptyset$. Then by Theorem 12.53, $\mathrm{cc}_X(x) \cup \mathrm{cc}_X(y)$ is connected. Since it contains both x and y, it must be the case that $\mathrm{cc}_X(x) \cup \mathrm{cc}_X(y) \subset \mathrm{cc}_X(x)$ and $\mathrm{cc}_X(x) \cup \mathrm{cc}_X(y) \subset \mathrm{cc}_X(y)$. It follows that

$$\mathrm{cc}_X(x) = \mathrm{cc}_X(x) \cup \mathrm{cc}_X(y) = \mathrm{cc}_X(y). \qquad \square$$

From now on let X be a cubical set. Let us observe first the following simple proposition.

Proposition 2.48 *For every $x \in X$ there exists a vertex $V \in \mathcal{K}_0(X)$ such that $\mathrm{cc}_X(x) = \mathrm{cc}_X(V)$.*

Proof. By Proposition 2.15(i), there exists an elementary cube Q such that $x \in \mathring{Q}$. Therefore, $\mathring{Q} \cap X \neq \emptyset$, and it follows from Proposition 2.15(vi) that $Q \subset X$. Let V be any vertex of Q. Since Q as a cube is connected (see Theorem 12.54), $Q \subset \mathrm{cc}_X(x)$ and consequently $V \in \mathrm{cc}_X(x)$. Hence $\mathrm{cc}_X(V) \cap \mathrm{cc}_X(x) \neq \emptyset$, and, by Theorem 2.47, $\mathrm{cc}_X(V) = \mathrm{cc}_X(x)$. $\square$

Corollary 2.49 *A cubical set can have only a finite number of connected components.*

Proof. By Proposition 2.48, every connected component of a cubical set is a connected component of one of its vertices, and a cubical set has only a finite number of vertices. $\square$

2.3 Connected Components and $H_0(X)$

Definition 2.50 A sequence of vertices $V_0, V_1, \ldots, V_n \in \mathcal{K}_0(X)$ is an *edge path in X* if there exist edges $E_1, E_2, \ldots, E_n \in \mathcal{K}_1(X)$ such that V_{i-1}, V_i are the two faces of E_i for $i = 1, 2, \ldots, n$. For $V, V' \in \mathcal{K}_0(X)$, we write $V \sim_X V'$ if there exists an edge path $V_0, V_1, \ldots, V_n \in \mathcal{K}_0(X)$ in X such that $V = V_0$ and $V' = V_n$. We say that X is *edge connected* if $V \sim_X V'$ for any $V, V' \in \mathcal{K}_0(X)$.

As Exercise 2.27 indicates, $\sim_X$ is an equivalence relation. Also left to the exercises is the proof of the following proposition.

Proposition 2.51 1. *Every elementary cube is edge-connected.*
2. *If X and Y are edge-connected cubical sets and $X \cap Y \neq \emptyset$, then $X \cup Y$ is edge-connected.*

Proposition 2.52 *Assume that $V \sim_X V'$ for some $V, V' \in \mathcal{K}_0(X)$. Then there exists a chain $c \in C_1(X)$ such that $|c|$ is connected and $\partial c = \widehat{V'} - \widehat{V}$.*

Proof. Let $V_0, V_1, \ldots, V_n \in \mathcal{K}_0(X)$ be an edge path from $V = V_0$ to $V' = V_n$ and let $E_1, E_2, \ldots, E_n \in \mathcal{K}_1(X)$ be the corresponding edges. Without loss of generality, we may assume that the edge path is minimal. Then any two edges as well as any two vertices in the path are different. We will show that for some coefficients $\alpha_i \in \{-1, 1\}$ the chain

$$c := \sum_{i=1}^{n} \alpha_i \widehat{E}_i$$

satisfies the conclusion of the proposition. We do so by induction in n. If $n = 1$, then $\partial E_1 = \pm(\widehat{V_1} - \widehat{V_0})$. Taking $c = \alpha_1 \widehat{E}_1$ with an appropriate coefficient $\alpha_1 \in \{-1, 1\}$, we get $\partial(c) = \widehat{V_1} - \widehat{V_0}$. Since $|c| = |\alpha_1 E_1| = E_1$, it is connected.

Consider in turn the second step of the induction argument. Let

$$c' := \sum_{i=1}^{n-1} \alpha_i \widehat{E}_i$$

with coefficients chosen so that $\partial c' = \widehat{V}_{n-1} - \widehat{V_0}$ and $|c'|$ is connected. Choose α_n so that $\partial(\alpha_n E_n) = \widehat{V_n} - \widehat{V}_{n-1}$. Then obviously $\partial c = \widehat{V_n} - \widehat{V_0}$. Since $|c| = |c'| \cup E_n$ and $|c'| \cap E_n \neq \emptyset$, it follows from Theorem 2.47 that $|c|$ is connected. □

For $x \in X$ we define the *edge-connected component of x in X* as the union of all edge-connected cubical subsets of X that contain x. We denote it by $\text{ecc}_X(x)$.

Since the number of cubical subsets of a given cubical set is finite, we may use an induction argument based on Proposition 2.51 to prove the following proposition.

Proposition 2.53 *For any $x \in X$, $\text{ecc}_X(x)$ is edge-connected.*

The same argument as in the case of connected components shows that we have the following counterpart of Theorem 2.47.

Proposition 2.54 *For any $x, y \in X$ either*

$$\mathrm{ecc}_X(x) = \mathrm{ecc}_X(y) \text{ or } \mathrm{ecc}_X(x) \cap \mathrm{ecc}_X(y) = \emptyset.$$

Theorem 2.55 *A cubical set X is connected if and only if it is edge-connected.*

Proof. Assume first that X is edge-connected. Let $V \in X$ be a vertex. It is enough to show that $\mathrm{cc}_X(V) = X$, because $\mathrm{cc}_X(V)$ is connected. Since $\mathrm{cc}_X(V) \subset X$, we need only to show the opposite inclusion. Thus let $x \in X$. By Proposition 2.48, we can select another vertex $W \in X$ such that $x \in \mathrm{cc}_X(x) = \mathrm{cc}_X(W)$. By Proposition 2.52, there exists a chain $c \in C_1(X)$ such that $\partial c = \widehat{V} - \widehat{W}$ and $|c|$ is connected. Since $V, W \in |c|$, it follows that $\mathrm{cc}_X(W) = \mathrm{cc}_X(V)$. Therefore, $x \in \mathrm{cc}_X(V)$, which we needed to prove.

To prove the opposite implication, assume that X is not edge-connected. Then there exist vertices $V_0, V_1 \in \mathcal{K}_0(X)$ such that $\mathrm{ecc}_X(V_0) \cap \mathrm{ecc}_X(V_1) = \emptyset$. Let $X_0 := \mathrm{ecc}_X(V_0)$ and

$$X_1 := \bigcup \{\mathrm{ecc}_X(V) \mid V \in \mathcal{K}_0(X) \text{ and } \mathrm{ecc}_X(V) \cap \mathrm{ecc}_X(V_0) = \emptyset\}.$$

The sets X_0 and X_1 are disjoint, nonempty closed subsets of X. We will show that $X = X_0 \cup X_1$. Let $x \in X$. Let Q be an elementary cube such that $x \in \overset{\circ}{Q}$. Then by Proposition 2.15(vi), $Q \subset X$. Let $V \in \mathcal{K}_0(Q)$ be any vertex of Q. Since by Proposition 2.51 Q is edge-connected, $Q \subset \mathrm{ecc}_X(x)$ and $Q \subset \mathrm{ecc}_X(V)$. Now if $\mathrm{ecc}_X(V) = \mathrm{ecc}_X(V_0) = X_0$, then $x \in X_0$. Otherwise, $x \in X_1$. This shows that $X = X_0 \cup X_1$, which implies that X is not connected, a contradiction. □

The proof of the next proposition is left as an exercise.

Proposition 2.56 *If X is cubical, then for every $x \in X$ its connected component $\mathrm{cc}_X(x)$ is a cubical set.*

From Theorem 2.55 and Proposition 2.56 we easily get the following corollary.

Corollary 2.57 *If X is cubical, then for every $x \in X$ its connected component and edge-connected component coincide.*

The following lemma is needed in the proof of the main theorem of this section.

Lemma 2.58 *Assume X is a cubical set and $X_1, X_2, \ldots, X_n$ are its connected components. If $c_i \in C_k(X_i)$ are k-dimensional chains, then*

$$\left| \sum_{i=1}^{n} c_i \right| = \bigcup_{i=1}^{n} |c_i|.$$

Proof. The left-hand side is contained in the right-hand side by Proposition 2.19(iv). To show the opposite inclusion, take $x \in \bigcup_{i=1}^{n} |c_i|$. Then for some $i_0 \in \{1, 2, \ldots, n\}$ there exists a $Q \in \mathcal{K}_k(X_{i_0})$ such that $x \in Q$ and $c_{i_0}(Q) \neq 0$. Since $Q \notin \mathcal{K}_k(X_j)$ for $j \neq i_0$, it must be $c_j(Q) = 0$ for $j \neq i_0$. It follows that

$$\left(\sum_{i=1}^{n} c_i\right)(Q) = c_{i_0}(Q) \neq 0,$$

that is, $x \in |\sum_{i=1}^{n} c_i|$. □

Finally, we are able to prove the main theorem of this section.

Theorem 2.59 *Let X be a cubical set. Then $H_0(X)$ is a free abelian group. Furthermore, if $\{P_i \mid i = 1, \ldots, n\}$ is a collection of vertices in X consisting of one vertex from each connected component of X, then*

$$\left\{[\widehat{P}_i] \in H_0(X) \mid i = 1, \ldots, n\right\}$$

forms a basis for $H_0(X)$.

Proof. Let $X_i := \mathrm{ccx}(P_i)$ and let $c \in Z_0(X)$. By Proposition 2.52, $[\widehat{P}] = [\widehat{P}_i]$ for any $P \in \mathcal{K}_0(X_i)$. Since $Z_0(X) = C_0(X)$, there exist integers α_P such that

$$[c] = \sum_{P \in \mathcal{K}_0(X)} \alpha_P [\widehat{P}] = \sum_{i=1}^{n} \sum_{P \in \mathcal{K}_0(X_i)} \alpha_P [\widehat{P}] = \sum_{i=1}^{n} \left(\sum_{P \sim_X P_i} \alpha_P\right) [\widehat{P}_i].$$

This shows that the classes $[\widehat{P}_i]$ generate $H_0(X)$.

It remains to show that the generators are free, that

$$\sum_{i=1}^{n} \alpha_i [\widehat{P}_i] = 0$$

implies that all $\alpha_i = 0$. To do so put $c := \sum_{i=1}^{n} \alpha_i \widehat{P}_i$. Since $[c] = 0$, we can select a $b \in C_1(X)$ such that $c = \partial b$. Let $b = \sum_{E \in \mathcal{K}_1(X)} \beta_E \widehat{E}$. Let

$$b_i := \sum_{E \in \mathcal{K}_1(X_i)} \beta_E \widehat{E}.$$

We have

$$\sum_{i=1}^{n} \alpha_i \widehat{P}_i = c = \partial b = \sum_{i=1}^{n} \partial b_i.$$

Therefore,

$$0 = \sum_{i=1}^{n} \left(\alpha_i \widehat{P}_i - \partial b_i\right).$$

But

$$|\alpha_i \widehat{P_i} - \partial b_i| \subset X_i.$$

Therefore, by Lemma 2.58,

$$\emptyset = |0| = \bigcup_{i=1}^{n} |\alpha_i \widehat{P_i} - \partial b_i|,$$

which shows that $|\alpha_i \widehat{P_i} - \partial b_i| = \emptyset$; that is, by Proposition 2.19(i), $\alpha_i \widehat{P_i} = \partial b_i$.

Let $\epsilon : C_0(X) \to \mathbf{Z}$ be the group homomorphism defined by $\epsilon(\widehat{P}) = 1$ for every vertex $P \in X$. Let E be an elementary edge. Then $\partial \widehat{E} = \widehat{V_1} - \widehat{V_0}$, where V_0 and V_1 are vertices of E. Observe that

$$\begin{aligned}\epsilon(\partial \widehat{E}) &= \epsilon(\widehat{V_1} - \widehat{V_0}) \\ &= \epsilon(\widehat{V_1}) - \epsilon(\widehat{V_0}) \\ &= 1 - 1 \\ &= 0.\end{aligned}$$

This implies that $\epsilon(\partial b_i) = 0$ and hence

$$0 = \epsilon(\partial b_i) = \epsilon(\alpha_i \widehat{P_i}) = \alpha_i \epsilon(\widehat{P_i}) = \alpha_i. \qquad \square$$

Exercises

2.27 Show that $\sim_X$ is an equivalence relation.

2.28 Find all minimal edge paths in the unit cube $Q = [0,1]^3$ connecting the vertex $V_0 = (0,0,0)$ to $V_1 = (1,1,1)$. For each edge path present the corresponding chain c such that $\partial c = \widehat{V_1} - \widehat{V_0}$.

2.29 Prove Proposition 2.51.

2.30 Prove Proposition 2.56.
Hint: First show that for every $x \in X$

$$\mathrm{cc}_X(\mathrm{x}) = \bigcup \{Q \in \mathcal{K} \mid Q \cap \mathrm{cc}(\mathrm{x}) \neq \emptyset\}.$$

2.4 Elementary Collapses

As the reader might have realized by now, even very simple cubical sets contain a large number of elementary cubes. We shall now discuss a method that allows us to reduce the number of elementary cubes needed to compute the homology of the set. For this end we need to refine the concepts of faces and proper faces (see Definition 2.7), but in the context of cubical sets rather than elementary cubes.

2.4 Elementary Collapses

Definition 2.60 Let X be a cubical set and let $Q \in \mathcal{K}(X)$. If Q is not a proper face of some $P \in \mathcal{K}(X)$, then it is a *maximal face* in X. $\mathcal{K}_{\max}(X)$ is the set of maximal faces in X. A face that is a proper face of exactly one elementary cube in X is a *free face in X*.

Example 2.61 Let $X = [0,1] \times [0,1] \times [0,1]$. Then $\mathcal{K}_0(X) \cup \mathcal{K}_1(X) \cup \mathcal{K}_2(X)$ is the set of proper faces. The set of free faces is given by $\mathcal{K}_2(X)$. For this simple case, $\mathcal{K}_{\max}(X) = \{X\}$.

Example 2.62 Referring to the cubical set $X \subset \mathbf{R}^2$ shown in Figure 2.5, the following elementary cubes are free faces:
$$[-1] \times [2],$$
$$[0,1] \times [0], \quad [0,1] \times [1], \quad [0] \times [0,1], \quad [1] \times [0,1].$$

Lemma 2.63 *Let X be a cubical set. Let $Q \in \mathcal{K}(X)$ be a free face in X and assume $Q \prec P \in \mathcal{K}(X)$. Then $P \in \mathcal{K}_{\max}(X)$ and $\dim Q = \dim P - 1$.*

Proof. Assume $P \prec R$. Then $Q \prec R$, contradicting the uniqueness of P.

Assume $\dim Q < \dim P - 1$. Then there exists $R \in \mathcal{K}(X)$ different from Q and P such that $Q \prec R \prec P$. □

Definition 2.64 Let Q be a free face in X and let P be the unique cube in $\mathcal{K}(X)$ such that Q is a proper face of P. Let $\mathcal{K}'(X) := \mathcal{K}(X) \setminus \{Q,P\}$. Define
$$X' := \bigcup_{R \in \mathcal{K}'(X)} R.$$
Then X' is a cubical space obtained from X via an *elementary collapse of P by Q*.

Proposition 2.65 *If X' is a cubical space obtained from X via an elementary collapse of P by Q, then*
$$\mathcal{K}(X') = \mathcal{K}'(X).$$

Proof. The inclusion $\mathcal{K}'(X) \subset \mathcal{K}(X')$ is obvious. To prove the opposite inclusion assume that there exists an elementary cube $S \in \mathcal{K}(X') \setminus \mathcal{K}'(X)$. It follows that $S \in \{P, Q\}$. Let $x \in \mathring{S} \subset S \subset X'$. Then $x \in R$ for some $R \in \mathcal{K}'(X)$ and $R \cap \mathring{S} \supset \{x\} \neq \emptyset$. By Proposition 2.15(vi), $S \subset R$. Since $S \notin \mathcal{K}'(X)$, S is a proper face of $R \in \mathcal{K}'(X) \subset \mathcal{K}(X)$. But neither $S = Q$ nor $S = P$ can be a proper face of such an R, a contradiction. □

Example 2.66 Let $X = [0,1] \times [0,1] \subset \mathbf{R}^2$ (see Figure 2.6). Then
$$\mathcal{K}_2(X) = \{[0,1] \times [0,1]\},$$
$$\mathcal{K}_1(X) = \{[0] \times [0,1], [1] \times [0,1], [0,1] \times [0], [0,1] \times [1]\},$$
$$\mathcal{K}_0(X) = \{[0] \times [0], [0] \times [1], [1] \times [0], [1] \times [1]\}.$$

72   2 Cubical Homology

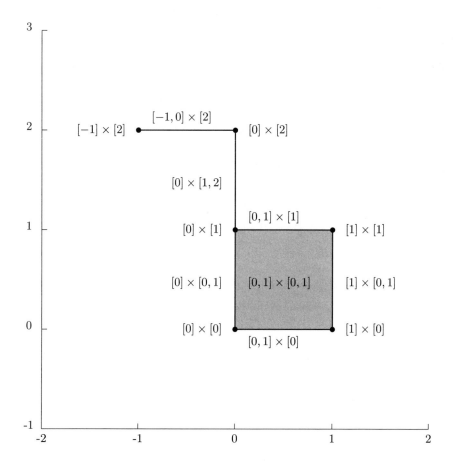

Fig. 2.5. Elementary cubes of $X \subset \mathbf{R}^2$.

There are four free faces, the elements of $\mathcal{K}_1(X)$. Let $Q = [0,1] \times [1]$, then $Q \prec P = [0,1] \times [0,1]$. If we let X' be the cubical space obtained from X via the elementary collapse of P by Q, then $X' = [0] \times [0,1] \cup [1] \times [0,1] \cup [0,1] \times [0]$ and

$$\mathcal{K}_1(X') = \{[0] \times [0,1], [1] \times [0,1], [0,1] \times [0]\},$$
$$\mathcal{K}_0(X') = \{[0] \times [0], [0] \times [1], [1] \times [0], [1] \times [1]\}.$$

Observe that the free faces of X' are different from those of X. In particular, $[0] \times [1]$ and $[1] \times [1]$ are free faces with $[0] \times [1] \prec [0] \times [0,1]$. Let X'' be the space obtained by collapsing $[0] \times [0,1]$ by $[0] \times [1]$. Then

$$\mathcal{K}_1(X'') = \{[1] \times [0,1], [0,1] \times [0]\},$$
$$\mathcal{K}_0(X'') = \{[0] \times [0], [1] \times [0], [1] \times [1]\}.$$

On X'' we can now perform an elementary collapse of $[1] \times [0,1]$ by $[1] \times [1]$ to obtain X''', where

$$\mathcal{K}_1(X'') = \{[0,1] \times [0]\},$$
$$\mathcal{K}_0(X'') = \{[0] \times [0], [1] \times [0]\}.$$

A final elementary collapse of $[0,1] \times [0]$ by $[1] \times [0]$ results in the single point $X'''' = [0] \times [0]$. Thus, through this procedure we have reduced a 2-cube to a single point.

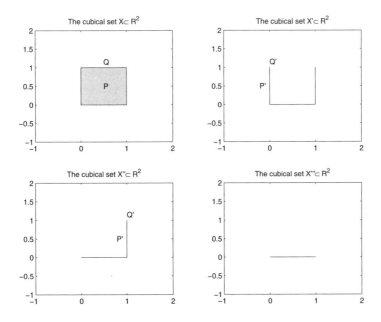

Fig. 2.6. Sequence of elementary collapses of $[0,1] \times [0,1] \subset \mathbf{R}^2$.

In the previous example, using elementary collapses, we reduce the cubical complex from that of a unit square containing nine elementary cubes to a cubical complex consisting of a single vertex. We now prove that two cubical complexes related by an elementary collapse have the same homology. Thus the homology of a square is the same as the homology of a vertex. The latter, of course, is trivial to compute. We begin with a lemma, which, when viewed from a geometrical point of view (see Figure 2.7), is fairly simple.

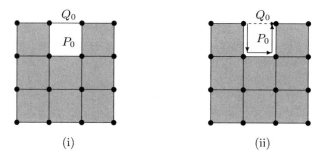

Fig. 2.7. (i) Gray indicates X' obtained from X by the elementary collapse of $P_0 \in \mathcal{K}_2(X)$ by $Q_0 \in \mathcal{K}_1(X)$. Observe that if $c \in C_2(X)$ and $\partial c \in C_{k-1}(X')$, then the support of c is entirely contained in the gray area. (ii) $c \in C_1(X)$ is indicated by the dashed arrow. $c' \in C_1(X')$ is indicated by the arrows that follow the sides and bottom of P_0. Observe that $c - c' \in B_1(X)$.

Lemma 2.67 *Assume X is a cubical set and X' is obtained from X via an elementary collapse of $P_0 \in \mathcal{K}_k(X)$ by $Q_0 \in \mathcal{K}_{k-1}(X)$. Then*

(i) $\{c \in C_k(X) \mid \partial c \in C_{k-1}(X')\} \subset C_k(X')$;
(ii) for every $c \in C_{k-1}(X)$ there exists $c' \in C_{k-1}(X')$ such that $c - c' \in B_{k-1}(X)$.

Proof. The proof of (i) begins with some simple observations. First, since P_0 is the unique element of $\mathcal{K}_k(X)$ of which Q_0 is a face, $\langle \partial \widehat{P}, \widehat{Q}_0 \rangle = 0$ if $P \neq P_0$. Similarly, by Exercise 2.19 $\langle \partial \widehat{P}_0, \widehat{Q}_0 \rangle = \pm 1$. Now assume that $c \in C_k(X)$ is such that $\partial c \in C_{k-1}(X')$. Then $|\partial c| \subset X'$. In particular, $Q_0 \not\subset |\partial c|$ and consequently $\langle \partial c, \widehat{Q}_0 \rangle = 0$. Since $c = \sum_{P \in \mathcal{K}_k(X)} \langle c, \widehat{P} \rangle \widehat{P}$, we have

$$0 = \langle \partial \sum_{P \in \mathcal{K}_k(X)} \langle c, \widehat{P} \rangle \widehat{P}, \widehat{Q}_0 \rangle = \sum_{P \in \mathcal{K}_k(X)} \langle c, \widehat{P} \rangle \langle \partial \widehat{P}, \widehat{Q}_0 \rangle = \pm \langle c, \widehat{P}_0 \rangle.$$

Therefore, $c \in C_k(X')$.

To prove (ii) assume that $c \in C_{k-1}(X)$. Let

$$c' := c - \langle c, \widehat{Q}_0 \rangle \langle \widehat{\partial P_0, \widehat{Q}_0} \rangle \widehat{\partial P_0}.$$

Then, obviously, $c - c' = \langle c, \widehat{Q}_0 \rangle \langle \widehat{\partial P_0, \widehat{Q}_0} \rangle \widehat{\partial P_0} \in B_{k-1}(X)$. Since

$$\langle c', \widehat{Q}_0 \rangle = \langle c, \widehat{Q}_0 \rangle - \langle c, \widehat{Q}_0 \rangle \langle \widehat{\partial P_0, \widehat{Q}_0} \rangle^2 = 0,$$

it follows that $c' \in C_{k-1}(X')$. □

2.4 Elementary Collapses

Theorem 2.68 *Assume X is a cubical set and X' is obtained from X via an elementary collapse of $P_0 \in \mathcal{K}_k(X)$ by $Q_0 \in \mathcal{K}_{k-1}(X)$. Then*

$$H_*(X') \cong H_*(X).$$

Proof. We begin the proof by making some obvious comparisons between the chains, cycles, and boundaries of the two complexes. Since $X' \subset X$, it follows that $C_n(X') \subset C_n(X)$ for all $n \in \mathbf{Z}$. Moreover, by Proposition 2.65, $\mathcal{K}_n(X') = \mathcal{K}_n(X)$ for $n \in \mathbf{Z} \setminus \{k-1, k\}$. Therefore,

$$C_n(X') = C_n(X) \quad \text{for all } n \in \mathbf{Z} \setminus \{k-1, k\}.$$

Since by (2.9) $Z_n(X) = C_n(X) \cap \ker \partial_n$ for any cubical set X and any $n \in \mathbf{Z}$, we have $Z_n(X') \subset Z_n(X)$ for all $n \in \mathbf{Z}$ and

$$Z_n(X') = Z_n(X) \quad \text{for all } n \in \mathbf{Z} \setminus \{k-1, k\}.$$

Similarly, since by (2.10) $B_n(X) = \partial_{n+1}(C_{n+1}(X))$ for any cubical set X and any $n \in \mathbf{Z}$, we have $B_n(X') \subset B_n(X)$ for all $n \in \mathbf{Z}$ and

$$B_n(X') = B_n(X) \quad \text{for all } n \in \mathbf{Z} \setminus \{k-2, k-1\}.$$

Observe that since the cycles and boundaries are the same,

$$H_k(X') = H_k(X) \quad \text{for all } n \in \mathbf{Z} \setminus \{k-2, k-1, k\}.$$

The only possible difference on the level of the k-chains is in the cycles. We now show that they are the same.

$$\begin{aligned}
Z_k(X) &= C_k(X) \cap \ker \partial_k \\
&= C_k(X) \cap \ker \partial_k \cap \ker \partial_k \\
&\stackrel{1}{\subset} C_k(X) \cap \{c \in C_k \mid \partial_k c \in C_{k-1}(X')\} \cap \ker \partial_k \\
&\stackrel{2}{\subset} C_k(X') \cap \ker \partial_k \\
&= Z_k(X').
\end{aligned}$$

Inclusion 1 is due to the fact that $\{0\} \subset C_{k-1}(X')$, while inclusion 2 follows from Lemma 2.67(i). Since also $Z_k(X') \subset Z_k(X)$, we conclude that $Z_k(X') = Z_k(X)$ and $H_k(X') = H_k(X)$.

On the level of the $(k-2)$-chains the only possible difference is in the boundaries. We will show they are the same by proving that $B_{k-2}(X) \subset B_{k-2}(X')$. To this end take $b \in B_{k-2}(X)$. Then $b = \partial c$ for some $c \in C_{k-1}(X)$. By Lemma 2.67(ii) there exists $c' \in C_{k-1}(X')$ such that $c - c' \in B_{k-1}(X)$. In particular, for some d,

$$b - \partial c' = \partial(c - c') = \partial^2 d = 0$$

and thus $b = \partial c' \in B_{k-2}(X')$. Therefore, we have shown that $B_{k-2}(X) = B_{k-2}(X')$. It follows that $H_{k-2}(X') = H_{k-2}(X)$.

It remains to consider the homology groups in dimension $k-1$. In this case the homology groups are not equal, but we will construct an isomorphism. Let $\xi \in H_{k-1}(X')$. Then $\xi = [z]_{X'}$ for some cycle $z \in Z_{k-1}(X')$. We put the subscript X' in $[z]_{X'}$ to emphasize that the homology class is taken in X'. This is important, because $Z_{k-1}(X') \subset Z_{k-1}(X)$, which means that we can also consider the homology class $[z]_X$ of z in $H_{k-1}(X)$. Since $B_{k-1}(X') \subset B_{k-1}(X)$, one easily verifies that if $[z]_{X'} = [z']_{X'}$ for some $z' \in Z_{k-1}(X')$, then $[z]_X = [z']_X$. This means that we have a well-defined map $\iota : H_{k-1}(X') \to H_{k-1}(X)$ given by $\iota([z]_{X'}) = [z]_X$. It is straightforward to verify that this map is a homomorphism of groups. To show that it is a monomorphism, assume that $[z]_X = 0$ for some $z \in Z_{k-1}(X')$. Then $z = \partial c$ for some $c \in C_k(X)$. It follows from Lemma 2.67(i) that $c \in C_k(X')$, which shows that $[z]_{X'} = 0$. Consequently the map is a monomorphism. To show that it is an epimorphism, consider $[z]_X$ for some $z \in Z_{k-1}(X)$. By Lemma 2.67(ii) there exists $c' \in C_{k-1}(X')$ such that $z - c' \in B_{k-1}(X)$. In particular, $\partial(z - c') = 0$, that is, $\partial c' = \partial z = 0$. This shows that $c' \in Z_{k-1}(X')$ and since $\iota([c']_{X'}) = [c']_X = [z]_X$, the map is surjective. □

Corollary 2.69 *Let $Y \subset X$ be cubical sets. Furthermore, assume that Y can be obtained from X via a series of elementary collapses. Then*

$$H_*(Y) \cong H_*(X).$$

Elementary collapses have been introduced in a combinatorial fashion. One begins with a cubical complex and removes two specially chosen elementary cubes. In particular, no topological motivation is provided. This is due to the fact that to provide a complete explanation requires understanding how continuous maps induce linear maps between homology groups—a topic covered in Chapter 6. Therefore, with this in mind, we shall limit ourselves to some simple examples that are meant to suggest that elementary collapses are induced by continuous maps called deformation retractions defined as follows.

Definition 2.70 Let $A \subset X$. A *deformation retraction* of X onto A is a continuous map $h : X \times [0,1] \to X$ such that

$$h(x, 0) = x \text{ for all } x \in X,$$
$$h(x, 1) \in A \text{ for all } x \in X,$$
$$h(a, 1) = a \text{ for all } a \in A.$$

If such an h exists, then A is called a *deformation retract* of X. The map h is called a *strong deformation retraction* and the set A a *strong deformation retract* if the third identity is reinforced as follows:

$$h(a, s) = a \text{ for all } a \in A \text{ and all } s \in [0,1].$$

Note that the map $r : X \to A$ defined by $r(x) = h(x, 1)$ has the property $r_{|A} = \mathrm{id}_A$. Any continuous map with this property is called a *retraction* and its image A is called a *retract*. Thus a deformation retract is a special case of a retract. We will come back to this topic later in Section 6.5.

Deformation retracts that are not strong deformation retracts only occur in complicated spaces. In our class of cubical sets, all examples of deformation retracts will be strong. The relation of deformation retraction to homology is discussed in Section 6.5.

Example 2.71 Let $P = [0, 1]$. The free faces in P are its vertices $Q_0 = [0]$ and $Q_1 = [1]$. Observe that Q_1 is obtained from P by the elementary collapse of P by Q_0. It is easy to show that the map $h : P \times [0, 1] \to P$ defined by

$$h(x, s) := (1 - s)x + s$$

is a strong deformation retraction of P onto Q_1. This construction can be easily carried over to any one-dimensional elementary cube and its vertex (this also is a particular case of Exercise 6.19). Thus, in the case of $P \in \mathcal{K}_1$, an elementary collapse corresponds to a deformation retraction.

Example 2.72 Let $P = [0, 1] \times [0, 1]$ and let $Q = [1] \times [0, 1]$. Observe that Q is a free face of P and that $P' = [0] \times [0, 1] \cup [0, 1] \times [0] \cup [0, 1] \times [1]$ is the result of the elementary collapse of P by Q. We shall now show that P' is a strong deformation retract of P.

Let $r : P \to P'$ be defined by

$$r(x, y) := \begin{cases} (3 - 3y + x, 1) & \text{if } y \geq \frac{1}{3}x + \frac{2}{3}, \\ (1, 3\frac{x+y}{2x+1} - 1) & \text{if } -\frac{1}{3}x + \frac{1}{3} \leq y \leq \frac{1}{3}x + \frac{2}{3}, \\ (3y + x, 0) & \text{if } y \leq -\frac{1}{3}x + \frac{1}{3}. \end{cases}$$

It is left as an exercise to show that r is well defined and continuous. The verification that $r|P' = \mathrm{id}_{P'}$ is straightforward. Now one can easily check that

$$h((x, y), s) := (1 - s)(x, y) + sr(x, y).$$

defines the required deformation retraction of P onto P'.

Exercises ─────────────

2.31 Determine the maximal and free faces of the cubical set presented in the file exC2d.txt in the folder CubTop/Examples by calling the CubTop program with the corresponding functions. Present your output in two bitmap files.

2.32 * Let X be a two-dimensional cubical subset of $\mathbf{R}^2$. Show that X has at least one free edge.

2.33 Use the elementary collapses to reduce the elementary cube $[0, 1]^3$ to its vertex $(0, 0, 0)$.

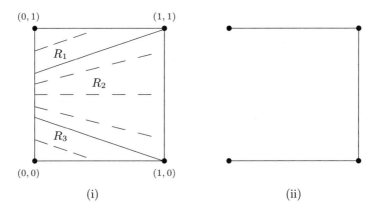

Fig. 2.8. Deformation retraction of a square. (i) The map r is defined by different formulas in the three regions R_1, R_2, and R_3 bounded by solid lines. The deformation discussed in Example 2.72 occurs along the dashed lines.(ii) The image of the collapsed square.

2.34 Let X be the solid cubical set discussed in Exercise 2.23. Here is an alternative way of computing the homology of X: Use the elementary collapses of X onto the simple closed curve Γ defined as the union of four line segments $[1,2] \times [1] \times [0]$, $[2] \times [1,2] \times [0]$, $[1,2] \times [2] \times [0]$, $[1] \times [1,2] \times [0]$. Compute the homology of Γ and deduce the homology of X.

2.35 Let X be a cubical set in $\mathbf{R}^2$ of dimension 2. In Exercise 2.32 we show that any such set must have a free edge. Prove that X can be reduced to a one-dimensional cubical set by elementary collapses. Conclude that $H_2(X) = 0$.

2.36 Let X be a cubical set and let $\mathcal{M} \subset \mathcal{K}_{\max}(X)$ be a family of selected maximal faces of X. Let

$$X' := \sum_{R \in \mathcal{K}(X) \setminus \mathcal{M}} R.$$

Show that

$$\mathcal{K}(X') = \mathcal{K}(X) \setminus \mathcal{M}.$$

2.37 Let $P = [0,1] \times [0,1] \times [0,1]$. Then $Q = [0] \times [0,1] \times [0,1]$ is a free face of P. Let P' be obtained via an elementary collapse of P by Q. Construct a deformation retraction of P to P'.

2.5 Acyclic Cubical Spaces

We shall now study an important class of cubical sets—those that have trivial homology, namely, the homology of a point.

Definition 2.73 A cubical set X is *acyclic* if

$$H_k(X) \cong \begin{cases} \mathbf{Z} & \text{if } k = 0, \\ 0 & \text{otherwise.} \end{cases}$$

Example 2.66 shows that the unitary cube $[0,1]^2$ in $\mathbf{R}^2$ is acyclic, and Exercise 2.33 shows that $[0,1]^3$ in $\mathbf{R}^3$ is acyclic. We may conjecture that any elementary cube is acyclic. Although the idea of collapsing cubes onto their faces is very transparent, writing it down precisely requires a lot of work. Therefore, we present another proof, based on a special feature of cycles located in elementary cubes.

Let $Q \in \mathcal{K}^d$ be an elementary cube. For some $i \in \{1, 2, \ldots, d\}$ let $I_i(Q)$ be nondegenerate, namely $I_i(Q) = [l, l+1]$, where $l \in \mathbf{Z}$. Fix $k > 0$. The family of k-dimensional faces of Q decomposes into

$$\mathcal{K}_k(Q) = \mathcal{K}_k([l], i) \cup \mathcal{K}_k([l, l+1], i) \cup \mathcal{K}_k([l+1], i),$$

where $\mathcal{K}_k(\Delta, i) := \{P \in \mathcal{K}_k(Q) \mid I_i(P) = \Delta\}$.

Example 2.74 Let $Q = [p, p+1] \times [l, l+1] \times [q]$. Then

$$\mathcal{K}_1([l], 2) = \{[p, p+1] \times [l] \times [q]\},$$
$$\mathcal{K}_1([l, l+1], 2) = \{[p] \times [l, l+1] \times [q], [p+1] \times [l, l+1] \times [q]\},$$
$$\mathcal{K}_1([l+1], 2) = \{[p, p+1] \times [l+1] \times [q]\}.$$

As indicated in the next lemma, k-cycles in elementary cubes have the following nice geometric feature. Let $z \in Z_k(Q)$ be with the property that $|z|$ contains no elementary cubes whose ith component is $[l+1]$. Then $|z|$ contains no elementary cubes whose ith component is $[l, l+1]$.

Lemma 2.75 *Assume $Q \in \mathcal{K}^d$ and $i \in \{1, 2, \ldots d\}$. If z is a k-cycle in Q such that $\langle z, \widehat{P} \rangle = 0$ for every $P \in \mathcal{K}_k([l+1], i)$, then $\langle z, \widehat{P} \rangle = 0$ for every $P \in \mathcal{K}_k([l, l+1], i)$.*

Proof. Since z is a chain in Q, we have

$$z = \sum_{P \in \mathcal{K}_k(Q)} \langle z, \widehat{P} \rangle \widehat{P}.$$

Therefore, for any $R \in \mathcal{K}_{k-1}(Q)$, we have

$$\langle \partial z, \widehat{R} \rangle = \langle \sum_{P \in \mathcal{K}_k(Q)} \langle z, \widehat{P} \rangle \partial \widehat{P}, \widehat{R} \rangle = \sum_{P \in \mathcal{K}_k(Q)} \langle z, \widehat{P} \rangle \langle \partial \widehat{P}, \widehat{R} \rangle.$$

80   2 Cubical Homology

Since z is a cycle it follows from our assumption, that for any $R \in \mathcal{K}_{k-1}(Q)$

$$0 = \langle \partial z, \widehat{R} \rangle = \sum_{P \in \mathcal{K}_k([l],i)} \langle z, \widehat{P} \rangle \langle \partial \widehat{P}, \widehat{R} \rangle + \sum_{P \in \mathcal{K}_k([l,l+1],i)} \langle z, \widehat{P} \rangle \langle \partial \widehat{P}, \widehat{R} \rangle. \quad (2.11)$$

Let $P_0 \in \mathcal{K}_k([l, l+1], i)$ and let R_0 be the elementary cube defined by

$$I_j(R_0) = \begin{cases} [l+1] & \text{if } j = i, \\ I_j(P_0) & \text{otherwise.} \end{cases}$$

Obviously, R_0 cannot be a face of P for $P \in \mathcal{K}_k([l], i)$, hence the first sum on the right-hand side of (2.11) disappears for $R = R_0$. Moreover, R_0 is a face of P for $P \in \mathcal{K}_k([l, l+1], i)$ if and only if $P = P_0$. This means that Eq. (2.11) reduces for $R = R_0$ to $0 = \langle z, \widehat{P_0} \rangle \langle \partial \widehat{P_0}, \widehat{R_0} \rangle$. Since $\langle \partial \widehat{P_0}, \widehat{R_0} \rangle \neq 0$, we get $\langle z, \widehat{P_0} \rangle = 0$. □

Theorem 2.76 *All elementary cubes are acyclic.*

Proof. Let Q be an elementary cube. Since Q is connected, it follows from Theorem 2.59 that $H_0(Q) = \mathbf{Z}$. Therefore, it remains to prove that $H_k(Q) = 0$ for $k > 0$, which is equivalent to showing that every k-cycle in Q is a boundary. We will show this fact by induction on $n := \dim Q$. If $n = 0$ and $k > 0$, then $Z_k(Q) = C_k(Q) = 0 = B_k(Q)$, which implies that $H_k(Q) = 0$.

Therefore, assume that $n > 0$ and $H_k(Q) = 0$ for all elementary cubes of dimension less than n. Since $n > 0$, we can choose some i such that $I_i(Q)$ is nondegenerate.

For every $P \in \mathcal{K}_k([l+1], i)$, let P^* denote the elementary cube given by

$$I_j(P^*) := \begin{cases} [l, l+1] & \text{if } j = i, \\ I_j(P) & \text{otherwise.} \end{cases}$$

Then obviously P is a face of P^*.

Let z be a k-cycle in Q. Define

$$c := \sum_{P \in \mathcal{K}_k([l+1],i)} \langle z, \widehat{P} \rangle \langle \widehat{\partial P^*}, \widehat{P} \rangle \widehat{P^*},$$

$$z' := z - \partial c.$$

For every $P_0 \in \mathcal{K}_k([l+1], i)$, we have

$$\langle \partial c, \widehat{P_0} \rangle = \sum_{P \in \mathcal{K}_k([l+1],i)} \langle z, \widehat{P} \rangle \langle \widehat{\partial P^*}, \widehat{P} \rangle \langle \widehat{\partial P^*}, \widehat{P_0} \rangle.$$

Since $I_i(P^*) = [l, l+1]$ and $I_i(P_0) = [l+1]$, we have $\langle \widehat{\partial P^*}, \widehat{P_0} \rangle \neq 0$ if and only if $P = P_0$. Therefore, $\langle \partial c, \widehat{P_0} \rangle = \langle z, \widehat{P_0} \rangle$ and $\langle z', \widehat{P_0} \rangle = 0$. It follows from Lemma 2.75 that $|z'| \subset Q'$, where Q' is an $(n-1)$-dimensional cube defined by

$$I_j(Q') := \begin{cases} [l] & \text{if } j = i, \\ I_j(Q) & \text{otherwise.} \end{cases}$$

By the induction assumption, $z' = \partial c'$. This shows that $z = \partial(c + c')$, that is z is a boundary. □

While Theorem 2.76 shows us that the building blocks of our theory are acylic, it sheds no light on the question of how to determine if a given cubical set is acyclic. Theorem 2.78 provides us with such a method. As we shall see in Chapter 6 this is a simple version of a much more general and powerful theorem called the Meyer–Vietoris sequence. Before stating and proving Theorem 2.78, we need the following proposition.

Proposition 2.77 *If $K, L \subset \mathbf{R}^n$ are cubical sets, then*

$$C_k(K \cup L) = C_k(K) + C_k(L).$$

Proof. Since $K \subset K \cup L$, obviously $C_k(K) \subset C_k(K \cup L)$. Similarly, $C_k(L) \subset C_k(K \cup L)$. Hence $C_k(K) + C_k(L) \subset C_k(K \cup L)$, because $C_k(K \cup L)$ is a group.

To prove the opposite inclusion, let $c \in C_k(K \cup L)$. In terms of the basis elements this can be written as

$$c = \sum_{i=1}^{m} \alpha_i \widehat{Q}_i, \quad \alpha_i \neq 0.$$

Let $A := \{i \mid Q_i \subset K\}$ and $B := \{1, 2, \ldots, m\} \setminus A$. Put $c_1 := \sum_{i \in A} \alpha_i \widehat{Q}_i$, $c_2 := \sum_{i \in B} \alpha_i \widehat{Q}_i$. Obviously, $|c_1| \subset K$. Let $i \in B$. Then $Q_i \subset K \cup L$ and $Q_i \not\subset K$. By Proposition 2.15(vi), $\overset{\circ}{Q}_i \cap K = \emptyset$. Therefore, $\overset{\circ}{Q}_i \subset L$ and by Proposition 2.15(iv), $Q_i = \operatorname{cl} \overset{\circ}{Q}_i \subset \operatorname{cl} L = L$ Hence $|c_2| \subset L$. It follows that $c = c_1 + c_2 \in C_k(K) + C_k(L)$. □

Theorem 2.78 *Assume $X, Y \subset \mathbf{R}^n$ are cubical sets. If X, Y, and $X \cap Y$ are acyclic, then $X \cup Y$ is acyclic.*

Proof. We will first prove that $H_0(X \cup Y) \cong \mathbf{Z}$. By Theorem 2.59, the assumption that X and Y are acyclic implies that X and Y are connected. $X \cap Y$ is acyclic implies that $X \cap Y \neq \emptyset$. Therefore, $X \cup Y$ is connected and hence by Theorem 2.59, $H_0(X \cup Y) \cong \mathbf{Z}$.

Now consider the case of $H_1(X \cup Y)$. Let $z \in Z_1(X \cup Y)$ be a cycle. We need to show that $z \in B_1(X \cup Y)$. By Proposition 2.77, $z = z_X + z_Y$ for some $z_X \in C_1(X)$ and $z_Y \in C_1(Y)$. Since z is a cycle, $\partial z = 0$. Thus

$$\begin{aligned} 0 &= \partial z \\ &= \partial(z_X + z_Y) \\ &= \partial z_X + \partial z_Y. \end{aligned}$$

Therefore

$$-\partial z_Y = \partial z_X.$$

Observe that $-\partial z_Y, \partial z_X \in C_0(X \cap Y) = Z_0(X \cap Y)$. From the assumption of acyclicity, $H_0(X \cap Y)$ is isomorphic to $\mathbf{Z}$.

Let P_0 be a vertex in $X \cap Y$. By Theorem 2.59, $H_0(X \cap Y)$ is generated by $[\widehat{P_0}]_{X \cap Y}$. Let $[\partial z_X]_{X \cap Y}$ denote the homology class of ∂z_X in $H_0(X \cap Y)$. Then $[\partial z_X]_{X \cap Y} = n[\widehat{P_0}]_{X \cap Y}$ for some $n \in \mathbf{Z}$.

We will now show that $n = 0$. The fact that $\partial z_X \in C_0(X \cap Y)$ implies that $\partial z_X = \sum a_i \widehat{P_i}$, where $P_i \in \mathcal{K}_0(X \cap Y)$. By Theorem 2.59, $[\partial z_X]_{X \cap Y} = n[\widehat{P_0}]_{X \cap Y}$ implies that $\sum a_i = n$. Define the group homomorphism $\epsilon : C_0(X \cap Y) \to \mathbf{Z}$ by $\epsilon(\widehat{P}) = 1$ for each $P \in \mathcal{K}_0(X \cap Y)$. Then $\epsilon(\partial \widehat{Q}) = 0$ for any $Q \in \mathcal{K}_1(X \cap Y)$. Therefore, $\epsilon(\partial z_X) = 0$, but

$$\epsilon(\partial z_X) = \sum a_i = n.$$

Hence, $n = 0$.

Since $[\partial z_X]_{X \cap Y} = 0$, there exists $c \in C_1(X \cap Y)$ such that $\partial c = \partial z_X$. Now observe that

$$\partial(-c + z_X) = -\partial c + \partial z_X = 0.$$

Therefore, $-c + z_X \in Z_1(X)$. But $H_1(X) = 0$, which implies that there exists $b_X \in C_2(X)$ such that $\partial b_X = -c + z_X$. The same argument shows that there exists $b_Y \in C_2(Y)$ such that $\partial b_Y = c + z_Y$. Finally, observe that $b_X + b_Y \in C_2(X \cup Y)$ and

$$\begin{aligned}\partial(b_X + b_Y) &= \partial b_X + \partial b_Y \\ &= (-c) + z_X + c + z_Y \\ &= z_X + z_Y \\ &= z.\end{aligned}$$

Therefore, $z \in B_1(X \cup Y)$, which implies that $[z]_{X \cap Y} = 0$. Thus $H_1(X \cup Y) = 0$.

We now show that $H_k(X \cup Y) \cong 0$ for all $k > 1$. Let $z \in Z_k(X \cup Y)$ be a cycle. Then by Proposition 2.77, $z = z_X + z_Y$ for some $z_X \in C_k(X)$ and $z_Y \in C_k(Y)$. Since z is a cycle, $\partial z = 0$. As in the previous case we show that

$$-\partial z_Y = \partial z_X. \tag{2.12}$$

Of course, this does not imply that $\partial z_X = 0$. However, since $z_Y \in C_k(Y)$ and $z_X \in C_k(X)$, we conclude that $b := \partial z_X \in C_{k-1}(X \cap Y)$.

Since $X \cap Y$ is acyclic, $H_{k-1}(X \cap Y) = 0$. Therefore, $b \in B_{k-1}(X \cap Y)$. This implies that there exists a $c \in C_k(X \cap Y)$ such that $b = \partial c$. It follows that $z_X - c \in Z_k(X)$ and from (2.12) that $z_Y + c \in Z_k(Y)$. By the acyclicity of X and Y, there exist $c_X \in C_{k+1}(X)$ and $c_Y \in C_{k+1}(Y)$ such that $z_X - c = \partial c_X$ and $z_Y - c = \partial c_Y$. Therefore,

$$z = z_X + z_Y = \partial(c_X + c_Y) \in B_k(X \cup Y). \qquad \square$$

2.5 Acyclic Cubical Spaces

Definition 2.79 A *rectangle* is a set of the form $X = [k_1, l_1] \times [k_2, l_2] \times \ldots \times [k_n, l_n] \subset \mathbf{R}^n$, where k_i, l_i are integers and $k_i \leq l_i$.

Any rectangle is a cubical set. It is an elementary cube if $k_i = l_i$ or $k_i + 1 = l_i$ for every i.

We leave the proof of the following proposition to the reader.

Proposition 2.80 *A cubical set is convex if and only if it is a rectangle.*

Proposition 2.81 *Any rectangle is acyclic.*

Proof. For $\Delta = [k, l] \subset \mathbf{R}$, an interval with integer endpoints, set

$$\mu(\Delta) := \begin{cases} l - k & \text{if } l > k, \\ 1 & \text{otherwise.} \end{cases}$$

Note that Δ is an elementary interval if and only if $\mu(\Delta) = 1$.

Let X be a rectangle, that is,

$$X = \Delta_1 \times \Delta_2 \times \cdots \times \Delta_d,$$

where $\Delta_i = [k_i, l_i]$ is an interval with integer endpoints. Put

$$\mu(X) := \mu(\Delta_1)\mu(\Delta_2)\ldots\mu(\Delta_d).$$

The proof will proceed by induction on $m := \mu(X)$.

If $m = 1$, then X is an elementary cube. Therefore, X is acyclic by Theorem 2.76. Thus assume that $m > 1$. Then $\mu(\Delta_{i_0}) = l_{i_0} - k_{i_0} \geq 2$ for some $i_0 \in \{1, 2, \ldots, d\}$. Let

$$X_1 := [k_1, l_1] \times \cdots \times [k_{i_0}, k_{i_0} + 1] \times \cdots \times [k_n, l_n]$$

and

$$X_2 := [k_1, l_1] \times \cdots \times [k_{i_0} + 1, l_{i_0}] \times \cdots \times [k_n, l_n].$$

Then X_1, X_2, and $X_1 \cap X_2$ are rectangles (see Figure 2.9) and

$$\mu(X_1) = \mu(X_1 \cap X_2) = \frac{\mu(X)}{\mu(\Delta_{i_0})} < m,$$

$$\mu(X_2) = \mu(X)\frac{\mu(\Delta_{i_0}) - 1}{\mu(\Delta_{i_0})} < m.$$

Therefore, by the induction assumption, X_1, X_2, and $X_1 \cap X_2$ are acyclic. The result now follows from Theorem 2.78. □

Since rectangles are always the products of intervals, they represent a small class of cubical sets. A slightly larger collection of acyclic spaces is the following.

Definition 2.82 A cubical set $X \subset \mathbf{R}^d$ is *star-shaped* with respect to a point $x \in \mathbf{Z}^d$ if X is the union of a finite number of rectangles each of which contains the point x.

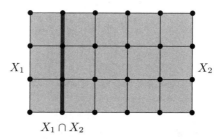

Fig. 2.9. X_0 is the two-dimensional rectangle on the left. X_2 is the two-dimensional rectangle on the right. $X_1 \cap X_2$ is the one-dimensional rectangle.

Proposition 2.83 *Let X_i, for $i = 1, \ldots, n$ be a collection of star-shaped sets with respect to the same point x. Then*

$$\bigcup_{i=1}^{n} X_i \quad \text{and} \quad \bigcap_{i=1}^{n} X_i$$

are star-shaped.

Proof. Since X_i is star-shaped, it can be written as $X_i = \cup R_{i,j}$, where $R_{i,j}$ is a rectangle and $x \in R_{i,j}$. Thus, if $X = \cup_i X_i$, then $X = \cup_{i,j} R_{i,j}$ and hence it is star-shaped.

On the other hand

$$\bigcap_i X_i = \bigcap_i \left(\bigcup_j R_{i,j} \right) = \bigcup_j \left(\bigcap_i R_{i,j} \right).$$

But, since $x \in R_{i,j}$ for every i, j, for each j the set $\bigcap_i R_{i,j}$ is a rectangle and contains x. This means that $\bigcap_i X_i$ is star-shaped. □

Proposition 2.84 *Every star-shaped set is acyclic.*

Proof. Let X be a star-shaped cubical set. Then $X = \bigcup_{i=1}^{n} R_i$, where each R_i is a rectangle and there exists $x \in X$ such that $x \in R_i$ for all $i = 1, \ldots, n$. The proof is by induction on n.

If $n = 1$, then X is a rectangle and hence by Proposition 2.81 is acyclic.

So assume that every star-shaped cubical set that can be written as the union of $n - 1$ rectangles containing the same point is acyclic. Let $Y = \bigcup_{i=1}^{n-1} R_i$. Then by the induction hypothesis, Y is acyclic. R_n is a rectangle and hence by Proposition 2.81 it is acyclic. Furthermore, $R_i \cap R_n$ is a rectangle for each $i = 1, \ldots, n - 1$ and

$$Y \cap R_n = \bigcup_{i=1}^{n-1} (R_i \cap R_n).$$

Therefore, $Y \cap R_n$ is a star-shaped set that can be written in terms of $n-1$ rectangles. By the induction hypothesis, it is also acyclic. Therefore, by Theorem 2.78, X is acyclic. □

The following simple result will be used later.

Proposition 2.85 *Assume that $\mathcal{Q}$ is a family of rectangles in $\mathbf{R}^d$ such that the intersection of any two of them is nonempty. Then $\bigcap \mathcal{Q}$ is nonempty.*

Proof. First consider the case when $d = 1$. Then rectangles become intervals. Let a denote the supremum of the set of left endpoints of the intervals and let b denote the infimum of the set of right endpoints. We cannot have $b < a$, because then one can find two disjoint intervals in the family. Therefore, $\emptyset \neq [a,b] \subset \bigcap \mathcal{Q}$.

If $d > 1$, then each rectangle is a product of intervals, the intersection of all rectangles is the product of the intersections of the corresponding intervals, and the conclusion follows from the previous case. □

Exercises

2.38 Prove Proposition 2.80.

2.39 Give an example where X and Y are acyclic cubical sets, but $X \cup Y$ is not acyclic.

2.40 Consider the capital letter **H** as a three-dimensional cubical complex. Compute its homology.

2.6 Homology of Abstract Chain Complexes

We now turn to a purely algebraic description of homology groups. The cubical chain complex is a particular case of what we present below.

Definition 2.86 A *chain complex* $\mathcal{C} = \{C_k, \partial_k\}_{k \in \mathbf{Z}}$ consists of abelian groups C_k, called *chains*, and homomorphisms $\partial_k : C_k \to C_{k-1}$, called *boundary operators*, such that

$$\partial_k \circ \partial_{k+1} = 0. \tag{2.13}$$

$\mathcal{C}$ is a *free chain complex* if C_k is free for all $k \in \mathbf{Z}$. The *cycles* of $\mathcal{C}$ are the subgroups

$$Z_k := \ker \partial_k,$$

while the *boundaries* are the subgroups

$$B_k := \operatorname{im} \partial_{k+1}.$$

Observe that (2.13) implies that

$$\operatorname{im} \partial_{k+1} \subset \ker \partial_k, \tag{2.14}$$

and hence the following definition makes sense.

Definition 2.87 The kth *homology group* of the chain complex $\mathcal{C}$ is
$$H_k(\mathcal{C}) := Z_k/B_k.$$
The *homology* of $\mathcal{C}$ is the sequence
$$H_*(\mathcal{C}) := \{H_k(\mathcal{C})\}_{k \in \mathbf{Z}}.$$

Observe that this is a purely algebraic definition. The inclusion (2.14) implies that for each $z \in Z_k$,
$$\partial_k(z + B_k) = \partial_k z + \partial_k(B_k) = 0 + \partial_k(\partial_{k+1}(C_{k+1})) = 0,$$
so ∂_k induces the trivial homomorphism
$$\bar{\partial}_k : H_k(\mathcal{C}) \to H_{k-1}(\mathcal{C}) , \ \bar{\partial}_k = 0,$$
on quotient groups. Thus the homology sequence $H_*(\mathcal{C})$ may also be viewed as a chain complex called the *homology complex* of $\mathcal{C}$ with the trivial boundary operators. Conversely, if in some chain complex $\mathcal{C}$ we have $\partial_k = 0$ for all n, then $H_*(\mathcal{C}) = \mathcal{C}$.

Definition 2.88 $\mathcal{C}$ is a *finitely generated free chain complex* if

1. each C_k is a finitely generated free abelian group,
2. $C_k = 0$ for all but finitely many k.

In this book we are only concerned with finitely generated free chain complexes. However, it should be mentioned that we have already seen an example of an infinitely generated free chain complex $\mathcal{C}^d$ with groups C_k^d of chains of $\mathbf{R}^d$ and boundary operators $\partial_k : C_k^d \to C_{k-1}^d$.

Example 2.89 Consider the cubical set consisting of two vertices P and Q. The cubical chain complex $\mathcal{C}(X)$ has only one nontrivial group
$$C_0(X) = \mathbf{Z}\widehat{P} \oplus \mathbf{Z}\widehat{Q} \cong \mathbf{Z}^2.$$
All boundary maps are necessarily 0. Consider now an abstract chain complex $\mathcal{C}$ given by
$$\mathcal{C}_k := \begin{cases} C_0(X) & \text{if } k = 0, \\ \mathbf{Z} & \text{if } k = -1, \\ 0 & \text{otherwise,} \end{cases}$$
where $\partial_k := 0$ if $k \neq 0$. The only nontrivial boundary map $\partial_0 : C_0 \to C_{-1}$ is defined on the generators as follows:
$$\partial_0 \widehat{P} = \partial_0 \widehat{Q} := 1.$$
It is clear that $\operatorname{im} \partial_0 = \mathbf{Z} = C_{-1} = \ker \partial_{-1}$. Therefore,

$$H_{-1}(\mathcal{C}) = 0,$$

We leave as an exercise the verification that

$$\ker \partial_0 = \mathbf{Z}(\widehat{P} - \widehat{Q}),$$

and, because $\partial_1 = 0$, we have

$$H_0(\mathcal{C}) = \ker \partial_0 = \mathbf{Z}(\widehat{P} - \widehat{Q}) \cong \mathbf{Z}.$$

The remaining homology groups are trivial. This is a special case of reduced homology, which is introduced in the next section.

Example 2.90 Define an abstract chain complex $\mathcal{C}$ as follows. The only non-trivial groups C_k of $\mathcal{C}$ are in dimensions $k = 0, 1, 2$. Let

$$C_0 := \mathbf{Z}v \cong \mathbf{Z},$$
$$C_1 := \mathbf{Z}e_1 \oplus \mathbf{Z}e_2 \cong \mathbf{Z}^2,$$
$$C_2 := \mathbf{Z}g \cong \mathbf{Z},$$

where v, e_1, e_2, and g are some fixed generators. Define $\partial_1 : C_1 \to C_0$ to be zero. Define $\partial_2 : C_2 \to C_1$ on the generator g by

$$\partial_2 g := 2e_1.$$

The remaining boundary maps must be zero because $C_k = 0$ for all $k \notin \{0, 1, 2\}$. It is easily seen that

$$H_0(\mathcal{C}) = C_0 = \mathbf{Z}v \cong \mathbf{Z}.$$

Computing the group H_1 requires some acquaintance with abelian groups (see Chapter 13 and Section 3.4).

$$H_1(\mathcal{C}) = \frac{\mathbf{Z}e_1 \oplus \mathbf{Z}e_2}{\mathbf{Z}2e_1 \oplus 0} = \frac{\mathbf{Z}e_1}{2\mathbf{Z}e_1} \oplus \mathbf{Z}e_2 \cong \mathbf{Z}_2 \oplus \mathbf{Z}.$$

Finally, since $\partial_3 = 0$ and $\ker \partial_2 = 0$, we have

$$H_2(\mathcal{C}) = 0.$$

Definition 2.91 Let $\mathcal{C} = \{C_n, \partial_n\}$ be a chain complex. A chain complex $\mathcal{D} = \{D_n, \partial'_n\}$ is a *chain subcomplex* of $\mathcal{C}$ if

1. D_n is a subgroup of C_n for all $n \in \mathbf{Z}$,
2. $\partial'_n = \partial_n |_{D_n}$.

The condition $\partial'_n = \partial_n |_{D_n}$ means that the boundary operator of the chain subcomplex is just the boundary operator of the larger complex restricted to its domain. For this reason we can simplify the notation by writing ∂ instead of ∂'.

Example 2.92 The following statements are immediate consequences of the definition of the cubical boundary operator.

(a) Let $X \subset Y$ be cubical sets. Then $\mathcal{C}(X)$ is a chain subcomplex of $\mathcal{C}(Y)$.
(b) Let $X \subset \mathbf{R}^d$ be a cubical set. Then $\mathcal{C}(X)$ is a chain subcomplex of $\mathcal{C}(\mathbf{R}^d) := \{C_k^d\}$.

Exercises ───

2.41 Let $\mathcal{C}$ be the abstract chain complex discussed in Example 2.90.

(a) Compute its homology modulo 2.
(b) Compute its homology modulo 3.
(c) Compute its homology modulo p, where p is a prime number.

2.42 Let $\mathcal{C} = \{C_k, \partial_k\}$ be a chain complex and let $\mathcal{D} = \{D_k, \partial_k'\}$ be a chain subcomplex. Consider a new complex whose chain groups are quotient groups C_k/D_k and whose boundary maps

$$\bar{\partial}_k : C_k/D_k \to C_{k+1}/D_{k+1}$$

are given by

$$[c + D_k] \mapsto [\partial_k c + D_{k-1}].$$

Prove that $\bar{\partial}_k$ is well defined and that it is a boundary map.
This new complex is called the *relative chain complex*. The *relative n-cycles* are $Z_k(\mathcal{C}, \mathcal{D}) := \ker \bar{\partial}_k$. The *relative n-boundaries* are $B_k(\mathcal{C}, \mathcal{D}) := \operatorname{im} \bar{\partial}_{k+1}$. The *relative homology groups* are

$$H_k(\mathcal{C}, \mathcal{D}) := Z_k(\mathcal{C}, \mathcal{D})/B_k(\mathcal{C}, \mathcal{D}).$$

(This topic is discussed in greater detail in Chapter 9.)

2.7 Reduced Homology

In the proofs of Theorem 2.59 and Theorem 2.78 we use a specific group homomorphism to deal with the fact that the 0th homology group is isomorphic to $\mathbf{Z}$. In mathematics, seeing a particular trick being employed to overcome a technicality in different contexts suggests the possibility of a general procedure to take care of the problem. In Theorem 2.78 we have to consider three cases. Essentially this is caused by the fact that acyclicity means the homology is $\mathbf{Z}$ in the dimension zero and zero in the higher dimensions. We would need only one case if acyclicity meant the homology being zero in every dimension. However, we know that the homology in dimension zero counts the number of connected components, that is, it must be nonzero. We can, therefore, ask the following question: Is there a different homology theory such that in the previous two examples we would have trivial 0th level homology?

2.7 Reduced Homology

Hopefully, this question does not seem too strange. We have spent most of Chapter 1 motivating the homology theory that we are using and as we do so we have to make choices of how to define our algebraic structures. From a purely algebraic point of view, given $\mathcal{K}(X)$, all we need in order to define homology groups is a chain complex $\{C_k(X), \partial_k\}_{k \in \mathbf{Z}}$. This means that if we change our chain complex, then we will have a new homology theory. The trick we employ in proving Theorem 2.59 and Theorem 2.78 involves the group homomorphism $\epsilon : C_0(X) \to \mathbf{Z}$ defined by sending each elementary cubical chain to 1. Furthermore, we show in each case that $\epsilon \circ \partial_1 = 0$, which means that

$$\operatorname{im} \partial_1 \subset \ker \epsilon.$$

With this in mind, we introduce the following definition.

Definition 2.93 Let X be a cubical set. The *augmented cubical chain complex* of X is given by $\{\tilde{C}_k(X), \tilde{\partial}_k\}_{k \in \mathbf{Z}}$, where

$$\tilde{C}_k(X) = \begin{cases} \mathbf{Z} & \text{if } k = -1, \\ C_k(X) & \text{otherwise}, \end{cases}$$

and

$$\tilde{\partial}_k := \begin{cases} \epsilon & \text{if } k = 0, \\ \partial_k & \text{otherwise}. \end{cases}$$

It is left as an exercise to show that the augmented cubical chain complex is an abstract chain complex in the sense of Definition 2.86. The added chain group $C_{-1} = \mathbf{Z}$ in the dimension -1 seems not to carry any geometric information. Nevertheless, it may be interpreted as follows. Since vertices have no faces, we have defined $C_{-1}(X)$ to be 0. But we may also adopt a convention that the empty set $\emptyset$ is the face of any vertex. Hence we define $\tilde{C}_{-1}$ to be the free abelian group generated by the singleton $\{\widehat{\emptyset}\}$, which is isomorphic to $\mathbf{Z}$ with $\{\widehat{\emptyset}\}$ corresponding to 1. The boundary of any dual vertex $\widehat{P}$ is $\tilde{\partial}_0(\widehat{P}) = \{\widehat{\emptyset}\}$, which precisely matches with the definition of ϵ.

Augmenting the chain complex $\mathcal{C}(X)$ by the group C_{-1} leads to a reduction of the 0-dimensional homology group. Indeed, all 0-chains are 0-cycles of $C_0(X)$, while the 0-cycles of $\tilde{C}_0(X)$ are only those in $\ker \epsilon$. This observation motivates the following terminology.

Definition 2.94 The homology groups $H_k(\tilde{\mathcal{C}}(X))$ are the *reduced homology groups* of X and are denoted by

$$\tilde{H}_k(X).$$

A chain $z \in C_k(X)$ that is a cycle in $\tilde{Z}_k$ is a *reduced cycle* in $\mathcal{C}(X)$. The homology class of a reduced cycle z with respect to the reduced homology is denoted by $[z]_\sim$.

The following theorem indicates the relationship between the two homology groups we now have at our disposal.

Theorem 2.95 *Let X be a cubical set. $\tilde{H}_0(X)$ is a free abelian group and*

$$H_k(X) \cong \begin{cases} \tilde{H}_0(X) \oplus \mathbf{Z} & \text{for } k = 0, \\ \tilde{H}_k(X) & \text{otherwise.} \end{cases}$$

Furthermore, if $\{P_i \mid i = 0, \ldots, n\}$ is a collection of vertices in X consisting of one vertex from each connected component of X, then

$$\{[P_i - P_0]_\sim \in \tilde{H}_0(X) \mid i = 1, \ldots, n\} \tag{2.15}$$

forms a basis for $\tilde{H}_0(X)$.

Proof. Notice that since we change the boundary operator only on the zero level, $\tilde{H}_k(X) = H_k(X)$ for $k \geq 1$. Therefore, it is enough to prove (2.15). We begin with showing that $\{[P_i - P_0]_\sim \mid i = 1, 2, \ldots, n\}$ is linearly independent. Assume

$$\sum_{i=1}^n \alpha_i [\widehat{P_i} - \widehat{P_0}]_\sim = 0.$$

Then there exists a chain $c \in \tilde{C}_1(X) = C_1(X)$ such that

$$\sum_{i=1}^n \alpha_i (\widehat{P_i} - \widehat{P_0}) = \partial c,$$

which can be rewritten as

$$\sum_{i=1}^n \alpha_i \widehat{P_i} - \left(\sum_{i=1}^n \alpha_i\right) \widehat{P_0} = \partial c.$$

Taking regular homology classes, we obtain

$$\sum_{i=1}^n \alpha_i [\widehat{P_i}] - \left(\sum_{i=1}^n \alpha_i\right) [\widehat{P_0}] = 0.$$

But by Theorem 2.59 the homology classes $[\widehat{P_i}]$, for $i = 0, 1, \ldots, n$, constitute the basis of $H_0(X)$. Therefore, $\alpha_i = 0$ for $i = 1, 2, \ldots, n$.

It remains to show that $[P_i - P_0]_\sim$ for $i = 1, 2, \ldots, n$ generates $\tilde{H}_0(X)$. Let c be a reduced cycle in dimension 0. Then, in particular, $c \in C_0(X)$ and by Theorem 2.59 there exists

$$c' = \sum_{i=0}^n \alpha_i \widehat{P_i}$$

such that $[c] = [c'] \in H_0(X)$. Therefore, there exists $b \in C_1(X)$ such that $c = c' + \partial_1 b$.

Since c is a reduced cycle, $\epsilon(c) = 0$. On the other hand,

$$\begin{aligned} \epsilon(c) &= \epsilon(c' + \partial_1 b) \\ &= \epsilon(c') + \epsilon(\partial_1 b) \\ &= \epsilon(\sum_{i=0}^{n} \alpha_i \widehat{P_i}) \\ &= \sum_{i=0}^{n} \alpha_i. \end{aligned}$$

Therefore, $\sum_{i=0}^{n} \alpha_i = 0$, which shows that c' is a reduced cycle, too. Then $0 = -\sum_{i=0}^{n} \alpha_i \widehat{P_0}$ and we can write

$$\begin{aligned} c' &= \sum_{i=0}^{n} \alpha_i \widehat{P_i} - \sum_{i=0}^{n} \alpha_i \widehat{P_0} \\ &= \sum_{i=0}^{n} \alpha_i (\widehat{P_i} - \widehat{P_0}). \end{aligned}$$

So, using the fact that $c - c'$ is a boundary, on the level of reduced homology we obtain

$$[c]_\sim = [c']_\sim = \sum_{i=0}^{n} \alpha_i [\widehat{P_i} - \widehat{P_0}]_\sim. \qquad \square$$

This theorem allows us to give an alternative characterization of acyclic spaces.

Corollary 2.96 *Let X be a nonempty cubical set. Then X is acyclic if and only if*

$$\tilde{H}_*(X) = 0.$$

Exercise _____

2.43 Verify that the augmented cubical chain complex $\tilde{\mathcal{C}}(X)$ is indeed an abstract chain complex in the sense of Definition 2.86.

2.8 Bibliographical Remarks

There are many approaches to homology and the best-known ones, which can be found in many classical textbooks, are *simplicial homology theory*, *singular homology theory*, and *cellular homology theory*. We refer the reader in particular to J. R. Munkres [69] and W. S. Massey [51]. A brief overview of the simplicial homology is provided in Chapter 11.

The notion of cubical homology has previously appeared in somewhat different contexts than the one chosen in this book. In [36] and [51] the notion of the *cubical singular complex* is introduced. It extends to a more general class of spaces than cubical sets, but it is far from the combinatorial spirit of this book because *singular cubes* are equivalence classes of continuous functions from the standard cube $[0,1]^d$ to a given topological space. In a series of papers by J. Blass and W. Holsztyński [8], a cellular cubical complex is introduced. It is, again, more general than the cubical complexes presented here, but it is more abstract and developed for different purposes. Perhaps the closest approach to our geometric cubical complexes is found in the paper by R. Ehrenborg and G. Hetyei [23].

The notion of elementary collapse studied in Section 2.4 comes from the *simple homotopy theory*. For example, see the book by M. M. Cohen [12].

3
Computing Homology Groups

In light of the discussion of the previous chapter, given a cubical set X we know that its homology groups $H_*(X)$ are well defined. We have also computed $H_*(X)$ for some simple examples and discussed the method of elementary collapse, which can be used in special cases to compute these groups. In this chapter we want to go further and argue that the homology groups of any cubical set are computable. In fact, we will derive Algorithm 3.78, which, given a cubical set X, takes as input the list of all elements of $\mathcal{K}_{\max}(X)$ and as output presents the associated homology groups $H_*(X)$. While the input is well defined, how to present the output may not be so clear at this moment. Obviously, we need to obtain a set of abelian groups. However, for this to be of use we need to know that we can present these groups in a finite and recognizable form. In particular, if X and Y are two cubical sets, it is desirable that our algorithm outputs the groups $H_*(X)$ and $H_*(Y)$ in such a way that it is evident whether or not they are isomorphic. With this in mind, by the end of this chapter we will prove the following result.

Corollary 3.1 *Any finitely generated abelian group G is isomorphic to a group of the form:*
$$\mathbf{Z}^r \oplus \mathbf{Z}_{b_1} \oplus \mathbf{Z}_{b_2} \oplus \cdots \oplus \mathbf{Z}_{b_k}, \tag{3.1}$$
where r is a nonnegative integer, $\mathbf{Z}_b$ denotes the group of integers modulo b, $b_i > 1$ provided $k > 0$, and b_i divides b_{i+1} for $i \in \{1, 2, \ldots, k-1\}$ provided $k > 1$. The numbers r and $b_1, b_2, \ldots, b_k$ are uniquely determined by G.

Clearly, $\mathbf{Z}^r$ is a free abelian group while $\mathbf{Z}_{b_i}$ are finite cyclic groups. However, what is important from the point of view of classification is that up to isomorphism every finitely generated abelian group is exactly determined by a unique finite set of numbers $\{r, b_1, b_2, \ldots, b_k\}$. The number r is the rank of the free subgroup $\mathbf{Z}^r$ and is called the *Betti number* of G, and the numbers $b_1, b_2, \ldots, b_k$ are called the *torsion coefficients* of G.

Algorithm 3.78 produces generators of the subgroups $\mathbf{Z}^r$ and $\mathbf{Z}_{b_i}$. In particular, the output determines the Betti numbers and the torsion coefficients.

Obviously, the generators are not needed for the purpose of classifying groups. However, we make use of this additional information in Chapter 7 when we discuss maps between homology groups.

Corollary 3.1 follows from *fundamental decomposition theorem* of abelian groups (Theorem 3.61). We shall prove this theorem by constructing Algorithm 3.58, which, in fact, produces generators of the subgroups $\mathbf{Z}^r$ and $\mathbf{Z}_{b_i}$ for a quotient group of two free subgroups of $\mathbf{Z}^p$.

As just mentioned, in this chapter we develop Algorithm 3.78 that takes $\mathcal{K}_{\max}(X)$ of a cubical set X as input and outputs generators of homology groups and the integers $\{r, b_1, b_2, \ldots, b_k\}$. This might surprise the reader who has carefully studied the table of contents, since much of the next chapter is also devoted to developing algorithms for computing homology. The justification of this is simple. There are two sides to computability—finding an algorithm and finding an efficient algorithm. In this chapter we produce an algorithm that computes homology groups. However, from the point of view of computations it is not practical, because it is not efficient. The algorithms in Chapter 4 are used to reduce the size of the chain complex so that the results presented here can be effectively applied.

3.1 Matrix Algebra over Z

Before we begin to develop the promised algorithms, we need a careful discussion of matrix algebra over the integers. As we already mentioned, in the case of arbitrary finitely generated free abelian groups there is no natural choice of a basis. To represent a homomorphism as a matrix we must always explicitly state which bases are chosen and, in fact, we will soon see the benefits of changing the bases. In particular, the canonical basis of the group of k-chains of a cubical set X consists of all elementary k-chains. However, in order to find a basis for the group of k-cycles in X, we have to choose another basis in $C_k(X)$. Therefore, if for no other reason than to keep track of the boundary operator, it is important to be able to tell how the matrix of a homomorphism changes when we change bases. To do so, assume G is a free, finitely generated, abelian group and $V = \{v_1, v_2, \ldots, v_n\}$ is a basis in G. Consider the homomorphism $\xi_V : G \to \mathbf{Z}^n$, defined on the basis V by

$$\xi_V(v_j) = \mathbf{e}_j. \tag{3.2}$$

One can easily check that ξ_V is the inverse of the isomorphism considered in Example 13.66. It is called the *coordinate isomorphism associated with the basis* V. The name is justified by the fact that if $g = \sum_{i=1}^{n} x_i v_i$ is an element of G expressed as the linear combination of the elements of the basis, then

$$\xi_V(g) = \mathbf{x} := (x_1, x_2, \ldots, x_n) \in \mathbf{Z}^n \,,$$

that is, the isomorphism ξ_V provides the coordinates of g in the basis V.

Example 3.2 Let $G := C_0(\{0,1\})$ be the group of 0-chains of the cubical two-point set $X := \{0,1\}$. The canonical basis of G consists of the elementary dual vertices
$$v_1 := \widehat{[0]} \text{ and } v_2 := \widehat{[1]}.$$
Any $g \in G$ can be expressed as $g = x_1 v_1 + x_2 v_2$ with unique $x_1, x_2 \in \mathbf{Z}$. Thus the coordinate isomorphism associated with this basis is given by
$$\xi_V(g) := (x_1, x_2) \in \mathbf{Z}^2.$$

Let $V' := \{v'_1, v'_2, \ldots, v'_n\}$ be another basis in G and let the coordinates of g in this basis be given by $\mathbf{x}' := \xi_{V'}(g)$. Then
$$\mathbf{x}' = \xi_{V'}(g) = \xi_{V'}(\xi_V^{-1}(\mathbf{x})),$$
which means that the coordinates in the new basis may be computed from the coordinates in the old basis by multiplying the coordinates $\mathbf{x}$ of g in the old basis by the matrix $A_{\xi_{V'}\xi_V^{-1}}$ of the isomorphism $\xi_{V'}\xi_V^{-1} : \mathbf{Z}^n \to \mathbf{Z}^n$ (compare with Section 13.3.3). This matrix is called the *change-of-coordinates matrix from V to V'*. Obviously, the matrix $A_{\xi_V \xi_{V'}^{-1}}$, the change-of-coordinates matrix from V' to V, is the inverse of the change-of-coordinates matrix $A_{\xi_{V'}\xi_V^{-1}}$ from V to V'. When the old basis and the new basis are clear from the context, we will just speak about the change-of-coordinates matrix and the inverse change-of-coordinates matrix.

The next example explains how to obtain these matrices.

Example 3.3 Let G and $g \in G$ be as in the previous example, and consider the new basis $V' = \{v'_1, v'_2\}$ in G given by
$$v'_1 := v_1, \qquad v'_2 := v_2 - v_1.$$
The inverse formulas are
$$v_1 = v'_1, \qquad v_2 = v'_1 + v'_2,$$
so $g = x_1 v_1 + x_2 v_2$ can be written in terms of the new basis as
$$g = x_1 v'_1 + x_2(v'_1 + v'_2) = (x_1 + x_2)v'_1 + x_2 v'_2.$$
Thus $\mathbf{x}' = (x'_1, x'_2) = \xi_{V'}(g) = (x_1 + x_2, x_2)$, so the change of coordinates is given by
$$A_{\xi_{V'}\xi_V^{-1}} = \begin{bmatrix} 1 & 1 \\ 0 & 1 \end{bmatrix}.$$

Note that the columns of this matrix represent coordinates of the old basis elements expressed in terms of the new basis. If we investigate its inverse change-of-coordinates matrix

$$A_{\xi_V \xi_{V'}^{-1}} = A_{\xi_{V'} \xi_V^{-1}}^{-1} = \begin{bmatrix} 1 & -1 \\ 0 & 1 \end{bmatrix},$$

we immediately notice that the columns are the coordinates of the new basis elements expressed in terms of the original basis. This observation will soon be generalized to any change of basis in $\mathbf{Z}^n$.

Assume now that H is another free, finitely generated, abelian group and $W = \{w_1, w_2, \ldots, w_m\}$ is a basis of H. Let $f : G \to H$ be a homomorphism of groups and let $h = f(g)$ for some fixed $g \in G$ and $h \in H$. If $\mathbf{x} := \xi_V(g)$ and $\mathbf{y} := \xi_W(h)$ are the coordinates of g and h, respectively, in bases V and W, then

$$\mathbf{y} = \xi_W(f(g)) = \xi_W f \xi_V^{-1}(\mathbf{x}).$$

Since $\xi_W f \xi_V^{-1} : \mathbf{Z}^n \to \mathbf{Z}^m$, the latter equation may be written in terms of the canonical matrix of $\xi_W f \xi_V^{-1}$ (compare Section 13.4.3) as

$$\mathbf{y} = A_{\xi_W f \xi_V^{-1}} \mathbf{x}.$$

Therefore, the matrix B of f in the bases V and W equals the canonical matrix of $\xi_W f \xi_V^{-1}$, namely

$$B = A_{\xi_W f \xi_V^{-1}}.$$

Assume that $V' := \{v_1', v_2', \ldots, v_n'\}$ is another basis in G and $W' := \{w_1', w_2', \ldots, w_n'\}$ is another basis in H.

Proposition 3.4 *Let $f : G \to H$ be a homomorphism and let B and B' denote the matrix of f, respectively, in the bases V, W and V', W'. Then*

$$B' = A_{\xi_{W'} \xi_W^{-1}} B A_{\xi_V \xi_{V'}^{-1}}.$$

Proof. We have [compare with (13.30)]

$$B' = A_{\xi_{W'} f \xi_{V'}^{-1}} = A_{\xi_{W'} \xi_W^{-1} \xi_W f \xi_V^{-1} \xi_V \xi_{V'}^{-1}}$$
$$= A_{\xi_{W'} \xi_W^{-1}} A_{\xi_W f \xi_V^{-1}} A_{\xi_V \xi_{V'}^{-1}} = A_{\xi_{W'} \xi_W^{-1}} B A_{\xi_V \xi_{V'}^{-1}}. \quad \square$$

The above proposition shows that in order to obtain the matrix of a homomorphism in a new basis, the matrix of this homomorphism in the old basis is multiplied on the right by the change-of-coordinates matrix from V' to V and on the left by the change-of-coordinates matrix from W to W'. An important special case is that of subgroups of $\mathbf{Z}^n$. First observe that if $V := \{\mathbf{v}_1, \mathbf{v}_2, \ldots, \mathbf{v}_n\} \subset \mathbf{Z}^m$ is a sequence of elements of $\mathbf{Z}^m$ and the elements of V are taken to be column vectors, then V may be identified with the matrix

$$[\mathbf{v}_1 \ \mathbf{v}_2 \ \cdots \ \mathbf{v}_n] \in M_{m,n}(\mathbf{Z}).$$

It is straightforward to verify that in this case

$$V\mathbf{e}_i = \mathbf{v}_i. \tag{3.3}$$

An integer matrix is called **Z**-*invertible* (compare Section 13.4.3) if it is invertible and its inverse is an integer matrix.

Proposition 3.5 *The columns* $\mathbf{v}_1, \mathbf{v}_2, \ldots, \mathbf{v}_n$ *of an* $n \times n$ *integer matrix* V *constitute a basis of* $\mathbf{Z}^n$ *if and only if* V, *is* **Z**-*invertible. In this case*

$$A_{\xi_V} = V^{-1}. \tag{3.4}$$

Proof. Assume that the columns of V constitute a basis in $\mathbf{Z}^n$. Since the coordinate isomorphism $\xi_V : \mathbf{Z}^n \to \mathbf{Z}^n$ satisfies $\xi_V(\mathbf{v}_i) = \mathbf{e}_i$, we see from (3.3) that A_{ξ_V} is the inverse of V. Therefore, V is **Z**-invertible and property (3.4) is satisfied.

It remains to be proved that if V is **Z**-invertible, then the columns of V constitute a basis of $\mathbf{Z}^n$. Let W denote the inverse of V. To show that the columns are linearly independent, assume that

$$\sum_{i=1}^{n} \alpha_i \mathbf{v}_i = 0.$$

Then

$$0 = \sum_{i=1}^{n} \alpha_i \mathbf{v}_i = \sum_{i=1}^{n} \alpha_i V \mathbf{e}_i = V \left(\sum_{i=1}^{n} \alpha_i \mathbf{e}_i \right)$$

and

$$0 = W0 = WV \left(\sum_{i=1}^{n} \alpha_i \mathbf{e}_i \right) = \sum_{i=1}^{n} \alpha_i \mathbf{e}_i.$$

Therefore, all $\alpha_i = 0$, because the canonical vectors $\mathbf{e}_i$ are linearly independent.

To show that the columns generate $\mathbf{Z}^n$, take $\mathbf{y} \in \mathbf{Z}^n$. Since W is an integer matrix, $\mathbf{x} := (x_1, x_2, \ldots, x_n) := W\mathbf{y} \in \mathbf{Z}^n$. Then

$$\mathbf{y} = V\mathbf{x} = \sum_{i=1}^{n} x_i \mathbf{v}_i. \qquad \square$$

The relationship between a change of coordinates and a change of basis in $\mathbf{Z}^n$ can also be explained by turning the reasoning around: Given a change-of-coordinates in $\mathbf{Z}^n$, what is the change of basis associated to it? It is often more convenient to express old coordinates as a function of the new ones rather than the new as a function of the old. So suppose

$$\mathbf{x} = Q\mathbf{u} \tag{3.5}$$

is a change-of-coordinates formula where $\mathbf{x}$ is the original coordinate vector with respect to the canonical basis in $\mathbf{Z}^n$, $\mathbf{u}$ is the new coordinate vector, and Q is a **Z**-invertible matrix. Equation (3.5) can be written as

$$\mathbf{x} = u_1\mathbf{q}_1 + u_2\mathbf{q}_2 + \cdots + u_n\mathbf{q}_n,$$

where $\mathbf{q}_1, \mathbf{q}_2, \ldots, \mathbf{q}_n$ are the columns of Q. However, this means that the new coordinates are taken with respect to the basis consisting of the columns of Q. Since the change-of-coordinates matrix expressing the new coordinates in the terms of the old is the inverse of Q, we arrive again at Eq. (3.4).

From Propositions 3.4 and 3.5, we get the following important corollary.

Corollary 3.6 *Let $f : \mathbf{Z}^n \to \mathbf{Z}^m$ be a homomorphism of groups. Let the columns of $V := [\mathbf{v}_1 \mathbf{v}_2 \ldots \mathbf{v}_n]$ and $W := [\mathbf{w}_1 \mathbf{w}_2 \ldots \mathbf{w}_m]$ be bases in $\mathbf{Z}^n$ and $\mathbf{Z}^m$, respectively. Then the matrix B of f in these bases may be obtained from the matrix of f in the canonical bases by the formula*

$$B = W^{-1} A_f V.$$

Example 3.7 Consider a homomorphism $f : \mathbf{Z}^3 \to \mathbf{Z}^3$ given in the canonical bases by the matrix

$$A = \begin{bmatrix} 3 & 2 & 3 \\ 0 & 2 & 0 \\ 2 & 2 & 2 \end{bmatrix}.$$

It can be verified that the following matrices

$$R = \begin{bmatrix} 1 & -2 & -1 \\ -1 & 3 & 0 \\ 0 & 0 & 1 \end{bmatrix} \quad \text{and} \quad Q = \begin{bmatrix} 1 & 0 & 0 \\ -2 & 3 & 1 \\ 0 & 1 & 0 \end{bmatrix}$$

are $\mathbf{Z}$-invertible; hence the columns of each form a basis for $\mathbf{Z}^3$ (see Exercise 3.3). Consider a new basis for the group $\mathbf{Z}^3$ viewed as the domain of f spanned by the columns of R and a new basis for the group $\mathbf{Z}^3$ viewed as the target space of f spanned by the columns of Q.

It is easy to verify that

$$B := Q^{-1} AR = \begin{bmatrix} 1 & 0 & 0 \\ 0 & 2 & 0 \\ 0 & 0 & 0 \end{bmatrix}.$$

By Corollary 3.6, B is the matrix of f with respect to the new bases. Note that we obtained a diagonal matrix, which is much simpler to study than the original matrix A. At present, the matrixes R and Q seem to be taken out of the blue, but we will come back to this topic and show how to obtain them.

There is another way of viewing Corollary 3.6 in the light of the previous discussion on the relation between the change-of-coordinate formula (3.5) and the change of basis. Suppose we have a change of coordinates in $\mathbf{Z}^n$ given by the formula

$$\mathbf{x} = R\mathbf{u}$$

and another one in $\mathbf{Z}^m$ given by

$$\mathbf{y} = Q\mathbf{v}.$$

If we substitute the old coordinates by the new ones in the equation

$$\mathbf{y} = A_f \mathbf{x},$$

we get

$$Q\mathbf{v} = A_f R\mathbf{u},$$

which is equivalent to

$$\mathbf{v} = Q^{-1} A_f R\mathbf{u}.$$

If we substitute $B := Q^{-1} A_f R$, we get the equation

$$\mathbf{v} = B\mathbf{u}.$$

Thus, in another way, we come to the same conclusion that B is the matrix of f with respect to the new coordinates. As we have already noticed, the columns of R and Q are the coordinates of the new basis vectors in $\mathbf{Z}^n$ and, respectively, $\mathbf{Z}^m$.

There are three elementary ways to obtain a new basis from an old basis. If $V := \{v_1, v_2, \ldots, v_n\}$ is a basis in G, then the new basis may be obtained by

(b1) exchanging elements v_i and v_j,
(b2) multiplying v_i by -1,[1]
(b3) replacing v_j by $v_j + qv_i$, where $q \in \mathbf{Z}$.

The fact that operations (b1) and (b2) transform a basis into a basis is straightforward. The same fact concerning (b3) is the contents of Exercise 3.1. Performing these operations on the canonical basis in $\mathbf{Z}^n$, one obtains the bases, which are the columns of the matrices

$$E_{i,j} := \begin{bmatrix} 1 & & & & & & \\ & \cdot & & & & & \\ & & 1 & & & & \\ & & & 0 \,..\, 0 \; 1 & & & (i) \\ & & & \,.\, 1 \quad\; 0 & & & \\ & & & \cdot \quad\;\; \cdot \;\; \cdot & & & \\ & & & 0 \quad\;\; 1 \,. & & & \\ & & & 1\, 0 \,..\, 0 & & & (j) \\ & & & \quad\quad\;\; 1 & & & \\ & & & & & \cdot & \\ & & & & & & 1 \end{bmatrix},$$

---

[1] In a linear algebra course the analogy of (b2) is multiplication by any nonzero real number. This is because every nonzero real number is invertible (i.e., $a\frac{1}{a} = 1$), but in the integers only ± 1 have multiplicative inverses.

$$E_i := \begin{bmatrix} 1 & & & & & \\ & \ddots & & & & \\ & & 1 & & & \\ & & & -1 & & \\ & & & & 1 & \\ & & & & & \ddots \\ & & & & & & 1 \end{bmatrix} \quad (i) \; ,$$

and

$$E_{i,j,q} := \begin{bmatrix} 1 & & & & & & \\ & \ddots & & & & & \\ & & 1 & 0 & \cdots & 0 & q & & (i)\\ & & & 1 & & 0 & & \\ & & & & \ddots & \vdots & & \\ & & & & & 1 & 0 & \\ & & & & & & 1 & (j)\\ & & & & & & & \ddots \\ & & & & & & & & 1 \end{bmatrix},$$

where all empty spaces denote zero. As we will see in the sequel, these matrices are very important. They are called *elementary matrices*.

Notice that all three elementary matrices are **Z**-invertible. Indeed,

$$E_{i,j}^{-1} = E_{i,j}, \quad E_i^{-1} = E_i, \quad \text{and} \quad E_{i,j,q}^{-1} = E_{i,j,-q}. \tag{3.6}$$

In particular, finding the inverse of an elementary matrix is extremely easy, unlike the case of a general integer matrix.

Let A be a fixed matrix. Denote by $\mathbf{r}_1, \mathbf{r}_2, \ldots, \mathbf{r}_m$ its rows and by $\mathbf{c}_1, \mathbf{c}_2, \ldots, \mathbf{c}_n$ its columns. One easily verifies that the multiplication of A from the left by one of the matrices $E_{i,j}, E_i, E_{i,j,q}$ is equivalent to performing, respectively, one of the following elementary operations on rows of A.

(r1) Exchange rows $\mathbf{r}_i$ and $\mathbf{r}_j$.
(r2) Multiply $\mathbf{r}_i$ by -1.
(r3) Replace $\mathbf{r}_i$ by $\mathbf{r}_i + q\mathbf{r}_j$, where $q \in \mathbf{Z}$.

Similarly, the multiplication of A from the right by one of the matrices $E_{i,j}, E_i, E_{i,j,q}$ is equivalent to performing, respectively, one of the following elementary operations on columns of A.

(c1) Exchange columns $\mathbf{c}_i$ and $\mathbf{c}_j$.
(c2) Multiply $\mathbf{c}_j$ by -1.
(c3) Replace $\mathbf{c}_j$ by $\mathbf{c}_j + q\mathbf{c}_i$, where $q \in \mathbf{Z}$.

These are, in fact, row operations on the transposed matrix A^T.

Example 3.8 Let A be a 5×3 matrix. If we wish to exchange the second and third columns, this can be done by the elementary matrix

$$E_{2,3} = \begin{bmatrix} 1 & 0 & 0 & 0 & 0 \\ 0 & 0 & 1 & 0 & 0 \\ 0 & 1 & 0 & 0 & 0 \\ 0 & 0 & 0 & 1 & 0 \\ 0 & 0 & 0 & 0 & 1 \end{bmatrix}$$

since

$$\begin{bmatrix} a_{11} & a_{12} & a_{13} & a_{14} & a_{15} \\ a_{21} & a_{22} & a_{23} & a_{24} & a_{25} \\ a_{31} & a_{32} & a_{33} & a_{34} & a_{35} \end{bmatrix} \begin{bmatrix} 1 & 0 & 0 & 0 & 0 \\ 0 & 0 & 1 & 0 & 0 \\ 0 & 1 & 0 & 0 & 0 \\ 0 & 0 & 0 & 1 & 0 \\ 0 & 0 & 0 & 0 & 1 \end{bmatrix} = \begin{bmatrix} a_{11} & a_{13} & a_{12} & a_{14} & a_{15} \\ a_{21} & a_{23} & a_{22} & a_{24} & a_{25} \\ a_{31} & a_{33} & a_{32} & a_{34} & a_{35} \end{bmatrix}.$$

Notice that the matrices $E_{i,j}$, E_i, $E_{i,j,q}$ themselves may be obtained by respectively performing the operations (r1), (r2), and (r3) or (c1), (c2), and (c3) on the identity $m \times m$ matrix $I_{m \times m}$.

Since elementary row operations on a matrix $A \in \mathrm{M}_{m,n}(\mathbf{Z})$ are equivalent to the multiplication of A on the left by elementary $m \times m$ matrices, they are related to changes of basis in $\mathbf{Z}^m$ viewed as the target space of the homomorphism from $\mathbf{Z}^n$ to $\mathbf{Z}^m$ defined by A. Similarly, elementary column operations are equivalent to the multiplication of A on the left by elementary $n \times n$ matrices, so they are related to changes of basis in $\mathbf{Z}^n$ viewed as the domain space of A. The relation between those matrices on the left or right of A and new bases of $\mathbf{Z}^m$ or, respectively, $\mathbf{Z}^n$ is exhibited in Corollary 3.6.

A natural question to ask is how elementary operations affect the image and kernel of A. We start from a proposition about the column operations because we need it in the sequel, and we leave the analogous statement for row operations as an exercise.

Proposition 3.9 *Let A be an $m \times n$ integer matrix and let B be obtained from A by elementary column operations. More precisely, let $B := AR$, where R is the product of elementary $n \times n$ matrices representing the column operations. Then*

$$\mathrm{im}\, A = \mathrm{im}\, B$$

and

$$\ker A = R(\ker B).$$

Proof. Since R is invertible, it is an epimorphism, so

$$\mathrm{im}\, B = \{B\mathbf{x} \mid \mathbf{x} \in \mathbf{Z}^m\} = \{AR\mathbf{x} \mid \mathbf{x} \in \mathbf{Z}^m\} = \{A\mathbf{y} \mid \mathbf{y} \in \mathbf{Z}^m\} = \mathrm{im}\, A.$$

Again, since R is invertible, for any $\mathbf{x} \in \mathbf{Z}^n$,

$$A\mathbf{x} = 0 \iff BR^{-1}\mathbf{x} = 0 \iff R^{-1}\mathbf{x} \in \ker B \iff \mathbf{x} \in R(\ker B). \qquad \square$$

Example 3.10 Consider the matrix

$$A = \begin{bmatrix} 3 & 2 & 0 \\ 2 & 0 & 3 \end{bmatrix}.$$

We want to perform column operations on A. It is customary to carry out this task using row operations on the transposed matrix

$$A^T = \begin{bmatrix} 3 & 2 \\ 2 & 0 \\ 0 & 3 \end{bmatrix}.$$

A practical way of keeping track of elementary operations is by performing elementary row operations on the augmented matrix:

$$[I_{3\times 3} | A^T] = \begin{bmatrix} 1 & 0 & 0 & 3 & 2 \\ 0 & 1 & 0 & 2 & 0 \\ 0 & 0 & 1 & 0 & 3 \end{bmatrix} \xmapsto{-r_2} \begin{bmatrix} 1 & -1 & 0 & 1 & 2 \\ 0 & 1 & 0 & 2 & 0 \\ 0 & 0 & 1 & 0 & 3 \end{bmatrix}$$

$$\xmapsto[-2r_1]{} \begin{bmatrix} 1 & -1 & 0 & 1 & 2 \\ -2 & 3 & 0 & 0 & -4 \\ 0 & 0 & 1 & 0 & 3 \end{bmatrix} \xmapsto{+r_3} \begin{bmatrix} 1 & -1 & 0 & 1 & 2 \\ -2 & 3 & 1 & 0 & -1 \\ 0 & 0 & 1 & 0 & 3 \end{bmatrix}$$

$$\xmapsto{+2r_2} \begin{bmatrix} -3 & 5 & 2 & 1 & 0 \\ -2 & 3 & 1 & 0 & -1 \\ 0 & 0 & 1 & 0 & 3 \end{bmatrix} \xmapsto{\times(-1)} \begin{bmatrix} -3 & 5 & 2 & 1 & 0 \\ 2 & -3 & -1 & 0 & 1 \\ 0 & 0 & 1 & 0 & 3 \end{bmatrix}$$

$$\xmapsto{-3r_2} \begin{bmatrix} -3 & 5 & 2 & 1 & 0 \\ 2 & -3 & -1 & 0 & 1 \\ -6 & 9 & 4 & 0 & 0 \end{bmatrix} =: [P|B].$$

Note that P is the product of the elementary matrices corresponding to row operations, thus $B = PA^T$. Furthermore, B is a very simple matrix. In the next section we present an algorithm performing such a simplification.

In order to see the result of column operations, consider

$$B^T = \begin{bmatrix} 1 & 0 & 0 \\ 0 & 1 & 0 \end{bmatrix}.$$

Then

$$B^T = (PA^T)^T = AP^T = AR,$$

where

$$R = P^T = \begin{bmatrix} -3 & 2 & -6 \\ 5 & -3 & 9 \\ 2 & -1 & 4 \end{bmatrix}$$

is the product of elementary matrices corresponding to column operations.
What information about A can we extract from the above calculations?

3.1 Matrix Algebra over $\mathbf{Z}$

First, by Corollary 3.6, the columns of R form the new basis for the domain $\mathbf{Z}^3$ with respect to which the corresponding homomorphism f_A is represented by the new matrix B^T.

Next, by Proposition 3.9, $\operatorname{im} A$ is equal to $\operatorname{im} B^T = \mathbf{Z}^2$; thus A is an epimorphism. This leads to an interesting observation. Since $\operatorname{im} A = \mathbf{Z}^2$ is generated by the columns $A\mathbf{e}_1$, $A\mathbf{e}_2$, $A\mathbf{e}_3$ of A, the elements $(3,2), (2,0)$, and $(0,3)$ of $\mathbf{Z}^2$ generate the whole group $\mathbf{Z}^2$. But the reader may verify that no two of them do. Therefore, a result known from linear algebra stating that from any set of generators of a vector space one can extract a basis does not generalize to free abelian groups.

Finally, we identify $\ker A$. It is easy to verify that $\ker B^T$ is generated by

$$\mathbf{e}_3 = \begin{bmatrix} 0 \\ 0 \\ 1 \end{bmatrix}.$$

So, by Proposition 3.9, $\ker A$ is generated by the third column of R denoted by $\mathbf{c}_3$:

$$\ker A = \langle \mathbf{c}_3 \rangle = \mathbf{Z} \begin{bmatrix} -6 \\ 9 \\ 4 \end{bmatrix}.$$

The following straightforward proposition is needed in the sequel.

Proposition 3.11 *Elementary row operations leave zero columns intact. Similarly, elementary column operations leave zero rows intact.*

Another useful feature is as follows.

Proposition 3.12 *Assume A is an integer matrix and A' results from applying a sequence of elementary row and/or column operations to A. If $p \in \mathbf{Z}$ divides all entries of A, then it divides all entries of A'.*

Proof. In the case where A' results from A by applying only one elementary row or column operation, the conclusion follows immediately from the form of elementary row and column operations. The general case follows by an easy induction argument. □

We finish this section with some algorithms. First some comments on notation used in this chapter are in order. Let A be a variable of type **matrix** storing an $m \times n$ matrix A with entries a_{ij}, where $i \in \{1, \ldots, m\}$ and $j \in \{1, \ldots, n\}$. In what follows we will need to be able to refer to submatrices of A consisting of all entries $a_{i,j}$ such that $i \in \{k, \ldots, l\}$ and $j \in \{k', \ldots, l'\}$ for some $k \leq l$ and $k' \leq l'$. In the algorithms these submatrices will be extracted from the variable A by writing $A[k:l, k':l']$ (see Section 14.2.4). To be consistent, in the discussion of the algorithms we will denote this submatrix by $A[k:l, k':l']$. In keeping with this notation we will often write $A[i,j]$ to refer to the entry $a_{i,j}$. In order to avoid separate analysis of special cases it will be convenient to use the notation for submatrices also when $k > l$ or $k' > l'$. In this case we will treat $A[k:l, k':l']$ as a zero matrix.

When presenting the algorithms we will use $*$ to denote the multiplication of matrices as well as the multiplication of a matrix by a number (see Section 14.4). Also, whenever we want to simultaneously store a matrix A and its inverse, we will use $\bar{A}$ as the variable name for the inverse.

The algorithms presented in this chapter are based on the following six simple algorithms:

(ar1) rowExchange(**matrix** B, **int** i, j)
(ar2) rowMultiply(**matrix** B, **int** i)
(ar3) rowAdd(**matrix** B, **int** i, j, q)
(ac1) columnExchange(**matrix** B, **int** i, j)
(ac2) columnMultiply(**matrix** B, **int** j)
(ac3) columnAdd(**matrix** B, **int** i, j, q)

which return the matrix B to which, respectively, the (r1), (r2), (r3) elementary row operations or (c1), (c2), (c3) elementary column operations have been applied. For instance, the implementation of rowAdd may appear as follows

Algorithm 3.13 Add a multiple of a row
    function rowAdd(**matrix** B, **int** i, j, q)
    n := **numberOfColumns**(B);
    $B[i, 1:n] := B[i, 1:n] + q * B[j, 1:n]$;
    return B;

and the implementation of columnAdd may appear as follows.

Algorithm 3.14 Add a multiple of a column
    function columnAdd(**matrix** B, **int** i, j, q)
    m := **numberOfRows**(B);
    $B[1:m, j] := B[1:m, j] + q * B[1:m, i]$;
    return B;

We leave the implementation of the remaining algorithms as an exercise.

Let $A \in M_{m,n}(\mathbf{Z})$ be the matrix of a homomorphism $f : \mathbf{Z}^n \to \mathbf{Z}^m$ in the canonical bases. Let some other bases in $\mathbf{Z}^n$ and $\mathbf{Z}^m$ be given, respectively, by the columns of $R \in M_{n,n}(\mathbf{Z})$ and $Q \in M_{m,m}(\mathbf{Z})$. We know that the matrix of f in these bases is given by

$$B = Q^{-1}AR.$$

Let $E \in M_{m,m}(\mathbf{Z})$ be a matrix of an elementary row operation. Letting

$$Q' := QE^{-1}$$

and

$$\bar{Q}' := EQ^{-1}.$$

we obtain

$$Q'\bar{Q}' = QE^{-1}EQ^{-1} = I_{m \times m}$$

and

$$\bar{Q}'Q' = EQ^{-1}QE^{-1} = I_{m \times m}$$

which shows that Q' and $\bar{Q}'$ are mutually inverse **Z**-invertible matrices and

$$B' := EB = EQ^{-1}AR = \bar{Q}'AR = (Q')^{-1}AR.$$

Thus the matrix B' obtained from B by an elementary row operation is the matrix of f in the bases given by the columns of Q' and R. The new matrix Q' may be obtained from Q by a respective column operation on Q. Moreover, the inverse $(Q')^{-1}$ of Q' may be obtained from the inverse of Q by applying to Q^{-1} the same row operation as to B. This gives us a convenient method of keeping track of the matrices of f and the respective bases by simultaneously performing elementary operations on B and Q. In the case of the row exchange, the method takes the form of the following algorithm.

Algorithm 3.15 Row exchange operation keeping track of bases
    function rowExchangeOperation(**matrix** $B, Q, \bar{Q}$, **int** i, j)
  $B := $ rowExchange(B, i, j);
  $\bar{Q} := $ rowExchange$(\bar{Q}, i, j)$;
  $Q := $ columnExchange(Q, i, j);
  return $(B, Q, \bar{Q})$;

Similarly, when multiplying a row by -1, we obtain the following algorithm.

Algorithm 3.16 Multiply a row by -1 keeping track of bases
    function rowMultiplyOperation(**matrix** $B, Q, \bar{Q}$, **int** i)
  $B := $ rowMultiply(B, i);
  $\bar{Q} := $ rowMultiply$(\bar{Q}, i)$;
  $Q := $ columnMultiply(Q, i);
  return $(B, Q, \bar{Q})$;

Finally, when adding to a row a multiple of another row, we obtain the following algorithm.

Algorithm 3.17 Add a multiple of a row keeping track of bases
    function rowAddOperation(**matrix** $B, Q, \bar{Q}$, **int** i, j, q)
  $B := $ rowAdd(B, i, j, q);
  $\bar{Q} := $ rowAdd$(\bar{Q}, i, j, q)$;
  $Q := $ columnAdd$(Q, i, j, -q)$;
  return $(B, Q, \bar{Q})$;

We may summarize the above discussion in the following straightforward proposition.

Proposition 3.18 *Let $A, B \in \mathrm{M}_{m,n}(\mathbf{Z})$. Let $R \in \mathrm{M}_{n,n}(\mathbf{Z})$ and $Q \in \mathrm{M}_{m,m}(\mathbf{Z})$ be $\mathbf{Z}$-invertible matrices such that*

$$B = Q^{-1}AR.$$

If $i, j, q \in \mathbf{Z}$, then Algorithm 3.15 applied to (B, Q, Q^{-1}, i, j), Algorithm 3.16 applied to (B, Q, Q^{-1}, i), and Algorithm 3.17 applied to (B, Q, Q^{-1}, i, j, q) return a matrix $B' \in \mathrm{M}_{m,n}(\mathbf{Z})$ and mutually inverse $\mathbf{Z}$-invertible matrices $Q', \bar{Q}' \in \mathrm{M}_{m,m}(\mathbf{Z})$ such that

$$B' = (Q')^{-1}AR = \bar{Q}'AR.$$

Moreover, the matrix B' is obtained from B by applying a respective elementary row operation.

Similarly, we can construct analogous algorithms

```
columnExchangeOperation(matrix B, R, R̄, int i, j)
columnMultiplyOperation(matrix B, R, R̄, int j)
columnAddOperation(matrix B, R, R̄, int i, j, q)
```

performing column operations and prove the following proposition.

Proposition 3.19 *Let $A, B \in \mathrm{M}_{m,n}(\mathbf{Z})$. Let $R \in \mathrm{M}_{n,n}(\mathbf{Z})$ and $Q \in \mathrm{M}_{m,m}(\mathbf{Z})$ be $\mathbf{Z}$-invertible matrices such that*

$$B = Q^{-1}AR.$$

If $i, j, q \in \mathbf{Z}$, then algorithm `columnExchangeOperation` *applied to (B, R, R^{-1}, i, j), algorithm* `columnMultiplyOperation` *applied to (B, R, R^{-1}, j), and algorithm* `columnAddOperation` *applied to (B, R, R^{-1}, i, j, q) return a matrix $B' \in \mathrm{M}_{m,n}(\mathbf{Z})$ and mutually inverse $\mathbf{Z}$-invertible matrices $R', \bar{R}' \in \mathrm{M}_{n,n}(\mathbf{Z})$ such that*

$$B' = Q^{-1}AR'.$$

Moreover, the matrix B' is obtained from B by applying a respective elementary column operation.

Exercises

3.1 Let $V := \{v_1, v_2, \ldots, v_n\}$ be a basis of a free abelian group G. Let $V' := \{v'_1, v'_2, \ldots, v'_n\}$, where

$$v'_k = \begin{cases} v_k & \text{if } k \neq j, \\ v_j + qv_i & \text{otherwise,} \end{cases}$$

for distinct $i, j \in \{1, 2, \ldots, n\}$ and $q \in \mathbf{Z}$. Show that V' is also a basis in G.

3.2 Let $f : \mathbf{Z}^n \to \mathbf{Z}^n$ be an isomorphism. Prove that $A_{f^{-1}} = A_f^{-1}$.

3.3 Let A be a $\mathbf{Z}$-invertible $n \times n$ matrix. Show that the columns of A form a basis of $\mathbf{Z}^n$.

3.4 Let $V = \{v_1, v_2, \ldots, v_n\}$ and $V' = \{v'_1, v'_2, \ldots, v'_n\}$ be two bases of a free abelian group G. This implies that every $v'_i \in V'$ can be written as a linear combination
$$v'_i = \sum_{j=1}^n p_{ij} v_j.$$
Let $P := [p_{ij}]$. Show that the inverse change-of-coordinates matrix is related to P by the formula
$$A_{\xi_V \xi_{V'}^{-1}} = P^T.$$

3.5 Let A be an $m \times n$ integer matrix and let B be obtained from A by elementary row operations. More precisely, let $B := PA$, where P is the product of elementary $m \times m$ matrices representing the row operations. Show that
$$\operatorname{im} B = \operatorname{im} PA$$
and
$$\ker B = \ker A.$$

3.6 Using Algorithms 3.15 and 3.17 as examples, write the algorithms `columnExchangeOperation` and `columnAddOperation` performing the tasks described in Proposition 3.19.

3.2 Row Echelon Form

We will use the elementary row and column operations from the previous section to transform arbitrary integer matrices into simpler matrices. The steps we present should remind the reader of *Gaussian elimination* [79, Chapter 1], with the constraint that we must always use integer coefficients. Our eventual goal is to produce "diagonal" matrices, since, as in Example 3.10, in this case the image and kernel of the map are immediately evident. We put the word "diagonal" in quotation marks to emphasize that we are dealing with $m \times n$ matrices. As a first step toward diagonalization, we introduce the following notion, which generalizes the concept of an upper triangular matrix.

Definition 3.20 The *pivot position* of a nonzero vector is the position of the first nonzero element of this vector. A matrix A is in *row echelon form* if for any two consecutive rows $\mathbf{r}_i$ and $\mathbf{r}_{i+1}$, if $\mathbf{r}_{i+1} \neq 0$, then $\mathbf{r}_i \neq 0$ and the pivot position of $\mathbf{r}_{i+1}$ is greater than the pivot position of $\mathbf{r}_i$.

Notice that if a matrix is in row echelon form, then all its nonzero rows come first, followed by zero rows, if there are any.

Example 3.21 The following matrix is in row echelon form:

$$B = \begin{bmatrix} 0 & 2 & 0 & 7 & 3 \\ 0 & 0 & 1 & 0 & 11 \\ 0 & 0 & 0 & 0 & 7 \\ 0 & 0 & 0 & 0 & 0 \\ 0 & 0 & 0 & 0 & 0 \end{bmatrix}.$$

The first three rows are the nonzero rows. The pivot positions of these three rows are, respectively, 2, 3, and 5.

The following proposition is straightforward.

Proposition 3.22 *Assume $A = [a_{ij}]$ is an $m \times n$ matrix in row echelon form and the nonzero rows are $\mathbf{r}_1, \mathbf{r}_2, \ldots, \mathbf{r}_k$. If l_i is the pivot position of $\mathbf{r}_i$, then $\{l_i\}_{i=1}^{k}$ is a strictly increasing sequence and $a_{i'l_i} = 0$ for $i' > i$.*

A matrix A is in *column echelon form* if its transpose A^T is in row echelon form. For our purposes, column echelon form is more useful than row echelon form but, as is observed in Example 3.10, row operations and row echelon form are more customary than the column ones. Therefore, we will often turn problems concerning a matrix A around by considering its transpose A^T, bringing it to row echelon form, and transposing the resulting matrix back.

As the following proposition indicates, if a matrix is in column echelon form, then one has an immediate description of the image and the kernel of the matrix.

Proposition 3.23 *Suppose that $A = [a_{ij}] \in M_{m,n}(\mathbf{Z})$ is in column echelon form. If $\mathbf{v}_1, \mathbf{v}_2, \ldots, \mathbf{v}_n$ are the columns of A and $\mathbf{v}_k$ is the last nonzero column, then $\{\mathbf{v}_1, \mathbf{v}_2, \ldots, \mathbf{v}_k\}$ is a basis of $\mathrm{im}\, A$ and $\{\mathbf{e}_{k+1}, \mathbf{e}_{k+2}, \ldots, \mathbf{e}_n\}$ is a basis of $\ker A$.*

Proof. By definition, A^T is in row echelon form. The rows of A^T are $\mathbf{v}_1^T, \mathbf{v}_2^T, \ldots, \mathbf{v}_n^T$ and since A^T is in row echelon form, the nonzero rows of A^T are $\mathbf{v}_1^T, \mathbf{v}_2^T, \ldots \mathbf{v}_k^T$. Let l_j be the first nonzero element in row $\mathbf{v}_j^T$. Then, by Proposition 3.22, $\{l_j\}_{j=1}^{k}$ is a strictly increasing sequence. We will show that $\mathbf{v}_1, \mathbf{v}_2, \ldots, \mathbf{v}_k$ are linearly independent. Let $\sum_{j=1}^{k} \alpha_j \mathbf{v}_j = 0$. Let $\pi_i : \mathbf{Z}^m \to \mathbf{Z}$ denote the projection onto the ith coordinate. Then for $s = 1, 2, \ldots, k$,

$$0 = \pi_{l_s}\left(\sum_{j=1}^{k} \alpha_j \mathbf{v}_j\right) = \sum_{j=1}^{k} \alpha_j \pi_{l_s}(\mathbf{v}_j) = \sum_{j=1}^{k} \alpha_j a_{l_s j} = \sum_{j=1}^{s} \alpha_j a_{l_s j}, \qquad (3.7)$$

because, by Proposition 3.22, $a_{l_s j} = a_{jl_s}^T = 0$ for $j > s$. In particular, $0 = \alpha_1 a_{l_1 1}$ and since $a_{l_1 1} \neq 0$, we get $\alpha_1 = 0$. Assuming that $\alpha_1 = \alpha_2 = \ldots = \alpha_{s-1} = 0$ for some $s \leq k$, we get, from (3.7), that $0 = \alpha_s a_{l_s s}$ and, consequently, $\alpha_s = 0$. This proves that $\mathbf{v}_1, \mathbf{v}_2, \ldots, \mathbf{v}_k$ are linearly independent. Obviously, they generate $\mathrm{im}\, A$. Therefore, they constitute a basis of $\mathrm{im}\, A$.

To show that $\{\mathbf{e}_{k+1}, \mathbf{e}_{k+2}, \ldots, \mathbf{e}_n\}$ constitute a basis of $\ker A$, take

$$\mathbf{x} = (x_1, x_2, \ldots, x_n) = \sum_{j=1}^{n} x_j \mathbf{e}_j \in \ker A.$$

Then

$$0 = A\mathbf{x} = \sum_{j=1}^{n} x_j A\mathbf{e}_j = \sum_{j=1}^{n} x_j \mathbf{v}_j = \sum_{j=1}^{k} x_j \mathbf{v}_j.$$

Since $\mathbf{v}_1, \mathbf{v}_2, \ldots, \mathbf{v}_k$ are linearly independent, $x_1 = x_2 = \ldots = x_k = 0$. Therefore, $\mathbf{x} = \sum_{j=k+1}^{n} x_j \mathbf{e}_j$, which shows that $\{\mathbf{e}_{k+1}, \mathbf{e}_{k+2}, \ldots, \mathbf{e}_n\}$ generate $\ker A$. Since, obviously, they are linearly independent, they form a basis of $\ker A$. □

Proposition 3.23 provides justification for the traditional terminology of referring to the group $\text{im}\, A$ as the *column space of* A.

Our goal in this section is to develop an algorithm that brings an integer matrix $A \in \text{M}_{m,n}(\mathbf{Z})$ to row echelon form by means of elementary row operations. We will achieve this goal gradually by working on the matrix row by row. Therefore, we will need the following concept. We say that a matrix A satisfies the (k, l) *criterion of row echelon form* if the submatrix $A[1:m, 1:l]$ consisting of the first l columns of A is in row echelon form and the nonzero rows of this submatrix are exactly the first k rows.

Example 3.24 The following matrix

$$\begin{bmatrix} 2 & 0 & 1 & 0 & 1 & 0 \\ 0 & 0 & 1 & 4 & 0 & 1 \\ 0 & 0 & 0 & 0 & 0 & 1 \\ 0 & 0 & 0 & 0 & 3 & 0 \\ 0 & 0 & 0 & 0 & 0 & 2 \end{bmatrix}$$

is not in row echelon form, but it satisfies the $(2, 3)$ criterion of row echelon form. It also satisfies the $(2, 4)$ criterion of row echelon form, but it satisfies neither the $(2, 2)$ nor the $(2, 5)$ criterion of row echelon form.

Assume $A \in \text{M}_{m,n}(\mathbf{Z})$ and $k \in \{0, 1, 2, \ldots, m\}$, $l \in \{0, 1, 2, \ldots, n\}$. The following four propositions will turn out to be useful in induction arguments based on the (k, l) criterion of row echelon form.

Proposition 3.25 *Every matrix satisfies the* $(0, 0)$ *criterion of row echelon form. If A satisfies the (k, n) criterion of row echelon form, then A is in row echelon form and the nonzero rows of A are exactly the first k rows.* □

Proposition 3.26 *Assume A satisfies the $(k-1, l-1)$ criterion of row echelon form, $A[k, l] \neq 0$, and $A[k+1:m, l] = 0$. Then A satisfies the (k, l) criterion of row echelon form.*

Proof. The pivot position of the $(k-1)$st row of A is at most $l-1$ and the pivot position of the kth row is exactly l. The last $m-k$ rows of $A[1:m, 1:l]$ are zero. The conclusion follows. □

Proposition 3.27 *Assume A satisfies the (k,l) criterion of row echelon form. If $A[k+1:m, l+1:l'] = 0$ for some $l' \in \{l+1, l+2, \ldots, n\}$, then A satisfies the (k, l') criterion of row echelon form.*

Proof. The pivot positions of the first k rows of $A[1:k, 1:l']$ and $A[1:k, 1:l]$ are the same. The last $m-k$ rows of $A[1:k, 1:l']$ are zero. The conclusion follows. □

Proposition 3.28 *Assume A satisfies the (k,l) criterion of row echelon form and A' results from applying an elementary row operation to the matrix A, which does not change the first k' rows for some $k' \geq k$. Then the first k' rows and the first l columns of A and A' coincide.*

Proof. Obviously, the first k rows of A cannot change. The first l columns can change only in the entries of the submatrix $A[k+1:m, 1:l]$, but this part of A is zero by the assumption that A satisfies the (k,l) criterion of row echelon form. □

The operation fundamental to most of the algorithms presented in this chapter is that of row reduction. It consists of applying suitable elementary row operations in order to achieve the situation where all elements in a given column below a selected position are zero. In the case of matrices with real entries, this is a simple procedure involving division. Unfortunately, dividing one integer by another does not necessarily produce an integer—a problem we need to circumvent.

To begin, consider the case of an integer matrix $B \in \mathrm{M}_{m,n}(\mathbf{Z})$ satisfying the $(k-1, l-1)$ criterion of row echelon form for some $k \in \{1, 2, \ldots, m\}$, $l \in \{1, 2, \ldots, n\}$ and such that $B[k, l] \neq 0$. Our modest goal is to decrease the magnitudes of the entries in the subcolumn $B[k+1:m, l]$.

Algorithm 3.29 Partial row reduction
    function partRowReduce(**matrix** $B, Q, \bar{Q}$, **int** k, l)
        for i := k + 1 **to** numberOfRows(B) **do**
            q := **floor**(B[i, 1]/B[k, 1]);
            $(B, Q, \bar{Q})$:= rowAddOperation$(B, Q, \bar{Q}, i, k, -q)$;
        endfor;
        return $(B, Q, \bar{Q})$;

Example 3.30 Let
$$A = \begin{bmatrix} 2 & 3 & 1 & -1 \\ 3 & 2 & 1 & 4 \\ 4 & 4 & -2 & -2 \end{bmatrix}$$

and let $(B, Q, \bar{Q}) = \mathtt{partRowReduce}(A, I_{3\times 3}, I_{3\times 3}, 1, 1)$. Then

$$B = \begin{bmatrix} 2 & 3 & 1 & -1 \\ 1 & -1 & 0 & 5 \\ 0 & -2 & -4 & 0 \end{bmatrix}, \quad Q = \begin{bmatrix} 1 & 0 & 0 \\ 1 & 1 & 0 \\ 2 & 0 & 1 \end{bmatrix}, \quad \bar{Q} = \begin{bmatrix} 1 & 0 & 0 \\ -1 & 1 & 0 \\ -2 & 0 & 1 \end{bmatrix}.$$

Before stating the properties of Algorithm 3.29 we introduce the following notation concerning the subcolumn $B[k:m,l]$ of a matrix $B \in \mathrm{M}_{m,n}(\mathbf{Z})$:

$$\alpha_{kl}(B) := \begin{cases} \min\{|B[i,l]| \mid i \in [k,m], B[i,l] \neq 0\} & \text{if } B[k:m,l] \neq 0, \\ 0 & \text{otherwise.} \end{cases}$$

Throughout this chapter this quantity will be useful for measuring the progress in decreasing the magnitudes of the entries in the subcolumns of B.

Proposition 3.31 *Let $A, B \in \mathrm{M}_{m,n}(\mathbf{Z})$ be such that B satisfies the $(k-1, l-1)$ criterion of row echelon form for some $k \in \{1, 2, \ldots, m\}$, $l \in \{1, 2, \ldots, n\}$ and*

$$B = Q^{-1}AR$$

for some $\mathbf{Z}$-invertible matrices $Q \in \mathrm{M}_{m,m}(\mathbf{Z})$ and $R \in \mathrm{M}_{n,n}(\mathbf{Z})$. If $B[k,l] = \alpha_{kl}(B) \neq 0$, then Algorithm 3.29 applied to (B, Q, Q^{-1}, k, l) returns a matrix $B' \in \mathrm{M}_{m,n}(\mathbf{Z})$ and mutually inverse $\mathbf{Z}$-invertible matrices $Q', \bar{Q}' \in \mathrm{M}_{m,m}(\mathbf{Z})$ such that

1. *the first k rows and $l-1$ columns of B and B' coincide,*
2. $\alpha_{kl}(B') < \alpha_{kl}(B) \quad \text{or} \quad B'[k+1:m,l] = 0$,
3. $B' = (Q')^{-1}AR = \bar{Q}'AR$.

Proof. The first property follows from Proposition 3.28. The outcome of applying `rowAddOperation` for $i = k+1, k+2, \ldots, m$ is that $B'[i,l]$ is the remainder from dividing $B[i,l]$ by $B[k,l]$. If at least one of the remainders is nonzero, then $\alpha_{kl}(B')$ is the smallest nonzero remainder, which obviously is less than $B[k,l] = \alpha_{kl}(B)$. Otherwise all remainders are zero, that is, $B[k+1:m,l] = 0$. This proves the second property. The final property follows from Proposition 3.18 by a straightforward induction argument. □

Note that replacing the row operations by column operations we can construct

function `partColumnReduce`(**matrix** B, R, R̄, **int** k, l)

with analogous properties.

Of course, Algorithm 3.29 works as desired only when $B[k,l] = \alpha_{kl}(B) \neq 0$, which, obviously, is a very special case. Thus the next step is to develop an algorithm that brings an integer matrix into this special form. We begin by a very simple procedure that (assuming it exists) identifies the first entry with the smallest nonzero magnitude in a part of a vector.

112 3 Computing Homology Groups

Algorithm 3.32 Smallest nonzero entry
    function smallestNonzero(**vector** v, **int** k)
    alpha := min $\{\text{abs}(\text{v}[i]) \mid i \text{ in } [k : \text{length}(\text{v})] \text{ and } \text{v}[i] \neq 0\}$;
    i_0 := min $\{i \mid i \text{ in } [k : \text{length}(\text{v})] \text{ and } \text{abs}(\text{v}[i]) = \text{alpha}\}$;
    return (alpha, i_0);

We make use of Algorithm 3.32 as follows.

Algorithm 3.33 Row preparation
    function rowPrepare(**matrix** B, Q, $\bar{Q}$, **int** k, l)
    m := **numberOfRows**(B);
    (α, i) := smallestNonzero(B[1 : m, l], k);
    (B, Q, $\bar{Q}$) := rowExchangeOperation(B, Q, $\bar{Q}$, k, i);
    return (B, Q, $\bar{Q}$);

Example 3.34 Let
$$A = \begin{bmatrix} 3 & 2 & 1 & 4 \\ 2 & 3 & 1 & -1 \\ 4 & 4 & -2 & -2 \end{bmatrix}.$$

Let $(B, Q, \bar{Q}) = \text{rowPrepare}(A, I_{3\times 3}, I_{3\times 3}, 1, 1)$. Then

$$B = \begin{bmatrix} 2 & 3 & 1 & -1 \\ 3 & 2 & 1 & 4 \\ 4 & 4 & -2 & -2 \end{bmatrix}, \quad Q = \begin{bmatrix} 0 & 1 & 0 \\ 1 & 0 & 0 \\ 0 & 0 & 1 \end{bmatrix}, \quad \bar{Q} = Q^{-1} = \begin{bmatrix} 0 & 1 & 0 \\ 1 & 0 & 0 \\ 0 & 0 & 1 \end{bmatrix}.$$

Proposition 3.35 *Let $A, B \in \mathrm{M}_{m,n}(\mathbf{Z})$ be such that B satisfies the $(k-1, l-1)$ criterion of row echelon form for some $k \in \{1, 2, \ldots, m\}$, $l \in \{1, 2, \ldots, n\}$ and*
$$B = Q^{-1}AR$$
for some $\mathbf{Z}$-invertible matrices $Q \in \mathrm{M}_{m,m}(\mathbf{Z})$ and $R \in \mathrm{M}_{n,n}(\mathbf{Z})$. If $B[k : m, l] \neq 0$, then Algorithm 3.33 applied to (B, Q, Q^{-1}, k, l) returns a matrix $B' \in \mathrm{M}_{m,n}(\mathbf{Z})$ and mutually inverse $\mathbf{Z}$-invertible matrices $Q', \bar{Q}' \in \mathrm{M}_{m,m}(\mathbf{Z})$ such that

1. *the first $k-1$ rows and $l-1$ columns of B and B' coincide,*
2. *$B'[k, l] = \alpha_{kl}(B') = \alpha_{kl}(B) \neq 0$,*
3. *$B' = (Q')^{-1}AR = \bar{Q}'AR.$*

Proof. The first property follows from Proposition 3.28. The remaining properties are straightforward. □

We can combine these algorithms to row-reduce the first column of an integer matrix.

Algorithm 3.36 Row reduction
  function rowReduce(**matrix** $B, Q, \bar{Q}$, **int** k, l)
    m := **numberOfRows**(B);
    while $B[k+1:m,l] \neq 0$ **do**
      $(B, Q, \bar{Q})$:= rowPrepare$(B, Q, \bar{Q}, k, l)$;
      $(B, Q, \bar{Q})$:= partRowReduce$(B, Q, \bar{Q}, k, l)$;
    endwhile ;
    return $(B, Q, \bar{Q})$;

Proposition 3.37 *Let* $A, B \in M_{m,n}(\mathbf{Z})$ *be such that* B *satisfies the* $(k-1, l-1)$ *criterion of row echelon form for some* $k \in \{1, 2, \ldots, m\}$, $l \in \{1, 2, \ldots, n\}$ *and*
$$B = Q^{-1}AR$$
for some $\mathbf{Z}$*-invertible matrices* $Q \in M_{m,m}(\mathbf{Z})$ *and* $R \in M_{n,n}(\mathbf{Z})$. *If* $B[k : m, l] \neq 0$, *then Algorithm 3.36 applied to* (B, Q, Q^{-1}, k, l) *returns a matrix* $B' \in M_{m,n}(\mathbf{Z})$ *and mutually inverse* $\mathbf{Z}$*-invertible matrices* $Q', \bar{Q}' \in M_{m,m}(\mathbf{Z})$ *such that*

1. $B' = (Q')^{-1}AR = \bar{Q}'AR$,
2. B' *satisfies the* (k, l) *criterion of row echelon form*.

Proof. Let $B^{(i)}$ denote the value of variable B after completing the ith pass of the **while** loop. First observe that on every call of rowPrepare the assumptions of Proposition 3.35 are satisfied and by this proposition on every call of partRowReduce the assumptions of Proposition 3.31 hold. By Proposition 3.35 rowPrepare does not change the value of α_{kl} and by Proposition 3.31 partRowReduce decreases it unless $B^{(i)}[k+1:m,l] = 0$, which can happen only in the last pass through the **while** loop. Therefore, the sequence

$$\alpha_i := \alpha_{kl}(B^{(i)})$$

is a decreasing sequence of positive integers. Hence, this sequence must be finite and the **while** loop must be completed. It follows that the algorithm always halts and $B' = B^{(i_*)}$, where i_* denotes the number of iterations through the **while** loop. The only way to leave the **while** loop is when the test fails, which implies

$$B'[k+1:m,l] = 0. \tag{3.8}$$

Since by Propositions 3.31 and 3.35 the first $l-1$ columns of $B^{(i)}$ do not depend on i, B' satisfies the $(k-1, l-1)$ criterion of row echelon form. Proposition 3.26, Eq. (3.8), and Proposition 3.35.2 imply the second property.

The first property follows as an immediate consequence of the corresponding properties of the rowPrepare and partRowReduce algorithms. □

Example 3.38 Let
$$A = \begin{bmatrix} 3 & 2 & 1 & 4 \\ 2 & 3 & 1 & -1 \\ 4 & 4 & -2 & -2 \end{bmatrix}.$$

After the first iteration of the while statement in Algorithm 3.36 called with arguments $(A, I_{3\times 3}, I_{3\times 3}, 1, 1)$,

$$B = \begin{bmatrix} 2 & 3 & 1 & -1 \\ 1 & -1 & 0 & 5 \\ 0 & -2 & -4 & 0 \end{bmatrix}, \quad Q = \begin{bmatrix} 1 & 1 & 0 \\ 1 & 0 & 0 \\ 2 & 0 & 1 \end{bmatrix}, \quad \bar{Q} = Q^{-1} = \begin{bmatrix} 0 & 1 & 0 \\ 1 & -1 & 0 \\ 0 & -2 & 1 \end{bmatrix}.$$

Thus, after another application of rowPrepare,

$$B = \begin{bmatrix} 1 & -1 & 0 & 5 \\ 2 & 3 & 1 & -1 \\ 0 & -2 & -4 & 0 \end{bmatrix}, \quad Q = \begin{bmatrix} 1 & 1 & 0 \\ 0 & 1 & 0 \\ 0 & 2 & 1 \end{bmatrix}, \quad \bar{Q} = Q^{-1} = \begin{bmatrix} 1 & -1 & 0 \\ 0 & 1 & 0 \\ 0 & -2 & 1 \end{bmatrix}.$$

After the application of partRowReduce,

$$B = \begin{bmatrix} 1 & -1 & 0 & 5 \\ 0 & 5 & 1 & -11 \\ 0 & -2 & -4 & 0 \end{bmatrix}, \quad Q = \begin{bmatrix} 3 & 1 & 0 \\ 2 & 1 & 0 \\ 4 & 2 & 1 \end{bmatrix}, \quad \bar{Q} = Q^{-1} = \begin{bmatrix} 1 & -1 & 0 \\ -2 & 3 & 0 \\ 0 & -2 & 1 \end{bmatrix}.$$

With regard to Example 3.38, observe that we have now reduced the problem of finding a row echelon reduction of A, to the problem of finding a row echelon reduction of

$$B[2:3, 2:4] = \begin{bmatrix} 5 & 1 & -11 \\ -2 & -4 & 0 \end{bmatrix}.$$

This motivates the following algorithm.

Algorithm 3.39 Row echelon
    function rowEchelon(**matrix** B)
    m := **numberOfRows**(B);
    n := **numberOfColumns**(B);
    Q := $\bar{Q}$:= **identityMatrix**(m);
    k := 0; l := 1;
    repeat
        while $1 \leq n$ **and** $B[k+1:m, l] = 0$ **do** l := l + 1;
        if l = n + 1 **then break endif**;
        k := k + 1;
        $(B, Q, \bar{Q})$:= **rowReduce**$(B, Q, \bar{Q}, k, l)$;
    until k = m;
    return $(B, Q, \bar{Q}, k)$;

Theorem 3.40 *Algorithm 3.39 always stops. Given a matrix $A \in \mathrm{M}_{m,n}(\mathbf{Z})$, on input it returns a matrix $B \in \mathrm{M}_{m,n}(\mathbf{Z})$, mutually inverse $\mathbf{Z}$-invertible matrices $Q, \bar{Q} \in \mathrm{M}_{m,m}(\mathbf{Z})$, and a number k such that B is in row echelon form. Furthermore, exactly the first k rows of B are nonzero and*

$$B = Q^{-1}A = \bar{Q}A. \tag{3.9}$$

Proof. To prove that the algorithm always halts, we need to show that every loop of the algorithm is exited. Clearly, the **repeat** loop is called at most m times. The **while** loop is always exited, because the l variable is increased at every pass through this loop. So even if the condition $\mathtt{B[k+1:m,l]} = 0$ is met at every pass of the loop, the variable l finally reaches $\mathtt{n}+1$.

Let $B^{(0)} := A$, $Q^{(0)} := \bar{Q}^{(0)} := I_{m \times m}$ and $l^{(0)} := 0$ denote the initial values of variables B, Q, Q̄, and l. For $i = 1, 2, \ldots, k$, let $B^{(i)}$, $Q^{(i)}$, $\bar{Q}^{(i)}$, and $l^{(i)}$ respectively denote the values of variables B, Q, Q̄, and l on completing the ith pass of the **repeat** loop. In particular, $B^{(k)} = B$, $Q^{(k)} = Q$, and $\bar{Q}^{(k)} = \bar{Q}$.

Observe that $l^{(i)}$ is set when the **while** loop is completed on the ith pass of the **repeat** loop. At this moment the k variable contains $i-1$ and the B variable contains $B^{(i-1)}$. Therefore, the test in the **while** loop implies that for every $i = 1, 2, \ldots, k$ we have

$$l^{(i)} > l^{(i-1)} \quad \text{implies} \quad B^{(i-1)}[i:m, l^{(i-1)}:l^{(i)}-1] = 0 \tag{3.10}$$

and

$$B^{(i-1)}[i:m, l^{(i)}] \neq 0. \tag{3.11}$$

We will show by induction in i that for $i = 0, 1, 2, \ldots, k$,

$$B^{(i)} = (Q^{(i)})^{-1}A = \bar{Q}^{(i)}A, \tag{3.12}$$

$$B^{(i)} \text{ satisfies the } (i, l^{(i)}) \text{ criterion of row echelon form.} \tag{3.13}$$

Indeed, for $i = 0$, Eq. (3.12) follows from the fact that $B^{(0)} = A$, $Q^{(0)} = \bar{Q}^{(0)} = I_{m \times m}$ and Eq. (3.13) is vacuously fulfilled. Therefore, assume that Eqs. (3.12) and (3.13) are satisfied for some $i-1 < k$. Then $B^{(i-1)}$ satisfies the $(i-1, l^{(i-1)})$ criterion of row echelon form. However, by (3.10) and Proposition 3.27, $B^{(i-1)}$ also satisfies the $(i-1, l^{(i)}-1)$ criterion of row echelon form. By (3.11) we may apply Proposition 3.37 to $B = B^{(i-1)}$, $Q = Q^{(i-1)}$, $k = i-1$, and $l = l^{(i)} - 1$, from which we get (3.12) and (3.13).

Since $B = B^{(k)}$, $Q^{(k)} = Q$, and $\bar{Q}^{(k)} = \bar{Q}$, it follows from (3.12) that (3.9) is satisfied.

If the **repeat** loop is exited when the k variable reaches m, then all rows of B are nonzero. Otherwise, it is exited via the **break** statement. In both cases

$$B[k+1:m, l^{(k)}:n] = 0.$$

Hence by Proposition 3.27, B satisfies the (k, n) criterion of row echelon form, and by Proposition 3.25, B is in row echelon form and exactly the first k rows of B are nonzero. □

3 Computing Homology Groups

Corollary 3.41 *Every matrix with integer coefficients can be brought to a row echelon form by means of elementary row operations over* $\mathbf{Z}$.

As pointed out in Example 3.10, the row echelon form of a matrix may be used to construct a basis of the kernel and image of that matrix.

Algorithm 3.42 Kernel-image algorithm
   function kernelImage(matrix B)
   m := numberOfRows(B);
   n := numberOfColumns(B);
   BT := transpose(B);
   $(\text{B}, \text{P}, \bar{\text{P}}, \text{k})$:= rowEchelon(BT);
   BT := transpose(B);
   PT := transpose(P);
   return (PT[1 : m, k + 1 : n], BT[1 : m, 1 : k]);

Theorem 3.43 *Given an* $m \times n$ *matrix* A *on input, Algorithm 3.42 returns an* $m \times (n-k)$ *matrix* W *and an* $m \times k$ *matrix* V *such that the columns of* W *constitute a basis of* $\ker A$ *and the columns of* V *constitute a basis of* $\operatorname{im} A$.

Proof. Let B, P and k denote respectively the values assigned to the variables B, P and k. By Theorem 3.40, $B = PA^T$, B is in row echelon form, and the first k rows of B are its nonzero rows. By Proposition 3.23, the first k columns of B^T constitute the basis of $\operatorname{im} B^T$ and the vectors $\mathbf{e}_{k+1}, \mathbf{e}_{k+2}, \ldots, \mathbf{e}_n \in \mathbf{Z}^n$ constitute a basis of $\ker B^T$. Since $B^T = AP^T$ and P^T is invertible, Proposition 3.9 implies that $\operatorname{im} B^T = \operatorname{im} A$. Therefore, the columns of V constitute a basis of $\operatorname{im} A$. Again, by Proposition 3.9, $\ker A = P^T(\ker B^T)$. Therefore, $P^T\mathbf{e}_{k+1}, P^T\mathbf{e}_{k+2}, \ldots, P^T\mathbf{e}_n \in \mathbf{Z}^n$ is a basis of $\ker A$. But these are exactly the columns of W. □

Example 3.44 Let $A \in \mathrm{M}_{4,3}(\mathbf{Z})$ be given by

$$A = \begin{bmatrix} 0 & 2 & 2 \\ 1 & 0 & -1 \\ 3 & 4 & 1 \\ 5 & 3 & -2 \end{bmatrix}.$$

We will find bases for $\ker A$ and $\operatorname{im} A$. Applying Algorithm 3.42 to A, we get the following values of the variables P, B and k:

$$P = \begin{bmatrix} 0 & 1 & 0 \\ 1 & 0 & 0 \\ 1 & -1 & 1 \end{bmatrix}, \quad B = \begin{bmatrix} 2 & 0 & 4 & 3 \\ 0 & 1 & 3 & 5 \\ 0 & 0 & 0 & 0 \end{bmatrix},$$

and $k = 2$. Therefore, the first two columns of B^T—$[2, 0, 4, 3]^T$ and $[0, 1, 3, 5]^T$—form a basis for $\operatorname{im} B = \operatorname{im} A$, whereas the third column of P^T—$[1, -1, 1]^T$—is a basis of $\ker A$.

Exercises

3.7 Prove Proposition 3.22.

3.8 Show that $l^{(i)}$, the value of the l variable on ith pass of the **repeat** loop in Algorithm 3.39, equals the pivot position of the ith row of the matrix B returned by this algorithm.

3.9 Use the algorithm rowEchelon to obtain an algorithm, columnEchelon, computing the column echelon form of a matrix.

3.3 Smith Normal Form

As indicated earlier, the action of a linear map is most transparent when it is presented in terms of bases in which it has a diagonal matrix. With this in mind we present an algorithm that produces a diagonal matrix with the property that the ith diagonal entry divides the $(i+1)$st diagonal entry. This latter property will be used to prove the classification theorem for finitely generated abelian groups. It should also be emphasized that the approach presented here does not lead to the most efficient algorithm. Instead we focus on expressing the essential ideas.

Since the algorithm is quite complex, we start with an example.

Example 3.45 Consider the matrix

$$A = \begin{bmatrix} 3 & 2 & 3 \\ 0 & 2 & 0 \\ 2 & 2 & 2 \end{bmatrix}.$$

In Example 3.10 we keep track of row operations by performing them on the matrix augmented by the identity matrix on the left. Now we need to keep track of both row and column operations, so we shall work with the following augmented matrix:

$$\left[\begin{array}{c|c} I & A \\ \hline 0 & I \end{array} \right].$$

When row operations are performed on A, only the upper blocks change. When column operations are performed on A, only the right-hand-side blocks change. The zero matrix in the lower left corner never changes; it is there only to complete the expression to a square matrix. At the final stage we obtain a matrix

$$\left[\begin{array}{c|c} P & B \\ \hline 0 & R \end{array} \right],$$

where B is in the above-mentioned diagonal form, R is a matrix of column operations, and P is a matrix of row operations. So, in particular, we have $B = PAR$.

The first step is to identify a nonzero entry of A with the minimal absolute value (we will use a bold character for it) and bring it, by row and column operations, to the upper left-hand corner of the matrix B obtained from A. By exchanging columns c_4 and c_5, we get

$$\left[\begin{array}{ccc|ccc} 1 & 0 & 0 & 3 & 2 & 3 \\ 0 & 1 & 0 & 0 & 2 & 0 \\ 0 & 0 & 1 & 2 & 2 & 2 \\ \hline 0 & 0 & 0 & 1 & 0 & 0 \\ 0 & 0 & 0 & 0 & 1 & 0 \\ 0 & 0 & 0 & 0 & 0 & 1 \end{array}\right] \mapsto \left[\begin{array}{ccc|ccc} 1 & 0 & 0 & \mathbf{2} & 3 & 3 \\ 0 & 1 & 0 & 2 & 0 & 0 \\ 0 & 0 & 1 & 2 & 2 & 2 \\ \hline 0 & 0 & 0 & 0 & 1 & 0 \\ 0 & 0 & 0 & 1 & 0 & 0 \\ 0 & 0 & 0 & 0 & 0 & 1 \end{array}\right].$$

Note that the first pivot entry containing 2 divides the entries in its column below it, but it does not divide the entries in its row on the right of it. By subtracting r_1 from r_2 and r_3, we get zero entries below the pivot, 2, and then by subtracting c_4 from c_5 and c_6, we reduce the value of the entries on the right of 2.

$$\mapsto \left[\begin{array}{ccc|ccc} 1 & 0 & 0 & 2 & 3 & 3 \\ -1 & 1 & 0 & 0 & -3 & -3 \\ -1 & 0 & 1 & 0 & -1 & -1 \\ \hline 0 & 0 & 0 & 0 & 1 & 0 \\ 0 & 0 & 0 & 1 & 0 & 0 \\ 0 & 0 & 0 & 0 & 0 & 1 \end{array}\right] \mapsto \left[\begin{array}{ccc|ccc} 1 & 0 & 0 & 2 & \mathbf{1} & 1 \\ -1 & 1 & 0 & 0 & -3 & -3 \\ -1 & 0 & 1 & 0 & -1 & -1 \\ \hline 0 & 0 & 0 & 0 & 1 & 0 \\ 0 & 0 & 0 & 1 & -1 & -1 \\ 0 & 0 & 0 & 0 & 0 & 1 \end{array}\right].$$

Now the minimal absolute value of nonzero entries of the matrix B is 1, so we repeat the procedure. By exchanging c_4 and c_5, we bring 1 to the upper left corner and then use a series of row and column subtractions to zero out the entries below and on the right of the pivot 1.

$$\mapsto \left[\begin{array}{ccc|ccc} 1 & 0 & 0 & \mathbf{1} & 2 & 1 \\ -1 & 1 & 0 & -3 & 0 & -3 \\ -1 & 0 & 1 & -1 & 0 & -1 \\ \hline 0 & 0 & 0 & 1 & 0 & 0 \\ 0 & 0 & 0 & -1 & 1 & -1 \\ 0 & 0 & 0 & 0 & 0 & 1 \end{array}\right] \mapsto \cdots \mapsto \left[\begin{array}{ccc|ccc} 1 & 0 & 0 & \mathbf{1} & 0 & 0 \\ 2 & 1 & 0 & 0 & 6 & 0 \\ 0 & 0 & 1 & 0 & 2 & 0 \\ \hline 0 & 0 & 0 & 1 & -2 & -1 \\ 0 & 0 & 0 & -1 & 3 & 0 \\ 0 & 0 & 0 & 0 & 0 & 1 \end{array}\right].$$

Now the row and column of the pivot entry 1 are in the desired form. It remains to continue reductions in the 2×2 matrix

$$B[2:3, 2:3] = \begin{bmatrix} 6 & 0 \\ \mathbf{2} & 0 \end{bmatrix}.$$

The minimal nonzero entry of this matrix is 2. By exchanging row r_2 with r_3 in the augmented matrix, we bring 2 to the upper right corner of $B[2:3, 2:3]$ and by subtracting $3r_2$ from r_3 we zero out the entry 6. Thus we obtain

$$\left[\begin{array}{c|c} P & B \\ \hline 0 & R \end{array}\right] = \left[\begin{array}{cc|ccc} 1\,0 & 0 & 1 & 0 & 0 \\ 0\,0 & 1 & 0 & 2 & 0 \\ 2\,1 & -3 & 0 & 0 & 0 \\ \hline 0\,0 & 0 & 1 & -2 & -1 \\ 0\,0 & 0 & -1 & 3 & 0 \\ 0\,0 & 0 & 0 & 0 & 1 \end{array}\right].$$

Therefore, the final matrix

$$B = \begin{bmatrix} 1 & 0 & 0 \\ 0 & 2 & 0 \\ 0 & 0 & 0 \end{bmatrix}$$

is diagonal with $b_1 = 1$ dividing $b_2 = 2$. We want to know the changes of basis corresponding to the row and column operations. We have obtained

$$R = \begin{bmatrix} 1 & -2 & -1 \\ -1 & 3 & 0 \\ 0 & 0 & 1 \end{bmatrix} \quad \text{and} \quad P = \begin{bmatrix} 1 & 0 & 0 \\ 0 & 0 & 1 \\ 2 & 1 & -3 \end{bmatrix}.$$

In order to apply Corollary 3.6, we need

$$Q = P^{-1} = \begin{bmatrix} 1 & 0 & 0 \\ -2 & 3 & 1 \\ 0 & 1 & 0 \end{bmatrix}.$$

Since $B = Q^{-1}AR$, the columns of R form the new basis of $\mathbf{Z}^3$, viewed as the domain of B, expressed in terms of the canonical basis and the columns of Q correspond to the new basis of $\mathbf{Z}^3$ viewed as the target space of B.

With the above example in mind, we now proceed with the formal description of the algorithm. Given an $m \times n$ matrix $A \in \mathrm{M}_{m,n}(\mathbf{Z})$, our ultimate aim is to produce bases given by columns of some matrices $Q \in \mathrm{M}_{m,m}(\mathbf{Z})$ and $R \in \mathrm{M}_{n,n}(\mathbf{Z})$ such that the matrix of the homomorphism f_A in the new bases, that is,

$$B := Q^{-1}AR,$$

satisfies $B[i,j] = 0$ if $i \neq j$ and $B[i,i]$ divides $B[i+1, i+1]$. We shall do this in a recursive manner. We say that $B \in \mathrm{M}_{m,n}(\mathbf{Z})$ is in Smith form up to the kth entry if the following three conditions are satisfied:

$$B = \left[\begin{array}{ccc|c} B[1,1] & & 0 & \\ & \ddots & & 0 \\ 0 & & B[k,k] & \\ \hline & 0 & & B[k+1:m, k+1:n] \end{array}\right];$$

$$B[i,i] \text{ divides } B[i+1, i+1] \quad \text{for } i = 1, 2, \ldots, k-1; \tag{3.14}$$

and

$$B[k,k] \text{ divides } B[i,j] \quad \text{for all } i, j > k. \tag{3.15}$$

For the moment let us ignore the conditions on divisibility. Then the problem is essentially the same as that solved by rowReduce, except that it needs to be solved not only for the kth column, but simultaneously for the kth column and the kth row.

First we extend smallestNonzero to find an entry that has the smallest nonzero magnitude for the entire submatrix $B[k:m, k:n]$.

Algorithm 3.46 Minimal nonzero entry
    function minNonzero(**matrix** B, **int** k)
    vector v, q;
    for i := 1 **to numberOfRows**(B)
      if i < k **then**
        v[i] := q[i] := 0
      else
        (v[i], q[i]) :=
        smallestNonzero(B[i, 1 : **numberOfColumns**(B)], k);
      endif;
    endfor;
    (alpha, i_0) := smallestNonzero(v, k);
    return (alpha, i_0, q(i_0));

Having found the entry with minimal nonzero magnitude, we now need to move it to the (k, k) position. We leave it to the reader to check that this is done by the following algorithm.

Algorithm 3.47 Move minimal nonzero entry
    function moveMinNonzero(**matrix** B, Q, $\bar{Q}$, R, $\bar{R}$, **int** k)
    (alpha, i, j) := minNonzero(B, k);
    (B, Q, $\bar{Q}$) := rowExchangeOperation(B, Q, $\bar{Q}$, k, i);
    (B, R, $\bar{R}$) := columnExchangeOperation(B, R, $\bar{R}$, k, j);
    return (B, Q, $\bar{Q}$, R, $\bar{R}$);

We now turn to the consideration of condition (3.15). The following algorithm checks if the (k, k) entry of a matrix B divides all entries in the submatrix $B[k+1:m, k+1:n]$.

Algorithm 3.48 Check for divisibility
    function checkForDivisibility(**matrix** B, **int** k)
    for i := k + 1 **to numberOfRows**(B)
      for j := k + 1 **to numberOfColumns**(B)

3.3 Smith Normal Form 121

```
            q := floor(B[i, j]/B[k, k]);
            if q * B[k, k] ≠ B[i, j] then
                return (false, i, j, q);
            endif;
        endfor;
    endfor;
    return (true, 0, 0, 0);
```

The first entry returned by this algorithm is true when the divisibility test succeeds. Otherwise, false is returned together with the coordinates of the first entry where the division test fails, followed by the integer part of the quotient, which indicated failure.

Combining these algorithms we obtain an algorithm that performs one step of the recursive procedure.

Algorithm 3.49 Partial Smith form algorithm
```
    function partSmithForm(matrix B, Q, Q̄, R, R̄, int k)
    m := numberOfRows(B);
    n := numberOfColumns(B);
    repeat
        (B, Q, Q̄, R, R̄) := moveMinNonzero(B, Q, Q̄, R, R̄, k);
        (B, Q, Q̄) := partRowReduce(B, Q, Q̄, k, k);
        if B[k + 1 : m, k] ≠ 0 then next; endif;
        (B, R, R̄) := partColumnReduce(B, R, R̄, k, k);
        if B[k, k + 1 : n] ≠ 0 then next; endif;
        (divisible, i, j, q) := checkForDivisibility(B, k);
        if not divisible then
            (B, Q, Q̄) := rowAddOperation(B, Q, Q̄, i, k, 1);
            (B, R, R̄) := columnAddOperation(B, R, R̄, k, j, −q);
        endif;
    until divisible;
    return (B, Q, Q̄, R, R̄);
```

Proposition 3.50 *Let $A, B \in M_{m,n}(\mathbf{Z})$ be integer matrices such that B is in Smith form up to the $(k-1)$st entry for some $k \in \{1, 2, \ldots, \min(m,n)\}$, $B[k:m, k:n] \neq 0$ and*
$$B = Q^{-1}AR$$
for some $\mathbf{Z}$-invertible matrices $Q \in M_{m,m}(\mathbf{Z})$ and $R \in M_{n,n}(\mathbf{Z})$. Algorithm 3.49 applied to $(B, Q, Q^{-1}, R, R^{-1}, k)$ always halts. It returns a matrix B' that is in Smith form up to the kth entry. It also returns two pairs $(Q', \bar{Q}')$ and $(R', \bar{R}')$ of mutually inverse $\mathbf{Z}$-invertible matrices such that
$$B' = Q'^{-1}AR' = \bar{Q}'AR'.$$

Proof. Let b_p denote the absolute value of B$[k,k]$ after the call of the function moveMinNonzero on the p-th pass of the **repeat** loop. We will show that the sequence $\{b_p\}$ is strictly decreasing.

Consider the pth pass of the **repeat** loop such that b_p is not the last value of the sequence. Then either one of the two **next** statements is executed or the divisibility test fails. When the first **next** statement is executed, the subcolumn B$[k+1:m,k]$ of B, returned by partRowReduce, is nonzero. Therefore, by Proposition 3.31, the subcolumn contains elements whose absolute value is less than b_p. It follows that $b_{p+1} < b_p$. When the other **next** statement is executed, the argument is similar. Thus consider the case when checkForDivisibility returns divisible equal to false. Let B, i, j, q be the values of variables B, i, j, q after completing the call to checkForDivisibility. Let

$$r := B[k,j] - qB[k,k].$$

It must be $r \neq 0$, because we assumed that checkForDivisibility returned false. Thus, as a nonzero remainder, r satisfies

$$0 < |r| < b_p = |B[k,k]|.$$

Moreover, $B[k,j] = B[i,k] = 0$, because checkForDivisibility is called only when no **next** statement is executed. After adding the kth row to the ith row, the new value of B$[i,k]$ equals $B[k,k]$. Therefore, after subtracting q-times the kth column from the jth column, the value of B$[i,j]$ becomes r and, consequently, $b_{p+1} \le |r| < b_p$. This shows that $\{b_p\}$ is a strictly decreasing sequence of positive integers. Therefore, it must be finite and the algorithm must stop.

Next, we will show that B' is in the Smith form up to the kth entry. First observe that by Proposition 3.31, Proposition 3.18 and their analogues for column operations the first $k-1$ rows and columns of B and B' do not differ. Since B' is the value of the B variable after the **repeat** loop is left, we must have $B'[k+1:m,k] = 0$, $B'[k,k+1:n] = 0$ and $B'[k,k]$ divides all $B'[i,j]$ for $i,j > k$. Therefore, by definition, B' is in Smith form up to the kth entry. In order to prove (3.14), we only need to show that $B'[k-1,k-1]$ divides $B'[k,k]$, because $B'[i,i] = B[i,i]$ for $i = 1,2,\ldots,k-1$. However, since $B[k-1,k-1]$ divides all entries of $B[k:m,k:n]$, by Proposition 3.12 $B[k-1,k-1] = B'[k-1,k-1]$ divides all entries of $B'[k:m,k:n]$. In particular, it divides $B'[k,k]$.

The last assertion is a straightforward consequence of the properties of functions used to modify the B variable inside the **repeat** loop. □

The following algorithm is obtained by inductively applying Algorithm 3.50.

Algorithm 3.51 Smith algorithm
    function smithForm(**matrix** B)
        m := **numberOfRows**(B);
        n := **numberOfColumns**(B);

```
Q := Q̄ := identityMatrix(m);
R := R̄ := identityMatrix(n);
s := t := 0;
while B[t + 1 : m, t + 1 : n] ≠ 0 do
    t := t + 1;
    (B, Q, Q̄, R, R̄) := partSmithForm(B, Q, Q̄, R, R̄, t);
    if B[t, t] < 0 then
        (B, Q, Q̄) := rowMultiplyOperation(B, Q, Q̄, t);
    endif;
    if B[t, t] = 1 then
        s := s + 1;
    endif;
endwhile ;
return (B, Q, Q̄, R, R̄, s, t);
```

Theorem 3.52 *Algorithm 3.51 always stops. Given a matrix $A \in \mathrm{M}_{m,n}(\mathbf{Z})$, on input it returns a matrix $B \in \mathrm{M}_{m,n}(\mathbf{Z})$, $\mathbf{Z}$-invertible, mutually inverse matrices $Q, \bar{Q} \in \mathrm{M}_{m,m}(\mathbf{Z})$, $R, \bar{R} \in \mathrm{M}_{n,n}(\mathbf{Z})$, and nonnegative integers s and t such that*

$$B = Q^{-1}AR = \bar{Q}AR.$$

Furthermore, B has the form

$$B = \begin{bmatrix} b_1 & & & & & \\ & b_2 & & 0 & & \\ & & \cdot & & & 0 \\ & 0 & & \cdot & & \\ & & & & b_t & \\ \hline & 0 & & & & 0 \end{bmatrix}, \qquad (3.16)$$

where b_i are positive integers, $b_i = 1$ for $i = 1, 2, \ldots, s$, and b_i divides b_{i+1} for $i = 1, 2, \ldots, t - 1$.

Proof. Observe that the variable t is increased on every pass of the **while** loop in Algorithm 3.51. Therefore, $B[t + 1 : m, t + 1 : n])$ must finally become zero, which means that the loop is left and the algorithm halts. Moreover, B is in the form (3.16), because, by Proposition 3.50, matrix B is in Smith form up to the kth entry and $B[t + 1 : m, t + 1 : n]) = 0$. The rest of the assertion is straightforward. □

Corollary 3.53 *Let A be an $n \times m$ matrix with integer coefficients. By means of elementary row and column operations over $\mathbf{Z}$, it is possible to bring A to the form (3.16), where the b_i are positive integers and b_i divides b_{i+1} for all* *i*. In particular, if $f : G \to G'$ is a homomorphism of finitely generated free abelian groups, then there are bases of G and G' such that the matrix of f with respect to those bases is in the form (3.16).

The matrix B returned by Algorithm 3.51 is called the *Smith normal form* of A.

It should be emphasized that the problem of reducing a matrix to its Smith normal form is different from the classical problem of diagonalizing a square matrix. In the first case we can have two bases, one for the domain and another for the image. In the latter, a single basis consisting of eigenvectors is required.

We finish this section with an application of the Smith normal form algorithm to the issue of solvability of linear systems of equations $Ax = b$ under the restriction that all the terms must be integers. Observe that even the question of whether a solution exists is nontrivial. Consider

$$A = \begin{bmatrix} 2 & 0 \\ 0 & 3 \end{bmatrix}.$$

This is a diagonal matrix, with nonzero entries in the diagonal. Therefore, $\det A \neq 0$. However, this does not mean that we can necessarily solve the equation $Ax = b$. As an example, consider $b = \begin{bmatrix} 1 \\ 0 \end{bmatrix}$. There is no integer-valued vector that solves the equation. Before proceeding, the reader might wish to consider what is the set of integer vectors b for which an integer vector solution exists.

However, even though we began on a pessimistic note, Algorithm 3.51 may be used to find integer-valued solutions of linear equations, if they exist.

Algorithm 3.54 Linear equation solver

```
function Solve(matrix A, vector b)
m := numberOfRows(A);
(B, Q, Q̄, R, R̄, s, t) := smithForm(A);
c := Q̄ * b;
vector u;
for i := 1 to t do
   if B[i, i] divides c[i] then
      u[i] := c[i]/B[i, i];
   else
      return "Failure";
   endif;
endfor;
for i := t + 1 to m do
   if c[i] ≠ 0 then
      return "Failure"
   else
```

```
        u[i] := 0;
    endif;
  endfor;
  return R * u;
```

Theorem 3.55 *If the equation*

$$Ax = b \qquad (3.17)$$

has a solution $x \in \mathbf{Z}^n$, then Algorithm 3.54 returns one such solution. Otherwise, it returns `"Failure"`.

Proof. By Theorem 3.52 variables $\bar{Q}$, B, and R, returned by `smithForm`, satisfy $B = \bar{Q}AR$. Since Q and R are $\mathbf{Z}$-invertible with their inverses given by $\bar{Q}$ and $\bar{R}$, Eq. (3.17) is equivalent to the equation

$$B\bar{R}x = \bar{Q}b. \qquad (3.18)$$

Let $c := \bar{Q}b$ and $u := \bar{R}x$. Then Eq. (3.18) has an integer solution if and only if

$$Bu = c$$

has an integer solution. But, since B is in normal Smith form, the latter has an integer solution if and only if $B[i,i]$ divides c_i for $i = 1, 2, \ldots, t$ and $c_i = 0$ for $i = t+1, t+2, \ldots, m$. □

Exercise _____

3.10 For each matrix A specified below, find its normal form B and two integer matrices R and Q, invertible over $\mathbf{Z}$, such that $QB = AR$. Use the information provided by R and Q for presenting (1) a pair of bases with respect to which the normal form is assumed, (2) a basis for $\ker A$, and (3) a basis for $\operatorname{im} A$.

(a) $A = \begin{bmatrix} 6 & 4 \\ 4 & 0 \\ 0 & 6 \end{bmatrix}$

(b) The matrix A in Example 3.44

(c) $A = \begin{bmatrix} 2 & 0 & 0 \\ 0 & 3 & 0 \\ 0 & 0 & 9 \end{bmatrix}$

3.4 Structure of Abelian Groups

By now we have essentially constructed all the tools necessary to compute quotients of finitely generated free abelian groups. Therefore, we will construct

Algorithm 3.58, which allows one to take finitely generated free abelian groups $H \subset G$ and produce the quotient group G/H. Furthermore, in the next section we use this algorithm to compute the homology of any finite chain complex. However, before doing so we prove the fundamental decomposition theorem of abelian groups, Theorem 3.61. As indicated in the introduction, this leads to Corollary 3.1, which provides us with a means of distinguishing the homology groups that will be computed. For the proof of Theorem 3.61 we do not need the full power of Algorithm 3.58. Instead we only need to understand the quotients $\mathbf{Z}^m/H$, where H is a subgroup of $\mathbf{Z}^m$. Observe (see Lemma 13.68) that $H \subset \mathbf{Z}^m$ implies that H is free of rank less than or equal to m. We start with the following example.

Example 3.56 Consider a subgroup H of $\mathbf{Z}^3$ generated by the columns of the matrix

$$A = \begin{bmatrix} 3 & 2 & 3 \\ 0 & 2 & 0 \\ 2 & 2 & 2 \end{bmatrix}$$

investigated earlier in Example 3.45. We want to describe the quotient group $\mathbf{Z}^3/H$, which is actually equal to $\mathbf{Z}^3/\operatorname{im} A$. Note that the columns are not linearly independent—the third one is equal to the first one. We compute the Smith normal form of A,

$$B = \begin{bmatrix} 1 & 0 & 0 \\ 0 & 2 & 0 \\ 0 & 0 & 0 \end{bmatrix},$$

in Example 3.45. This normal form is obtained using the bases consisting of the columns of the matrices

$$R = \begin{bmatrix} 1 & -2 & -1 \\ -1 & 3 & 0 \\ 0 & 0 & 1 \end{bmatrix} \quad \text{and} \quad Q = \begin{bmatrix} 1 & 0 & 0 \\ -2 & 3 & 1 \\ 0 & 1 & 0 \end{bmatrix}$$

in $\mathbf{Z}^3$ viewed, respectively, as the domain and the target space. We will denote the columns of R and Q, respectively, by $\mathbf{r}_1, \mathbf{r}_2, \mathbf{r}_3$, and $\mathbf{q}_1, \mathbf{q}_2, \mathbf{q}_3$. Since $AR = QB$, we get equations on column vectors:

$$A\mathbf{r}_1 = \mathbf{q}_1,$$
$$A\mathbf{r}_2 = 2\mathbf{q}_2,$$
$$A\mathbf{r}_3 = 0.$$

Thus $H = \operatorname{im} A = \mathbf{Z}\mathbf{q}_1 \oplus \mathbf{Z}2\mathbf{q}_2$. By using the basis $\{\mathbf{q}_1, \mathbf{q}_2, \mathbf{q}_3\}$ for $\mathbf{Z}^3$, we get

$$\frac{\mathbf{Z}^3}{H} = \frac{\mathbf{Z}\mathbf{q}_1 \oplus \mathbf{Z}\mathbf{q}_2 \oplus \mathbf{Z}\mathbf{q}_3}{\mathbf{Z}\mathbf{q}_1 \oplus \mathbf{Z}2\mathbf{q}_2}.$$

Consider the equivalence classes of the generators $\mathbf{q}_i$ of $\mathbf{Z}^3$. Due to the presence of $\mathbf{q}_1$ in the denominator, $[\mathbf{q}_1] = 0$. Due to the presence of $2\mathbf{q}_2$, $[\mathbf{q}_2]$

is a cyclic element of order 2, so it generates a cyclic subgroup isomorphic to Z_2. The class $[\mathbf{q}_3]$ generates an infinite cyclic group isomorphic to $\mathbf{Z}$. This suggests that
$$\frac{\mathbf{Z}^3}{H} = \langle [\mathbf{q}_2] \rangle \oplus \mathbf{Z}[\mathbf{q}_3] \cong \mathbf{Z}_2 \oplus \mathbf{Z}.$$
A formal argument for this guess is a part of the proof of the next theorem.

In the above example we investigate a quotient group of the form $\mathbf{Z}^n/\mathrm{im}\, A$. We don't want to limit ourselves to this case because the ultimate goal will be computing homology groups $H_k(X) = \ker \partial_k / \mathrm{im}\, \partial_{k+1}$. Even if we identify $C_k(X)$ with a power $\mathbf{Z}^p$, we still have a quotient G/H of two different subgroups of $\mathbf{Z}^p$ to consider: $G := \ker \partial_k$ and $H = \mathrm{im}\, \partial_{k+1} \subset G$. Thus we will assume that we have subgroups $H \subset G \subset \mathbf{Z}^p$ with, possibly, unrelated bases V for H and W for G. We will need to know how to express the elements of V as linear combinations of the elements of W. The algorithm Linear equation solver helps us to deal with this more complex situation. As usual, we shall identify the bases V and W with matrices containing their elements as columns.

Before presenting the quotient group algorithm, we need the following result, which generalizes Proposition 3.5.

Proposition 3.57 *Assume $H \subset \mathbf{Z}^p$ has as a basis $V = \{\mathbf{v}_1, \mathbf{v}_2, \ldots, \mathbf{v}_m\}$. Let $\iota_H : H \to \mathbf{Z}^p$ denote the inclusion map. Then $\eta_V : \mathbf{Z}^m \to H$ defined for $\mathbf{x} = (x_1, x_2, \ldots, x_m) \in \mathbf{Z}^m$ by*

$$\eta_V(\mathbf{x}) := V\mathbf{x} = \sum_{i=1}^{m} x_i \mathbf{v}_i$$

is an isomorphism of groups, whose inverse is ξ_V. In particular, for every $\mathbf{x} \in \mathbf{Z}^m$

$$\xi_V^{-1}(\mathbf{x}) = V\mathbf{x} \tag{3.19}$$

and

$$V = A_{\iota_H \xi_V^{-1}}. \tag{3.20}$$

Moreover, if R is a $\mathbf{Z}$-invertible $m \times m$ matrix, then the columns of $V' := VR$ constitute another basis of H and the change-of-coordinates matrix is

$$A_{\xi_{V'} \xi_V^{-1}} = R^{-1}. \tag{3.21}$$

Proof. Because $V = \{\mathbf{v}_1, \mathbf{v}_2, \ldots \mathbf{v}_m\}$ is a basis in H, one can easily verify that η_V is an isomorphism. To prove that its inverse is ξ_V it is enough to verify 3.19 on the canonical bases. Indeed, by the definition of the coordinates isomorphism and (3.3), we have

$$\xi_V^{-1}(\mathbf{e}_i) = \mathbf{v}_i = V\mathbf{e}_i$$

for every canonical vector $\mathbf{e}_i \in \mathbf{Z}^m$. Similarly, the equality

$$A_{\iota_H \xi_V^{-1}} \mathbf{e}_i = \iota_H(\mathbf{v}_i) = \mathbf{v}_i = V \mathbf{e}_i$$

proves (3.20).

Let $\mathbf{r}_1, \mathbf{r}_2, \ldots, \mathbf{r}_m$ denote the columns of R. Because R is $\mathbf{Z}$-invertible, $\{\mathbf{r}_1, \mathbf{r}_2, \ldots, \mathbf{r}_m\}$ is a basis in $\mathbf{Z}^m$. The fact, that η_V is an isomorphism, implies that $\{\eta_V(\mathbf{r}_1), \eta_V(\mathbf{r}_2), \ldots, \eta_V(\mathbf{r}_m)\}$ is a basis in H. But $\eta_V(\mathbf{r}_i)$ is the ith column of V'. Therefore, the columns of V' constitute another basis of H.

To prove (3.21) take $\mathbf{v} \in H$. Let $\mathbf{x} := \xi_V(\mathbf{v})$ and $\mathbf{y} := \xi_{V'}(\mathbf{v})$ be the coordinates of $\mathbf{v}$ respectively in the bases V and V'. Then, by (3.19), $\mathbf{v} = V\mathbf{x}$ and $\mathbf{v} = V'\mathbf{y} = VR\mathbf{y}$. Therefore, $V(\mathbf{x} - R\mathbf{y}) = 0$. Since the columns of V are linearly independent, $\ker V = 0$ and consequently we get $\mathbf{x} = R\mathbf{y}$ and $\mathbf{y} = R^{-1}\mathbf{x}$. By the definition of the change-of-coordinates matrix, we get (3.21). □

Algorithm 3.58 Quotient group finder
  function quotientGroup(matrix W, V)
  n := numberOfColumns(V);
  matrix A;
  for i := 1 to numberOfColumns(V)
    A[i] := Solve(W, V[i]);
  endfor;
  $(B, Q, \bar{Q}, R, \bar{R}, s, t)$:= smithForm(A);
  U := W * Q;
  return (U, B, s);

Theorem 3.59 *Assume H and G are subgroups of $\mathbf{Z}^p$, $H \subset G$, and*

$$\{\mathbf{v}_1, \mathbf{v}_2, \ldots, \mathbf{v}_n\} \subset \mathbf{Z}^p, \{\mathbf{w}_1, \mathbf{w}_2, \ldots, \mathbf{w}_m\} \subset \mathbf{Z}^p$$

constitute bases in H and G, respectively. Algorithm 3.58, when started with $W = [\mathbf{w}_1 \ \mathbf{w}_2 \ \ldots \ \mathbf{w}_m]$ and $V = [\mathbf{v}_1 \ \mathbf{v}_2 \ \ldots \ \mathbf{v}_n]$ returns a $p \times m$ matrix U, an $m \times n$ matrix B and a nonnegative integer s such that if $U = [\mathbf{u}_1 \ \mathbf{u}_2 \ \ldots \ \mathbf{u}_m]$, then $\{\mathbf{u}_1 \ \mathbf{u}_2 \ \ldots \ \mathbf{u}_m\}$ is another basis in G and

$$[\mathbf{u}_1] = [\mathbf{u}_2] = \ldots = [\mathbf{u}_s] = 0, \quad (3.22)$$

$[\mathbf{u}_i]$ is of order B_{ii} for $i = s+1, s+2, \ldots, n$ and of infinite order for $i = n+1, n+2, \ldots, m$. Moreover,

$$G/H = \bigoplus_{i=s+1}^{n} \langle [\mathbf{u}_i] \rangle \oplus \bigoplus_{i=n+1}^{m} \mathbf{Z}[\mathbf{u}_i]. \quad (3.23)$$

Proof. Let Q denote the matrix assigned to variable Q, when smithForm is called. Since $U = WQ$, it follows from Proposition 3.57 that $\{\mathbf{u}_1 \ \mathbf{u}_2 \ \ldots \ \mathbf{u}_m\}$ is a basis in G.

Let $\iota_H : H \to \mathbf{Z}^p$, $\iota_G : G \to \mathbf{Z}^p$, and $\iota : H \to G$ denote the inclusion maps. Applying (3.20) twice, we get

$$W A_{\xi_W \iota \xi_V^{-1}} = A_{\iota_G \xi_W^{-1}} A_{\xi_W \iota \xi_V^{-1}} = A_{\iota_G \iota \xi_V^{-1}} = A_{\iota_H \xi_V^{-1}} = V.$$

Therefore, the matrix equation

$$WX = V \tag{3.24}$$

has a solution $X = A_{\xi_W \iota \xi_V^{-1}}$, which is the matrix of ι in bases V and W. The solution is unique, because the columns of W are linearly independent, and hence $\ker W = 0$.

Let A denote the matrix stored in variable A after completing the **for** loop. Let R denote the matrix assigned to variable R, when smithForm is called. Since A solves Eq. (3.24), it is the matrix of the inclusion map $\iota : H \to G$ in bases V and W. Put $V' := VR$ and let $\mathbf{v}'_1, \mathbf{v}'_2, \ldots, \mathbf{v}'_n$ denote the columns of V'. By Proposition 3.57 and 3.4, the inclusion map ι is represented by matrix B in bases $\{\mathbf{v}'_1, \mathbf{v}'_2, \ldots, \mathbf{v}'_n\}$ and $\{\mathbf{u}_1, \mathbf{u}_2, \ldots, \mathbf{u}_m\}$. Thus,

$$B_{ii} \mathbf{u}_i = \iota(\mathbf{v}'_i) = \mathbf{v}'_i \in H \tag{3.25}$$

for $i = 1, 2, \ldots, n$. Observe that, as a basis vector, $\mathbf{v}'_i \neq 0$, which implies that $B_{ii} \neq 0$ for $i = 1, 2, \ldots n$. Thus the value of the variable t, which counts the number of the nonzero diagonal entries in matrix B, is exactly n. Since $|B_{ii}| = 1$ for $i = 1, 2, \ldots, s$, we see from (3.25) that in this case $\mathbf{u}_i \in H$, which shows (3.22).

Consider, in turn, $[\mathbf{u}_i]$ for $i \in \{s+1, s+2, \ldots, m\}$. If $k[\mathbf{u}_i] = 0$ for some $k \in \mathbf{Z}$, then $k\mathbf{u}_i \in H$. Thus,

$$k\mathbf{u}_i = \sum_{j=1}^n \alpha_j \mathbf{v}'_j = \sum_{j=1}^n \alpha_j B_{jj} \mathbf{u}_j$$

for some $\alpha_j \in \mathbf{Z}$. This implies that either $i > n$ and then $k = 0$ or $i \in \{s+1, s+2, \ldots, n\}$ and $k = \alpha_i B_{ii}$. As a consequence, $[\mathbf{u}_i]$ is of infinite order in the first case and of order $|B_{ii}|$ in the other case.

It remains to prove (3.23). Let $g \in G$. Then, for some $\alpha_i \in \mathbf{Z}$,

$$g = \sum_{i=1}^m \alpha_i \mathbf{u}_i$$

and by (3.22)

$$[g] = \sum_{i=1}^m \alpha_i [\mathbf{u}_i] = \sum_{i=s+1}^m \alpha_i [\mathbf{u}_i].$$

Therefore, $[g]$ belongs to the right-hand side of (3.23). To show the uniqueness of the decomposition, assume that also

$$[g] = \sum_{i=s+1}^{m} \beta_i[\mathbf{u}_i]$$

for some $\beta_i \in \mathbf{Z}$. Then

$$\sum_{i=s+1}^{m} (\alpha_i - \beta_i)\mathbf{u}_i \in H.$$

Thus, for some $\gamma_i \in \mathbf{Z}$,

$$\sum_{i=s+1}^{m} (\alpha_i - \beta_i)\mathbf{u}_i = \sum_{i=1}^{n} \gamma_i \mathbf{v}'_i.$$

By (3.25),

$$\sum_{i=s+1}^{m} (\alpha_i - \beta_i)\mathbf{u}_i = \sum_{i=1}^{n} \gamma_i B_{ii} \mathbf{u}_i.$$

This implies $\alpha_i = \beta_i$ for $i = n+1, n+2, \ldots, m$ and $|B_{ii}|$ divides $\alpha_i - \beta_i$ for $i = s+1, s+2, \ldots, n$ and the uniqueness is proved. □

The following corollary is obtained from the previous theorem by shifting indices of $[\mathbf{u}_i]$ so to eliminate the first s trivial equivalence classes and by switching the order of finite cyclic with infinite cyclic elements.

Corollary 3.60 *Consider subgroups $H \subset G \subset \mathbf{Z}^p$. Then there exist elements $\mathbf{u}_1, \mathbf{u}_2, \ldots, \mathbf{u}_q \in G$ and an integer $r \geq 0$ and $k := q - r$ such that*

(a) the quotient group G/H is generated by equivalence classes $[\mathbf{u}_i]$, $i = 1, 2, \ldots, q$;
(b) if $r > 0$, $[\mathbf{u}_i]$ is cyclic of infinite order for $i \in \{1, 2, \ldots, r\}$;
(c) if $k = q - r > 0$, $[\mathbf{u}_{r+i}]$ is cyclic of order b_i for $i \in \{1, 2, \ldots, k\}$, where $b_i > 1$ and b_i divides b_{i+1} for $i \in \{1, 2, \ldots, k-1\}$ provided $k > 1$;
(d) the group G/H can be decomposed as

$$G/H = \bigoplus_{i=1}^{r} \mathbf{Z}[\mathbf{u}_i] \oplus \bigoplus_{i=r+1}^{q} \langle [\mathbf{u}_i] \rangle.$$

By means of this corollary we can characterize the general structure of a finitely generated abelian group.

Theorem 3.61 (Fundamental decomposition theorem) *Let G be a finitely generated abelian group. Then G can be decomposed as a direct sum of cyclic groups. More explicitly, there exist generators, $g_1, g_2, \ldots, g_q$, of G and an integer, $0 \leq r \leq q$, such that*

(a) $G = \bigoplus_{i=1}^{q} \langle g_i \rangle$;
(b) if $r > 0$, then $g_1, g_2, \ldots, g_r$ are of infinite order;
(c) if $k = q - r > 0$, then $g_{r+1}, g_{r+2}, \ldots, g_{r+k}$ have finite orders $b_1, b_2, \ldots, b_k$, respectively, $b_i > 1$, and b_i divides b_{i+1} for $i \in \{1, 2, \ldots, k-1\}$ provided $k > 1$.

The numbers r and $b_1, b_2, \ldots, b_k$ are uniquely determined by G, although the generators $g_1, g_2, \ldots, g_q$ are not.

Before proving Theorem 3.61 let us formulate the following corollary.

Corollary 3.62 *Given a finitely generated abelian group G, there exists a free subgroup F such that*
$$G = F \oplus T(G). \tag{3.26}$$
In particular, $G/T(G)$ is free.

Proof. Using the notation of Theorem 3.61, let
$$F := \bigoplus_{i=1}^{r} \mathbf{Z} g_i, \qquad T := \bigoplus_{i=1}^{k} \langle g_{r+i} \rangle.$$
It is easy to verify that $T = T(G)$, thus $G = F \oplus T(G)$. Therefore, $G/T(G) \cong F$, which is free. □

The presentation of G given by (3.26) is called the *normal decomposition* of G. It is worth remarking that F is unique only up to an isomorphism and depends on the choice of generators.

Proof of Theorem 3.61. Let $S := \{s_1, \ldots, s_m\}$ be a set of generators for G. Define $f : \mathbf{Z}^m \to G$ on the canonical basis of $\mathbf{Z}^m$ by $f(\mathbf{e}_i) = s_i$. This is a group homomorphism and so $H := \ker f$ is a subgroup of $\mathbf{Z}^m$. By Theorem 13.43,
$$\bar{f} : \mathbf{Z}^m / H \to G$$
is an isomorphism. Thus to prove the theorem it is sufficient to obtain the desired decomposition for the group $\mathbf{Z}^m/H$. Therefore, the conclusions (a), (b) and (c) follow immediately from Corollary 3.60.

The statement about the uniqueness of the Betti number r and torsion coefficients b_i is not immediate, but the outline of the proof is provided in Exercise 3.14. □

We conclude this section by reminding the reader to check that Corollary 3.1 does indeed follow from Theorem 3.61.

Exercises

3.11 For each matrix A in Exercise 3.10, find the normal decomposition of the group $\mathbf{Z}^m / \operatorname{im} A$, where m is the number of rows of A.

3.12 If $G \simeq \mathbf{Z}_6 \oplus \mathbf{Z}_{20} \oplus \mathbf{Z}_{25}$, find the torsion coefficients and prime power factors of G.

3.13 Find two different free subgroups of $\mathbf{Z} \oplus \mathbf{Z}_2$ complementing the torsion subgroup $\mathbf{Z}_2$.

3.14 (a) Let p be a prime and let $d_1, d_2, \ldots, d_m$ be nonnegative integers. Show that if

$$G \simeq (\mathbf{Z}_p)^{d_1} \oplus (\mathbf{Z}_{p^2})^{d_2} \oplus \ldots \oplus (\mathbf{Z}_{p^m})^{d_m},$$

then the integers d_i are uniquely determined by G.
[**Hint:** Consider the kernel of the homomorphism $f_i : G \to G$ given by $f_i(a) := p^i a$. Show that f_1 and f_2 determine d_1. Proceed by induction.]
(b) Let $p_1, p_2, \ldots, p_s$ be a sequence of primes. Generalize (a) to a finite direct sum of terms of the form $(\mathbf{Z}_{p_i}^j)^{d_{ij}}$.
(c) Derive the last conclusion of Theorem 5.1 on the uniqueness of the Betti number and torsion coefficients.

3.5 Computing Homology Groups

The previous sections provided us with sufficient background to formulate an algorithm for computing the kth homology group of a free chain complex. As usual we start from a simple example.

Example 3.63 Consider the abstract chain complex $\mathcal{C}$ defined as follows. The chain groups are $C_0 := \mathbf{Z}^3$, $C_1 := \mathbf{Z}^3$, and $C_k = 0$ for all $k \neq 0, 1$. The only nontrivial boundary map $\partial_1 : C_1 \to C_0$ is given by

$$\partial_1 \mathbf{x} := A\mathbf{x},$$

where A is the matrix from Example 3.56:

$$A = \begin{bmatrix} 3 & 2 & 3 \\ 0 & 2 & 0 \\ 2 & 2 & 2 \end{bmatrix}.$$

Recall that in Example 3.45 we have found that the Smith normal form of A is

$$B = \begin{bmatrix} 1 & 0 & 0 \\ 0 & 2 & 0 \\ 0 & 0 & 0 \end{bmatrix}$$

when considered in the bases given by the columns of

$$R = \begin{bmatrix} 1 & -2 & -1 \\ -1 & 3 & 0 \\ 0 & 0 & 1 \end{bmatrix}$$

in C_0 and the columns of

$$Q = \begin{bmatrix} 1 & 0 & 0 \\ -2 & 3 & 1 \\ 0 & 1 & 0 \end{bmatrix}$$

in C_1. The group $H_0(\mathcal{C})$ has already been computed in Example 3.56. Indeed,
$$H_0(\mathcal{C}) := \ker \partial_0 / \operatorname{im} \partial_1 = \mathbf{Z}^3 / \operatorname{im} A = \langle [\mathbf{q}_2] \rangle \oplus \mathbf{Z}[\mathbf{q}_3] \cong \mathbf{Z}_2 \oplus \mathbf{Z},$$
where $\mathbf{q}_1, \mathbf{q}_2, \mathbf{q}_3$ are the columns of Q. Next, by Proposition 3.9.
$$H_1(\mathcal{C}) =: \ker \partial_1 / \operatorname{im} \partial_2 = \ker A = R(\ker B).$$
Since $\ker B$ is generated by $\mathbf{e}_3$, $R(\ker B)$ is generated by the last column $\mathbf{r}_3$ of R. Thus
$$H_1(\mathcal{C}) = \mathbf{Z}\mathbf{r}_3 \cong \mathbf{Z}.$$

We shall now formulate a general algorithm for finding the homology group of any finitely generated free complex $\mathcal{C} = \{C_k, \partial\}$. We begin by assuming that for each C_k we are given a particular basis that we refer to as the canonical basis. This permits us, via the coordinate isomorphism, to identify C_k with $\mathbf{Z}^{p_k}$ for some p_k and to identify the boundary homomorphisms $\partial_k : C_k \to C_{k-1}$ with matrices $D_k := A_{\partial_k}$.

We assume that these matrices are stored in an array of matrices D, constituting the input of the following algorithm.

Algorithm 3.64 Homology group of a chain complex
    function homologyGroupOfChainComplex(array[0 :] of matrix D)
    array[−1 :] of matrix V, W;
    for k := 0 to lastIndex(D) do
        (W[k], V[k − 1]) := kernelImage(D[k]);
    endfor;
    V[lastIndex(D)] := 0;
    array[0 :] of list H;
    for k := 0 to lastIndex(D) do
        H[k] := quotientGroup(W[k], V[k]);
    endfor;
    return H;

Theorem 3.65 *Assume $\mathcal{C} = (C_k, \partial_k)$ is a chain complex and algorithm homologyGroupOfChainComplex is called with an array D such that D[k] is the matrix of the boundary operator $\partial_k : C_k \to C_{k-1}$ in some fixed bases in C_k and C_{k-1}. Then it returns an array H such that H[k] represents the kth homology group of $\mathcal{C}$ in the sense of Theorem 3.59.*

Proof. This is a straightforward consequence of the definition of the homology group, Theorems 3.43 and 3.59. □

Exercise ───

3.15 For each matrix A in Exercise 3.10, find the homology groups (with identification of generators) of the abstract chain complex $\mathcal{C}$ defined the same way as in Example 3.63, whose only nontrivial boundary map is $\partial_1 = A$.

3.6 Computing Homology of Cubical Sets

Algorithm 3.64 computes the homology groups of an abstract chain complex, and we would like to make use of it to compute the homology groups of a cubical set. What is needed is an algorithm that, given a cubical set, computes the associated cubical chain complex, that is the generators of the groups of chains, and the boundary maps.

The first step is to decide on the data structures needed in such an algorithm. For an elementary interval we use

typedef interval := **hash**{endpoint} **of int**

where endpoint is defined by

typedef endpoint := (left, right)

The representation of an elementary cube in R^d is given by an array of d elementary intervals.

typedef cube := **array**[1 :] **of** interval

A cubical set X is stored as an array of elementary cubes in $\mathcal{K}_{\max}(X)$.

typedef cubicalSet := **array**[1 :] **of** cube

A chain $c \in C_k(X)$ is represented as a hash whose keys are the elementary cubes in $\mathcal{K}_k(|c|)$ and the value of the hash on an elementary cube $Q \in \mathcal{K}_k(|c|)$ is just $c(Q)$. Of course, in the hash we store only those elementary cubes for which the chain has a nonzero value.

typedef chain := **hash**{cube} **of int**

Finally, the boundary map $\partial_k : C_k(X) \to C_{k-1}(X)$ is stored as a hash whose keys are the elementary cubes in $\mathcal{K}_k(X)$ and the value of the hash on an elementary cube $Q \in \mathcal{K}_k(X)$ is the chain $\partial_k(\widehat{Q})$ stored again as a hash.

typedef boundaryMap := **hash**{cube} **of** chain

To be able to switch easily between a hash representation and a vector representation of a chain, we need two algorithms that implement the coordinate isomorphism and its inverse.

Our first algorithm computes the vector of canonical coordinate of a chain.

Algorithm 3.66 Finding the coordinate vector of a chain
    function canonicalCoordinates(**chain** c, **array**[1 :] **of cube** K)
    vector v;
    for i := 1 **to** lastIndex(K) **do**
      if defined(c{K[i]}) **then**
        v[i] := c{K[i]};
      else
        v[i] := 0;
      endif;
    endfor;
    return v;

The following proposition is straightforward.

3.6 Computing Homology of Cubical Sets   135

Proposition 3.67 *Assume Algorithm 3.66 is called with K containing an array of all elementary cubes in $\mathcal{K}_k(X)$ and c containing a hash representation of a chain $c \in C_k(X)$. Then it returns an array, v, of all coefficients of c in the canonical basis $\widehat{\mathcal{K}}_k(X)$, that is the coordinate vector of c in this basis.*

The next algorithm does the inverse: Given canonical coordinates of a chain, it computes the chain itself.

Algorithm 3.68 Reconstructing a chain from its canonical coordinates
    function chainFromCanonicalCoordinates(
                vector v, array[1 :] **of** cube K)
    chain c := ();
    for i := 1 **to** lastIndex(K) **do**
      if v[i] $\neq$ 0 **then**
        c{K[i]} := v[i];
      endif;
    endfor;
    return c;

The following proposition is straightforward.

Proposition 3.69 *Assume Algorithm 3.68 is called with K containing an array of all elementary cubes in $\mathcal{K}_k(X)$ and v containing the vector of canonical coordinates of a chain $c \in C_k(X)$. Then it returns the hash representation of c.*

Our next task is to compute the chain groups $C_k(X)$. Since these are free groups, we can represent them as lists of generators. Therefore, we need to compute the families $\mathcal{K}_k(X)$. We will first present an algorithm that computes the primary faces of an elementary cube. In the case of an elementary interval, $I = [k, k+1]$, these are just the two degenerate intervals $[k]$ and $[k+1]$.

Algorithm 3.70 Primary faces of an elementary cube
    function primaryFaces(cube Q)
    set of cube L := $\emptyset$;
    for i := 1 **to** lastIndex(Q) **do**
      if Q[i]{left} $\neq$ Q[i]{right} **then**
        R := Q;
        R[i]{left} := R[i]{right} := Q[i]{left};
        L := **join**(L, R);
        R[i]{left} := R[i]{right} := Q[i]{right};
        L := **join**(L, R);
      endif;
    endfor;
    return L;

It is straightforward to check that, given an elementary cube as input, the algorithm returns a list of all its primary faces. The next algorithm accepts as input a cubical set represented as a list of its maximal faces and uses Algorithm 3.70 to produce the generators of $C_k(X)$. It uses an algorithm dim, which, given an elementary cube, returns the dimension of this cube (see Exercise 3.16).

Algorithm 3.71 The groups of cubical chains of a cubical set
```
    function cubicalChainGroups(cubicalSet K)
    cube Q;
    array[ −1 : ] of list of cube E;
    while K ≠ ∅ do
        (Q, K) := cutFirst(K);
        k := dim(Q);
        L := primaryFaces(Q);
        K := union(K, L);
        E[k − 1] := union(E[k − 1], L);
        E[k] := join(E[k], Q);
    endwhile ;
    return E;
```

Theorem 3.72 *Let X be a cubical set. Assume Algorithm 3.71 is called with K containing a list of elementary cubes in $\mathcal{K}_{\max}(X)$. Then it returns an array E such that for $k = 0, 1, 2, \ldots, \dim(X)$ the element E[k] is a list that contains all elements of $\mathcal{K}_k(X)$, namely the generators of the group $C_k(X)$.*

Proof. We will show first that the algorithm always stops. Let $n := \dim X$ and, for a family $\mathcal{X} \subset \mathcal{K}(X)$, let

$$p(\mathcal{X}) := (p_n(\mathcal{X}), p_{n-1}(\mathcal{X}), \ldots, p_0(\mathcal{X})),$$

where

$$p_i(\mathcal{X}) := \operatorname{card} \{Q \in \mathcal{X} \mid \dim Q = i\}.$$

Let $\mathcal{K}_j$ denote the family of cubes represented by the value of the K variable on entering the jth pass of the **while** loop. Let Q_j denote the elementary cube removed from $\mathcal{K}_j$ on the call to **cutFirst** and put $m_j := \dim Q_j$. The family $\mathcal{K}_{j+1}$ differs from $\mathcal{K}_j$ by one m_j-dimensional cube removed from $\mathcal{K}_j$ and several $(m_j - 1)$-dimensional cubes added to it. This means that, in lexicographical order,

$$p(\mathcal{K}_{j+1}) < p(\mathcal{K}_j).$$

Since all vectors $p(\mathcal{K}_j)$ have only nonnegative entries, the sequence $\{p(\mathcal{K}_j)\}$ cannot be infinite and the algorithm must stop.

It remains to prove that the kth element of the returned array E contains exactly the elementary cubes in $\mathcal{K}_k(X)$. Obviously, it contains nothing else.

3.6 Computing Homology of Cubical Sets

To show that it contains all such elementary cubes, take $Q \in \mathcal{K}_k(X)$. Then we can choose a sequence of elementary cubes

$$Q_0 \supset Q_1 \supset Q_2 \supset \ldots \supset Q_r = Q$$

such that $Q_0 \in \mathcal{K}_{\max}(X)$ and Q_{i+1} is a primary face of Q_i. We will show by induction that Q_i is in the list $\mathrm{E}[\dim Q + r - i]$. Since $Q_0 \in \mathcal{K}_{\max}(X)$, it is added to $\mathrm{E}[\dim Q_0] = \mathrm{E}[\dim Q + r]$ on the pass of the **while** loop in which Q_0 is removed from K. Now assume that Q_j is in the list $\mathrm{E}[\dim Q + r - j]$ for some $j < r$. Then Q_{j+1} is added to $\mathrm{E}[\dim Q_{j+1}] = \mathrm{E}[\dim Q + r - (j+1)]$ on the pass of the **while** loop in which Q_j is removed from K. This proves that $Q = Q_r$ is in the list $\mathrm{E}[\dim Q]$. □

The next thing we need is an algorithm computing the matrix of the boundary map in the canonical bases. First we present an algorithm that computes the boundary of an elementary chain.

Algorithm 3.73 The boundary operator of an elementary cube
```
function boundaryOperator(cube Q)
sgn := 1;
chain c := ();
for i := 1 to lastIndex(Q) do
    if Q[i]{left} ≠ Q[i]{right} then
        R := Q;
        R[i]{left} := R[i]{right} := Q[i]{left};
        c{R} := -sgn;
        R[i]{left} := R[i]{right} := Q[i]{right};
        c{R} := sgn;
        sgn := -sgn;
    endif;
endfor;
return c;
```

Theorem 3.74 *Assume Algorithm 3.73 is called with the representation Q of an elementary cube Q. Then it returns a hash c representing the chain $\partial \widehat{Q}$.*

Proof. The conclusion follows easily from Proposition 2.36. □

Using Algorithm 3.73 we obtain an algorithm for computing the matrix of the boundary map in a cubical chain complex. Its input is a cubical chain complex represented by the generators of the cubical chain groups and stored in the following data structure.

    typedef cubicalChainComplex = **array**[0 :] **of array**[1 :] **of** cube;

Algorithm 3.75 The matrix of the boundary operator
```
function boundaryOperatorMatrix(cubicalChainComplex E)
array[ 0 : ] of matrix D;
for k := 0 to lastIndex(E) do
```

```
        m := lastIndex(E[k − 1]);
        for j := 1 to lastIndex(E[k]) do
            c := boundaryOperator(E[k][j])
            D[k][1 : m, j] := canonicalCoordinates(c, E[k − 1]);
        endfor;
    endfor;
    return D;
```

The proof of the following proposition is straightforward.

Proposition 3.76 *Let X be a cubical set. Assume Algorithm 3.75 is called with an array E such that E[k] is a list that contains all elements of $\mathcal{K}_k(X)$, that is, the generators of the group $C_k(X)$. Then it returns an array D such that D[k] is the matrix of $\partial_k^X : C_k(X) \to C_{k-1}(X)$ in the canonical bases.*

We are now almost ready to present the algorithm that, given a cubical set, computes its homology. It is sufficient to combine Algorithms 3.71, 3.75, and 3.64. However, the output of Algorithm 3.64 gives the information about homology only in terms of the coordinates and we would prefer to write down exactly the cycles generating the homology groups. To this end we need the following algorithm, which will translate the output of Algorithm 3.64 to concrete chains.

Algorithm 3.77 Generators of homology
```
    function generatorsOfHomology(
                array[ 0 : ] of list H, cubicalChainComplex E)
    for k := 1 to lastIndex(H) do
        m := lastIndex(E[k]);
        (U, B, s, t) := H[k];
        array[ 0 : ] hash{generators, orders} HG;
        for j := s + 1 to lastIndex(E[k]) do
            if j ≤ t then order := "infinity";
            else order := B[j, j] endif;
            c := chainFromCanonicalCoordinates(U[1:m, j], E[k]);
            HG[k]{generators}[j − s] := c;
            HG[k]{orders}[j − s] := order;
        endfor;
    endfor;
    return HG;
```

Finally, we are in the position to state an algorithm that computes the homology of *any* cubical set.

Algorithm 3.78 Cubical homology algorithm
```
    function homology(cubicalSet K)
    E := cubicalChainGroups(K);
```

```
D := boundaryOperatorMatrix(E);
H := homologyGroupOfChainComplex(D);
H := generatorsOfHomology(H, E);
return H;
```

Theorem 3.79 *Let X be a cubical set. Assume Algorithm 3.78 is called with K containing a list of elementary cubes in $\mathcal{K}_{\max}(X)$. Then it returns an array of hashes H for which H[k]{generators} is a list $(c_1, c_2, \ldots, c_n)$ and H[k]{orders} is a list $(b_1, b_2, \ldots, b_n)$ such that $c_i \in Z_k(X)$ is a cycle of order b_i and*

$$H_k(X) = \bigoplus_{i=1}^{n} \langle [c_i] \rangle. \tag{3.27}$$

Proof. The theorem follows from Proposition 3.76 and Theorems 3.72 and 3.65. □

In principle Algorithm 3.78 can perform the homology calculations when the cubical set is relatively small, for instance in the examples described in Section 1.1 and in Chapter 8. However, as mentioned in the introduction, while Theorem 3.79 clearly demonstrates that the homology of a cubical set is computable, it does not guarantee that Algorithm 3.78 is always efficient. In fact, this algorithm performs poorly on large complexes, an issue that is discussed, and at least partially resolved, in Chapter 4.

3.7 Preboundary of a Cycle—Algebraic Approach

In Chapter 6 we discuss the construction of chain selectors of multivalued maps which relies on the solution to the following problem:

> Given an acyclic cubical set X, an integer $k \geq 0$, and a k-dimensional reduced cycle $z \in C_k(X)$, construct a $(k+1)$-dimensional chain $c \in C_{k+1}$ such that
> $$\partial_{k+1} c = z. \tag{3.28}$$

Recall from Section 2.7 that the chain groups and boundary maps in all dimensions $k \geq 1$ of the augmented complex

$$\mathcal{C} := \tilde{\mathcal{C}}(X)$$

coincide with those of the cubical complex $\mathcal{C}(X)$, so the only difference between cycles and reduced cycles is in dimension 0. Any 0-dimensional chain in $C_0(X)$ is a cycle, whereas the reduced cycles are precisely those in the kernel of the augmentation map ϵ. Recall that the group of reduced cycles is generated by the differences $\widehat{P} - \widehat{Q}$ of elementary 0-chains in $C_0(X)$.

Observe that acyclicity of X means that $H_*(\mathcal{C}) = \tilde{H}_k(X) = 0$ for all $k \geq 0$ and consequently $Z_k(\mathcal{C}) = B_k(\mathcal{C})$. Thus every k-cycle in $\mathcal{C}$ is a boundary, which implies that a solution to (3.28) exists (of course, it need not be unique). Any such solution will be called a *preboundary* of z.

In this section we use Theorem 3.55 to present a purely algebraic approach to the solution of this problem. In order to convert Eq. (3.28) to a matrix equation, let us rewrite it in the form

$$(\xi_k \partial \xi_{k+1}^{-1})(\xi_{k+1} c) = \xi_k z, \qquad (3.29)$$

where $\xi_k : C_k(X) \to \mathbf{Z}^{m_k}$ stands for the coordinate isomorphism and m_k is the number of elementary k-dimensional cubes in X. Letting

$$D := A_{\xi_k \partial \xi_{k+1}^{-1}}, \quad \mathbf{x} := \xi_{k+1} c, \quad \text{and} \quad \mathbf{y} := \xi_k z,$$

we obtain the equation

$$D\mathbf{x} = \mathbf{y}.$$

Combining Algorithms 3.66 and 3.68 with Algorithms 3.54 and 3.75, we obtain a solution to the preboundary problem in the form of the following algorithm. Note that we assume that the function dim accepts as its argument a variable of type chain and returns the dimension of the chain stored in this variable (see Exercise 3.17).

Algorithm 3.80 Algebraic preboundary algorithm
    function preBoundary(chain z, cubicalSet X)
    if z = () then return (); endif;
    k := dim(z);
    E := cubicalChainGroups(X);
    D := boundaryOperatorMatrix(E);
    y := canonicalCoordinates(z, E[k]);
    x := Solve(D, y);
    return chainFromCanonicalCoordinates(x, E[k + 1]);

Theorem 3.81 *Assume Algorithm 3.80 is called with X containing a list representation of an acyclic cubical set X and z containing a hash representation of a reduced cycle $z \in \tilde{Z}_k(X)$ for some $k \geq 1$. Then it returns the hash representation of a preboundary of z, that is, a chain $c \in C_{k+1}(X)$ such that $\partial c = z$.*

Proof. The theorem follows immediately from Propositions 3.67, 3.69, and 3.76 and Theorem 3.55. □

Note that the above algorithm could be easily generalized to cycles of any abstract finitely generated chain complex $\mathcal{C}$ that is acyclic, such as $H_*(\mathcal{C}) = 0$. Although the algebraic methods are very general and theoretically correct, in practice they may lead to unacceptably expensive computations.

In Chapter 7 we present a more efficient algorithm, which relies on the geometric structure of cubical chain complexes.

Exercises ───

3.16 Write an algorithm that, given an elementary cube represented as an array in data structure `cube`, returns the dimension of the cube.

3.17 Write an algorithm that, given a chain represented as a hash in data structure `chain`, returns the dimension of the chain.
Hint: Let `c` be a variable of type `chain`. Then `c` is a hash. If `c` contains a correct chain, then all keys of `c` are elementary cubes of the same dimension and the dimension of these cubes is the dimension of the chain.

3.8 Bibliographical Remarks

The algorithms presented in Sections 3.1 through 3.5 are based on material from standard introductory courses in abstract algebra (see, e.g., [47]). The Smith normal form Algorithm 3.51 is based on the outline of the construction given in the first chapter of [69]. The matrix algebra in integer coefficients is an important tool in operations research and optimization; see, for example, [31, Section 7.4] for another presentation of Smith normal form. The algorithms of Sections 3.6 and 3.7 represent new material.

4
Chain Maps and Reduction Algorithms

Continuous maps are used to compare topological spaces, linear maps play the same role for vector spaces, and homomorphisms are the tool for comparing abelian groups. It is, therefore, natural to introduce *chain maps*, which are maps between chain complexes. This notion permits us to compare different chain complexes in the same fashion that homomorphisms allow us to compare abelian groups. A homomorphism of abelian groups is required to preserve group addition. In a chain complex we have an additional operation: taking the boundary of a chain. Therefore, the definition of a chain map will also require that it preserves boundaries.

As we shall see, this gives a new insight into the concept of elementary collapse introduced in Chapter 2 and leads to a more general class of elementary reductions. This, in turn, leads to additional algorithms for the computation of homology groups. However, the full picture of how topology, algebra, and, also, analysis tie together will start to emerge in the next two chapters, where we discuss continuous maps $f : X \to Y$ between cubical sets X, Y and their effect on homology groups. Chain maps become a tool in defining a map $f_* : H_*(X) \to H_*(Y)$ induced by f in homology.

4.1 Chain Maps

As indicated above, we want to introduce homomorphisms between chain complexes. Since our goal is to compare the homologies of these complexes, it is important that these homomorphisms induce maps on homology groups. Our discussion, at this level, is purely algebraic so we can begin with two abstract chain complexes $\mathcal{C} = \{C_k, \partial_k\}$ and $\mathcal{C}' = \{C'_k, \partial'_k\}$. For each $k \in \mathbf{Z}$, let $\varphi_k : C_k \to C'_k$ be a linear map. Since φ_k is intended to induce a map on homology, $\varphi_{k*} : H_k(\mathcal{C}) \to H_k(\mathcal{C}')$, an obvious question we should ask is: Are there any necessary conditions that φ_k must satisfy?

To answer this, we need to begin with the fundamental question of how, given φ_k, can we define φ_{k*}? There does not appear to be much choice. Elements of homology are equivalence classes of cycles. Thus we should focus our attention on cycles rather than on arbitrary chains and consider φ_k applied to cycles. Let us choose $z \in Z_k$, which gives rise to $[z] \in H_k(\mathcal{C})$. If φ_k is to induce a map φ_{k*}, then there must be a relationship between $\varphi_k(z)$, which is an element of C'_k, and $\varphi_{k*}([z])$, which is an element of $H_k(\mathcal{C}')$. On this level of generality the only option seems to be the following: $\varphi_{k*}([z])$ must be the homology class generated by $\varphi_k(z)$. Formally, this can be written as

$$\varphi_{k*}([z]) := [\varphi_k(z)]. \tag{4.1}$$

But this formula makes sense only if $\varphi_k(z)$ actually generates a homology class, that is, if $\varphi_k(z)$ is a cycle. More generally, φ_k must map cycles to cycles.

This observation leads to an easily expressed restriction on the φ_k. By definition, $z \in Z_k$ implies that $\partial_k z = 0$. In order for (4.1) to make sense, we need $\varphi_k(z) \in Z'_k$, or equivalently $\partial'_k \varphi_k(z) = 0$. Since φ_{k-1} is a linear map, $\varphi_{k-1}(\partial_k z) = \varphi(0) = 0$. Combining these statements gives

$$\partial'_k \varphi_k(z) = 0 = \varphi_{k-1}(\partial_k z).$$

The homology groups $H_k(\mathcal{C})$ and $H_k(\mathcal{C}')$ are abelian groups and therefore we should insist that $\varphi_{k*} : H_k(\mathcal{C}) \to H_k(\mathcal{C}')$ be an abelian group homomorphism, namely a linear map. Thus, at the very least, $\varphi_{k*}(0) = 0$. But $0 \in H_k(\mathcal{C})$ and $0 \in H_k(\mathcal{C}')$ are really the equivalence classes consisting of boundaries. Thus φ_k must map boundaries to boundaries.

Again, this observation can be expressed in terms of a simple formula. Let $b \in B_k$ (i.e., $b = \partial_{k+1} c$ for some chain $c \in C_{k+1}$). Since boundaries must map to boundaries, $\varphi_k(b) \in B'_k$. But what chain is $\varphi_k(b)$ to be the boundary of? The only one we have at our disposal is c, so the easiest constraint is to ask that $\varphi_k(b) = \partial'_{k+1}(\varphi_{k+1} c)$. But observe that this is equivalent to

$$\partial'_{k+1} \varphi_{k+1}(c) = \varphi_k(b) = \varphi_k(\partial_{k+1} c).$$

As one might have guessed from the time spent discussing it, the relationship

$$\partial' \varphi = \varphi \partial$$

is extremely important, and thus we make the following formal definition.

Definition 4.1 Let $\mathcal{C} = \{C_k, \partial_k\}$ and $\mathcal{C}' = \{C'_k, \partial'_k\}$ be chain complexes. A sequence of homomorphisms $\varphi_k : C_k \to C'_k$ is a *chain map* if, for every $k \in \mathbf{Z}$,

$$\partial'_k \varphi_k = \varphi_{k-1} \partial_k. \tag{4.2}$$

We will use the notation $\varphi : \mathcal{C} \to \mathcal{C}'$ to represent the collection of homomorphisms $\{\varphi_k : C_k \to C'_k\}$.

A chain map $\varphi : \mathcal{C} \to \mathcal{C}'$ is called a *chain isomorphism* if for each k the homomorphism φ_k is an isomorphism. The collection of inverse homomorphisms

$\varphi_k^{-1} : C_k' \to C_k$ is denoted by $\varphi^{-1} : \mathcal{C}' \to \mathcal{C}$. We leave it to the reader to check that φ^{-1} is a chain map.

Another way to describe an equality such as (4.2) is through the language of *commutative diagrams*. More precisely, to say that the diagram

$$\begin{array}{ccc} C_k & \xrightarrow{\varphi_k} & C_k' \\ \downarrow{\partial_k} & & \downarrow{\partial_k'} \\ C_{k-1} & \xrightarrow{\varphi_{k-1}} & C_{k-1}' \end{array}$$

commutes is equivalent to saying that $\partial_k' \varphi_k = \varphi_{k-1} \partial_k$.

The following result is essentially a restatement of the introductory comments.

Proposition 4.2 *If $\varphi : \mathcal{C} \to \mathcal{C}'$ is a chain map, then*

$$\varphi_k(Z_k) \subset Z_k' := \ker \partial_k'$$

and

$$\varphi_k(B_k) \subset B_k' := \operatorname{im} \partial_{k+1}'$$

for all $k \in \mathbf{Z}$.

Definition 4.3 Let $\varphi : \mathcal{C} \to \mathcal{C}'$ be a chain map. Define the homomorphism of groups $\varphi_{k*} : H_k(\mathcal{C}) \to H_k(\mathcal{C}')$ by

$$\varphi_{k*}([z]) := [\varphi_k(z)],$$

where $z \in Z_k$, and put $\varphi_* := \{\varphi_{k*}\} : H_*(\mathcal{C}) \to H_*(\mathcal{C}')$.

That this homomorphism is well defined follows from Proposition 4.2. More precisely, the elements of $H_k(\mathcal{C})$ are the equivalence classes $[z]$ of elements $z \in Z_k$. We need to show that the definition does not depend on the choice of z. By Proposition 4.2, $\varphi_k(z) \in Z_k'$, and so $[\varphi_k(z)] \in H_k(\mathcal{C}')$. Now assume that $[z] = [z']$ for some $z' \in Z_k$. Then $z' = z + b$, where $b \in B_k$. Since φ is a chain map, $\varphi_k b \in B_k(\mathcal{C}')$, and hence,

$$\varphi_{k*}[z'] = [\varphi_k z'] = [\varphi_k(z+b)] = [\varphi_k z + \varphi_k b] = [\varphi_k z] = \varphi_{k*}[z].$$

We now present some simple, but important, examples of chain maps.

Example 4.4 Consider a chain complex $\mathcal{C} = \{C_k, \partial_k\}$ with a chain subcomplex $\mathcal{C}' = \{C_k', \partial_k\}$. Let $\iota_k : C_k' \to C_k$ be the inclusion map given by

$$\iota_k c' = c'$$

for every $c' \in C_k'$. Since

146   4 Chain Maps and Reduction Algorithms

$$\partial_k \iota_k c' = \partial_k c' = \iota_{k-1} \partial_k c',$$

ι is a chain map. Therefore,

$$\iota_* : H_*(\mathcal{C}') \to H_*(\mathcal{C})$$

is defined.

On the level of abstract chain complexes this example may appear trite. However, consider the case where $A \subset X$ are cubical sets in $\mathbf{R}^d$. Then the inclusion map $i : A \hookrightarrow X$ that maps elementary cubes to elementary cubes by $i(Q) = Q$ gives rise to an inclusion map of the chain complexes, $\iota : \mathcal{C}(A) \hookrightarrow \mathcal{C}(X)$, defined by

$$\iota_k \widehat{Q} = \widehat{Q} \quad \text{for all } Q \in \mathcal{K}_k(A).$$

Thus, beginning with a continuous map between two topological spaces, $i : A \hookrightarrow X$, we have arrived at a map between their homology groups $\iota_* : H_*(A) \to H_*(X)$.

Given this rapid success, the reader might wonder at this point why we wait until Chapter 6 to formalize the process of passing from a continuous map to a map on homology. The reason is simple. The inclusion map sends cubes to cubes; however, in the class of continuous maps this is an extremely rare property.

The reader might notice that ι is a monomorphism, and ask whether or not $\iota_* : H_*(A) \to H_*(X)$ must also be a monomorphism. This is false because a nontrivial cycle in $\mathcal{C}(A)$ might become a boundary in $\mathcal{C}(X)$. For example, $\Gamma^1 \hookrightarrow [0,1]^2$ but $H_1(\Gamma^1) \cong \mathbf{Z}$ and $H_1([0,1]^2) = 0$. Therefore, ι_* is not a monomorphism in this case.

We leave the proof of the following proposition as an exercise.

Proposition 4.5 *Let $\mathcal{C} = \{C_k, \partial_k\}$ and $\mathcal{C}' = \{C'_k, \partial'_k\}$ be chain complexes. If $\varphi : \mathcal{C} \to \mathcal{C}'$ is a chain isomorphism, then*

$$\varphi_* : H_*(\mathcal{C}) \to H_*(\mathcal{C}')$$

is an isomorphism.

Example 4.6 Let $X \subset \mathbf{R}^{d-1}$ be a cubical set and $Y := \{0\} \times X \subset \mathbf{R}^d$. Define $\kappa : \mathcal{C}(X) \to \mathcal{C}(Y)$ by

$$\kappa_k(c) := \widehat{[0]} \diamond c.$$

Since any chain in Y must be of the form $\widehat{[0]} \diamond c$, where $c \in \mathcal{C}(X)$, κ is invertible with the inverse $\kappa^{-1}(\widehat{[0]} \diamond c) = c$. Since

$$\partial_k \kappa_k(c) = \widehat{[0]} \diamond \partial_k c = \kappa_{k-1} \partial_k(c),$$

κ is a chain map. Thus, by Proposition 4.5,

$$\kappa_* : H_*(X) \to H_*(Y)$$

is an isomorphism.

Observe that, as in Example 4.4, the chain map κ can be viewed as being generated by a continuous map $j : X \to Y$ given by

$$j(x) = (0, x).$$

We can easily generalize Example 4.6 as follows. Again, let $X \subset \mathbf{R}^{d-1}$. Define $j : X \to \mathbf{R}^d$ by

$$j(x_1, \ldots, x_{d-1}) = (x_1, \ldots, x_l, m, x_{l+1}, \ldots, x_{d-1}).$$

If $m \in \mathbf{Z}$, then $j(X)$ is a cubical subset of $\mathbf{R}^d$. It is left to the reader to check that it gives rise to a chain isomorphism between $C(X)$ and $C(j(X))$.

Example 4.7 In the previous example we have taken a cubical set in $\mathbf{R}^{d-1}$ and immersed it in the higher-dimensional space $\mathbf{R}^d$. We can also consider translating cubical sets within the same space. To be more precise, let $X \subset \mathbf{R}^d$ be cubical and let $(m_1, \ldots, m_d) \in \mathbf{Z}^d$. Define $t : X \to \mathbf{R}^d$ by

$$t(x_1, \ldots, x_d) = (x_1 + m_1, \ldots, x_d + m_d).$$

Observe that $t(X) \subset \mathbf{R}^d$ is a cubical set.

To obtain a map on homology we need to generate a chain map. Let $Q \in \mathcal{K}_k(X)$. Then Q is the product of elementary intervals,

$$Q = [a_1, b_1] \times [a_2, b_2] \times \cdots \times [a_d, b_d],$$

where $b_i - a_i = 1$ or $b_i = a_i$. Define $\tau : \mathcal{C}(X) \to \mathcal{C}(t(X))$ by $\tau_k(\widehat{Q}) := \widehat{P}$, where

$$P := [a_1 + m_1, b_1 + m_1] \times [a_2 + m_2, b_2 + m_2] \times \cdots \times [a_d + m_d, b_d + m_d].$$

It is left to the reader to check that τ is a chain isomorphism and therefore, by Proposition 4.5, to conclude that

$$\tau_* : H_*(X) \to H_*(t(X))$$

is an isomorphism.

In the previous examples, there is a fairly transparent correspondence between the continuous map and the associated chain map. We now turn to a fundamental example where this is not the case.

Example 4.8 Consider the elementary cube $Q = [0, 1]^d$ and the projection $p : Q \to Q$ given by

$$p(x_1, x_2, x_3, \ldots, x_d) := (0, x_2, x_3, \ldots, x_d).$$

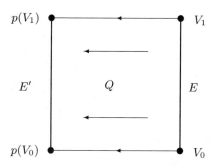

Fig. 4.1. Projection of $Q = [0,1]^2$ and the right edge $E = [1] \times [0,1]$ to the left edge $Q' = E' = [0] \times [0,1]$ described in Example 4.8. Horizontal edges project to vertices so corresponding chains project to 0.

This is illustrated in Figure 4.1 for the case $d = 2$. We want to associate with p a chain map $\pi : \mathcal{C}(Q) \to \mathcal{C}(Q)$ and begin by mimicking the procedure of the previous examples. Any face E of Q can be written as

$$E = E_1 \times E_2 \times E_3 \times \cdots \times E_d,$$

where E_i can be $[0,1]$, $[0]$, or $[1]$. The image of E under p is the elementary cube

$$E' := p(E) = [0] \times E_2 \times E_3 \times \cdots \times E_d.$$

If we were to blindly follow what has been done before, it would be natural to define $\pi \widehat{E} := \widehat{p(E)}$. However, by definition, $\pi_k : C_k(Q) \to C_k(Q)$. But, if $E \in \mathcal{K}_k(Q)$ and $E_1 = [0,1]$, then $p(E) \in \mathcal{K}_{k-1}(Q)$. This violates the constraint that k-dimensional chains must be mapped to k-dimensional chains. Therefore, if a nondegenerate interval is projected to a point, the only natural choice for the value of the associated chain map is the trivial chain 0. Thus we define

$$\pi_k(\widehat{E}) := \begin{cases} \widehat{E'} & \text{if } E_1 = [0] \text{ or } E_1 = [1], \\ 0 & \text{otherwise.} \end{cases} \quad (4.3)$$

We will show that π is a chain map. Let $E \in \mathcal{K}_k(Q)$ and decompose it as $E = E_1 \times P$, where $P := E_2 \times E_3 \times \ldots \times E_d$. Using Eq. (2.7), we obtain

$$\partial \widehat{E} = \partial \widehat{E_1} \diamond \widehat{P} + (-1)^{\dim(E_1)} \widehat{E_1} \diamond \partial \widehat{P}.$$

If $E_1 = [0]$ or $E_1 = [1]$,

$$\pi \partial \widehat{E} = \pi \left(\partial \widehat{E_1} \diamond \widehat{P} + \widehat{E_1} \diamond \partial \widehat{P} \right)$$
$$= \pi(\widehat{E_1} \diamond \partial \widehat{P})$$
$$= \widehat{[0]} \diamond \partial \widehat{P}$$

and, consequently,

$$\partial \pi \widehat{E} = \partial(\widehat{[0]} \diamond \widehat{P}) = \widehat{[0]} \diamond \partial \widehat{P} = \pi \partial \widehat{E}.$$

If $E_1 = [0,1]$, then $\pi \widehat{E} = 0$, by definition, so $\partial \pi \widehat{E} = 0$. On the other hand,

$$\begin{aligned}\pi \partial \widehat{E} &= \pi((\widehat{[1]} - \widehat{[0]}) \diamond \widehat{P} - \widehat{E_1} \diamond \widehat{\partial P}) \\ &= \widehat{[0]} \diamond \widehat{P} - \widehat{[0]} \diamond \widehat{P} - 0 \\ &= 0.\end{aligned}$$

Thus $\pi : \mathcal{C}(Q) \to \mathcal{C}(Q)$ is a chain map. However, it has additional structure. Observe that on the level of chains it is a *projection* (that is, $\pi_k^2 = \pi_k$). To see this, observe that if $E \in \mathcal{K}_k(p(Q))$, then $E = [0] \times E_2 \times \cdots \times E_d$, and hence $\pi_k(\widehat{E}) = \widehat{E}$.

There is another way to interpret the map p. Let $Q' := \{0\} \times [0,1]^{d-1}$. Then $Q' = p(Q)$. For reasons that become transparent later, we have been careful to distinguish p from the projection $p' : Q \to Q'$ given by the same formula but with Q' as the target space. Similarly, we can define $\pi' : \mathcal{C}(Q) \to \mathcal{C}(Q')$ by the same formula as π, with the only difference being that $\mathcal{C}(Q')$ is the range. The same argument, as above, shows that π' is a chain projection.

The relation between the two projection maps is $p = ip'$, where $i : Q' \hookrightarrow Q$ is the inclusion map. This brings us back to Example 4.6 and the reader can check that on the chain level

$$\pi_k = \iota_k \circ \pi'_k.$$

Exercises

4.1 Suppose that a chain map $\varphi : \mathcal{C} \to \mathcal{C}'$ is a chain isomorphism, that is, for each k, the homomorphism φ_k is an isomorphism. Show that φ^{-1} is a chain map.

4.2 Prove Proposition 4.2.

4.3 Show that reflection of $\Gamma^{d-1} = \mathrm{bd}\,[0,1]^d$ about an axis in $\mathbf{R}^d$ gives rise to a chain map on $\mathcal{C}(\Gamma^{d-1})$ that, moreover, is a chain isomorphism.

4.2 Chain Homotopy

We now know that chain maps $\varphi, \psi : \mathcal{C} \to \mathcal{C}'$ generate homology maps $\varphi_*, \psi_* : H_*(\mathcal{C}) \to H_*(\mathcal{C}')$. It is natural to ask what conditions guarantee that $\varphi_* = \psi_*$. To motivate an answer, we return to the setting of cubical complexes. In particular, let X and Y be cubical sets and let $\varphi, \psi : \mathcal{C}(X) \to \mathcal{C}(Y)$.

We shall present a method of linking the two chain maps as restrictions of a chain map on a bigger space. Let us make two copies of X, namely, $\{0\} \times X$ and $\{1\} \times X$. Let $\iota^0 : \mathcal{C}(X) \to \mathcal{C}(\{0\} \times X)$ and $\iota^1 : \mathcal{C}(X) \to \mathcal{C}(\{1\} \times X)$ be the inclusion maps described in Example 4.6. Consider the cubical chain map $\chi : \mathcal{C}(\{0,1\} \times X) \to \mathcal{C}(Y)$ defined by

$$\chi_k(\widehat{[0] \times Q}) = \varphi_k(\widehat{Q}), \tag{4.4}$$
$$\chi_k(\widehat{[1] \times Q}) = \psi_k(\widehat{Q}), \tag{4.5}$$

for all $Q \in \mathcal{K}_k(X)$, and observe that $\varphi = \chi \circ \iota^0$ and $\psi = \chi \circ \iota^1$. Let us assume that χ can be extended to a chain map from $\mathcal{C}([0,1] \times X)$ to $\mathcal{C}(Y)$ that satisfies (4.4) and (4.5). For example, in the trivial case that $\psi = \varphi$ the desired extension can be obtained by setting $\chi_{k+1}(\widehat{[0,1] \times Q}) = 0$ for each $Q \in \mathcal{K}_k(X)$. We leave the verification that this is a chain map as an exercise.

Let us now make use of the fact that χ is a chain map, that is, that $\partial \chi = \chi \partial$;

$$\begin{aligned}
\partial_{k+1} \chi_{k+1}(\widehat{[0,1]} \diamond \widehat{Q}) &= \chi_k \partial_{k+1}(\widehat{[0,1]} \diamond \widehat{Q}) \\
&= \chi_k \left(\widehat{[1]} \diamond \widehat{Q} - \widehat{[0]} \diamond \widehat{Q} - \widehat{[0,1]} \diamond \partial_k \widehat{Q} \right) \\
&= \chi_k \left(\widehat{[1] \times Q} - \widehat{[0] \times Q} - \widehat{[0,1]} \diamond \partial_k \widehat{Q} \right) \\
&= \chi_k \left(\widehat{[1] \times Q} \right) - \chi_k \left(\widehat{[0] \times Q} \right) - \chi_k \left(\widehat{[0,1]} \diamond \partial_k \widehat{Q} \right).
\end{aligned}$$

Therefore, by (4.4) and (4.5),

$$\partial_{k+1} \chi_{k+1}(\widehat{[0,1]} \diamond \widehat{Q}) = \psi_k(\widehat{Q}) - \varphi_k(\widehat{Q}) - \chi_k \left(\widehat{[0,1]} \diamond \partial_k \widehat{Q} \right). \tag{4.6}$$

A geometric interpretation of Eq. (4.6) is indicated in Figure 4.2. If we let

$$D_k(\widehat{Q}) := \chi_{k+1}(\widehat{[0,1]} \diamond \widehat{Q}), \tag{4.7}$$

then (4.6) becomes

$$\partial_{k+1} D_k(\widehat{Q}) = \psi_k(\widehat{Q}) - \phi_k(\widehat{Q}) - D_{k-1}\left(\partial_k \widehat{Q}\right),$$

where $D_k : C_k(X) \to C_{k+1}(Y)$ is a homomorphism of groups. This expression is more commonly written as

$$\psi_k - \phi_k = \partial_{k+1} D_k + D_{k-1} \partial_k.$$

Now assume that this relationship is satisfied for some homomorphisms $D_k : C_k(X) \to C_{k+1}(Y)$ not necessarily given by (4.7) and apply it to a cycle $z \in C_k(X)$:

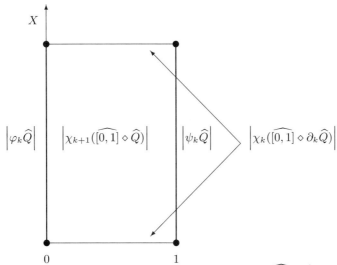

Fig. 4.2. Interpretation of Eq. (4.6). The boundary of $\chi_{k+1}(\widehat{[0,1]} \diamond \widehat{Q})$ is composed of three parts: The support of $\varphi_k(\widehat{Q})$ is viewed on the left; the support of $\psi_k(\widehat{Q})$ is viewed on the right; and the support of $\chi_k(\widehat{[0,1]} \diamond \partial_k \widehat{Q})$ is viewed as the lower and upper faces.

$$\psi_k(z) - \phi_k(z) = \partial_{k+1} D_k(z) + D_{k-1} \partial_k(z)$$
$$= \partial_{k+1} D_k(z) \in B_k(Y).$$

Hence, we see that the difference of the values of ψ and ϕ on any cycle is a boundary, and therefore, they induce the same map on homology. This observation leads to the following definition, which is presented in the more general setting of abstract chain complexes.

Definition 4.9 Let $\varphi, \psi : \mathcal{C} \to \mathcal{C}'$ be chain maps. A collection of group homomorphisms
$$D_k : C_k \to C'_{k+1}$$
is a *chain homotopy* between φ and ψ if, for all $k \in \mathbf{Z}$,
$$\partial'_{k+1} D_k + D_{k-1} \partial_k = \psi_k - \varphi_k. \tag{4.8}$$

We say that φ is *chain homotopic* to ψ if there exists a chain homotopy between φ and ψ.

The usefulness of chain homotopy may be summarized in the following, just-proved theorem.

Theorem 4.10 *If φ and ψ are chain homotopic, then $\varphi_* = \psi_*$.*

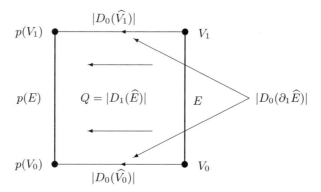

Fig. 4.3. Chain homotopy between the projection π and the identity on $\mathcal{C}(Q)$ in Example 4.11. $\widehat{E} - \pi\widehat{E}$ represents a part of $\partial \widehat{Q}$. $D_{k-1}\partial_k\widehat{E}$ represents the remaining lower and upper faces obtained as images of ∂E under the continuous deformation h.

Example 4.11 We will show that the chain projection $\pi : \mathcal{C}(Q) \to \mathcal{C}(Q)$, in Example 4.8, is chain homotopic to the identity map on $\mathcal{C}(Q)$. Any $E \in \mathcal{K}_k(Q)$ can be written, as before, as $E = E_1 \times P$, where E_1 is $[0]$, $[1]$, or $[0,1]$ (the last case is excluded if $k = 0$).

Define $D_k : C_k(Q) \to C_{k+1}(Q)$ by

$$D_k(\widehat{E}) := \begin{cases} \widehat{[0,1]} \diamond \widehat{P} & \text{if } E_1 = [1], \\ 0 & \text{if } E_1 = [0], \\ 0 & \text{if } E_1 = [0,1]. \end{cases}$$

There is a simple geometric interpretation of this definition. View the corresponding projection $p : Q \to Q$ discussed in Example 4.8 as the final stage $t = 1$ of the deformation $h : Q \times [0,1] \to Q$ given by the formula

$$h(x,t) := ((1-t)x_1, x_2, x_3, \ldots, x_d).$$

Recall that E_1 is $[1]$, $[0]$, or $[0,1]$. In the first case, when $E_1 = [1]$, the support of $D_k(\widehat{E})$ is precisely the image $h(E, [0,1])$ of E under the deformation and it represents the face through which E is projected to $E' = [0] \times P$. In the second case, when $E_1 = [0]$, the deformation does not move E, so there is no $(k+1)$-dimensional face involved. In the third case, when $E_1 = [0,1]$, the deformation contracts E_1 to its vertex $[0]$ so, as in the second case, no $(k+1)$-dimensional face is involved.

We need to verify that condition (4.8) holds for $\psi = \pi$ and $\varphi = \text{id}$. In other words, we must show that for every $E \in \mathcal{K}_k(Q)$

$$\partial_{k+1} D_k \widehat{E} + D_{k-1} \partial_k \widehat{E} = \widehat{E} - \pi \widehat{E}. \tag{4.9}$$

Consider the first case in which $E_1 = [1]$. Then

$$\begin{aligned}
\partial_{k+1} D_k \widehat{E} + D_{k-1} \partial_k \widehat{E} &= \partial_{k+1}(\widehat{[0,1]} \diamond \widehat{P}) + D_{k-1}(\widehat{[1]} \diamond \partial_k \widehat{P}) \\
&= \widehat{[1]} \diamond \widehat{P} - \widehat{[0]} \diamond \widehat{P} - \widehat{[0,1]} \diamond \partial_k \widehat{P} + \widehat{[0,1]} \diamond \partial_k \widehat{P} \\
&= \widehat{E} - \pi \widehat{E}.
\end{aligned}$$

Next let $E_1 = [0]$. Then each term in the left-hand side of Eq. (4.9) is zero by the definition of D_k and the right-hand side is zero because $\pi \widehat{E} = \widehat{E}$. Finally, let $E_1 = [0,1]$. Then $D_k \widehat{E} = 0$ so we get

$$\begin{aligned}
\partial_{k+1} D_k \widehat{E} + D_{k-1} \partial_k \widehat{E} &= 0 + D_{k-1}(\widehat{[1]} \diamond \widehat{P} - \widehat{[0]} \diamond \widehat{P} - \widehat{[0,1]} \diamond \partial_k \widehat{P}) \\
&= \widehat{[0,1]} \diamond \widehat{P} - 0 - 0 \\
&= \widehat{E} \\
&= \widehat{E} - \pi \widehat{E}
\end{aligned}$$

because $\pi \widehat{E} = 0$.

We have shown that chain maps induce maps on homology. We also have a condition under which different chain maps induce the same map on homology. We now turn to the composition of maps. The proof of the following proposition is left as an exercise.

Proposition 4.12 *Let $\mathcal{C}$, $\mathcal{C}'$, and $\mathcal{C}''$ be chain complexes. If $\varphi : \mathcal{C} \to \mathcal{C}'$ and $\psi : \mathcal{C}' \to \mathcal{C}''$ are chain maps, then $\psi\varphi : \mathcal{C} \to \mathcal{C}''$ is a chain map and*

$$(\psi\varphi)_* = \psi_* \varphi_*.$$

Eventually we will use homology maps to compare topological spaces. With this in mind it is natural to ask: When does a chain map induce an isomorphism in homology?

Definition 4.13 A chain map $\varphi : \mathcal{C} \to \mathcal{C}'$ is called a *chain equivalence* if there exists a chain map $\psi : \mathcal{C}' \to \mathcal{C}$ such that $\psi\varphi$ is chain homotopic to $\mathrm{id}_\mathcal{C}$ and $\varphi\psi$ is chain homotopic to $\mathrm{id}_{\mathcal{C}'}$.

The following proposition is an easy consequence of Theorem 4.10 and Proposition 4.12.

Proposition 4.14 *If $\varphi : \mathcal{C} \to \mathcal{C}'$ is a chain equivalence, then $\varphi_* : H_*(\mathcal{C}) \to H_*(\mathcal{C}')$ is an isomorphism.*

Example 4.15 We continue Examples 4.8 and 4.11 by showing that the projection $\pi' : \mathcal{C}(Q) \to \mathcal{C}(Q')$ and inclusion $\iota : \mathcal{C}(Q') \to \mathcal{C}(Q)$ are chain equivalences. This will imply that $H_*(Q) \cong H_*(Q')$.

Indeed, $\pi'\iota = \mathrm{id}_{\mathcal{C}(Q')}$. So, in particular, $\pi'\iota$ is chain homotopic to $\mathrm{id}_{\mathcal{C}(Q')}$. The reverse composition is $\iota\pi' = \pi$. But Example 4.11 shows that π is chain homotopic to $\mathrm{id}_{\mathcal{C}(Q)}$.

154   4 Chain Maps and Reduction Algorithms

From the above we immediately get another way of proving Theorem 2.76. Since we already have one proof, we leave some verifications as exercises.

Corollary 4.16 *The elementary cube $Q = [0,1]^d$ is acyclic.*

Proof. We argue by induction on d. If $d = 0$, Q is a single vertex so it is acyclic by definition. Suppose we know the result for $P = [0,1]^{d-1}$ and consider $Q = [0,1]^d$. By Example 4.15, $H_*(Q) \cong H_*(Q')$, where $Q' = [0] \times P$ is a face of Q. By Example 4.6, $H_*([0] \times P) \cong H_*(P)$, so the conclusion follows. □

Corollary 4.17 *All elementary cubes are acyclic.*

Proof. Let $Q = I_1 \times I_2 \times \cdots \times I_d$ be an elementary cube. Example 4.6 shows that inserting a degenerate interval on a given coordinate does not change homology. Similarly, homology remains the same when a degenerate interval is removed. Hence we may assume that all I_i are nondegenerate. Let $I_i = [a_i, a_i + 1]$. Consider the translation map $\tau : \mathcal{C}([0,1]^d) \to \mathcal{C}(Q)$ defined as follows. If $P \in \mathcal{K}_k([0,1]^d)$, then $P = P_1 \times P_2 \times \cdots \times P_d$, where the P_i are elementary intervals and exactly k of them are nondegenerate. For each k, τ_k is given by

$$\tau_k \widehat{P} := \widehat{(a_1 + P_1)} \diamond \widehat{(a_2 + P_2)} \diamond \cdots \diamond \widehat{(a_d + P_d)},$$

where $a_i + P_i$ is the translation of the interval P_i by a_i. This chain map is discussed in Example 4.7. As commented there, τ is an isomorphism of chain complexes, so it induces the isomorphism in homology and the conclusion follows from the previous corollary. □

Exercises _____

4.4 Consider a chain map, $\varphi : \mathcal{C}(\Gamma^1) \to \mathcal{C}(\Gamma^1)$, which one can think of as being generated by rotating Γ^1 by 90 degrees in the clockwise direction. More precisely, let $\varphi_0 : C_0(\Gamma^1) \to C_0(\Gamma^1)$ be given on the canonical generators by

$$\varphi_0(\widehat{[0] \times [0]}) = \widehat{[0] \times [1]},$$
$$\varphi_0(\widehat{[0] \times [1]}) = \widehat{[1] \times [1]},$$
$$\varphi_0(\widehat{[1] \times [1]}) = \widehat{[1] \times [0]},$$
$$\varphi_0(\widehat{[1] \times [0]}) = \widehat{[0] \times [0]},$$

and let $\varphi_1 : C_1(X) \to C_1(X)$ be given by

$$\varphi_1(\widehat{[0,1] \times [0]}) = -\widehat{[0] \times [0,1]},$$
$$\varphi_1(\widehat{[0] \times [0,1]}) = \widehat{[0,1] \times [1]},$$
$$\varphi_1(\widehat{[0,1] \times [1]}) = -\widehat{[1] \times [0,1]},$$
$$\varphi_1(\widehat{[1] \times [0,1]}) = \widehat{[0,1] \times [0]}.$$

Construct a chain homotopy between φ_* and the identity map on $\mathcal{C}(\Gamma^1)$.

4.5 Prove Proposition 4.12.

4.6 Prove Proposition 4.14.

4.7 Consider the cubical cylinder $X = [0,1] \times \Gamma^1 \subset \mathbf{R}^3$, where Γ^1 is the boundary of the unit square. Let $X' = [0] \times \Gamma^1$ be obtained from X by projecting to the plane given by $x_1 = 0$ as in Example 4.8.

(a) Define the corresponding chain projection $\pi : \mathcal{C}(X) \to \mathcal{C}(X)$ with the image $\mathcal{C}(X')$.
(b) Show that π is chain homotopic to the identity.
(c) Derive the conclusion that $H_*(X) \cong H_*(\Gamma^1)$.

4.3 Internal Elementary Reductions

In Chapter 3 we have presented a purely algebraic algorithm for computing the homology of a chain complex. Unfortunately, if the chain complex is large, then the run time of the algorithm may be very long. On the other hand, in Section 2.4 we have introduced the notion of elementary collapse, in part so that we can efficiently compute the homology of some special complexes. Using chain maps we are in a position to generalize the idea of elementary collapse, which has a completely cubical interpretation, to that of an *elementary reduction*, which is based on algebraic considerations. As we shall see this reduction takes the form of a projection of chain groups.

4.3.1 Elementary Collapses Revisited

Our immediate goal is to reinterpret the elementary collapses introduced in Section 2.4 in terms of chain maps.

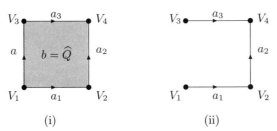

Fig. 4.4. The elementary collapse of a square results in removing a pair of generators (a, b) in dimensions 1 and 2.

Consider $\mathcal{C}(Q)$, the cubical complex of the elementary cube $Q = [0,1]^2$. When we perform the elementary collapse of Q by the edge $[0] \times [0,1]$, two generators are removed from the bases of the chain groups: The generator

$a := [0] \widehat{\times [0,1]}$ is removed from the basis $W_1 := \widehat{\mathcal{K}}_1(Q)$ of $C_1(Q)$ and $b := \widehat{Q}$ is removed from the basis $W_2 := \widehat{\mathcal{K}}_2(Q) = \{b\}$. The basis $W_0 := \widehat{\mathcal{K}}_0(Q) = \{\widehat{V}_1, \widehat{V}_2, \widehat{V}_3, \widehat{V}_4\}$ of dual vertices is left untouched. Let us denote by a_1, a_2, and a_3 the remaining generators in W_1. These bases are displayed in Figure 4.4(i).

The chain complex $\mathcal{C}(Q')$ of the collapsed cubical set Q' is indicated in Figure 4.4(ii). Again, explicitly writing out bases for the chain groups, we have that $C_0(Q')$ is generated by the basis $W_0' := W_0$, $C_1(Q')$ is generated by the basis $W_1' := W_1 \setminus \{a\} = \{a_1, a_2, a_3\}$, and $C_2(Q')$ is trivial so $W_2' := \emptyset$.

To define a chain map from the complex $\mathcal{C}(Q)$ to $\mathcal{C}(Q')$ we return to topology. Recall Example 2.72, which presents a deformation retraction of the square onto three of its edges. In this example the edge $|a|$ is mapped onto the three edges $|a_1|$, $|a_2|$, and $|a_3|$. Keeping track of orientations, this suggests that we want to map

$$a \mapsto a' := a_1 + a_2 - a_3.$$

So far we have focused on a single chain. Can we describe this operation on the entire chain complex? In other words, can we find a chain map $\pi' : \mathcal{C}(Q) \to \mathcal{C}(Q')$ that captures the collapse? The answer is yes and we claim that we have essentially done it. Clearly, $\pi_2' = 0$. The matrices of π_1' and π_0' are, respectively,

$$\begin{bmatrix} 1 & 0 & 0 & 1 \\ 0 & 1 & 0 & 1 \\ 0 & 0 & 1 & -1 \end{bmatrix} \quad \text{and} \quad \begin{bmatrix} 1 & 0 & 0 & 0 \\ 0 & 1 & 0 & 0 \\ 0 & 0 & 1 & 0 \\ 0 & 0 & 0 & 1 \end{bmatrix},$$

where the bases of the first matrix are W_1 ordered as (a_1, a_2, a_3, a) and W_1' as (a_1, a_2, a_3). We leave it to the reader to check that π' is a chain map and, therefore, induces maps on the associated homology groups.

We now show that π' induces an isomorphism in homology. If we demonstrate that π' is a chain equivalence, the conclusion will follow from Proposition 4.14. Observe that $\mathcal{C}(Q')$ is a chain subcomplex of $\mathcal{C}(Q)$ and thus the inclusion map $\iota : \mathcal{C}(Q') \to \mathcal{C}(Q)$ is a chain map. It is easy to check that

$$\pi' \circ \iota = \mathrm{id}_{\mathcal{C}(Q')}$$

so, a fortiori, $\pi' \circ \iota$ is chain homotopic to $\mathrm{id}_{\mathcal{C}(Q')}$. It remains to show that the map $\pi : \mathcal{C}(Q) \to \mathcal{C}(Q)$ defined by $\pi = \iota \circ \pi'$ is chain homotopic to $\mathrm{id}_{\mathcal{C}(Q)}$. Obviously, $\pi_2 = 0$. The matrices of π_1 and π_0 in the bases W_k are

$$\begin{bmatrix} 1 & 0 & 0 & 1 \\ 0 & 1 & 0 & 1 \\ 0 & 0 & 1 & -1 \\ 0 & 0 & 0 & 0 \end{bmatrix} \quad \text{and} \quad \begin{bmatrix} 1 & 0 & 0 & 0 \\ 0 & 1 & 0 & 0 \\ 0 & 0 & 1 & 0 \\ 0 & 0 & 0 & 1 \end{bmatrix}. \quad (4.10)$$

The maps π_1 and π_2 are clearly not equal to the identity. Nevertheless, we shall see that on the level of homology they induce the identity maps.

Consider the maps $D_k : C_k(Q) \to C_{k+1}(Q)$ given by $D_k = 0$, if $k \neq 1$, and

$$D_1 = \begin{bmatrix} 0 & 0 & 0 & 1 \end{bmatrix}.$$

Simple matrix calculations show that for $k = 0, 1, 2$

$$\pi_k - \mathrm{id}_{C_k} = \partial_{k+1} \circ D_k + D_{k-1} \circ \partial_k.$$

Thus the collection of maps D_k is a chain homotopy between π and $\mathrm{id}_{\mathcal{C}(Q)}$. This proves that π' and ι are chain equivalences.

4.3.2 Generalization of Elementary Collapses

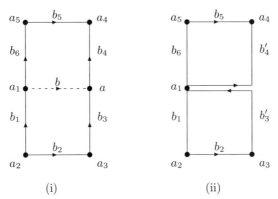

Fig. 4.5. A reduction where an elementary collapse is not possible. The pair of generators (a, b) in dimensions 0 and 1 is removed, but also the generators b_3 and b_4 are replaced by new ones.

As indicated in the introduction to this section, we want to be able to reduce the size of the chain complex used to compute homology. In the previous subsection this is done geometrically using an elementary collapse. Consider, however, an example such as that shown in Figure 4.5(i). This cubical set is made up only of vertices and edges and has no free faces. Therefore, no elementary collapse is possible.

On the other hand, we have seen that when an elementary collapse is written as a chain map, it can be interpreted as an algebraic operation that takes the form of a projection. With this in mind let us try to mimic the ideas of the previous section to somehow eliminate a pair of k- and $(k-1)$-dimensional chains by directly constructing a projection. To be more precise, let $\mathcal{C} = \{C_k, \partial_k\}$ be an abstract chain complex. Let $b \in C_m$ and $a \in C_{m-1}$. We want to construct a chain map $\pi : \mathcal{C} \to \mathcal{C}$ that is a projection with the property that $\pi_m(b) = 0$ and $a \notin \mathrm{im}\,\pi_{m-1}$.

As an analogy to motivate the formula, we will consider Figure 4.6(a). We want a projection that sends b to 0. V represents the subspace that is perpendicular to b, and hence can be taken to be the image of our projection. What

happens to a vector q under this projection? As indicated in Figure 4.6(b), it is sent to the vector q' in V. Observe that the relationship among the three vectors is
$$q = q' + \lambda b$$
for some λ. Thus, the projection takes the form
$$q \mapsto q' := q - \lambda b,$$
where λ is a scalar depending on q.

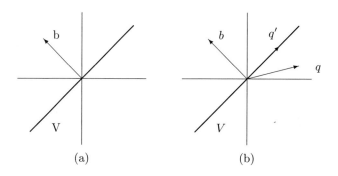

Fig. 4.6. Geometry of a projection.

Returning to the setting of the desired projection $\pi : \mathcal{C} \to \mathcal{C}$, let us begin with π_m. For $c \in C_m$ set
$$\pi_m(c) := c - \lambda b. \tag{4.11}$$
Of course, we still need to determine $\lambda \in \mathbf{Z}$. Recall that we also want to eliminate $a \in C_{m-1}$ from the image of π_{m-1}. In particular, this implies that $0 = \langle \pi_{m-1} \partial c, a \rangle$. Since we want π to be a chain map, we must have
$$\begin{aligned} 0 &= \langle \pi_{m-1} \partial c, a \rangle \\ &= \langle \partial \pi_m c, a \rangle \\ &= \langle \partial (c - \lambda b), a \rangle \\ &= \langle \partial c, a \rangle - \lambda \langle \partial b, a \rangle. \end{aligned}$$
Hence, it must be the case that
$$\lambda = \frac{\langle \partial c, a \rangle}{\langle \partial b, a \rangle}.$$
Observe that since π_k is a group homomorphism, λ must be an integer. Thus we can only define this projection if $\langle \partial b, a \rangle$ divides $\langle \partial c, a \rangle$ for every $c \in C_k$. The easiest way to guarantee this is to choose b and a such that $\langle \partial b, a \rangle = \pm 1$.

To determine π_{m-1}, we begin with the restriction that π is a chain map. In particular, $\pi_m b = 0$ implies that $\partial \pi_m b = 0$, and hence, $\pi_{m-1} \partial b = 0$. Clearly, ∂b needs to be sent to zero under the projection. Thus, following (4.11), we set

$$\pi_{m-1} c := c - \lambda \partial b$$

for any $c \in C_{m-1}$ and again we need to solve for λ. As before, a is not to be in the image of π_{m-1} and hence

$$0 = \langle \pi_{m-1} c, a \rangle = \langle c - \lambda \partial b, a \rangle = \langle c, a \rangle - \lambda \langle \partial b, a \rangle.$$

Solving for λ, we obtain

$$\lambda = \frac{\langle c, a \rangle}{\langle \partial b, a \rangle}.$$

Again, the simplest way to ensure that λ is an integer is to choose a pair b and a such that $\langle \partial b, a \rangle = \pm 1$.

Putting these calculations together leads to the following definition.

Definition 4.18 Let $\mathcal{C} = \{C_k, \partial_k\}$ be an abstract chain complex. A pair of generators (a, b) such that $a \in C_{m-1}$, $b \in C_m$ and

$$\langle \partial b, a \rangle = \pm 1$$

is called a *reduction pair*. It induces a collection of group homomorphisms

$$\pi_k c := \begin{cases} c - \frac{\langle c, a \rangle}{\langle \partial b, a \rangle} \partial b & \text{if } k = m-1, \\ c - \frac{\langle \partial c, a \rangle}{\langle \partial b, a \rangle} b & \text{if } k = m, \\ c & \text{otherwise}, \end{cases} \quad (4.12)$$

where $c \in C_k$.

Theorem 4.19 *The map $\pi : \mathcal{C} \to \mathcal{C}$ defined by the homomorphisms $\{\pi_k\}_{k \in \mathbb{Z}}$ of (4.12) is a chain map. Furthermore, π_k is a projection of C_k onto its image C'_k, that is, $\pi_k c = c$ for all $c \in C'_k$.*

Proof. The linearity of π_k is obvious since the maps ∂ and $\langle \cdot, a \rangle$ are linear.

We need to show that $\partial_k \pi_k = \pi_{k-1} \partial_k$ for all k. This is obvious for $k \notin \{m-1, m, m+1\}$.

Let $k = m - 1$ and $c \in C_{m-1}$. Then

$$\partial \pi_{m-1} c = \partial \left(c - \frac{\langle c, a \rangle}{\langle \partial b, a \rangle} \partial b \right) = \partial c - \frac{\langle c, a \rangle}{\langle \partial b, a \rangle} \partial^2 b = \partial c$$

because $\partial^2 = 0$. On the other hand, $\pi_{m-2} \partial c = \partial c$ because $\pi_{m-2} = \text{id}$.

Let $k = m$ and $c \in C_m$. Then

$$\partial \pi_m c = \partial \left(c - \frac{\langle \partial c, a \rangle}{\langle \partial b, a \rangle} b \right) = \partial c - \frac{\langle \partial c, a \rangle}{\langle \partial b, a \rangle} \partial b = \pi_{m-1} \partial c.$$

160   4 Chain Maps and Reduction Algorithms

Let $k = m+1$ and $c \in C_{m+1}$. Then

$$\pi_m \partial c = \partial c - \frac{\langle \partial^2 c, a \rangle}{\langle \partial b, a \rangle} b = \partial c$$

because $\partial^2 = 0$. On the other hand, $\partial \pi_{m+1} c = \partial c$, because $\pi_{m+1} = \text{id}$.

The second conclusion is equivalent to the identity $(\pi_k)^2 = \pi_k$. This is trivial if $k \notin \{m-1, m\}$. Let $k = m-1$ and $c \in C_{m-1}$. Then

$$(\pi_{m-1})^2 c = \pi_{m-1}\left(c - \frac{\langle c, a \rangle}{\langle \partial b, a \rangle} \partial b\right)$$

$$= \pi_{m-1} c - \frac{\langle c, a \rangle}{\langle \partial b, a \rangle} \pi_{m-1} \partial b$$

$$= \pi_{m-1} c - \frac{\langle c, a \rangle}{\langle \partial b, a \rangle}\left(\partial b - \frac{\langle \partial b, a \rangle}{\langle \partial b, a \rangle} \partial b\right)$$

$$= \pi_{m-1} c.$$

Let $k = m$ and $c \in C_m$.

$$(\pi_m)^2 c = \pi_m\left(c - \frac{\langle \partial c, a \rangle}{\langle \partial b, a \rangle} b\right)$$

$$= \pi_m c - \frac{\langle \partial c, a \rangle}{\langle \partial b, a \rangle} \pi_m b$$

$$= \pi_m c - \frac{\langle \partial c, a \rangle}{\langle \partial b, a \rangle}\left(b - \frac{\langle \partial b, a \rangle}{\langle \partial b, a \rangle} b\right)$$

$$= \pi_m c. \qquad \square$$

Example 4.20 Consider the cubical set X presented in Figure 4.5(i). More precisely, let X be the one-dimensional set consisting of all edges and vertices of the rectangle $R = [0,1] \times [0,2]$. We have mentioned earlier that an elementary collapse is not possible in X because it has no free face. Choose a and b, the remaining dual vertices a_1 to a_5, and the remaining dual edges b_1 to b_6 as shown in Figure 4.5(i). The only nontrivial boundary map of $\mathcal{C}(X)$ is ∂_1 defined on the bases $W_0 := \{a_1, a_2, \ldots, a_5, a\}$ and $W_1 := \{b_1, b_2, \ldots, b_6, b\}$ as follows:

$$\partial b_1 = a_1 - a_2,$$
$$\partial b_2 = a_3 - a_2,$$
$$\partial b_3 = a - a_3,$$
$$\partial b_4 = a_4 - a,$$
$$\partial b_5 = a_4 - a_5,$$
$$\partial b_6 = a_5 - a_1,$$
$$\partial b = a - a_1.$$

4.3 Internal Elementary Reductions   161

Since $\langle \partial b, a \rangle = 1$, (a,b) is a reduction pair. In light of the previous discussion it should be possible to eliminate it from the chain complex. Rather than immediately applying Definition 4.12, we shall go through arguments that lead to it and directly construct the projection. We start by defining π_1 on the generators. To begin with, we want to eliminate b, so we define

$$\pi_1 b := 0.$$

The definition of π_1 on the remaining generators should permit eliminating a from the boundaries of one-dimensional chains. This can be done directly by adding or subtracting the above equations. Indeed, a only appears in the equations for the boundary of b_3, b_4, and b. Thus, we get

$$\partial(b_3 - b) = (a - a_3) - (a - a_1) = a_1 - a_3,$$
$$\partial(b_4 + b) = (a_4 - a) + (a - a_1) = a_4 - a_1.$$

This suggests that we should set

$$\pi b_3 := b'_3 := b_3 - b,$$
$$\pi b_4 := b'_4 := b_4 + b.$$

The remaining generators may be left unchanged:

$$\pi b_i := b'_i := b_i, \quad i \notin \{3, 4\}.$$

We now define π_0 on generators. We want to have

$$\pi_0 a_i := a_i, \quad \text{for } i = 1, 2, 3, 4, 5,$$

and to define $\pi_0 a$ so to obtain a chain map. Since $\pi_1 b = 0$, we must have

$$0 = \partial \pi_1 b = \pi_0 \partial b = \pi_0(a) - \pi_0(a_1) = \pi_0(a) - a_1.$$

Therefore, the only choice is

$$\pi_0 a := a' := a_1.$$

It is easily verified that $\partial \pi_1 b_i = \pi_0 \partial b_i$ for all i.

The above formulas coincide with (4.12). As in the case of the elementary collapse discussed in the previous section, we want to look at the reduced complex $\mathcal{C}' := \pi(\mathcal{C}(X))$. Its chain groups, C'_0 and C', are generated, respectively, by

$$W'_0 := \{a_1, a_2, \ldots, a_5\}$$

and

$$W'_1 := \{b'_1, b'_2, \ldots, b'_6\}.$$

The reduced complex $\mathcal{C}'$ is a chain subcomplex of $\mathcal{C}(X))$, but it is not a cubical chain complex because its generators are not elementary dual cubes.

Its geometrical interpretation is presented in Figure 4.5(ii). The projection of b to 0 is interpreted as a deformation retraction of the edge $|b|$ to the single point, namely the vertex $|a_1|$. While we deform the edge, the vertex $|a|$ is mapped to $|a_1|$, which makes the definition $\pi(a) := a_1$ clear. While $|a|$ is mapped to $|a_1|$, the adjacent edges b_3 and b_4 can be viewed as being dragged along the edge b, thus producing the new generators b'_3 and b'_4.

We shall now formalize the reduction step discussed in the above example. Let $(\mathcal{C}, \partial) = (\{C_k\}_{k \in \mathbf{Z}}, \{\partial_k\}_{k \in \mathbf{Z}})$ be a finitely generated free chain complex. Let

$$k_{min}(\mathcal{C}) := \min\{k \mid C_k \neq 0\},$$
$$k_{max}(\mathcal{C}) := \max\{k \mid C_k \neq 0\}.$$

For each $k \in \{k_{min}(\mathcal{C}), k_{min}(\mathcal{C})+1, \ldots, k_{max}(\mathcal{C})\}$, let W_k be a fixed basis for C_k. Let $d_k + 1$ be the number of elements in W_k.

Fix $m \in \mathbf{Z}$ and assume that

$$W_{m-1} = \{a_1, a_2, \ldots, a_{d_{m-1}}, a\},$$
$$W_m = \{b_1, b_2, \ldots, b_{d_m}, b\}$$

where (a, b) is a reduction pair. Recall that this means that $\langle \partial b, a \rangle = \pm 1$. Let $\pi : \mathcal{C} \to \mathcal{C}$ be the associated chain map and let $c' \in C'_k$ denote the image of the chain $c \in C_k$ under π (i.e., $c' := \pi c$). Similarly,

$$C'_k := \pi_k(C_k).$$

To simplify the formulas that follow, we let

$$r := \partial b - \langle \partial b, a \rangle a.$$

Equation (4.12) is explicit and convenient for some purposes but, since π is a linear map, we would like to express it in terms of generators. Since π_k is the identity in all dimensions other than $m-1$ and m, it is enough to see the images of elements of W_{m-1} and W_m under π_{m-1} and π_m, respectively. As before we let $c' := \pi c$ for simplicity. It is straightforward to check that

$$a'_i = a_i - \frac{\langle a_i, a \rangle}{\langle \partial b, a \rangle} \partial b = a_i - 0 = a_i,$$

$$a' := a - \frac{\langle a, a \rangle}{\langle \partial b, a \rangle} \partial b = a - \frac{1}{\langle \partial b, a \rangle}(\langle \partial b, a \rangle a + r) = -\frac{r}{\langle \partial b, a \rangle},$$

$$b'_i = b_i - \frac{\langle \partial b_i, a \rangle}{\langle \partial b, a \rangle} b,$$

and

$$b' = b - \frac{\langle \partial b, a \rangle}{\langle \partial b, a \rangle} b = b - b = 0.$$

Thus we get the formulas

$$a'_i = a_i, \quad a' = -\frac{r}{\langle \partial b, a \rangle} \quad (4.13)$$

and

$$b'_i = b_i - \frac{\langle \partial b_i, a \rangle}{\langle \partial b, a \rangle} b, \quad b' = 0. \quad (4.14)$$

Define

$$W'_k := \begin{cases} \{b'_1, b'_2, \ldots, b'_{d_m}\} & \text{if } k = m, \\ \{a'_1, a'_2, \ldots, a'_{d_{m-1}}\} & \text{if } k = m-1, \\ W_k & \text{otherwise.} \end{cases} \quad (4.15)$$

We leave as an exercise the proof of the following proposition.

Proposition 4.21 W'_k is a basis for C'_k for all $k \in \mathbf{Z}$ and $W'_m \cup \{b\}$ is a basis for C_m.

Theorem 4.22 $H_*(C') \cong H_*(C)$.

Proof. We will show that $\pi' : C \to C'$ is a chain equivalence with the inclusion $\iota : C' \hookrightarrow C$ as a homotopical inverse. Indeed, by Theorem 4.19, π' is a projection so $\pi'\iota = \text{id}_{C'}$. Hence, it is sufficient to find a chain homotopy between $\iota\pi' = \pi$ and id_C. Let $D_k : C_k \to C_{k+1}$ be given by

$$D_k c = \begin{cases} \frac{\langle c, a \rangle}{\langle \partial b, a \rangle} b & \text{if } k = m-1, \\ 0 & \text{otherwise} \end{cases} \quad (4.16)$$

for any $c \in C_k$. We need to show the identity

$$\text{id}_{C_k} - \pi_k = \partial_{k+1} D_k + D_{k-1} \partial_k. \quad (4.17)$$

This is obvious if $k \notin \{m-1, m\}$, because in that case both sides are 0. Let $k = m-1$ and $c \in C_{m-1}$. Then

$$c - \pi_{m-1} c = c - \left(c - \frac{\langle c, a \rangle}{\langle \partial b, a \rangle} \partial b \right) = \frac{\langle c, a \rangle}{\langle \partial b, a \rangle} \partial b.$$

On the other hand,

$$\partial D_{m-1} c + D_{m-2} \partial c = \partial D_{m-1} c = \partial \left(\frac{\langle c, a \rangle}{\langle \partial b, a \rangle} b \right) = \frac{\langle c, a \rangle}{\langle \partial b, a \rangle} \partial b,$$

so the identity holds.

Let $k = m$ and $c \in C_m$. Then

$$c - \pi_m c = c - \left(c - \frac{\langle \partial c, a \rangle}{\langle \partial b, a \rangle} b\right) = \frac{\langle \partial c, a \rangle}{\langle \partial b, a \rangle} b.$$

On the other hand,

$$\partial D_m c + D_{m-1} \partial c = D_{m-1} \partial c = \frac{\langle \partial c, a \rangle}{\langle \partial b, a \rangle} b,$$

so again the identity holds. □

Exercises

4.8 Obtain the projection map π and the reduced complex $\mathcal{C}'$ if $\mathcal{C} := \mathcal{C}(\Gamma^2)$ and the reduction pair (a, b) is given by

$$a := \widehat{[0,1]} \diamond \widehat{[0]} \diamond \widehat{[0]},$$
$$b := \widehat{[0,1]} \diamond \widehat{[0,1]} \diamond \widehat{[0]}.$$

4.9 Compute the homology of the complex $\mathcal{C}(X)$ discussed in Example 4.20 by continuing to simplify the complex with elementary reductions.

4.10 Compute the homology groups of $\mathcal{C} := \mathcal{C}(\Gamma^2)$ by iterated elementary reductions. A reduction pair to start from is given in Exercise 4.8.

4.11 Consider the cubical chain complex $\mathcal{C}(R)$ where R is the rectangle $[0,1] \times [0,2]$. Let $a = \widehat{[1]} \times \widehat{[1]}$ and $b = \widehat{[0,1]} \times \widehat{[1]}$. Show that (a, b) is a reduction pair. Let $\pi : \mathcal{C}(R) \to \mathcal{C}(R)$ be the induced chain map. Write down a basis for $C'_k(R) = \pi_k(C_k(R))$ for each k.

4.12 Prove Proposition 4.21.

4.13 Let $\mathcal{C}'$ be the complex obtained from $\mathcal{C}$ by the reduction step. Let $A = (A_{i,j})$ be the matrix of ∂_k with respect to the bases W_k and W_{k-1} and $B = (B_{i,j})$ the matrix of $\partial_k |_{\mathcal{C}'}$ with respect to the bases W'_k and W'_{k-1}. Prove the following statements.

(i) If $k \notin \{m-1, m, m+1\}$, then $B = A$.
(ii) If $k = m-1$, then $B = A[1 : d_{m-2} + 1, 1 : d_{m-1}]$, that is, B is obtained from A by removing its last column.
(iii) If $k = m$, then B is a $d_{m-1} \times d_m$ matrix (one column and one row less than A) and its coefficients are given by the formula

$$B[i, j] = A[i, j] - \frac{\langle \partial b_j, a \rangle \langle \partial b, a_i \rangle}{\langle \partial b, a \rangle}.$$

(iv) If $k = m+1$, then $B = A[1 : d_m, 1 : d_{m+1} + 1]$, that is, B is obtained from A by removing its last row.

4.4 CCR Algorithm

From the viewpoint of programming, the internal reductions studied in the previous section are more complicated than elementary collapses. This is due to the fact that an elementary collapse can be presented as a removal of a pair of generators from the lists of bases W_{m-1} and W_m while leaving the remaining elements in the list. A general internal reduction requires replacing a basis by a different one. The manipulation with a data structure could be simpler if we could work with a fixed list of generators as is the case with collapses. However, if we just remove a pair (a, b) of generators of $\mathcal{C}$, where a is not a free face, the resulting collection of chain groups would not be a chain complex because the boundary operator may no longer be well defined. A solution to this problem is to define a new boundary operator. These ideas lead to a procedure for reducing the complex called the *chain complex reduction algorithm (CCR algorithm)*.

Assume, as before, that $(\mathcal{C}, \partial)$ is a finitely generated free abelian chain complex with a fixed basis W_k of C_k, and let (a, b) be a reduction pair, where $a \in W_{m-1}$ and $b \in W_m$. In the previous section, we define a subcomplex $\mathcal{C}'$ of $\mathcal{C}$ having fewer generators but the same homology as $\mathcal{C}$. The complication comes in dimension m where the new basis, $W'_m = \{b'_1, b'_2, \ldots, b'_{d_m}\}$, is related to the basis $W_m = \{b_1, b_2, \ldots, b_{d_m}, b\}$ by the formulas

$$b'_i = b_i - \frac{\langle \partial b_i, a \rangle}{\langle \partial b, a \rangle} b. \tag{4.18}$$

We want to define a new chain complex $\bar{\mathcal{C}}$ whose basis is a subbasis of $\mathcal{C}$. Define

$$\bar{W}_k := \begin{cases} \{b_1, b_2, \ldots, b_{d_m}\} & \text{if } k = m, \\ \{a_1, a_2, \ldots, a_{d_{m-1}}\} & \text{if } k = m-1, \\ W_k & \text{otherwise.} \end{cases} \tag{4.19}$$

Let $\bar{C}_k$ be the free abelian group generated by the set $\bar{W}_k$. Since $\bar{W}_k \subset W_k$, this is a subgroup of C_k but, as we have mentioned above, the collection $\bar{\mathcal{C}}$ of those subgroups is not necessarily a subcomplex of $\mathcal{C}$. We want to define a new boundary operator $\bar{\partial}$ so as to give $\bar{\mathcal{C}}$ the structure of a chain complex. The knowledge of the complex $\mathcal{C}'$ from the previous section is useful. We want to assign to the generators b_i of $\bar{C}_m$ the same geometric role that the generators b'_i had in the complex $\mathcal{C}'$. We start with the following example.

Example 4.23 We will review Example 4.20 in the new setting. The complex $\mathcal{C}(X)$ illustrated in Figure 4.5(i) has two nontrivial groups, $C_0(X)$ generated by $W_0 = \{a_1, a_2, \ldots, a_5, a\}$, $C_1(X)$ generated by $W_1 = \{b_1, b_2, \ldots, b_6, b\}$, and one nontrivial boundary map ∂_1 defined on generators by

$$\partial b_1 = a_1 - a_2,$$
$$\partial b_2 = a_3 - a_2,$$

$$\partial b_3 = a - a_3,$$
$$\partial b_4 = a_4 - a,$$
$$\partial b_5 = a_4 - a_5,$$
$$\partial b_6 = a_5 - a_1,$$
$$\partial b = a - a_1.$$

Let $\bar{C}_0$ be the free abelian group generated by $\bar{W}_0 := \{a_1, a_2, \ldots, a_5\}$, and let $\bar{C}_1$ be the free abelian group generated by $\bar{W}_1 := \{b_1, b_2, \ldots, b_6\}$. We have already observed that a can be eliminated from the boundaries by subtracting the equation for ∂b from the equation for ∂b_3 and adding it to the equation for ∂b_4. In this way we obtain

$$\partial(b_3 - b) = (a - a_3) - (a - a_1) = a_1 - a_3,$$
$$\partial(b_4 + b) = (a_4 - a) + (a - a_1) = a_4 - a_1.$$

Instead of using those equations to define the new generators b_3' and b_4', we use them for defining the new boundary map:

$$\bar{\partial}_1 b_3 := \partial_1(b_3 - b) = a_1 - a_3,$$
$$\bar{\partial}_1 b_4 := \partial_1(b_4 + b) = a_4 - a_1.$$

We complete the formula by setting

$$\bar{\partial}_1 b_i := \partial_1 b_i, \qquad i \notin \{3, 4\}.$$

This defines the new complex $\bar{C}$. As will be seen shortly, the complexes $\bar{C}$ and C' are isomorphic.

In order to generalize the above example, consider the collection of homomorphisms $\eta = \{\eta_k : \bar{C}_k \to C'_k\}$ given on any $c \in \bar{C}_k$ by the formula

$$\eta_k(c) := \begin{cases} c - \frac{\langle \partial c, a \rangle}{\langle \partial b, a \rangle} b & \text{if } k = m, \\ c & \text{otherwise.} \end{cases} \qquad (4.20)$$

Observe that η_m is given on generators by

$$\eta_m(b_i) = b_i', \qquad (4.21)$$

so it is an isomorphism of groups sending the basis $\bar{W}_m$ to the basis W'_m. When $k \neq m$, η_k is the identity on $\bar{C}_k = C'_k$. The definition of ∂ on C' may now be transported to $\bar{C}$ by the conjugacy formula

$$\bar{\partial} := \eta^{-1} \partial \eta \qquad (4.22)$$

or, more explicitly,

$$\bar{\partial}_k := \eta_{k-1}^{-1} \partial_k \eta_k. \qquad (4.23)$$

Theorem 4.24 *Equation (4.22) defines a boundary map on $\bar{\mathcal{C}}$, thus giving it a structure of chain complex. Moreover, $\eta : \bar{\mathcal{C}} \to \mathcal{C}'$ is a chain isomorphism. In particular, $H_*(\bar{\mathcal{C}}) = H_*(\mathcal{C}') = H_*(\mathcal{C})$.*

Proof. Since each η_k is an isomorphism, we get

$$\bar{\partial}_{k-1}\bar{\partial}_k = \eta_{k-2}^{-1}\partial_{k-1}\eta_{k-1}\eta_{k-1}^{-1}\partial_k\eta_k = \eta_{k-2}^{-1}\partial_{k-1}\partial_k\eta_k = 0.$$

Hence, $\bar{\partial}$ is a boundary map. Next,

$$\eta_{k-1}\bar{\partial}_k = \eta_{k-1}\eta_{k-1}^{-1}\partial_k\eta_k = \partial_k\eta_k,$$

so η is a chain map. Since each η_k is an isomorphism, the last conclusion is obvious. □

For the purpose of computation we want to have an explicit formula for $\bar{\partial}$.

Proposition 4.25 *The following formula holds:*

$$\bar{\partial}_k(c) = \begin{cases} \partial c - \frac{\langle \partial c, a \rangle}{\langle \partial b, a \rangle} \partial b & \text{if } k = m, \\ \partial c - \langle \partial c, b \rangle b & \text{if } k = m+1, \\ \partial c & \text{otherwise.} \end{cases} \quad (4.24)$$

Proof. It is clear from (4.20) and (4.23) that the only dimensions where $\bar{\partial}_k$ may differ from ∂_k are $k = m$ and $k = m+1$.

Let $k = m$. Then $\eta_{m-1} = \text{id}$ so

$$\bar{\partial}_m c = \partial_m \eta_m c = \partial\left(c - \frac{\langle \partial c, a \rangle}{\langle \partial b, a \rangle} b\right) = \partial c - \frac{\langle \partial c, a \rangle}{\langle \partial b, a \rangle} \partial b.$$

Let $k = m+1$. Then $\eta_{m+1} = \text{id}$ so

$$\bar{\partial}_{m+1} c = \eta_m^{-1}\partial_{m+1} c.$$

Since $c \in \bar{C}_{m+1} = C'_{m+1}$, we see that $\partial c \in C'_m$. Therefore ∂c can be written as

$$\partial_{m+1} c = \sum_i \alpha_i b'_i.$$

But $\eta_m^{-1} b'_i = b_i$, so

$$\bar{\partial}_{m+1} c = \sum_i \alpha_i b_i.$$

On the other hand, letting

$$\alpha := \sum_i \alpha_i \frac{\langle \partial b_i, a \rangle}{\langle \partial b, a \rangle},$$

we obtain

$$\partial_{m+1}c = \sum_i \alpha_i b'_i = \sum_i \alpha_i \left(b_i - \frac{\langle \partial b_i, a \rangle}{\langle \partial b, a \rangle} b \right) = \sum_i \alpha_i b_i - \alpha b.$$

From this we get
$$\langle \partial c, b \rangle = -\alpha$$
and
$$\partial c = \bar{\partial} c - \alpha b.$$
So, by combining the above formulas we get
$$\bar{\partial} c = \partial c + \alpha b = \partial c - \langle \partial c, b \rangle b. \qquad \square$$

In order to construct the reduction algorithm we need to rewrite (4.24) in terms of the bases.

Proposition 4.26 *If $e \in \bar{W}_k$, $e' \in \bar{W}_{k-1}$, then*

$$\langle \bar{\partial}_k(e), e' \rangle = \begin{cases} \langle \partial e, e' \rangle - \langle \partial e, a \rangle \langle \partial b, a \rangle \langle \partial b, e' \rangle & \text{if } k = m, \\ \langle \partial e, e' \rangle & \text{otherwise.} \end{cases} \qquad (4.25)$$

Proof. The formulas follow easily from (4.24) and the fact that $1/\langle \partial b, a \rangle = \langle \partial b, a \rangle$. $\square$

In the previous section, the reduced complex $\mathcal{C}'$ is described in terms of the chain projection map $\pi : \mathcal{C} \to \mathcal{C}$ having $\mathcal{C}'$ as the image. It does not make much sense to speak about the chain projection from $\mathcal{C}$ to $\bar{\mathcal{C}}$ because the latter one is not a subcomplex of the first. We may, however, define an analogous chain map $\bar{\pi} : \mathcal{C} \to \bar{\mathcal{C}}$ by the formula

$$\bar{\pi} := \eta^{-1} \pi. \qquad (4.26)$$

Since π induces an isomorphism in homology and η is a chain isomorphism, the composition $\bar{\pi}$ also induces an isomorphism in homology. Here is an explicit formula for that map.

Proposition 4.27

$$\bar{\pi}_k(c) := \begin{cases} c - \frac{\langle c, a \rangle}{\langle \partial b, a \rangle} \partial b & \text{if } k = m-1, \\ c - \langle \partial c, a \rangle b & \text{if } k = m, \\ c & \text{otherwise.} \end{cases} \qquad (4.27)$$

Proof. If $k \neq m$, η_k is the identity, so the conclusion follows from (4.12). The proof of the case $k = m$ is left as an exercise. $\square$

Note that the $\bar{\pi}$ gives the mathematical meaning to the elementary reduction step as an operation on chain complexes, but it does not explicitly serve to perform the reduction step. This step consists of two operations:

- Given a reduction pair (a, b) remove a from the list of generators W_{m-1} and b from the list of generators W_m.

- Modify the formula for the boundary map on the generators remaining in W_m.

Now let
$$\mathcal{C}^0 \xrightarrow{\bar{\pi}^1} \mathcal{C}^1 \xrightarrow{\bar{\pi}^2} \mathcal{C}^2 \cdots \quad (4.28)$$
be a sequence of chain complexes and projections obtained as follows. Initially set
$$\mathcal{C}^0 := \mathcal{C}, \ \partial^0 := \partial, \text{ and } W_k^0 := W_k.$$
Then iterate the elementary reduction and define
$$\mathcal{C}^{j+1} := \bar{\mathcal{C}}^j, \ W_k^{j+1} := \bar{W}_k^j, \ \partial^{j+1} := \bar{\partial}^j$$
as long as it is possible to find a reduction pair in $\bar{\mathcal{C}}^j$. Denote by $M(j) := \sum_k \operatorname{card}(W_k^j)$, for $j = 0, 1, 2, \ldots$. Since $\mathcal{C}$ is finitely generated, $M(j) < \infty$ and $M(j+1) = M(j) - 2$. Therefore, there exists a final element of that sequence denoted by $(\mathcal{C}^f, \partial^f)$, beyond which the construction cannot be extended.

Corollary 4.28
$$H_*(\mathcal{C}) \cong H_*(\mathcal{C}^f).$$

Proof. The identity follows from Theorem 4.24 by induction. □

Corollary 4.29 *Suppose that the elementary reductions can be successfully performed until $\partial^f = 0$. Then*
$$H(\mathcal{C}) \cong H(\mathcal{C}^f) = \mathcal{C}^f.$$

Proof. The first identity is given by the previous corollary and the second follows from the assumption $\partial^f = 0$ because then
$$H_k(\mathcal{C}) := \ker \partial_k^f / \operatorname{im} \partial_k^f = C_k^f / 0 = C_k^f.$$
□

Observe that if, through a reduction, we achieve $\partial^f = 0$, then we have computed $H(\mathcal{C})$. However, even if this is not attained, reduction in the number of generators of $\mathcal{C}^f$ speeds up the application of the algorithms presented in Chapter 3.

We are now ready to describe the algorithm based on the described construction. Since we study the case of an abstract chain complex, the generators need not be elementary cubes. Therefore, we assume that there is given a data type **generator** and the abstract generators may be any elements of this data type. The chain complex itself will be stored in
    typedef chainComplex = **array**[0 :] **of list of** generator;
For the chains we will use the data structure
    typedef chain = **hash**{generator} **of** int ;
and for the boundary map
    typedef boundaryMap = **array**[0 :] **of hash**{generator} **of** chain;

We first present the one-step reduction algorithm.

Algorithm 4.30 Reduction of a pair
  function reduce(
         chainComplex E, boundaryMap bd, int i, generator a, b)
  for each e in E[i + 1] do
     remove(b, bd[i + 1]{e});
  endfor;
  for each e in E[i] do
     if a in keys(bd[i]{e}) then
         for each e' in keys(bd[i]{b}) do
            bd[i]{e}{e'} :=
               bd[i]{e}{e'} − bd[i]{e}{a} ∗ bd[i]{b}{a} ∗ bd[i]{b}{e'};
         endfor;
     endif;
  endfor;
  remove(b, E[i]);
  remove(a, E[i − 1]);
  remove(b, bd[i]);
  remove(a, bd[i − 1]);
  return (E, bd);

Proposition 4.31 *Assume Algorithm 4.30 is called with a chain complex (C, ∂) represented in E and bd, an integer i, and a pair of generators in dimensions $i - 1$ and i constituting a reduction pair. Then it always stops and returns a new chain complex with the same homology as the original one.*

Proof. Obviously, the algorithm always stops. From Proposition 4.26 it follows that the returned chain complex is $(\bar{C}, \bar{\partial})$. The conclusion follows now from Theorem 4.24. □

Finally, we are ready to present the general CCR algorithm.

Algorithm 4.32 Reduction of a chain complex
  function reduceChainComplex(chainComplex E, boundaryMap bd)
  for i := lastIndex(E) downto 1 do
     repeat
         found := false;
         LOOP:
         for each b in E[i] do
            for each a in E[i − 1] do
               if abs(bd[i]{b}a) = 1 then
                  (E, bd) := reduce(E, bd, i, a, b);
                  found := true;
                  break LOOP;
               endif;
            endfor;
         endfor;

            until not found;
        endfor;
        return (E, bd);

Theorem 4.33 *Assume Algorithm 4.32 is called with a chain complex represented in E and bd. Then it always stops and returns a new chain complex with the same homology as the original one.*

Proof. To see that the algorithm always stops, observe that the only loop that might not be completed is the **repeat** loop. For this loop to complete it is necessary that no reduction pair is found. This must be the case because on every path of this loop the number of generators, where the reduction pair is searched, is smaller. Hence in the extreme case there are no generators at all. The rest of the theorem follows from Proposition 4.31. □

Exercises _____

4.14 Redo Exercise 4.10 in the setting of CCR algorithm.

4.15 Complete the proof of Proposition 4.27.

4.16 Verify that the matrix of $\bar{\partial}_k$ with respect to the bases $\bar{W}_k$ and $\bar{W}_{k-1}$ is identical to the matrix of ∂_k with respect to the bases W'_k and W'_{k-1} discussed in Exercise 4.13.

4.5 Bibliographical Remarks

Chain maps and chain homotopy presented in Sections 4.1 and 4.2 are standard tools in any homology theory. The main result of this chapter is the CCR algorithm (Algorithms 4.30 and 4.32) originally presented in [45]. The discussion in Section 4.3 is based on [42].

5
Preview of Maps

Consider two cubical sets X and Y. In Chapter 2 we have studied the associated homology groups $H_*(X)$ and $H_*(Y)$. Now assume that we are given a continuous map $f : X \to Y$. It is natural to ask if f induces a group homomorphism $f_* : H_*(X) \to H_*(Y)$. If so, do we get useful information out of it? The answer is yes and we will spend the next three chapters explaining how to define and compute f_*. It is worth noting, even at this very preliminary stage, that since $H_*(X)$ and $H_*(Y)$ are abelian groups, f_* is essentially a linear map and therefore, from the algebraic point of view, easy to use.

In Chapter 4 we have already seen that certain simple maps such as the inclusion map $j : A \hookrightarrow X$ of cubical sets induce chain maps and, consequently, maps in homology. However, those maps are very special in the sense that they map cubes to cubes. The typical continuous map will not have this property. Consider the simple example of $j : [0,1] \to [0,2]$ given by $j(x) = 2x$. Clearly, $H_*([0,1]) \cong H_*([0,2])$ and j is a homeomorphism so one would like to derive an isomorphism j_*. But it is not at all clear how one can pass from j to a chain map $\varphi : C([0,1]) \to C([0,2])$ that, in turn, induces an isomorphism $\varphi_* : H_*([0,1]) \to H_*([0,2])$. This fundamental question is dealt with in the following two chapters.

As we shall make clear in Chapter 6, constructing the desired chain maps and maps on homology can be done on a fairly general level. In Chapter 7 we provide the corresponding algorithms. A necessary step is the transformation of a continuous map to a cubical representation. There is a variety of ways in which this can be done. However, since the focus of this book is on algebraic topology and not numerical analysis, we have chosen to present the simplest approximation method, known as interval arithmetic. This is discussed in Section 5.1 and used in the remaining sections.

Even with the question of deriving a cubical representation of a continuous map resolved, there is a variety of technical issues that need to be overcome before the homology map can be defined. With this in mind we return, in Sections 5.2 through 5.4, to the style of Chapter 1. The emphasis is on the big picture. Here we introduce ideas and terminology that are explained and justified in Chapters 6 and 7.

5.1 Rational Functions and Interval Arithmetic

Let $X \subset \mathbf{R}^d$ and $Y \subset \mathbf{R}^{d'}$ be cubical sets and let $f : X \to Y$ be a continuous function. Our goal is to obtain a group homomorphism $f_* : H_*(X) \to H_*(Y)$ and eventually construct an algorithm computing it. The first obvious difficulty is that in general f is not a finite combinatorial object, hence it is not clear how it can be entered into a computer. We need some means of extracting a finite amount of data from f in such a way that the information, although finite, may still be used to construct a chain map and a map in homology.

To achieve this goal in a relatively simple way, in this chapter we restrict our attention to the set of *rational functions*, that is, maps $f : \mathbf{R}^d \to \mathbf{R}^{d'}$ such that
$$f = (f_1, f_2, \ldots, f_{d'}),$$
where
$$f_i(x_1, x_2, \ldots, x_d) = \frac{p_i(x_1, x_2, \ldots, x_d)}{q_i(x_1, x_2, \ldots, x_d)}$$
for some polynomials $p_i, q_i : \mathbf{R}^d \to \mathbf{R}$ with integer coefficients. This simple class of maps meets our needs in that it allows us to provide examples for all the essential ideas of maps induced in homology.

The second issue is approximating rational maps in a manner that leads to cubical representations. Given two intervals $[k_1, l_1]$ and $[k_2, l_2]$, where k_1, k_2, l_1, l_2 are integers, we define the arithmetic operations

$$[k_1, l_1] + [k_2, l_2] := [k_1 + k_2, l_1 + l_2], \tag{5.1}$$

$$[k_1, l_1] - [k_2, l_2] := [k_1 - l_2, l_1 - k_2], \tag{5.2}$$

$$[k_1, l_1] * [k_2, l_2] := [\min A, \max A], \tag{5.3}$$

$$[k_1, l_1] / [k_2, l_2] := [\text{floor}(\min B), \text{ceil}(\max B)], \tag{5.4}$$

where
$$A := \{k_1 * k_2, k_1 * l_2, l_1 * k_2, l_1 * l_2\},$$
$$B := \{k_1 / k_2, k_1 / l_2, l_1 / k_2, l_1 / l_2\},$$

and in the last operation it is assumed that $0 \notin [k_2, l_2]$. We also identify every integer k with the interval $[k, k]$. This is a simple example of *interval arithmetic*. Note that we use floor and ceiling, when defining the division of intervals, to ensure that the resulting interval has integer endpoints. This process, being an example of *outwards rounding*, is not needed for the other arithmetic operations, because the sum, difference and product of integers are integers. In real computations more sophisticated interval arithmetic, based on

5.1 Rational Functions and Interval Arithmetic

floating point numbers instead of integers, is used. As a consequence outwards rounding is necessary in every arithmetic operation (see [59]).

A useful feature of interval arithmetic is the following straightforward-to-prove proposition.

Proposition 5.1 *Assume I and J are two intervals and $\diamond \in \{+, -, *, /\}$ is an arithmetic operation. If $x \in I$ and $y \in J$, then $x \diamond y \in I \diamond J$.*

If $f : \mathbf{R}^d \to \mathbf{R}^{d'}$ is a rational function, it may be extended to intervals by replacing every arithmetic operation on numbers by the corresponding operation on intervals. To be more precise, recall that we are letting $f = (f_1, f_2, \ldots, f_{d'})$, where each $f_i : \mathbf{R}^d \to \mathbf{R}$ is a rational function. Since any rectangle $I_1 \times I_2 \times \cdots \times I_d$ can be identified with a sequence of d intervals $(I_1, I_2, \ldots, I_d)$, we can extend f_i to a function $\bar{f}_i$ that maps cubical rectangles to intervals with integer endpoints by replacing every arithmetic operation on numbers by the corresponding operation on intervals. Now, putting

$$\bar{f} := (\bar{f}_1, \bar{f}_2, \ldots, \bar{f}_{d'}), \tag{5.5}$$

we obtain a map that takes rectangles to rectangles or more precisely, a map

$$\bar{f} : \operatorname{CRect}(\mathbf{R}^d) \to \operatorname{CRect}(\mathbf{R}^{d'}),$$

where

$$\operatorname{CRect}(\mathbf{R}^d) := \{ X \subset \mathbf{R}^d \mid X \text{ is a cubical set and a rectangle} \}.$$

Theorem 5.2 *If $f : \mathbf{R}^d \to \mathbf{R}^{d'}$ is a rational function and $R \in \operatorname{CRect}(\mathbf{R}^d)$ is a rectangle, then*

$$f(R) \subset \bar{f}(R). \tag{5.6}$$

Proof. The theorem follows from Proposition 5.1 by an induction argument on the number of arithmetic operations in the rational function f. □

Equation (5.6) may look strange at first glance. So it is worth commenting that the left-hand side of this formula is the image of a rectangle R under a rational function f. This image in general is not a rectangle and it may be difficult to determine the image exactly. On the other hand, the right-hand side of the formula is the value of the extension of f to intervals on the intervals whose product is R. Unlike the left-hand side, the right-hand side is always a rectangle and it may be easily computed by applying the interval versions of arithmetic operations to arguments. Therefore, Eq. (5.6) may be viewed as an algorithm for finding rectangular upper estimates of the images of rectangles under a rational function.

Exercises

5.1 Consider the rational functions $f(x) := 1/(x+1) + 1/(x-1)$ and $g(x) := 2x/(x^2-1)$. Let $I = [5, 6]$. Find $\bar{f}(I)$ and $\bar{g}(I)$. Compare $\bar{f}$ and $\bar{g}$.

5.2 Given a rational number a, consider the set of numbers of the form $\mathbf{R}_a := \{na \mid n \in \mathbf{Z}\}$. Define counterparts of formulas (5.1)-(5.4) for intervals whose endpoints are not integers but elements of $\mathbf{R}_a$. Prove a counterpart of Proposition 5.1.

5.2 Maps on an Interval

As indicated in the introduction, the goal of this chapter is to introduce the homology of maps. To keep the technicalities to an absolute minimum, we begin our discussion with maps of the form $f : [a, b] \to [c, d]$. We do this for one very simple reason; we can draw pictures of these functions to help us develop our intuition. In practice we will want to apply these ideas to cubical sets where it is not feasible to visualize the maps, either because the map is too complicated or because the dimension is too high.

With this in mind consider the cubical sets $X = [-2, 2] \subset \mathbf{R}$ and $Y = [-3, 4] \subset \mathbf{R}$, and let $f : X \to Y$ be defined by

$$f(x) := \frac{(3x - 4)(5x + 4)}{15}.$$

Thus, we have two topological spaces and a continuous map between them. To treat these combinatorially it should come as no surprise that we will use the cubical structure for X and Y and $\bar{f}$ introduced earlier, but it remains to be seen how to view $\bar{f}$ as a combinatorial object.

The first observation we make is that in passing from a continuous function to a combinatorial object we lose information. Consider Figure 5.1, which shows two functions f and g that are close to one another. Since computing $\bar{f}$ involves rounding to nearest integers, it may happen that $\bar{f} = \bar{g}$. Thus the information that we will eventually obtain from f_* and g_* is much coarser than the information contained in the maps themselves.

Recall that elementary cubes are our fundamental objects for determining homology. Clearly,

$$X = [-2, -1] \cup [-1, 0] \cup [0, 1] \cup [1, 2].$$

So let us do our computations in terms of edges. From the combinatorial point of view, this suggests trying to map elementary intervals to sets of intervals. Since $f(-2) = 4$, $f(-1) = 7/15$, and f is monotone over the edge $[-2, -1]$, it is clear that

$$f([-2, -1]) \subset [0, 4].$$

Of course, this strategy of looking at the endpoints does not work for the edge $[0, 1]$ since f is not monotone here.

In dealing with this problem, we make use of the interval arithmetic introduced in the previous section. To be more precise, Theorem 5.2 implies that

5.2 Maps on an Interval

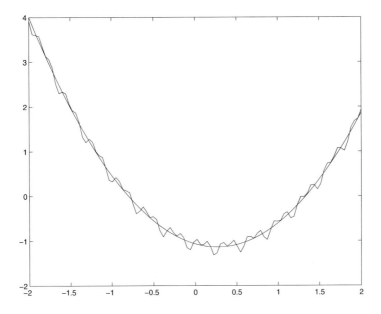

Fig. 5.1. The function $f(x) = (3x - 4)(5x + 4)/15$ and a function g that is close in the sense that $|f(x) - g(x)|$ is small for all $x \in [-2, 2]$.

$$
\begin{aligned}
f([0,1]) &\subset \bar{f}([0,1]) \\
&= ((3 * [0,1] - 4) * (5 * [0,1] + 4))/15 \\
&= (([3,3] * [0,1] - [4,4]) * ([5,5] * [0,1] + [4,4]))/[15,15] \\
&= (([0,3] - [4,4]) * ([0,5] + [4,4]))/[15,15] \\
&= ([-4,-1] * [4,9])/[15,15] \\
&= [-36,-4]/[15,15] \\
&= [\text{floor}\,(-\tfrac{36}{15}), \text{ceil}\,(-\tfrac{4}{15})] \\
&= [-3, 0].
\end{aligned}
$$

Performing these computations on all the elementary intervals in X leads to Table 5.1.

We can think of Table 5.1 as defining a map from edges to sets of edges. For example,

$$[0,1] \mapsto \{[-3,-2], [-2,-1], [-1,0]\}.$$

More precisely, Table 5.1 defines a multivalued map

$$\mathcal{F} : \mathcal{K}_{\max}(X) \rightrightarrows \mathcal{K}_{\max}(Y).$$

This leads to the following concept.

Table 5.1. Elementary Edges and Their Bounding Images in Terms of Sets of Edges for $f : X \to Y$

Edge of X	Bounds on the image	Image edges
$[-2, -1]$	$0 \leq f(x) \leq 4$	$\{[0,1],\ [1,2],\ [2,3],\ [3,4]\}$
$[-1, 0]$	$-2 \leq f(x) \leq 1$	$\{[-2,-1],\ [-1,0],\ [0,1]\}$
$[0, 1]$	$-3 \leq f(x) \leq 0$	$\{[-3,-2],\ [-2,-1],\ [-1,0]\}$
$[1, 2]$	$-1 \leq f(x) \leq 2$	$\{[-1,0],\ [0,1],\ [1,2]\}$

Definition 5.3 Let X and Y be cubical sets. A *combinatorial cubical multivalued map* $\mathcal{F} : \mathcal{K}_{\max}(X) \rightrightarrows \mathcal{K}_{\max}(Y)$ is a function from $\mathcal{K}_{\max}(X)$ to subsets of $\mathcal{K}_{\max}(Y)$.

We call $\mathcal{F}$ a combinatorial map, because only finite sets are involved in its definition. This will let us store $\mathcal{F}$ easily in the computer. On the other hand, as a purely combinatorial object, this map is in contrast to f, which acts on topological spaces X and Y. Therefore we need something extra, which would serve as a bridge between $\mathcal{F}$ and f. To achieve this represent the relation

$$\mathcal{F}([0,1]) = \{[-3,-2], [-2,-1], [-1,0]\}$$

graphically by means of the rectangle

$$[0,1] \times [-3,0] \subset [-2,2] \times [-3,4] = X \times Y.$$

Doing this for all the edges in the domain gives the region shown in Figure 5.2. Observe that the graph of $f : X \to Y$ is a subset of this region and therefore we can think of the region as representing an outer bound on the function f. The point is that we can think of this region as a graph of a *multivalued map* $F : X \rightrightarrows Y$, that is a map, which sends individual elements of X to subsets of Y. Consider for instance the points $x = 0.5$ and $x = 1.5$ in X. We can interpret the region as saying that $F(0.5) = [-3, 0]$, while $F(1.5) = [-1, 2]$.

Since our goal is to use the computer to perform the necessary computations, we do not need to, nor want to, consider arbitrary multivalued maps. Instead, we are satisfied with a very special class of such maps. To derive this class let us think about the essential elements of cubical homology. The first point is that our theory is built on cubical sets. Therefore, it seems reasonable to restrict the images of our multivalued maps to be cubical sets, that is, given $x \in X$, the image of x should be a cubical set.

Observe, however, that in using our cubical approach we do not think of the topological space X as being made up of points, but rather of elementary

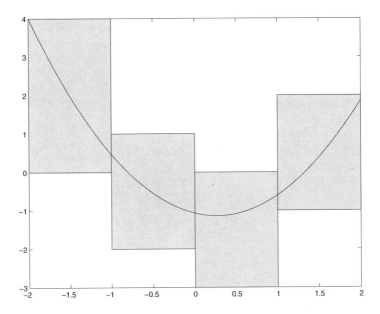

Fig. 5.2. The graph of the map produced by sending edges to sets of edges. Observe that the graph of the function $f(x) = (3x - 4)(5x + 4)/15$ lies inside the graph of this edge map.

cubes. Thus one might be tempted to add the condition that if x and x' belong to the same elementary cube, then their images should be the same. Unfortunately, this would lead to all maps being constant maps. To see why this is the case, consider our example of $X = [-2, 2]$. It contains the elementary cubes $[l, l + 1]$ for $l = -2, -1, \ldots, 1$. Since $-1 \in [-2, -1] \cap [-1, 0]$, imposing the above-mentioned condition would force the images of all points in $[-2, 0]$ to be the same. By induction we would get that the image of every point in $[-2, 2]$ would have to be the same.

We can, however, avoid this problem by using elementary cells rather than cubes. The elementary cells that make up $[-2, 2]$ are $(l, l + 1)$ for $l = -2, -1, 0, 1$ and $[l]$ for $l = -2, \ldots, 2$ and so different cells do not intersect.

Combining these ideas leads to the following definition.

Definition 5.4 Let X and Y be cubical sets. A multivalued map $F : X \rightrightarrows Y$ is *cubical* if

1. For every $x \in X$, $F(x)$ is a cubical set.
2. For every $Q \in \mathcal{K}(X)$, $F \vert_{\overset{\circ}{Q}}$ is constant; that is, if $x, x' \in \overset{\circ}{Q}$, then $F(x) = F(x')$.

Using elementary cells and the region in $X \times Y$ obtained from the combinatorial mapping $\mathcal{F}$ represented in Table 5.1, we see that the multivalued

180   5 Preview of Maps

map
$$F : [-2, 2] \rightrightarrows [-3, 4]$$
satisfies
$$F(x) = \begin{cases} [0, 4] & \text{if } x = -2, \\ [0, 4] & \text{if } x \in (-2, -1), \\ [0, 1] & \text{if } x = -1, \\ [-2, 1] & \text{if } x \in (-1, 0), \\ [-2, 0] & \text{if } x = 0, \\ [-3, 0] & \text{if } x \in (0, 1), \\ [-1, 0] & \text{if } x = 1, \\ [-1, 2] & \text{if } x \in (1, 2), \\ [-1, 2] & \text{if } x = 2. \end{cases}$$

There are two observations to be made at this point. First, for every $x \in X$,
$$f(x) \in F(x). \tag{5.7}$$

Thus the cubical multivalued map F acts as an outer approximation for the continuous function f. Geometrically this means that the graph of $f : X \to Y$ is a subset of the graph of F. Note that the direct comparison (5.7) wouldn't be possible with the map $\mathcal{F}$, because $\mathcal{F}$ acts on edges, not on individual points. We will refer to F as a *representation* of f.

Second, even though $F : X \rightrightarrows Y$ is a map defined on uncountably many points, it is completely characterized by the combinatorial multivalued map $\mathcal{F} : \mathcal{K}_1([-2, 2]) \rightrightarrows \mathcal{K}_1([-3, 4])$. This shouldn't be a surprise, because we derived F from the region defined by means of $\mathcal{F}$. To see it better it is convenient to introduce the following notation. If $\mathcal{E} := \{E_1, E_2, \ldots E_n\}$ is a set of edges in X, then we write
$$|\mathcal{E}| := E_1 \cup E_2 \cup \ldots \cup E_n.$$

This lets us easily compare F and $\mathcal{F}$. For instance, if $x \in (-2, -1)$, then
$$F(x) = [0, 4] = [0, 1] \cup [1, 2] \cup [2, 3] \cup [3, 4] = |\mathcal{F}([-2, -1])|.$$

If x is a vertex in X, that is a face of edges E_1 and E_2, then for arbitrarily chosen $x_i \in \overset{\circ}{E}_i$
$$F(x) = F(x_1) \cap F(x_2) = |\mathcal{F}(E_1)| \cap |\mathcal{F}(E_2)|, \tag{5.8}$$

therefore again $F(x)$ may be entirely characterized by $\mathcal{F}$. The two observations explain why we consider F as a bridge between f and $\mathcal{F}$.

The multivalued map F that we have constructed above is fairly coarse. If we want a tighter approximation, then one approach is to use finer graphs to describe X and Y. For example, let us write
$$X = \bigcup_{i=0}^{8} [-2 + \frac{i}{2}, -1.5 + \frac{i}{2}] \quad \text{and} \quad Y = \bigcup_{i=0}^{14} [-3 + \frac{i}{2}, -2.5 + \frac{i}{2}].$$

Table 5.2. Edges and Vertices for the Graphs of $X = [-2, 2]$ and $Y = [-3, 4]$

Edge of X	Bounds on the image	Image edges
$[-2, -1.5]$	$1.5 \leq f(x) \leq 4$	$\{[1.5, 2], [2, 2.5], [2.5, 3], [3, 3.5], [3.5, 4]\}$
$[-1.5, -1]$	$0 \leq f(x) \leq 2$	$\{[-0.5, 0], [0, 0.5], [0.5, 1], [1, 1.5], [1.5, 2]\}$
$[-1, -0.5]$	$-1 \leq f(x) \leq 0.5$	$\{[-1, -0.5], [-0.5, 0], [0, 0.5]\}$
$[-0.5, 0]$	$-1.5 \leq f(x) \leq 0$	$\{[-1.5, -1], [-1, -0.5], [-0.5, 0]\}$
$[0, 0.5]$	$-2 \leq f(x) \leq -0.5$	$\{[-2, -1.5], [-1.5, 0], [0, 0.5]\}$
$[0.5, 1]$	$-1.5 \leq f(x) \leq 0$	$\{[-1.5, -1], [-1, -0.5], [-0.5, 0]\}$
$[1, 1.5]$	$-1 \leq f(x) \leq 0.5$	$\{[-1, -0.5], [-0.5, 0], [0, 0.5]\}$
$[1.5, 2]$	$0 \leq f(x) \leq 2$	$\{[0, 0.5], [0.5, 1], [1, 1.5], [1.5, 2]\}$

Modifying our interval arithmetic so that it acts on intervals with endpoints being multiples of 1/2 (see Exercise 5.2), we obtain the data described in Table 5.2.

The graph of the corresponding multivalued map is shown on the left side of Figure 5.3. Observe that this is a tighter approximation to the function than what was obtained with intervals of unit length. In fact, one can obtain as good an approximation as one likes by choosing the edge lengths sufficiently small. In the second graph of Figure 5.3 one sees the graph of the multivalued map when the length of the edges is 0.05.

Of course, this method of subdividing the domain produces tighter and tighter approximations, but unfortunately it takes us out of the class of cubical sets. To avoid this we will adopt an alternative approach. Observe that subdividing has the effect of producing intervals that as individual sets make up a smaller fraction of the domain. We can obtain the same effect by scaling the size of the domain. To be more precise notice that a unit interval in $X = [-2, 2]$ represents a quarter of the domain. Now consider the map

$$\Lambda^{(2)} : \mathbf{R} \to \mathbf{R}$$

given by $\Lambda^{(2)}(x) = 2x$. Define

$$X^{(2)} = \Lambda^{(2)}(X) = [-4, 4],$$

and observe that a unit interval now makes up only an eighth of $X^{(2)}$. Of course, topologically $X^{(2)}$ and X are equivalent since $\Lambda^{(2)}$ has an inverse $\Omega_X^{(2)} : X^{(2)} \to X$ given by

$$\Omega_X^{(2)}(x) = \frac{1}{2}x.$$

It should not be forgotten that our interest is in obtaining a tighter approximation to the map $f : X \to Y$. With this in mind consider $f^{(2)} : X^{(2)} \to Y$ defined by

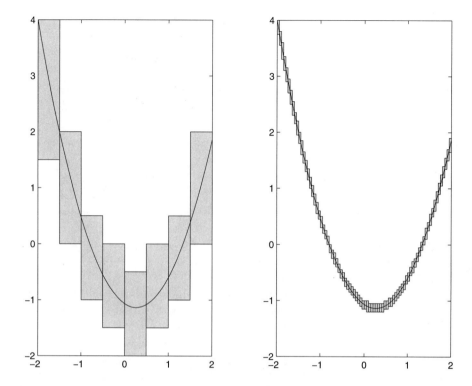

Fig. 5.3. Graphs of the multivalued approximation to $f(x) = (3x - 4)(5x + 4)/15$ obtained by means of interval arithmetic based on different basic lengths: 0.5 for the left graph and 0.05 for the right graph.

$$f^{(2)} := f \circ \Omega_X^{(2)}$$

or equivalently $f^{(2)} : [-4, 4] \to Y$ given by

$$f^{(2)}(x) = \frac{(3(\frac{x}{2}) - 4)(5(\frac{x}{2}) + 4)}{15} = \frac{(3x - 8)(5x + 8)}{60}.$$

The graph of the associated cubical multivalued map $F^{(2)}$ constructed using $\bar{f}^{(2)}$ is indicated in Figure 5.4.

There is no reason that one has to limit the scaling to a factor of 2. More generally, for any integer α we can define $\Lambda^{(\alpha)} : \mathbf{R} \to \mathbf{R}$ by

$$\Lambda^{(\alpha)}(x) = \alpha x.$$

By insisting that α is an integer we are guaranteed that elementary intervals are sent to cubical sets. Under this scaling $X^{(\alpha)} := \Lambda^{(\alpha)}(X) = [-2\alpha, 2\alpha]$. The rescaling of the domain X of f leads to a rescaling of f by means of the inverse map $\Omega_X^{(\alpha)} : X^{(\alpha)} \to X$ defined by $\Omega_X^{(\alpha)}(x) = x/\alpha$. We set

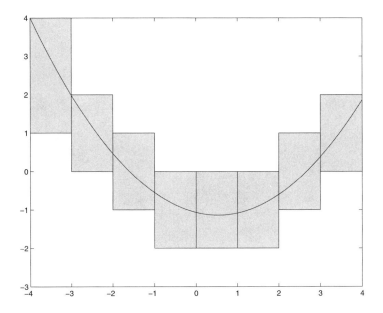

Fig. 5.4. The graph of the cubical multivalued representation of $f^{(2)}(x) = (3x - 8)(5x + 8)/60$. The size of the domain has been doubled, but the intervals are all elementary cubes.

$$f^{(\alpha)} := f \circ \Omega_X^{(\alpha)} : X^{(\alpha)} \to Y.$$

More explicitly, given $x \in X^{(\alpha)}$,

$$f^{(\alpha)}(x) = f\left(\frac{x}{\alpha}\right).$$

Note that we can recover the original map f from $f^{(\alpha)}$ by the formula

$$f(x) = f^{(\alpha)} \circ \Lambda^{(\alpha)}(x) = f^{(\alpha)}(\alpha x).$$

Figure 5.5 shows the graph of the associated cubical multivalued map $F^{(20)}$ constructed using $f^{(20)}$.

It may seem, at first, that the approximation given in the right-hand-side graph of Figure 5.3 is more accurate than that of Figure 5.5. After all, the sets $F(x)$ are smaller and hence provide more accurate bounds on $f(x)$. However, it must be kept in mind that we are interested in the images of homology classes rather than particular points. Thus, as shown in Chapter 6, modulo some technical conditions, it is sufficient for the homology of the set $F(x)$ to be the same as the homology of a point. In other words, the values of F need to be acyclic. Note that this is satisfied in both cases since $F(x)$ is an interval for any x.

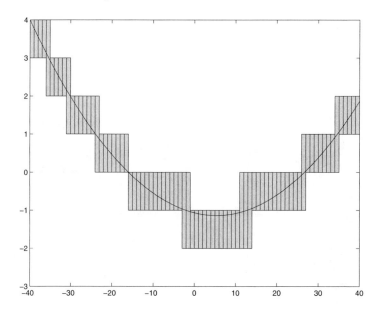

Fig. 5.5. The graph of the cubical multivalued representation of $f^{(20)}(x)$: $[-40, 40] \to [-3, 4]$.

Furthermore, while both the second graph in Figure 5.3 and Figure 5.5 have been obtained by dividing the domain into 80 intervals, in the first the range consists of approximately 100 intervals while in the latter only six intervals are used. As one might expect, shorter lists require fewer calculations. Therefore, from the computational point of view the latter approximation is preferable.

The reader might ask a question: If so, why don't we just consider the simplest multivalued "approximation" $F : [-2, 2] \rightrightarrows [-3, 4]$ of f given by

$$F(x) := [-3, 4]$$

for all x? It has all the desired properties, its only value is acyclic, and it does not require any calculation at all! We could. However, this is an artifact of the trivial topology of the range $Y = [-3, 4]$. As we shall see shortly, when the homology of the range is nontrivial, assuring that the acyclicity condition is satisfied, it is related to the size of the values.

Exercise

5.3 Consider the function $f : [-2, 2] \to [-3, 3]$ given by $f(x) = x(x^2 - 1)$.

(a) Use interval arithmetic to construct a cubical multivalued map F that is a representation of f.
(b) Write an explicit formula for $f^{(2)} : [-4, 4] \to [-3, 3]$.

(c) Write a program that draws the graph of a cubical multivalued map $F^{(\alpha)}$ that is an outer representation for a scaling of f by α.

5.3 Constructing Chain Selectors

In the previous section we consider the problem of approximating maps from one interval to another. Of course, the real goal is to use such an approximation to induce a map between homology groups. So in this section we begin with this question: How can we use the information in Figure 5.2 to construct a group homomorphism $f_* : H_*(X) \to H_*(Y)$?

By now we know that maps on homology are induced by chain maps. Thus we need to address the following issue. Given a multivalued representation $F : X \rightrightarrows Y$ of $f : X \to Y$, how can we produce an appropriate chain map $\varphi : C(X) \to C(Y)$? Our approach is to use the geometry of F.

By Proposition 2.15 for any $x \in X$ there exists a unique elementary cube $Q \in \mathcal{K}(X)$ such that $x \in \overset{\circ}{Q}$. In analogy to the assumption that $f(x) \in F(x)$ we will require inclusion on the level of the support of images of the chain map, that is,

$$\left|\varphi(\widehat{Q})\right| \subset F(\overset{\circ}{Q}). \tag{5.9}$$

Chain maps satisfying this property are called *chain selectors* of F.

To see what this means in practice, let us return to the example of the previous section where F is indicated by Figure 5.2. Recall that the canonical basis of $C_0([-2, 2])$ consists of dual vertices

$$\{\widehat{[-2]}, \widehat{[-1]}, \widehat{[0]}, \widehat{[1]}, \widehat{[2]}\}.$$

Similarly, the canonical basis of $C_0(Y)$ consists of

$$\{\widehat{[-3]}, \widehat{[-2]}, \widehat{[-1]}, \widehat{[0]}, \widehat{[1]}, \widehat{[2]}, \widehat{[3]}, \widehat{[4]}\}.$$

We begin by defining
$$\varphi_0 : C_0(X) \to C_0(Y)$$
so that it satisfies (5.9) on vertices. For lack of a better idea let us set

$$\varphi_0(\widehat{[v]}) := \widehat{[\max F(v)]}.$$

For example, $\varphi_0\widehat{[-2]} = \widehat{[4]}$, $\varphi_0\widehat{[-1]} = \widehat{[1]}$, $\varphi_0\widehat{[0]} = \widehat{[0]}$, etc.

We identify the dual vertices in $C_0(X)$ and, respectively, $C_0(Y)$ with the column vectors

$$e_j^n := \begin{bmatrix} 0 \\ \vdots \\ 0 \\ 1 \\ 0 \\ \vdots \\ 0 \end{bmatrix}$$

of the canonical bases of $\mathbf{Z}^5$, respectively, $\mathbf{Z}^8$, in the following manner:

$$\widehat{[-2]} = e_1^5, \quad \widehat{[-1]} = e_2^5, \quad \widehat{[0]} = e_3^5, \ldots, \widehat{[2]} = e_5^5,$$

for $C_0(X)$ and

$$\widehat{[-3]} = e_1^8, \quad \widehat{[-2]} = e_2^8, \quad \widehat{[-1]} = e_3^8, \ldots, \widehat{[4]} = e_8^8.$$

for $C_0(Y)$.

The matrix of φ_0 with respect to these bases is

$$\varphi_0 = \begin{bmatrix} 0 & 0 & 0 & 0 & 0 \\ 0 & 0 & 0 & 0 & 0 \\ 0 & 0 & 0 & 0 & 0 \\ 0 & 0 & 1 & 1 & 0 \\ 0 & 1 & 0 & 0 & 0 \\ 0 & 0 & 0 & 0 & 1 \\ 0 & 0 & 0 & 0 & 0 \\ 1 & 0 & 0 & 0 & 0 \end{bmatrix}.$$

It is natural to ask what would happen if we make a different choice; for example,

$$\varphi_0(\widehat{[v]}) := [\widehat{\min F(v)}].$$

We could get an entirely different chain map, but as proved in the next chapter, both choices lead to the same map in homology. When defining $\varphi_1 : C_1(X) \to C_1(Y)$, on dual edges in $C_1(X)$, we cannot choose an arbitrary dual edge of $C_1(Y)$ as we did for vertices because we face an additional constraint from the condition

$$\partial_1 \varphi_1 = \varphi_0 \partial_1$$

for a chain map. We will attempt to "lift" the definition of φ_0 to dimension 1 so that this condition is satisfied. Consider the interval $[-2, -1] \subset [-2, 2]$. How should we define $\varphi_1(\widehat{[-2, -1]})$? Let us return to the level of topology to find an answer. We know that

$$\left|\varphi_0(\widehat{[-2]})\right| = \left|\widehat{[4]}\right| = [4] \quad \text{and} \quad \left|\varphi_0(\widehat{[-1]})\right| = \left|\widehat{[1]}\right| = [1].$$

Since we began with a continuous map and since -2 and -1 are the endpoints of the interval $[-2, -1]$, it seems reasonable that we want the image of $[-2, -1]$

to stretch from the image of -2 to the image of -1, namely to go from 4 to 1. Figure 5.6 is an attempt to graphically indicate this idea. The polygonal curve indicated by the dashed line segments joins the pairs of vertices of the form $([v], |\varphi_0(\widehat{[v]})|)$. In particular, the projection onto the y-axis of the first segment is the interval extending from the image of -2 to the image of -1.

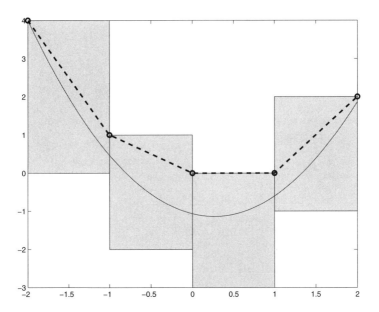

Fig. 5.6. The graph of a cubical map F that is a representation of f. Given a vertex v, $\varphi_0(\widehat{[v]}) := [\widehat{\max F(v)}]$. The circles indicate pairs of vertices $([v], |\varphi_0(\widehat{[v]})|)$ while the dashed segments connecting them can be used to localize, on the y-axis, the images of the corresponding edges under the chain map φ_1.

Of course, this needs to be done algebraically, so let

$$\varphi_1(\widehat{[-2,-1]}) := -\widehat{[1,2]} - \widehat{[2,3]} - \widehat{[3,4]}$$

and observe that

$$\left|\varphi_1(\widehat{[-2,-1]})\right| = [1,4] \subset F((-2,-1))$$

so (5.9) is satisfied. As in the case of the zero-dimensional level, we do not claim that this is a unique choice. Why the minus signs? As we proceed in the direction of the x-axis (i.e., from -2 to -1), the image of the interval goes in the opposite direction of the y-axis (i.e., from 4 to 1).

Similarly, we can set

$$\varphi_1([\widehat{-1,-0}]) := -[\widehat{0,1}].$$

But what about $\varphi_1([\widehat{0,1}])$, where $\varphi_1(\widehat{0}) = \varphi_1(\widehat{1}) = \widehat{0}$? Since the two endpoints are the same, let us just declare that $\varphi_1([\widehat{0,1}])$ does not map to any intervals, namely that $\varphi_1([\widehat{0,1}]) := 0$.

Again, we want to express φ_1 as a matrix. So let us identify dual edges with canonical basis vectors as we have done for vertices. For $C_1([-2,2])$ we set

$$[\widehat{-2,-1}] = \mathbf{e}_1^4, \quad [\widehat{-1,0}] = \mathbf{e}_2^4, \quad [\widehat{0,1}] = \mathbf{e}_3^4, \quad [\widehat{1,2}] = \mathbf{e}_4^4,$$

and for $C_1([-3,4])$ we set

$$[\widehat{-3,-2}] = \mathbf{e}_1^7, \quad [\widehat{-2,-1}] = \mathbf{e}_2^7, \ldots, [\widehat{3,4}] = \mathbf{e}_7^7.$$

Applying the reasoning described above to each of the intervals, we obtain the following matrix:

$$\varphi_1 = \begin{bmatrix} 0 & 0 & 0 & 0 \\ 0 & 0 & 0 & 0 \\ 0 & 0 & 0 & 0 \\ 0 & -1 & 0 & 1 \\ -1 & 0 & 0 & 1 \\ -1 & 0 & 0 & 0 \\ -1 & 0 & 0 & 0 \end{bmatrix}.$$

The reader should check that, as we wanted, φ is a chain map.

Recall that chain maps generate maps on homology, so we define $f_* : H_*(X) \to H_*(Y)$ by

$$f_* := \varphi_*.$$

Since X and Y are acyclic, we know that

$$H_k(X) \cong H_k(Y) \cong \begin{cases} \mathbb{Z} & \text{if } k = 0, \\ 0 & \text{otherwise.} \end{cases}$$

Thus the only interesting map is

$$f_0 : H_0(X) \to H_0(Y).$$

We know from Section 2.3 that all dual vertices in X are homologous and similarly for Y. So, let us take, for example, the equivalence class $\zeta := \left[\widehat{[-2]}\right]$ of $\widehat{[-2]}$ as a generator of $H_0(X)$ and the equivalence class $\xi := \left[\widehat{[3]}\right]$ of $\widehat{[3]}$ as a generator of $H_0(Y)$. Since $\varphi(\widehat{[-2]}) = \widehat{[4]}$ and $\widehat{[4]} \sim \widehat{[3]}$, f_0 is defined on the generator by the formula

$$f_0(\zeta) := \left[\varphi_0(\widehat{[-2]})\right] = \left[\widehat{[4]}\right] = \left[\widehat{[3]}\right] = \xi.$$

This is probably a good place to restate the caveat given at the beginning of the chapter. We are motivating the ideas behind homology maps at this point.

If you do not find these definitions and constructions completely rigorous, that is good—they are not. We will fill in the details later. For the moment we are just trying to get a feel for how we can obtain group homomorphisms from continuous maps.

Exercise

5.4 Construct another chain selector of F given in this section by starting from the definition
$$\varphi_0(\widehat{[v]}) := [\widehat{\min F(v)}].$$
Give an argument why your chain map induces the same homomorphism in homology as the one discussed in this text.

5.4 Maps of Γ^1

Up to now we have considered maps from one interval to another. Since intervals are acyclic spaces it is not surprising that the resulting maps on homology are not very interesting. So let us consider a space with nontrivial homology such as Γ^1. Unfortunately, it is rather difficult to draw the graph of a function $f : \Gamma^1 \to \Gamma^1$. So we will employ a trick. In order to draw simple pictures we will think of Γ^1 as the interval $[0, 4]$ but identify the endpoints.

More precisely, the identification of a point $x = (x_1, x_2) \in \Gamma^1$ with a number $t \in [0, 4]$ is made via the function $t \mapsto (x_1(t), x_2(t))$ defined as follows.

$$x(t) := (x_1(t), x_2(t)) := \begin{cases} (0,0) + t(1,0) & \text{if } t \in [0,1], \\ (1,0) + (t-1)(0,1) & \text{if } t \in [1,2], \\ (1,1) + (t-2)(-1,0) & \text{if } t \in [2,3], \\ (0,1) + (t-3)(0,-1) & \text{if } t \in [3,4]. \end{cases} \quad (5.10)$$

Note that the numbers $t \in (0, 4)$ correspond bijectively to points $x \in \Gamma^1 \setminus \{(0,0)\}$ but $x(0) = x(4) = (0,0)$. Also note that $t = 0, 1, 2, 3$ and $4 \sim 0$ correspond to elementary vertices of Γ^1 while the intervals $[0, 1]$, $[1, 2]$, $[2, 3]$, $[3, 4]$ correspond to respective elementary edges by going counterclockwise around the square $[0, 1]^2$. In fact, it is useful to go a step further and think of Γ^1 as the real line where one identifies each point t with its translate $t + 4$.

To see how this works in practice, consider the function $f : [0, 4] \to \mathbf{R}$ given by $f(t) = 2t$. We want to think of f as a map from $\Gamma^1 \to \Gamma^1$ and do this via the identification of $t \sim t + 4$ (see Figure 5.7).

While this process allows us to draw nice figures it must be kept in mind that what we are really interested in is f as a continuous mapping from Γ^1 to Γ^1. How should we interpret the drawing in Figure 5.7(b)? Observe that as we move across the interval $[0, 2]$ the graph of f covers all of $[0, 4]$. So going halfway around Γ^1 in the domain corresponds to going once around Γ^1 in the image. Thus, going all the way around Γ^1 in the domain results in going twice

190   5 Preview of Maps

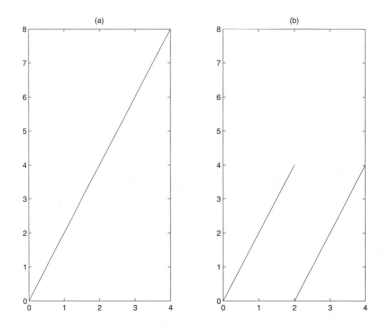

Fig. 5.7. Two versions of the graph of $f(t) = 2t$. The left-hand drawing indicates $f : [0, 4] \to \mathbf{R}$. In the right-hand drawing we have made the identification of $y \sim y+4$ and so can view $f : [0, 4] \to [0, 4]$. It is important to keep in mind that on both the x- and y-axes we make the identification of $0 \sim 4 \sim 8$. Thus $f(0) \sim 0 \sim f(2) \sim 4 \sim f(4) \sim 8$.

around Γ^1 in the image. In other words, f wraps Γ^1 around itself twice. In Figure 5.8 we show a variety of different maps and indicate how many times they wrap Γ^1 around itself. Our goal in this section is to see if we can detect the differences in these maps algebraically.

Recall that
$$H_k(\Gamma^1) \cong \begin{cases} \mathbf{Z} & \text{if } k = 0, 1, \\ 0 & \text{otherwise.} \end{cases}$$
We will focus our attention on $f_1 : H_1(\Gamma^1) \to H_1(\Gamma^1)$.

Since $H_1(\Gamma^1) \cong \mathbf{Z}$, it is generated by one element $\zeta := [z]$ where z is a counterclockwise cycle expressed, under the discussed identification of Γ^1 with $[0, 4]$, as
$$z := \widehat{[0, 1]} + \widehat{[1, 2]} + \widehat{[2, 3]} + \widehat{[3, 4]}.$$
If the construction of F and its chain selector φ for a map $f : \Gamma^1 \to \Gamma^1$ does not require rescaling, then it is enough to determine the value of φ_1 on z. What happens if we do need a rescaling, say, by $\alpha = 2$? Formally, the rescaling by 2 of $\Gamma^1 \subset \mathbf{R}^2$ should be defined as a map $\Lambda^{(2,2)} : \mathbf{R}^2 \to \mathbf{R}^2$ defined by
$$\Lambda^{(2,2)}(x_1, x_2) := (2x_1, 2x_2).$$

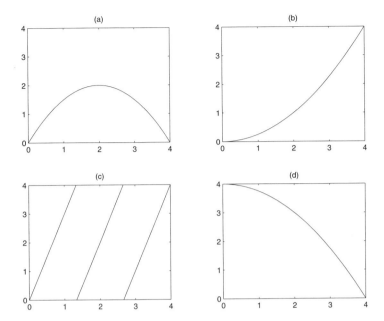

Fig. 5.8. Four different maps $f : \Gamma^1 \to \Gamma^1$. How do these different f's wrap Γ^1 around Γ^1? (a) $f(t) = t(4-t)/2$ wraps the interval $[0,2]$ halfway around Γ^1 and then over the interval $[2,4]$ f unwraps it. Thus we could say that the total amount of wrapping is 0. (b) $f(t) = t^2/4$ wraps Γ^1 once around Γ^1. (c) $f(t) = 3t$ wraps Γ^1 three times around Γ^1. (d) $f(t) = (16-t^2)/4$ wraps Γ^1 once around Γ^1, but in the opposite direction from the example in (b).

Thus the image of $X := \Gamma^1$ under this rescaling is the boundary of the square $[0,2]^2$. This is the way we develop the formal definitions in the next chapter. However, for the moment we are content to use the simpler presentation of X as the interval $[0,4]$ with identified endpoints. An analogy of that presentation for the rescaled set $X^{(2)}$ is viewing it as the interval $X^{(2)} = [0,8]$ with identified endpoints. In other words, we consider the rescaling $\Lambda^{(2)}(t) = 2t$ in $\mathbf{R}$, as in the first section. The reader is encouraged to check that the homology of $X^{(2)}$ is the same as that of Γ^1, in particular, $H_1(X^{(2)}) \cong \mathbf{Z}$. Moreover, this homology group is generated by one element $\zeta^{(2)} := [z^{(2)}]$, where $z^{(2)}$ is a counterclockwise cycle expressed, under the discussed identification with $[0,8]$, as

$$z^{(2)} := \widehat{[0,1]} + \widehat{[1,2]} + \widehat{[2,3]} + \widehat{[3,4]} + \widehat{[4,5]} + \widehat{[5,6]} + \widehat{[6,7]} + \widehat{[7,8]}. \quad (5.11)$$

If the map f requires rescaling by $\alpha = 2$, we will produce a map $f_1^{(2)} : H_1(X^{(2)}) \to H_1(X)$. Our goal is to determine the map $f_1 : H_1(X) \to H_1(X)$, so we will need to compose $f_1^{(2)}$ with the homology map $\Lambda_1^{(2)} : H_1(X) \to H_1(X^{(2)})$ of our rescaling. Observe that $\Lambda^{(2)}$ sends the support of z to the support of $z^{(2)}$ and it preserves orientation on the contour, so it can be guessed that we should have the formula

$$\Lambda_1^{(2)}(\zeta) = \zeta^{(2)}.$$

However, we omit the verification at this time.

After this preliminary discussion, we are now ready to show the construction of the homology map for some specific functions.

Example 5.5 Let us begin by considering the map $f : \Gamma^1 \to \Gamma^1$ given by

$$f(t) = \frac{t(4-t)}{2},$$

which is drawn in Figure 5.8(a). The first step is to obtain an approximation for f, which is done using the interval approximation $\bar{f}$ defined by (5.5). In Figure 5.9 we indicate the resulting cubical multivalued map F that is a representation for f.

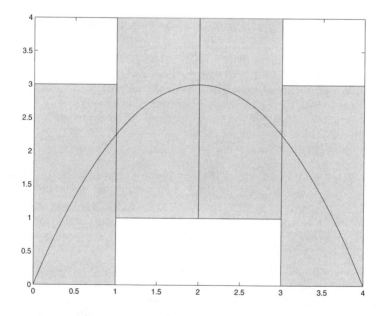

Fig. 5.9. The representation for the map $f(t) = t(4-t)/2$. Because we are identifying 0 and 4, $F(1) = F(3) = [1,3] \cup [4]$. Thus for this representation the image of a point need not be an acyclic set.

Recalling the relation (5.8) between F and $\mathcal{F}$, we see that

$$F([1]) = |\mathcal{F}([0,1])| \cap |\mathcal{F}([1,2])|$$
$$= [1,3] \cup [4].$$

5.4 Maps of Γ^1 193

This is troubling. What we are saying is that with this procedure the representation of a point is the union of two disjoint sets. This, in particular, is not an acyclic set, which is what we wanted to obtain and thus we turn to the scaling technique introduced at the end of Section 5.2.

To avoid this problem we can make use of the scaling $\alpha = 2$ to Γ^1. Recall that the scaling is only applied to the domain. To keep track of this let us write $f : X \to Y$, where $X = Y = \Gamma^1$. Then under the scaling $\Lambda^{(2)} : \mathbf{R} \to \mathbf{R}$, $X^{(2)} := \Lambda^{(2)}(X)$. Note that $X^{(2)}$ can be viewed as the interval $[0, 8]$ in $\mathbf{R}$ with 0 and 8 identified. Since the range is not being changed, we continue to view Y as the interval $[0, 4]$ with the identification of 0 with 4. Performing the same computations as before, we can compute a representation for $f^{(2)}$ and obtain the result indicated in Figure 5.10.

Using the same rules as before, we end up with the multivalued map

$$F^{(2)}(t) = \begin{cases} [0,2] \cup [3,4] & \text{if } t = 0, \\ [0,2] \cup [3,4] & \text{if } t \in (0,1), \\ [0,2] & \text{if } t = 1, \\ [0,2] & \text{if } t \in (1,2), \\ [1,2] & \text{if } t = 2, \\ [1,3] & \text{if } t \in (2,3), \\ [1,3] & \text{if } t = 3, \\ [1,3] & \text{if } t \in (3,4), \\ [1,3] & \text{if } t = 4, \\ [1,3] & \text{if } t \in (4,5), \\ [1,3] & \text{if } t = 5, \\ [1,3] & \text{if } t \in (5,6), \\ [1,2] & \text{if } t = 6, \\ [0,2] & \text{if } t \in (6,7), \\ [0,2] & \text{if } t = 7, \\ [0,2] \cup [3,4] & \text{if } t \in (7,8), \\ [0,2] \cup [3,4] & \text{if } t = 8. \end{cases}$$

We shall skip the verification that all the values are now acyclic.

Having determined $F^{(2)}$ we will construct its chain selector $\varphi_0 : C_0(X^{(2)}) \to C_0(X)$ in the same manner as in Section 5.3. Set $\varphi_0(v) = \max F^{(2)}(v)$ for any vertex v. Thus for example, $\varphi_0(\widehat{[0]}) = \widehat{[4]}$, $\varphi_0(\widehat{[1]}) = \widehat{[2]}$, etc. Having defined φ_0, the construction of $\varphi_1 : C_1(X^{(2)}) \to C_1(X)$ also follows as in Section 5.3. Of course, because of the identification, it is a little more subtle in this case. In particular, $\varphi_0(\widehat{[0]}) = \widehat{[4]} = \widehat{[0]}$.

Again, we assign canonical vectors to the duals of elementary intervals as follows. In $C_1(X^{(2)})$, set

$$\widehat{[0,1]} = \mathbf{e}_1^8, \quad \widehat{[1,2]} = \mathbf{e}_2^8, \ldots, \widehat{[7,8]} = \mathbf{e}_8^8,$$

and in $C_1(Y)$, set

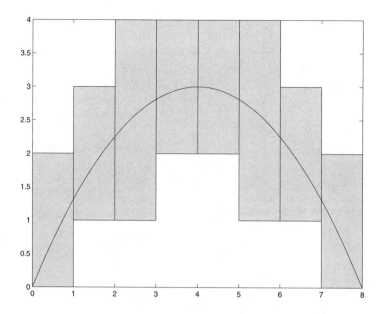

Fig. 5.10. The representation for the map $f^{(2)}(t) = t(8-t)/8$.

$$\widehat{[0,1]} = \mathbf{e}_1^4, \quad \widehat{[1,2]} = \mathbf{e}_2^4, \quad \widehat{[2,3]} = \mathbf{e}_3^4, \quad \widehat{[2,3]} = \mathbf{e}_4^4.$$

Using these bases we can write

$$\varphi_1 = \begin{bmatrix} 1 & 0 & 0 & 0 & 0 & 0 & 0 & -1 \\ 1 & 0 & 0 & 0 & 0 & 0 & 0 & -1 \\ 0 & 0 & 1 & 0 & 0 & -1 & 0 & 0 \\ 0 & 0 & 0 & 0 & 0 & 0 & 0 & 0 \end{bmatrix}.$$

As before, the negative signs arise because the image of the interval goes in the opposite direction from the ordering of the y-axis.

In order to understand the induced map from $H_1(X^{(2)})$ to $H_1(X)$, we need to see how φ_1 acts on $z^{(2)}$ defined by (5.11).

In vector notation, we have

$$z^{(2)} = \begin{bmatrix} 1 \\ 1 \\ \vdots \\ 1 \end{bmatrix}.$$

It is instantly checked that $\varphi_1(z^{(2)}) = 0$. Hence $f_1^{(2)} = 0$ and, consequently, $f_1 : H_1(X^{(2)}) \to H_1(\Gamma^1)$ is the zero homomorphism. A topological interpretation can be obtained by observing that wrapping does not occur here.

5.4 Maps of Γ^1

Example 5.6 Let's repeat this calculation for the map

$$f(t) = \frac{t^2}{4}.$$

We proceed exactly as before using the interval map $\bar{f}$ defined by (5.5). Figure 5.11 shows the resulting multivalued map and indicates how the chain maps[1] are defined. Of course, for purposes of calculation we want to view the chain map as a matrix

$$\varphi_1 = \begin{bmatrix} 1 & 0 & 0 & 0 \\ 0 & 0 & 1 & 0 \\ 0 & 0 & 1 & 0 \\ 0 & 0 & 0 & 1 \end{bmatrix}.$$

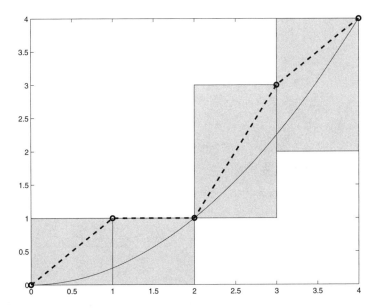

Fig. 5.11. The representation for the map $f(t) = t^2/4$. The circles indicate pairs of vertices $([v], |\varphi_0(\widehat{[v]})|)$ while the dashed segments connecting them can be used to localize, on the y-axis, the images of the corresponding edges under the chain map φ_1.

To determine the image of z under φ_1, we evaluate

$$\begin{bmatrix} 1 & 0 & 0 & 0 \\ 0 & 0 & 1 & 0 \\ 0 & 0 & 1 & 0 \\ 0 & 0 & 0 & 1 \end{bmatrix} \begin{bmatrix} 1 \\ 1 \\ 1 \\ 1 \end{bmatrix} = \begin{bmatrix} 1 \\ 1 \\ 1 \\ 1 \end{bmatrix}.$$

---
[1] The reader is encouraged to define those chain maps explicitly.

Thus $f_1 : H_1(X) \to H_1(X)$ is given by $f_1(\zeta) = \zeta$. Observe that this corresponds to the fact that $f(t) = t^2/4$ wraps Γ^1 around itself once in the clockwise direction.

Example 5.7 We shall do one more example, that of

$$f(t) = 2t.$$

Figure 5.12 shows the multivalued map is a representation when a scaling $\alpha = 2$ is used.

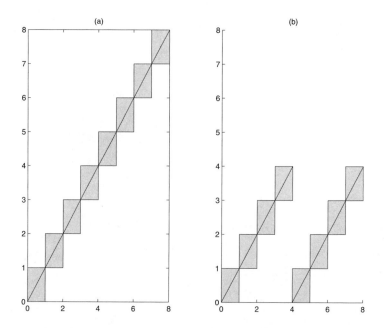

Fig. 5.12. The representation for the rescaling by 2 of the map $f(t) = 2t$. (a) The graph of the multivalued map with domain $[0,4]$ and image $[0,8]$. (b) The same graph. However, the interval $[0,4]$ has been identified with the interval $[4,8]$.

The rescaling of $[0,4]$ by 3 gives the interval $[0,12]$, which is composed of 12 elementary intervals. Following exactly the same process as in the previous cases we obtain

$$\varphi_1 = \begin{bmatrix} 1 & 0 & 0 & 0 & 0 & 1 & 0 & 0 & 0 & 0 & 0 & 0 \\ 0 & 1 & 0 & 0 & 0 & 0 & 0 & 1 & 0 & 0 & 0 & 0 \\ 0 & 0 & 1 & 0 & 0 & 0 & 0 & 0 & 0 & 1 & 0 & 0 \\ 0 & 0 & 0 & 0 & 1 & 0 & 0 & 0 & 0 & 0 & 1 & 0 \end{bmatrix}.$$

Again viewing how this acts on the generator $\zeta^{(3)} = [z^{(3)}]$ of $H_1(X^{(3)})$ (the construction of this generator, by analogy to previous cases, is left to the reader), we have

$$\varphi_1(z^{(3)}) = \begin{bmatrix} 1&0&0&0&0&1&0&0&0&0&0&0 \\ 0&1&0&0&0&0&0&1&0&0&0&0 \\ 0&0&1&0&0&0&0&0&0&1&0&0 \\ 0&0&0&0&1&0&0&0&0&0&1&0 \end{bmatrix} \begin{bmatrix} 1 \\ 1 \\ 1 \\ \vdots \\ 1 \end{bmatrix} = \begin{bmatrix} 2 \\ 2 \\ 2 \\ 2 \end{bmatrix} = 2z.$$

Finally, we use the homology of rescaling to get

$$f_1(\zeta) = f_1^{(3)} \Lambda_1^{(3)}(\zeta) = f_1^{(3)}(\zeta^{(3)}) = 2\zeta.$$

Thus the homology map on the first level is multiplication by 2. Observe that again it corresponds to the fact that $f(t) = 2t$ wraps Γ^1 around itself twice.

Exercises

5.5 Compute $f_1 : H_1(\Gamma^1) \to H_1(\Gamma^1)$ for $f(t) = -t$.

5.6 To motivate the use of computers, try to compute $f_1 : H_1(\Gamma^1) \to H_1(\Gamma^1)$ for $f(t) = 3t$.

6
Homology of Maps

In Chapter 5 we have provided a preview of the issues involved in defining homology of maps. In this chapter we revisit this material, but in a rigorous and dimension-independent manner. We begin with the introduction of representable sets. These are sets that can be constructed using elementary cells and represent a larger class than that of cubical sets. This extra flexibility is used in Section 6.2 to construct cubical multivalued maps. As described in Section 5.2, these multivalued maps provide representations of the continuous function for which we wish to compute the homology map. Section 6.3 describes the process by which one passes from the cubical map to a chain map from which one can define a map on homology. Section 6.4 shows that applying the above-mentioned steps (plus perhaps rescaling) to a continuous function leads to a well-defined map on homology. Finally, in the last section, we address the question of when do two different continuous maps give rise to the same map on homology.

6.1 Representable Sets

The goal of this chapter is to define homology maps induced by continuous functions on cubical sets. As mentioned above, an intermediate step is the introduction of cubical multivalued maps. These maps are required to have a weak form of continuity that is most easily expressed using sets defined in terms of elementary cells. For this reason we introduce the notion of a representable set, which is more general than that of a cubical set.

Definition 6.1 A set $Y \subset \mathbf{R}^d$ is *representable* if it is a finite union of elementary cells. The family of representable sets in $\mathbf{R}^d$ is denoted by $\mathcal{R}^d$.

Example 6.2 The set

$$X = ((0,1) \times (0,1)) \cup ([0] \times (0,1)) \cup ([0,2] \times [1]) \cup ([3] \times [0])$$

200   6 Homology of Maps

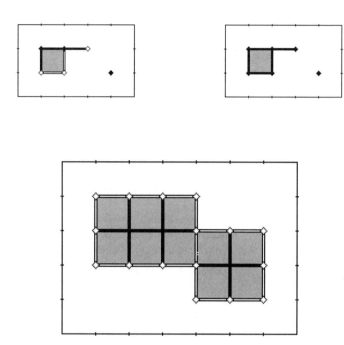

Fig. 6.1. The images generated by the CubTop program applied to the representable set X discussed in Example 6.2. Top left: The image of X. The elementary cells in X are displayed either in black (vertices and edges) or in gray (the square). Those in the closure of X but not in X are marked white. Top right: The image of $\operatorname{ch}(X) = \operatorname{cl} X$. Bottom: The image of $\operatorname{oh}(X)$ discussed in Example 6.9.

is representable. It is the union of three elementary vertices,

$$[0] \times [1], \quad [1] \times [1], \quad [3] \times [0],$$

three elementary edges,

$$[0] \times (0,1), \quad (0,1) \times [1], \quad (1,2) \times [1],$$

and one elementary square, $(0,1) \times (0,1)$. Figure 6.1, top left, shows the image of X generated by the CubTop program.

We begin with a series of results that characterize representable sets.

Proposition 6.3 *Representable sets have the following properties:*

(i) Every elementary cube is representable.
(ii) If $A, B \subset \mathbf{R}^d$ are representable, then $A \cup B$, $A \cap B$, and $A \setminus B$ are representable.

(iii) $X \subset \mathbf{R}^d$ *is a cubical set if and only if it is closed and representable.*

Proof. The first statement follows from Proposition 2.15(v), while the second is a consequence of Proposition 2.15(iii).

To prove the third statement, assume X is a cubical set. Then as a finite union of elementary cubes it is closed, and by (i) and (ii) it is representable. Thus assume that X is closed and representable. Then

$$X = \mathring{Q}_1 \cup \mathring{Q}_2 \cup \ldots \cup \mathring{Q}_m$$

for some $Q_i \in \mathcal{K}^d$, and by Proposition 2.15(iv)

$$X = \operatorname{cl} X = \operatorname{cl} \mathring{Q}_1 \cup \operatorname{cl} \mathring{Q}_2 \cup \ldots \cup \operatorname{cl} \mathring{Q}_m = Q_1 \cup Q_2 \cup \ldots \cup Q_m,$$

which shows that X is cubical. □

Proposition 6.4 *The set $A \subset \mathbf{R}^d$ is representable if and only if the following two conditions are satisfied:*

(i) $\operatorname{cl} A$ *is bounded;*

(ii) for every $Q \in \mathcal{K}^d$, $\mathring{Q} \cap A \neq \emptyset$ *implies* $\mathring{Q} \subset A$.

Proof. Assume A is representable. Then $A = \mathring{Q}_1 \cup \mathring{Q}_2 \cup \ldots \cup \mathring{Q}_m$ for some $Q_i \in \mathcal{K}^d$, and by Proposition 2.15(iv)

$$\operatorname{cl} A = \operatorname{cl} \mathring{Q}_1 \cup \operatorname{cl} \mathring{Q}_2 \cup \ldots \cup \operatorname{cl} \mathring{Q}_m = Q_1 \cup Q_2 \cup \ldots \cup Q_m$$

is bounded. Also, if $\mathring{Q} \cap A \neq \emptyset$ for some $Q \in \mathcal{K}^d$, then $\mathring{Q} \cap \mathring{Q}_i \neq \emptyset$ for some $i \in \{1, 2, \ldots, m\}$. It follows from Proposition 2.15(iii) that $Q = Q_i$ and consequently $\mathring{Q} \subset A$. Hence (i) and (ii) are satisfied.

On the other hand, if properties (i) and (ii) are satisfied, then by Proposition 2.15(i),

$$A = A \cap \mathbf{R}^d = \bigcup \{A \cap \mathring{Q} \mid Q \in \mathcal{K}^d\} = \bigcup \{\mathring{Q} \mid Q \in \mathcal{K}^d, \mathring{Q} \subset A\}. \quad (6.1)$$

Since $\operatorname{cl} A$ is bounded, it follows from Proposition 2.15(ii) that the last union in (6.1) is finite. Hence A is a representable set. □

Proposition 6.5 *Assume $A \in \mathcal{R}^d$. Then A is closed if and only if for every $Q \in \mathcal{K}^d$*

$$\mathring{Q} \subset A \Rightarrow Q \subset A. \quad (6.2)$$

Similarly, A is open if and only if for every $Q \in \mathcal{K}^d$

$$\mathring{Q} \cap A = \emptyset \Rightarrow Q \cap A = \emptyset. \quad (6.3)$$

Proof. If A is closed and $\overset{\circ}{Q} \subset A$ for some $Q \in \mathcal{K}$, then by Proposition 2.15(iv) $Q = \text{cl}\,\overset{\circ}{Q} \subset A$. To prove that this condition is sufficient, take the decomposition $A = \overset{\circ}{Q}_1 \cup \overset{\circ}{Q}_2 \cup \ldots \cup \overset{\circ}{Q}_m$ of A into open elementary cubes. We have

$$\text{cl}\,A = \text{cl}\,\overset{\circ}{Q}_1 \cup \text{cl}\,\overset{\circ}{Q}_2 \cup \ldots \cup \text{cl}\,\overset{\circ}{Q}_m = Q_1 \cup Q_2 \cup \ldots \cup Q_m \subset A,$$

which proves that A is closed.

Assume, in turn, that A is open. Then $\overset{\circ}{Q} \cap A = \emptyset$ implies $\text{cl}\,\overset{\circ}{Q} \cap A = \emptyset$ and by Proposition 2.15(iv) we get $Q \cap A = \emptyset$.

We will show in turn that Eq. (6.3) is sufficient. Since by Proposition 6.4(i) A is bounded, the following definitions make sense.

$$k_i := \max\{n \in \mathbf{N} \mid (x_1, x_2 \ldots x_d) \in A \Rightarrow n \leq x_i\} - 1,$$
$$l_i := \min\{n \in \mathbf{N} \mid (x_1, x_2 \ldots x_d) \in A \Rightarrow n \geq x_i\} + 1.$$

Now let

$$R := [k_1, l_1] \times [k_2, l_2] \times \ldots \times [k_d, l_d].$$

Then R is representable and $A \subset \text{int}\,R$. We will verify condition (6.3) in Proposition 6.5 in order to prove that $R \setminus A$ is closed. To this end take $Q \in \mathcal{K}^d$ such that $\overset{\circ}{Q} \subset R \setminus A$. Then $\overset{\circ}{Q} \cap A = \emptyset$ and by (6.3) $Q \cap A = \emptyset$, that is, $Q \subset R \setminus A$. Thus $R \setminus A$ is closed, and consequently A is open in R. But $A \subset \text{int}\,R$, hence A is open. □

Definition 6.6 Let $A \subset \mathbf{R}^d$ be a bounded set. Then the *open hull* of A is

$$\text{oh}\,(A) := \bigcup \{\overset{\circ}{Q} \mid Q \in \mathcal{K}, Q \cap A \neq \emptyset\}, \tag{6.4}$$

and the *closed hull* of A is

$$\text{ch}\,(A) := \bigcup \{Q \mid Q \in \mathcal{K}, \overset{\circ}{Q} \cap A \neq \emptyset\}. \tag{6.5}$$

Example 6.7 Consider the vertex $P = [0] \times [0] \in \mathbf{R}^2$. Then

$$\text{oh}\,(P) = \{(x_1, x_2) \in \mathbf{R}^2 \mid -1 < x_i < 1\}.$$

Generalizing this example leads to the following result.

Proposition 6.8 *Let $P = [a_1] \times \cdots \times [a_d] \in \mathbf{R}^d$ be an elementary vertex. Then*

$$\text{oh}\,(P) = (a_1 - 1, a_1 + 1) \times \cdots \times (a_d - 1, a_d + 1).$$

6.1 Representable Sets

Example 6.9 Let X be the representable set discussed in Example 6.2. Then
$$\operatorname{ch}(X) = \operatorname{cl} X = ([0,1] \times [0,1]) \cup ([1,2] \times [1]) \cup ([3] \times [0])$$
and
$$\operatorname{oh}(X) = ((-1,2) \times (1,2)) \cup ((2,4) \times (-1,1)).$$
Figure 6.1, top right and bottom, displays images of closed and open hulls generated by the CubTop program.

The names chosen for $\operatorname{oh}(A)$ and $\operatorname{ch}(A)$ are justified by the following proposition.

Proposition 6.10 *Assume $A \subset \mathbf{R}^d$. Then*

(i) $A \subset \operatorname{oh}(A)$ and $A \subset \operatorname{ch}(A)$.
(ii) *The set $\operatorname{oh}(A)$ is open and representable.*
(iii) *The set $\operatorname{ch}(A)$ is closed and representable.*
(iv) $\operatorname{oh}(A) = \bigcap \{U \in \mathcal{R}^d \mid U \text{ is open and } A \subset U\}$.
(v) $\operatorname{ch}(A) = \bigcap \{B \in \mathcal{R}^d \mid B \text{ is closed and } A \subset B\}$. *In particular, if K is a cubical set such that $A \subset K$, then $\operatorname{ch}(A) \subset K$.*
(vi) $\operatorname{oh}(\operatorname{oh}(A)) = \operatorname{oh}(A)$ and $\operatorname{ch}(\operatorname{ch}(A)) = \operatorname{ch}(A)$.
(vii) *If $x, y \in \mathbf{R}^d$ and $y \in \operatorname{oh}(x)$, then $\operatorname{ch}(x) \subset \operatorname{ch}(y)$.*
(viii) $Q \in \mathcal{K}^d$ and $x \in \overset{\circ}{Q}$ *implies that* $\operatorname{ch}(x) = Q$.
(ix) *Let $Q \in \mathcal{K}^d$ and let $x, y \in \overset{\circ}{Q}$. Then, $\operatorname{oh}(x) = \operatorname{oh}(y)$ and $\operatorname{ch}(x) = \operatorname{ch}(y)$.*

Proof. (i) Let $x \in A$. By Proposition 2.15(i) there exists an elementary cube Q such that $x \in \overset{\circ}{Q}$. It follows that $\overset{\circ}{Q} \cap A \neq \emptyset \neq Q \cap A$. Hence $x \in \operatorname{oh}(A)$ and $x \in \operatorname{ch}(A)$.

(ii) By Proposition 2.15(ii) the union in (6.4) is finite. Therefore, the set $\operatorname{oh}(A)$ is representable. To prove that $\operatorname{oh}(A)$ is open, we will show that it satisfies (6.3). Let $P \in \mathcal{K}^d$ be such that $\overset{\circ}{P} \cap \operatorname{oh}(A) = \emptyset$. Assume that $P \cap \operatorname{oh}(A) \neq \emptyset$. Then there exists a $Q \in \mathcal{K}$ such that $Q \cap A \neq \emptyset$ and $P \cap \overset{\circ}{Q} \neq \emptyset$. Since P is representable, it follows from Proposition 6.4 that $\overset{\circ}{Q} \subset P$. Therefore, $Q = \operatorname{cl} \overset{\circ}{Q} \subset P$, that is, $P \cap A \neq \emptyset$. This means that $\overset{\circ}{P} \subset \operatorname{oh}(A)$, a contradiction. It follows that $\operatorname{oh}(A)$ is open.

(iii) The set $\operatorname{ch}(A)$ is closed since it is the finite union of closed sets. By Proposition 6.3, $\operatorname{ch}(A)$ is representable.

(iv) Observe that since $\operatorname{oh}(A)$ is open, is representable, and contains A,
$$\bigcap \{U \in \mathcal{R}^d \mid U \text{ is open and } A \subset U\} \subset \operatorname{oh}(A).$$
To show the opposite inclusion, take an open set $U \in \mathcal{R}^d$ such that $A \subset U$. Let $x \in \operatorname{oh}(A)$. Then there exists a $Q \in \mathcal{K}$ such that $A \cap Q \neq \emptyset$ and $x \in \overset{\circ}{Q}$. It

follows that $\emptyset \neq Q \cap U = \mathrm{cl}\ \overset{\circ}{Q} \cap U$, that is, $\overset{\circ}{Q} \cap U \neq \emptyset$. By Proposition 6.4, $\overset{\circ}{Q} \subset U$; hence $x \in U$. This shows that $\mathrm{oh}\,(A) \subset U$ and since U is arbitrary,

$$\mathrm{oh}\,(A) \subset \bigcap \{U \in \mathcal{R}^d \mid U \text{ is open and } A \subset U\}.$$

(v) Since $\mathrm{ch}\,(A)$ is closed, is representable, and contains A,

$$\bigcap \{B \in \mathcal{R}^d \mid B \text{ is closed and } A \subset B\} \subset \mathrm{ch}\,(A).$$

Let $B \in \mathcal{R}^d$ be a closed set that contains A. We will show that $\mathrm{ch}\,(A) \subset B$. For this end take an $x \in \mathrm{ch}\,(A)$. Then there exists a $Q \in \mathcal{K}$ such that $\overset{\circ}{Q} \cap A \neq \emptyset$ and $x \in Q$. It follows that $\overset{\circ}{Q} \cap B \neq \emptyset$ and consequently $\overset{\circ}{Q} \subset B$. Hence $Q \subset B$ and $x \in B$. This shows that $\mathrm{ch}\,(A) \subset B$ and since B is arbitrary,

$$\mathrm{ch}\,(A) \subset \bigcap \{B \in \mathcal{R}^d \mid B \text{ is closed and } A \subset B\}.$$

(vi) This follows immediately from (iv) and (v).

(vii) Observe that since $y \in \mathrm{oh}\,(x)$, there exists a $P \in \mathcal{K}$ such that $y \in \overset{\circ}{P}$ and $x \in P$. Take $z \in \mathrm{ch}\,(x)$. Then there exists a $Q \in \mathcal{K}$ such that $z \in Q$ and $x \in \overset{\circ}{Q}$. It follows that $\overset{\circ}{Q} \subset P$ and hence also $Q \subset P$. Consequently $z \in P$, which proves (vii).

(viii) This is straightforward.

(ix) Let $z \in \mathrm{oh}\,(x)$. Then there exists a $P \in \mathcal{K}$ such that $z \in \overset{\circ}{P}$ and $x \in P$. It follows that $\overset{\circ}{Q} \subset P$, that is, $y \in P$. Consequently, $z \in \mathrm{oh}\,(y)$ and $\mathrm{oh}\,(x) \subset \mathrm{oh}\,(y)$. The same way one proves that $\mathrm{oh}\,(y) \subset \mathrm{oh}\,(x)$. The equality $\mathrm{ch}\,(x) = \mathrm{ch}\,(y)$ follows from (viii). □

Proposition 6.11 *Assume $x, y \in \mathbf{R}$, $x \leq y$ are two arbitrary real numbers. Then*

$$\mathrm{ch}\,([x, y]) = [\mathrm{floor}\,(x), \mathrm{ceil}\,(y)].$$

Proof. Let $p := \mathrm{floor}\,(x)$ and $q := \mathrm{ceil}\,(y)$. Obviously, $[p, q]$ is a cubical set and $[x, y] \subset [p, q]$. Therefore, $\mathrm{ch}\,([x, y]) \subset [p, q]$. To show the opposite inclusion, take $z \in [p, q]$. If $z \in [x, y]$, the conclusion is obvious. Otherwise,

$$z \in [p, x) \cup (y, q] \subset [p, p+1] \cup [q-1, q] \subset \mathrm{ch}\,([x, y]),$$

because $(p, p+1) \cap [x, y] \neq \emptyset \neq (q-1, q) \cap [x, y]$. □

Proposition 6.12 *Let $\Delta_1, \Delta_2, \ldots, \Delta_d$ be bounded intervals. Then*

$$\mathrm{ch}\,(\Delta_1 \times \Delta_2 \times \ldots \times \Delta_d) = \mathrm{ch}\,(\Delta_1) \times \mathrm{ch}\,(\Delta_2) \times \ldots \times \mathrm{ch}\,(\Delta_d).$$

Proof. Since $\operatorname{ch}(\Delta_1) \times \operatorname{ch}(\Delta_2) \times \ldots \times \operatorname{ch}(\Delta_d)$ is a closed representable set, which contains $\Delta_1 \times \Delta_2 \times \ldots \times \Delta_d$, it must also contain $\operatorname{ch}(\Delta_1 \times \Delta_2 \times \ldots \times \Delta_d)$. To show the opposite inclusion, take $y := (y_1, y_2, \ldots, y_d) \in \operatorname{ch}(\Delta_1) \times \operatorname{ch}(\Delta_2) \times \ldots \times \operatorname{ch}(\Delta_d)$. Then there exist elementary intervals I_i such that $y_i \in I_i$ and $\overset{\circ}{I_i} \cap \Delta_i \neq \emptyset$ for $i \in \{1, 2, \ldots, n\}$. Hence $\overset{\circ}{I_1} \times \overset{\circ}{I_2} \times \ldots \times \overset{\circ}{I_d}$ is an elementary cell that has a nonempty intersection with $\Delta_1 \times \Delta_2 \times \ldots \times \Delta_d$ and $y \in I_1 \times I_2 \times \ldots \times I_d$. Therefore, $y \in \operatorname{ch}(\Delta_1 \times \Delta_2 \times \ldots \times \Delta_d)$. □

As an immediate corollary of the last two propositions we obtain the following result.

Corollary 6.13 *Let $\Delta_1, \Delta_2, \ldots, \Delta_d$ be bounded intervals. Then $\operatorname{ch}(\Delta_1 \times \Delta_2 \times \ldots \times \Delta_d)$ is a rectangle.*

Exercises

6.1 Write the input file for the elementary cell $A = (2, 3) \subset \mathbf{R}$ for the CubTop program. Then alternate the functions `closedhull` and `openhull` to see the representable sets $\operatorname{ch}(A)$, $\operatorname{oh}(\operatorname{ch}(A))$, $\operatorname{ch}(\operatorname{oh}(\operatorname{ch}(A)))$, and so on. This experiment should help you to answer the question in Exercise 6.4.

6.2 The file exR2d.txt in the folder CubTop/Examples contains a representable set A. Run CubTop on the file exR2d.txt with the option -g to see what it looks like. Run CubTop `closedhull` to see $\operatorname{ch}(A)$. Then run CubTop `openhull` on the files of A and of $\operatorname{ch}(A)$ to see the effect.

6.3 Given a cubical set $X \subset \mathbf{R}^d$ and a positive integer r, define the *closed cubical ball* of radius r about X by the formula

$$\bar{B}_r(X) := \{y \in R^d \mid \operatorname{dist}(y, X) \leq r\},$$

where

$$\operatorname{dist}(y, X) := \inf\{\|y - x\|_0 \mid x \in X\}$$

and $\|y - x\|_0$ is the supremum norm of $y - x$ [see Eq. (12.3) in Chapter 12]. Give, with a proof, a formula for $\bar{B}_r(X)$ in terms of open and closed hulls of X. This will show, in particular, that $\bar{B}_r(X)$ is a cubical set.

6.4 Given a cubical set $X \subset R^d$ and a positive integer r, define the *open cubical ball* about X by the formula

$$B_r(X) := \{y \in R^d \mid \operatorname{dist}(y, X) < r\}.$$

Give, with a proof, a formula for $B_r(X)$ in terms of open and closed hulls of X (you will need them both). This will show, in particular, that $B_r(X)$ is a representable set. What happens if we extend the definition to open balls about bounded, not necessarily closed sets? You may use the experiments in Exercises 6.1 and 6.2 to make the right guess.

6.2 Cubical Multivalued Maps

There are a variety of ways in which one can pass from a continuous function to a chain map. Each has its advantages and disadvantages. The approach we will adopt involves using multivalued maps to approximate the continuous map. The motivation for this is given in Chapter 5. We now want to formalize those ideas.

Let X and Y be cubical sets. A *multivalued map* $F : X \rightrightarrows Y$ from X to Y is a function from X to subsets of Y, that is, for every $x \in X$, $F(x) \subset Y$. This class is very broad and, since we are interested in computations, we recall Definition 5.4 of a cubical multivalued map.

Let X and Y be cubical sets. A multivalued map $F : X \rightrightarrows Y$ is *cubical* if

1. For every $x \in X$, $F(x)$ is a cubical set.
2. For every $Q \in \mathcal{K}(X)$, $F|_{\overset{\circ}{Q}}$ is constant, that is, if $x, x' \in \overset{\circ}{Q}$, then $F(x) = F(x')$.

Cubical multivalued maps have the nice feature that, on one hand, they are maps between topological spaces and may be used to enclose single-valued continuous maps (which in general have infinitely many values), while on the other hand, they may be easily encoded with a finite amount of information. This makes them convenient to handle with a computer.

With our goal of using multivalued maps to enclose the image of a continuous function, there is a canonical choice for constructing cubical maps. To make this clear, let us return to the discussion in Section 5.2 where the use of multivalued maps is first presented. We have considered there the function $f(x) = (x - \frac{4}{3})(x + \frac{4}{5})$ as a map from $X = [-2, 2] \subset \mathbf{R}$ to $Y = [-3, 4] \subset \mathbf{R}$. Using interval arithmetic we have derived bounds on f that apply to each $Q \in \mathcal{K}_1(X)$ (see Table 5.1 and Figure 5.2). Then we have extend these bounds to vertices by intersecting the bounds of the adjacent edges.

Assume now that for a continuous map $f : X \to Y$, bounds on $f(Q)$ for $Q \in \mathcal{K}_{\max}(X)$ are given in the form of a combinatorial multivalued map $\mathcal{F} : \mathcal{K}_{\max}(X) \rightrightarrows \mathcal{K}(Y)$ (see Definition 5.3). The approach used in Section 5.2 may be easily generalized in the following way. First, given a collection $\mathcal{X}$ of elementary cubes in $\mathbf{R}^d$, we set

$$|\mathcal{X}| := \bigcup \mathcal{X}. \tag{6.6}$$

Obviously, $|\mathcal{X}|$ is a cubical set. Next, we define the multivalued maps $\lfloor \mathcal{F} \rfloor, \lceil \mathcal{F} \rceil : X \rightrightarrows Y$ by

$$\lfloor \mathcal{F} \rfloor(x) := \bigcap \{|\mathcal{F}(Q)| \mid x \in Q \in \mathcal{K}_{\max}(X)\}, \tag{6.7}$$

$$\lceil \mathcal{F} \rceil(x) := \bigcup \{|\mathcal{F}(Q)| \mid x \in Q \in \mathcal{K}_{\max}(X)\}. \tag{6.8}$$

Theorem 6.14 *The multivalued maps $\lfloor \mathcal{F} \rfloor$ and $\lceil \mathcal{F} \rceil$ are cubical.*

Proof. The first property of Definition 5.4 follows from the straightforward fact that the union and intersection of cubical sets is a cubical set. To prove the other observe that, by Proposition 2.15(vi), if $x, x' \in \overset{\circ}{P}$ for some $P \in \mathcal{K}(X)$, then the union in (6.8) and intersection in (6.7) are taken over the same family of maximal cubes for x' as for x. □

There is another property of multivalued maps which we need. Before we introduce it, let us consider the following example.

Example 6.15 Let $X = [0, 2]$ and let $Y = [-5, 5]$. Define $F : X \rightrightarrows Y$ by

$$F(x) := \begin{cases} [-5] & \text{if } x = 0, \\ [-4, -1] & \text{if } x \in (0, 1), \\ [0] & \text{if } x = 1, \\ [1, 4] & \text{if } x \in (1, 2), \\ [5] & \text{if } x = 2. \end{cases}$$

The graph of this function is given in Figure 6.2. Observe that F is a cubical map. However, from several points of view this is not satisfactory for our needs. The first is that intuitively it should be clear that a map of this type cannot be thought of as being continuous. The second is that we are interested in multivalued maps because we use them as representations of continuous maps. But it is obvious that there is no continuous map $f : X \to Y$ such that $f(x) \in F(x)$ for all $x \in X$.

To overcome these problems we need to introduce a notion of continuity for multivalued maps. Recall that for single-valued functions, continuity is defined in terms of the preimages of open sets. We want to do something similar for multivalued maps; however, as we will see, there are at least two reasonable ways to define a preimage. As a consequence, we have two different ways of defining the continuity of multivalued maps.

Let $F : X \rightrightarrows Y$, $A \subset X$, and $B \subset Y$. The *image of A* is defined by

$$F(A) := \bigcup_{x \in A} F(x).$$

The *weak preimage* of B under F is

$$F^{*-1}(B) := \{x \in X \mid F(x) \cap B \neq \emptyset\},$$

while the , *preimage* of B is

$$F^{-1}(B) := \{x \in X \mid F(x) \subset B\}.$$

Definition 6.16 A multivalued map F is *upper semicontinuous* if $F^{-1}(U)$ is open for any open set $U \subset Y$, and it is *lower semicontinuous* if the set $F^{*-1}(U)$ is open for any open set $U \subset Y$.

208   6 Homology of Maps

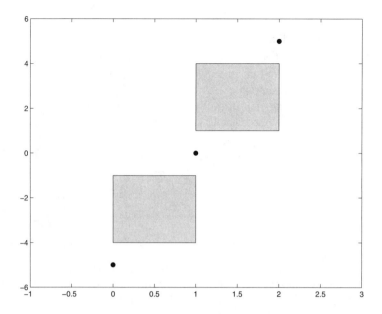

Fig. 6.2. The graph of the cubical map F.

Proposition 6.17 *Assume $F : X \rightrightarrows Y$ is a cubical map. Then F is lower semicontinuous if and only if the following property is satisfied:*

$$\text{If } P, Q \in \mathcal{K}(X) \text{ are such that } P \prec Q, \text{ then } F(\overset{\circ}{P}) \subset F(\overset{\circ}{Q}). \tag{6.9}$$

Proof. Suppose that F is lower semicontinuous. Since $F(\overset{\circ}{Q}) = F(x)$ for $x \in \overset{\circ}{Q}$, the set $F(\overset{\circ}{Q})$ is cubical and consequently closed. Thus the set $U := Y \setminus F(\overset{\circ}{Q})$ is open. By the lower semicontinuity of F,

$$V := F^{*-1}(U) = \{z \in X \mid F(z) \cap U \neq \emptyset\} \tag{6.10}$$

is open.

Now consider $x \in \overset{\circ}{P}$. Since F is cubical, $F(x) = F(\overset{\circ}{P})$. Therefore, it is sufficient to prove that $F(x) \subset F(\overset{\circ}{Q})$. This is equivalent to showing that $x \notin V$, which will be proved by contradiction.

So, assume that $x \in V$. Since $x \in \overset{\circ}{P}$ and $P \prec Q$, it follows that $x \in Q = \text{cl}(\overset{\circ}{Q})$. Thus, $V \cap \text{cl}(\overset{\circ}{Q}) \neq \emptyset$. But V is open, hence $V \cap \overset{\circ}{Q} \neq \emptyset$. Let $z \in V \cap \overset{\circ}{Q}$. Then, because F is cubical, $F(z) = F(\overset{\circ}{Q})$, and hence, $F(z) \cap U = \emptyset$. Thus $z \notin V$, a contradiction, which proves (6.9).

Suppose now that F has property (6.9). Let U be open in Y and let V be defined as in (6.10). We need to show that V is open. It is enough to show that for every point $x \in V$ the open hull $\operatorname{oh}(x)$ is contained in V. Indeed, $\operatorname{oh}(x)$ is the union of elementary cells $\mathring{Q}$ such that $x \in Q$. Fix such a Q. Let P be the unique elementary cube such that $x \in \mathring{P}$. Then $P \subset Q$. Since $F(x) = F(\mathring{P}) \subset F(\mathring{Q})$, it follows that $F(\mathring{Q}) \cap U \neq \emptyset$. Thus $\mathring{Q} \subset V$. Since this is true for all $\mathring{Q} \subset \operatorname{oh}(x)$, the conclusion follows. □

Here is a dual property of upper semicontinuous maps, the proof of which is left as an exercise.

Proposition 6.18 *Assume $F : X \rightrightarrows Y$ is a cubical map. Then F is upper semicontinuous if and only if it has the following property:*

$$\text{For any } P, Q \in \mathcal{K}(X) \text{ such that } P \text{ is a face of } Q, F(\mathring{Q}) \subset F(\mathring{P}). \quad (6.11)$$

It is not difficult to verify that $\lfloor \mathcal{F} \rfloor$ given by (6.7) and $\lceil \mathcal{F} \rceil$ given by (6.8) satisfy respectively the conditions (6.9) and (6.11). The details are left as an exercise. Therefore we get the following corollary.

Corollary 6.19 *The map $\lfloor \mathcal{F} \rfloor$ is lower semicontinuous and the map $\lceil \mathcal{F} \rceil$ is upper semicontinuous.*

The above propositions show how to construct lower semicontinuous and upper semicontinuous maps if their values on elementary cells are given. Indeed, if $F(\mathring{Q})$ is defined for all maximal Q and we let

$$F(\mathring{P}) = \bigcap \{F(\mathring{Q}) \mid P \prec Q \text{ and } Q \text{ is maximal}\},$$

we get a lower semicontinuous map. Similarly, if we let

$$F(\mathring{P}) = \bigcup \{F(\mathring{Q}) \mid P \prec Q \text{ and } Q \text{ is maximal}\},$$

we get an upper semicontinuous map.

We use lower semicontiuous maps for computing homology maps. However, by the following proposition (its proof is left as an exercise), the graphs of upper semicontinous maps are easier to draw.

Proposition 6.20 *Assume $F : X \rightrightarrows Y$ is a cubical map. Then F is upper semicontinuous if and only if its graph*

$$\operatorname{graph}(F) := \{(x, y) \in X \times Y \mid y \in F(x)\}$$

is closed.

210   6 Homology of Maps

Exercises

6.5 Let $X = [-1,1]^2 \subset \mathbf{R}^2$. Let $Y = [-2,2]^2 \subset \mathbf{R}^2$. Consider the map $A : X \to Y$ given by
$$A = \begin{bmatrix} 0.5 & 0 \\ 0 & 2 \end{bmatrix}.$$
Find a lower semicontinuous multivalued map $F : X \rightrightarrows Y$ with the property that $Ax \in F(x)$ for every $x \in X$. Try to find such a map with the smallest possible values.

6.6 Prove Proposition 6.18.

6.7 Prove Corollary 6.19.

6.8 Prove Proposition 6.20.

6.9 Let $a < b$ and $c < d$ be integers in $\mathbf{R}$ and let $F : [a,b] \rightrightarrows [c,d]$ be a cubical lower semicontinuous map with acyclic values. Prove that there exists a continuous single-valued map $f : [a,b] \to [c,d]$ such that $f(x) \in F(x)$ for all x.

6.3 Chain Selectors

As has been indicated since Chapter 5, multivalued maps are used to provide representation for continuous functions. Of course, we still need to indicate how this representation can be used to generate homology. By Section 4.1 it is sufficient to indicate how a multivalued map induces a chain map. In general, this problem is difficult and may not have a satisfactory solution. But there is a special class of multivalued maps for which the construction is relatively simple.

Definition 6.21 A cubical multivalued map $F : X \rightrightarrows Y$ is called *acyclic-valued* if for every $x \in X$ the set $F(x)$ is acyclic.

Theorem 6.22 (Chain selector) *Assume $F : X \rightrightarrows Y$ is a lower semicontinuous acyclic-valued cubical map. Then there exists a chain map $\varphi : C(X) \to C(Y)$ satisfying the following two conditions:*

$$\left|\varphi(\widehat{Q})\right| \subset F(\overset{\circ}{Q}) \text{ for all } Q \in \mathcal{K}(X), \tag{6.12}$$

$$\varphi(\widehat{Q}) \in \widehat{\mathcal{K}}_0(F(Q)) \text{ for any vertex } Q \in \mathcal{K}_0(X). \tag{6.13}$$

Proof. We will construct the homomorphisms $\varphi_k : C_k(X) \to C_k(Y)$ by induction on k.

For $k < 0$, $C_k(X) = 0$; therefore, there is no choice but to define $\varphi_k := 0$.

Consider $k = 0$. For each $Q \in \mathcal{K}_0$, choose $P \in \mathcal{K}_0(F(Q))$ and set

$$\varphi_0(\widehat{Q}) := \widehat{P}. \tag{6.14}$$

Clearly, $\left|\varphi_0\widehat{Q}\right| = P \in F(Q)$. Since $Q \in \mathcal{K}_0$, $\mathring{Q} = Q$ and hence $F(Q) = F(\mathring{Q})$. Therefore,
$$\left|\varphi_0\widehat{Q}\right| \subset F(\mathring{Q}).$$

Furthermore,
$$\varphi_{-1}\partial_0 = 0 = \partial_0\varphi_0.$$

To continue the induction, suppose now that the homomorphisms $\varphi_i : C_i(X) \to C_i(Y)$, $i = 0, 1, 2, \ldots, k-1$, are constructed in such a way that
$$\left|\varphi_i\widehat{Q}\right| \subset F(\mathring{Q}) \text{ for all } Q \in \mathcal{K}_i(X), \tag{6.15}$$

and
$$\varphi_{i-1}\partial_i = \partial_i\varphi_i. \tag{6.16}$$

Let $\widehat{Q} \in \widehat{\mathcal{K}}_k(X)$. Then $\partial\widehat{Q} = \sum_{j=1}^m \alpha_j\widehat{Q}_j$ for some $\alpha_j \in \mathbf{Z}$ and $\widehat{Q}_j \in \widehat{\mathcal{K}}_{k-1}(Q)$. Since F is lower semicontinuous, we have, by the induction assumption (6.15) and Proposition 6.17,
$$\left|\varphi_{k-1}\widehat{Q}_j\right| \subset F(\mathring{Q}_j) \subset F(\mathring{Q})$$

for all $j = 1, \ldots, m$. Thus
$$\left|\varphi_{k-1}\partial\widehat{Q}\right| \subset F(\mathring{Q}).$$

Since $F(\mathring{Q}) = F(x)$ for any $x \in \mathring{Q}$, the set $F(\mathring{Q})$ is acyclic. By the induction assumption (6.16),
$$\partial_{k-1}\varphi_{k-1}\partial_k\widehat{Q} = \varphi_{k-2}\partial_{k-1}\partial_k\widehat{Q} = 0,$$

that is, $\varphi_{k-1}\partial\widehat{Q}$ is a cycle.

We should distinguish $k = 1$. In this case, Q is an interval and
$$\partial\widehat{Q} = \widehat{B} - \widehat{A},$$

where A and B are vertices of Q. We show above that the vertices $\varphi_0(A)$ and $\varphi_0(B)$ are supported in $F(\mathring{Q})$. Since $F(\mathring{Q})$ is acyclic, its reduced homology $\tilde{H}_*(F(\mathring{Q}))$ is trivial. In particular, $\ker \epsilon = \operatorname{im} \partial_1$ in $\tilde{C}(F(\mathring{Q}))$. Thus there exists $c \in C_1(F(\mathring{Q}))$ such that $\partial_1 c = \varphi_0(B) - \varphi_0(A) \in \ker \epsilon$. We let
$$\varphi_1\widehat{Q} := c.$$

When $k > 1$, the acyclicity of $F(\overset{\circ}{Q})$ implies that there exists a chain $c \in C_k(F(\overset{\circ}{Q}))$ such that $\partial c = \varphi_{k-1}\partial \widehat{Q}$. Define
$$\varphi_k \widehat{Q} := c.$$
By definition, the homomorphism φ_k satisfies the property
$$\partial_k \varphi_k = \varphi_{k-1}\partial_k.$$
Also, since $\varphi_k \widehat{Q} = c \in C_k(F(\overset{\circ}{Q}))$, we get
$$\left|\varphi_k \widehat{Q}\right| \subset F(\overset{\circ}{Q}).$$
Therefore, the chain map $\varphi = \{\varphi_k\}_{k \in \mathbf{Z}} : C(X) \to C(Y)$ satisfying (6.12) is well defined. □

Observe that in the first nontrivial step (6.14) of the inductive construction of φ we are allowed to choose any $P \in \mathcal{K}_0(F(Q))$. Thus this procedure allows us to produce many chain maps of the type described in Theorem 6.22. This leads to the following definition.

Definition 6.23 Let $F : X \rightrightarrows Y$ be lower semicontinuous acyclic-valued cubical multivalued map. A chain map $\varphi : C(X) \to C(Y)$ satisfying the conditions (6.12) and (6.13) in Theorem 6.22:

$$\left|\varphi(\widehat{Q})\right| \subset F(\overset{\circ}{Q}) \text{ for all } Q \in \mathcal{K}(X),$$
$$\varphi(\widehat{Q}) \in \widehat{\mathcal{K}}_0(F(Q)) \text{ for any vertex } Q \in \mathcal{K}_0(X),$$

is called a *chain selector* of F.

Proposition 6.24 *Assume $F : X \rightrightarrows Y$ is a lower semicontinuous cubical map and φ is a chain selector for F. Then, for any $c \in C(X)$,*
$$|\varphi(c)| \subset F(|c|).$$

Proof. Let $c = \sum_{i=1}^m \alpha_i \widehat{Q}_i$, where $\alpha_i \in \mathbf{Z}$, $\alpha_i \neq 0$. Then
$$|\varphi(c)| = \left|\sum_{i=1}^m \alpha_i \varphi(\widehat{Q}_i)\right|$$
$$\subset \bigcup_{i=1}^m \left|\varphi(\widehat{Q}_i)\right|$$
$$\subset \bigcup_{i=1}^m F(\overset{\circ}{Q}_i)$$
$$\subset \bigcup_{i=1}^m F(Q_i)$$
$$= F(\bigcup_{i=1}^m Q_i) = F(|c|). \quad □$$

Theorem 6.22 indicates that for an appropriate multivalued map we can produce a chain selector, but as observed above this chain selector need not be unique. Of course, our primary interest is not in the chain selector, but rather in the homology maps that it induces. The following theorem shows that every chain selector leads to the same map on homology.

Theorem 6.25 *Let $\varphi, \psi : C(X) \to C(Y)$ be chain selectors for the lower semicontinuous acyclic-valued cubical map $F : X \rightrightarrows Y$. Then φ is chain homotopic to ψ and hence they induce the same homomorphism in homology.*

Proof. A chain homotopy $D = \{D_k : C_k(X) \to C_{k+1}(Y)\}_{k \in \mathbf{Z}}$ joining φ to ψ can be constructed by induction.

For $k < 0$, there is no choice but to set $D_k := 0$. Thus assume $k \geq 0$ and D_i is defined for $i < k$ in such a way that

$$\partial_{i+1} \circ D_i + D_{i-1} \circ \partial_i = \psi_i - \varphi_i, \tag{6.17}$$

and for all $Q \in \mathcal{K}_i(X)$ and $c \in C_i(Q)$,

$$|D_i(c)| \subset F(\mathring{Q}). \tag{6.18}$$

Let $\widehat{Q} \in C_k(X)$ be an elementary k-cube. Let

$$c := \psi_k(\widehat{Q}) - \varphi_k(\widehat{Q}) - D_{k-1}\partial_k(\widehat{Q}).$$

It follows easily from the induction assumption (6.17) that c is a cycle. Moreover, if $\partial \widehat{Q} = \sum_{i=1}^m \alpha_i \widehat{P}_i$ for some $\alpha_i \neq 0$ and $P_i \in \mathcal{K}_{k-1}(X)$, then P_i are faces of Q and, by the induction assumption (6.18) and Proposition 6.17,

$$\left|D_{k-1}\partial(\widehat{Q})\right| \subset \bigcup_{i=1}^m \left|D_{k-1}(\widehat{P}_i)\right| \subset \bigcup_{i=1}^m F(\mathring{P}_i) \subset F(\mathring{Q}).$$

Consequently,

$$|c| \subset \left|\psi_k(\widehat{Q})\right| \cup \left|\varphi_k(\widehat{Q})\right| \cup \left|D_{k-1}\partial_k(\widehat{Q})\right| \subset F(\mathring{Q}).$$

It follows that $c \in Z_k(F(\mathring{Q}))$.

Now, since $F(\mathring{Q})$ is acyclic, we conclude for $k > 0$ that there exists a $c' \in C_{k+1}(F(\mathring{Q}))$ such that $\partial c' = c$. In the case $k = 0$, the same conclusion follows (as in the proof of Theorem 6.22) from the identity $\tilde{H}_*(F(\mathring{Q})) = 0$ and from the equation

$$c := \psi_0(\widehat{Q}) - \varphi_0(\widehat{Q}) - D_{-1}\partial_0(\widehat{Q}) = \psi_0(\widehat{Q}) - \varphi_0(\widehat{Q}).$$

We put $D_k(\widehat{Q}) := c'$. One easily verifies that the induction assumptions are satisfied. Therefore, the construction of the required homotopy is completed. □

The above theorem lets us make the following fundamental definition.

Definition 6.26 Let $F : X \rightrightarrows Y$ be a lower semicontinuous acyclic-valued cubical map. Let $\varphi : C(X) \to C(Y)$ be a chain selector of F. The *homology map* of F, $F_* : H_*(X) \to H_*(Y)$ is defined by

$$F_* := \varphi_*.$$

Given $k \in \mathbf{Z}$, the restriction of F_* to the kth dimensional homology is denoted by F_{*k}.

Example 6.27 Let X and Y be connected cubical sets, $F : X \rightrightarrows Y$ be a lower semicontinuous acyclic-valued cubical map, and $\varphi : C(X) \to C(Y)$ be a chain selector of F. By Theorem 2.59, all vertices in a connected set are homologous, so $H_0(X) = \mathbf{Z}[\widehat{V}]$ and $H_0(Y) = \mathbf{Z}[\widehat{W}]$ for arbitrarily chosen $V \in \mathcal{K}_0(X)$ and $W \in \mathcal{K}_0(Y)$. Since $\varphi_0(\widehat{V}) \in \widehat{\mathcal{K}_0}(Y)$ and $\varphi_0(\widehat{V}) \sim \widehat{W}$, we get

$$F_{*0}([\widehat{V}]) = \varphi_{*0}([\widehat{V}]) = [\varphi_0(\widehat{V})] = [\widehat{W}].$$

In particular, if $X = Y$ is connected, then

$$F_{*0} = \operatorname{id}_{H_0(X)}.$$

Keep in mind that one of the purposes of introducing multivalued maps is to be able to compute the homology of a continuous map by some systematic method of representation. Obviously, and we have seen this in Chapter 5, what procedure one uses or the amount of computation one is willing to do determines how sharp a representation one obtains. A natural question is how much does this matter? In other words, given an acyclic-valued cubical map, does an acyclic-valued map with smaller values produce a different map on homology? To answer this we need to make the question precise.

Definition 6.28 Let X and Y be cubical spaces and let $F, G : X \rightrightarrows Y$ be lower semicontinuous cubical maps. F is a *submap* of G if

$$F(x) \subset G(x)$$

for every $x \in X$. This is denoted by $F \subset G$.

Proposition 6.29 *If $F, G : K \rightrightarrows L$ are two lower semicontinuous acyclic-valued cubical maps and F is a submap of G, then $F_* = G_*$.*

Proof. Let φ be a chain selector of F. Then φ is also a chain selector of G. Hence, by definition,

$$F_* = \varphi_* = G_*. \qquad \square$$

The above proposition may lead the reader to believe that the size of the values of a representation of a continuous map does not matter at all. But this is not true. As we have seen in Example 5.5, when the values are too large, they may be not acyclic.

A fundamental property of maps is that they can be composed. In the case of multivalued maps $F : X \rightrightarrows Y$ and $G : Y \rightrightarrows Z$, we will construct the multivalued map $G \circ F : X \rightrightarrows Z$, by setting

$$G \circ F(x) := G(F(x))$$

for every $x \in X$.

Proposition 6.30 *If $F : X \rightrightarrows Y$, $G : Y \rightrightarrows Z$, and $H : X \rightrightarrows Z$ are lower semicontinuous acyclic-valued cubical maps and $G \circ F \subset H$, then $H_* = G_* \circ F_*$.*

Proof. Let φ be a chain selector of F and ψ be a chain selector of G. Then by Proposition 6.24, for any $Q \in \mathcal{K}(X)$

$$\left|(\psi(\varphi(\widehat{Q})))\right| \subset G(\left|\varphi(\widehat{Q})\right|) \subset G(F(\overset{\circ}{Q})) \subset H(\overset{\circ}{Q}).$$

Hence $\psi \circ \varphi$ is a chain selector of H. But we can compose chain maps and hence

$$H_* = (\psi\varphi)_* = \psi_* \varphi_* = G_* F_*. \qquad \square$$

Corollary 6.31 *If $F : X \rightrightarrows Y$ and $G : Y \rightrightarrows Z$ are lower semicontinuous acyclic-valued cubical maps and $G \circ F$ is acyclic-valued, then $(G \circ F)_* = G_* \circ F_*$.*

6.4 Homology of Continuous Maps

We are finally in the position to discuss the homology of continuous maps. Recall the discussion of maps in Chapter 5. There we consider continuous functions of one variable and use interval arithmetic to get bounds on images of elementary cubes. While this provides a conceptually easy approach to getting bounds, there is a variety of methods for approximating functions over a given region. Furthermore, for different problems, different methods will be the most appropriate. Therefore, we would like our discussion of the construction of the multivalued map to be independent of the particular method of approximation employed. This leads us to make the following definition.

Definition 6.32 Let X and Y be cubical sets and let $f : X \to Y$ be a continuous function. A *cubical representation* of f, or simply a *representation* of f, is a lower semicontinuous multivalued cubical map $F : X \rightrightarrows Y$ such that

$$f(x) \in F(x) \tag{6.19}$$

for every $x \in X$.

Assume that f has a cubical representation $F : X \rightrightarrows Y$ that is acyclic-valued. Then the natural candidate for the definition of the associated homology map, $f_* : H_*(X) \to H_*(Y)$, is

216   6 Homology of Maps

$$f_* := F_*. \tag{6.20}$$

At least two questions need to be answered before we can be content with this approach to defining the homology of a continuous map. First, observe that given a continuous function, there may be many acyclic-valued cubical representations. Thus we need to show that all acyclic-valued cubical representations of a given function give rise to the same homomorphism on homology. This is the content of Section 6.4.1. Second, given cubical sets and a continuous map between them, it need not be the case that there exists a cubical representation, which is acyclic-valued. We use scaling to deal with this problem in Section 6.4.2.

6.4.1 Cubical Representations

From the viewpoint of computations, one often wants a cubical representation whose images are as small as possible. As will be made clear shortly, the following construction does this.

Proposition 6.33 *Let X and Y be cubical sets and let $f : X \to Y$ be a continuous function. The map $M_f : X \rightrightarrows Y$ defined by*

$$M_f(x) := \operatorname{ch}(f(\operatorname{ch}(x))) \tag{6.21}$$

is a cubical representation of f.

Proof. The fact that M_f is a cubical map follows from Proposition 6.10(iii) and (ix). Since obviously $f(x) \in M_f(x)$, M_f is a representation of f. To prove that M_f is lower semicontinuous it is enough to verify condition (6.9). Thus let $P, Q \in \mathcal{K}(X)$ be such that P is a face of Q. Choose $x \in \overset{\circ}{P}$ and $y \in \overset{\circ}{Q}$. Then
$$\operatorname{ch}(x) = P \subset Q = \operatorname{ch}(y).$$
Hence
$$M_f(\overset{\circ}{P}) = \operatorname{ch}(f(\operatorname{ch}(x))) \subset \operatorname{ch}(f(\operatorname{ch}(x))) = M_f(\overset{\circ}{Q}),$$
which, by Proposition 6.17, proves the lower semicontinuity of M_f. □

Definition 6.34 The map M_f is called the *minimal representation* of f. If M_f is acyclic-valued, then M_f is referred to as the *cubical carrier* of f.

Example 6.35 Consider the continuous function $f : [0,3] \to [0,3]$ given by $f(x) = x^2/3$. Figure 6.3 indicates the graph of f and its minimal cubical representation M_f. To verify that M_f really is the minimal cubical representation just involves checking the definition on all the elementary cells of $[0,3]$. To begin with, consider $[0] \in \mathcal{K}_0$. $\operatorname{ch}[0] = [0]$ and $f(0) = 0$; therefore, $M_f(0) = 0$. On the other hand, while $\operatorname{ch}[1] = [1]$, $f(1) = 1/3 \in (0,1)$ and hence $\operatorname{ch} f(1) = [0,1]$. Therefore, $M_f(1) = [0,1]$. Now consider the elementary cell $(0,1)$. $f((0,1)) \subset (0,1)$ so, as before, $M_f(x) = [0,1]$ for all $x \in (0,1)$. By similar arguments, $M_f(x) = [0,2]$ for all $x \in (1,2)$. The rest of the elementary cells can be checked in a similar manner.

Observe that if any cube from the graph of M_f were removed, then the graph of f would no longer be contained in the graph of M_f. In this sense M_f is minimal.

More precisely, we have the following proposition.

Proposition 6.36 *Let X and Y be cubical sets, let $f : X \to Y$ be a continuous function, and let $F : X \rightrightarrows Y$ be a cubical representation of f. Then M_f is a submap of F.*

Proof. Let $x \in X$ and let $Q \in \mathcal{K}(X)$ be such that $x \in \overset{\circ}{Q}$. Then $F(x) = F(\overset{\circ}{Q})$ and $F(x)$ is closed. Let $\bar{x} \in Q$. By Proposition 2.15(iv), there exists a sequence (x_n) in $\overset{\circ}{Q}$ such that $x_n \to \bar{x}$. By continuity of f, $f(\bar{x}) \in \operatorname{cl} F(\overset{\circ}{Q}) = F(\overset{\circ}{Q})$, which implies that $f(Q) \subset F(\overset{\circ}{Q})$. Therefore also $\operatorname{ch} f(Q) \subset F(\overset{\circ}{Q})$, because $F(\overset{\circ}{Q})$ is a cubical set. Since $x \in \overset{\circ}{Q}$, $\operatorname{ch}(x) = Q$. Thus

$$M_f(x) = \operatorname{ch}(f(\operatorname{ch}(x))) = \operatorname{ch}(f(Q)) \subset F(\overset{\circ}{Q}) = F(x). \qquad \square$$

Example 6.37 As the previous definition suggests, it is not true that a minimal representation is necessarily acyclic-valued. Let $X := \Gamma^1$, where Γ^1 is the boundary of the unit circle considered in Section 5.4. Note that X is the union of the elementary cubes:

$$K_1 := [0] \times [0,1], \qquad K_2 := [0,1] \times [1],$$
$$K_3 := [1] \times [0,1], \qquad K_4 := [0,1] \times [0].$$

Define the map $\lambda : [0,1] \to X$ for $s \in [0,1]$ by

$$\lambda(s) := \begin{cases} (0, 4s) & \text{if } s \in [0, 1/4], \\ (4s - 1, 1) & \text{if } s \in [1/4, 1/2], \\ (1, 3 - 4s) & \text{if } s \in [1/2, 3/4], \\ (4 - 4s, 0) & \text{if } s \in [3/4, 1], \end{cases}$$

and the map $f : X \to X$ for $(x_1, x_2) \in X$ by

$$f(x_1, x_2) := \begin{cases} \lambda(x_2) & \text{if } (x_1, x_2) \in K_1 \cup K_3, \\ \lambda(x_1) & \text{if } (x_1, x_2) \in K_2 \cup K_4. \end{cases}$$

Since the local formulas for $\lambda(t)$ and $f(x_1, x_2)$ agree at the points where the consecutive intervals intersect, f is continuous. For any $(x_1, x_2) \in \overset{\circ}{K_i}$,

$$M_f(x_1, x_2) = \operatorname{ch}(f(\operatorname{ch}(x_1, x_2))) = \operatorname{ch}(f(K_i)) = \operatorname{ch}(X) = X.$$

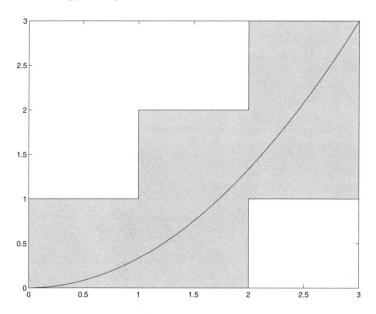

Fig. 6.3. The graph of a continuous function $f : [0,3] \to [0,3]$ and the closure of the graph of M_f. The identification of the interval $M_f(x)$ depending on weather x is a vertex or a point in an open elementary interval is discussed in Example 6.35.

Since X is not acyclic, it follows that M_f is not acyclic-valued.

In order to better visualize the function f, we rephrase the formula in the language of Section 5.4. Recall from Section 5.4 that any point $x \in \Gamma^1$ can be identified with $t \in [0,4]$ by the function

$$t \mapsto x(t) = (x_1(t), x_2(t)),$$

defined in Eq. (5.10), upon the identification $t \sim t + 4$. Also recall that $t = 0, 1, 2, 3$ and $4 \sim 0$ correspond to elementary vertices of Γ^1 while the intervals $[0,1], [1,2], [2,3], [3,4]$ correspond to, respectively, K_4, K_3, K_2, K_1, by going counterclockwise around the square $[0,1]^2$.

Now our map $f : X \to X$ can be given an alternative presentation in terms of the parameter $t \in [0,4]$:

$$t \mapsto \begin{cases} 4t & \text{if } t \in [0,2], \\ 16 - 4t & \text{if } t \in [2,4], \end{cases}$$

which looks like a tent map of slope 4. Of course, the values in the above formula fall outside the interval $[0,4]$, but the identification $t \sim t + 4$ breaks down the formula to four cases:

$$f(x(t)) := \begin{cases} x(4t) & \text{if } t \in [0,1], \\ x(4t-4) & \text{if } t \in [1,2], \\ x(12-4t) & \text{if } t \in [2,3], \\ x(16-4t) & \text{if } t \in [3,4]. \end{cases}$$

The graph of f as a function of t is illustrated in Figure 6.4.

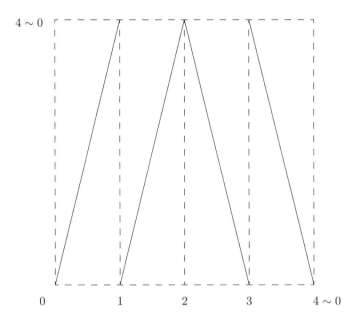

Fig. 6.4. The graph of f in Example 6.37 presented as a function of the parameter $t \in [0,4]$. It might not, at first sight, look like a graph of a continuous function, but keep in mind that the endpoints of the vertical dotted lines are identified by the relation $t \sim t+4$.

Proposition 6.38 *Let X and Y be cubical sets, and let $f : X \to Y$ be a continuous function such that its minimal representation is acyclic-valued. If $F, G : X \rightrightarrows Y$ are any other acyclic-valued representations of f, then $F_* = G_*$.*

Proof. By Proposition 6.36, M_f is a submap of F and a submap of G. Hence it follows from Proposition 6.29 that

$$F_* = (M_f)_* = G_*. \qquad \square$$

The last proposition implies that if the minimal representation of f is acyclic-valued, then (6.20) is well defined. Now that we have a formula to work with, we can, using some very simple examples, try to reinforce the intuition about homology of maps that we began to develop in Chapter 5. We begin with a technical result.

Proposition 6.39 *Let $X \subset \mathbf{R}^d$ be a cubical set. Let $A \subset X$ be such that diam $A < 1$. Then $\text{ch}(A)$ is acyclic.*

Proof. Let
$$\mathcal{Q} := \{Q \in \mathcal{K}(X) \mid \mathring{Q} \cap A \neq \emptyset\}.$$
Since X is cubical,
$$\text{ch}(A) = \bigcup_{Q \in \mathcal{Q}} Q.$$
Observe that for any two elementary cubes $P, Q \in \mathcal{Q}$ the intersection $P \cap Q$ is nonempty, because otherwise diam $A \geq 1$. Therefore, by Proposition 2.85, $\bigcap \mathcal{Q}$ is nonempty. It follows that $\text{ch}(A)$ is star-shaped and consequently acyclic by Proposition 2.84. □

The simplest nontrivial map is the identity.

Proposition 6.40 *Let X be a cubical set. Consider the identity map $\text{id}_X : X \to X$. Then M_{id_X} is acyclic-valued and*
$$(\text{id}_X)_* = \text{id}_{H_*(X)}.$$

Proof. By Proposition 6.10(vi),
$$M_{\text{id}_X}(x) = \text{ch}(\text{id}_X(\text{ch}(x))) = \text{ch}(\text{ch}(x)) = \text{ch}(x),$$
which, by Proposition 6.39, is acyclic. Therefore, M_{id_X} is a cubical carrier of id_X and
$$(\text{id}_X)_* = \left(M_{\text{id}_X}\right)_*.$$
Let $Q \in \mathcal{K}(X)$. Then
$$\left|\text{id}_{C(X)}(\widehat{Q})\right| = Q = \text{ch}(\mathring{Q}) = M_{\text{id}_X}(\mathring{Q}).$$
Therefore, $\text{id}_{C(X)}$ is a chain selector for M_{id_X}. Finally, it is easy to check that $\text{id}_{C(X)}$ induces the identity map $\text{id}_{H_*(X)}$ on homology. □

Lemma 6.41 *Let X, Y, and Z be cubical sets. Assume $f : X \to Y$ and $g : Y \to Z$ are continuous maps such that M_f, M_g, $M_{g \circ f}$, and $M_g \circ M_f$ are acyclic-valued. Then*
$$(g \circ f)_* = g_* \circ f_*.$$

Proof. Observe that

$$M_{g\circ f}(x) = \operatorname{ch}(g(f(\operatorname{ch}(x)))) \subset \operatorname{ch}(g(\operatorname{ch}(f(\operatorname{ch}(x))))) = M_g(M_f(x)),$$

that is, $M_{g\circ f} \subset M_g \circ M_f$. Therefore, from Proposition 6.29 and Corollary 6.31,

$$(g \circ f)_* = (M_{g\circ f})_* = (M_g \circ M_f)_* = (M_g)_* \circ (M_f)_* = g_* \circ f_*. \qquad \square$$

The last lemma is a temporary tool. In the next section we present a general result, Theorem 6.58, in which the technical assumptions about the acyclicity of the values of M_f, M_g, $M_{g\circ f}$, and $M_g \circ M_f$ is no longer needed. However, before we prove Theorem 6.58, we have to make sure that these four assumptions are satisfied whenever we want to compose homology maps.

Example 6.37 indicates that M_g need not be acyclic-valued. Therefore, we are not yet finished with the definition of the homology of a map. There are, however, some simple, but useful, maps whose minimal representations are acyclic-valued and so we already know their homology. These maps are studied in Chapter 4 and we are now in the position to conclude that the related chain maps presented there are in fact chain selectors of minimal representations. We start from the following example.

Example 6.42 Let $A \subset X$ be cubical sets in $\mathbf{R}^d$. In Example 4.4 we point out that the inclusion map $i : A \hookrightarrow X$ gives rise to an inclusion map of the chain complexes, $\iota : \mathcal{C}(A) \hookrightarrow \mathcal{C}(X)$, defined by

$$\iota_k \widehat{Q} = \widehat{Q} \quad \text{for all } Q \in \mathcal{K}_k(A).$$

We are now able to state precisely what we mean by "gives rise to." Since $i(Q) = Q$ for every elementary cube $Q \in A$, M_i is defined by

$$M_i(x) = \operatorname{ch}(\operatorname{ch}(x)) = \operatorname{ch}(x)$$

for all $x \in A$. From this,

$$M_i(\overset{\circ}{Q}) = Q$$

for all $Q \in \mathcal{K}(A)$. Hence the chain inclusion ι satisfies the two conditions defining a chain selector of M_i and $i_* := (M_i)_* = \iota_*$.

Example 6.43 Let f be any of the following maps:

(a) the immersion $j : X \to \{0\} \times X$ discussed in Example 4.6;
(b) the translation $t : X \to \mathbf{m} + X$ discussed in Example 4.7;
(b) the projection $p : Q \to Q$ discussed in Example 4.8.

We leave it to the reader to check that the minimal representation of f has the property

$$M_f(x) = f(\operatorname{ch}(x))$$

and to verify that the previously defined associated chain map, κ, τ, and π, respectively, is indeed a chain selector of M_f.

6.4.2 Rescaling

So far we are able to define the homology map of a continuous function when its minimal representation is acyclic-valued. Unfortunately, as indicated in Example 6.37 this is not true for every continuous function. We have encountered this problem before, in Section 5.4. In Section 5.2 we consider the procedure of subdividing the intervals. We could do the same thing here, namely we could try to make the images of all elementary cubes acyclic by subdividing the domain of the map into smaller cubes. However, that would require developing the homology theory for cubical sets defined on fractional grids. Obviously, this could be done, but it is not necessary. Instead we take the equivalent path, based on rescaling the domain of the function, already considered in Section 5.2. Observe that if we make the domain large, then as a fraction of the size of the domain the elementary cubes become small. This leads to the following notation.

Definition 6.44 A *scaling vector* is a vector of positive integers

$$\alpha = (\alpha_1, \alpha_2, \ldots, \alpha_d) \in \mathbf{Z}^d$$

and gives rise to the *scaling* $\Lambda^\alpha : \mathbf{R}^d \to \mathbf{R}^d$ defined by

$$\Lambda^\alpha(x) := (\alpha_1 x_1, \alpha_2 x_2, \ldots, \alpha_d x_d).$$

If $\beta = (\beta_1, \beta_2, \ldots, \beta_d)$ is another scaling vector, then set

$$\alpha\beta := (\alpha_1\beta_1, \alpha_2\beta_2, \ldots, \alpha_d\beta_d).$$

The following properties of scalings are straightforward and left as an exercise.

Proposition 6.45 *Let α and β be a scaling vector. Then Λ^α maps cubical sets onto cubical sets and $\Lambda^\beta \circ \Lambda^\alpha = \Lambda^{\alpha\beta}$.*

Definition 6.46 Let $X \subset \mathbf{R}^d$ be a cubical set and let $\alpha \in \mathbf{Z}^d$ be a scaling vector. Define $\Lambda_X^\alpha := \Lambda^\alpha|_X$. The *scaling of X by α* is

$$X^\alpha := \Lambda_X^\alpha(X) = \Lambda^\alpha(X).$$

Example 6.47 Recall that Example 6.37 describes a function f for which M_f was not a cubical representation. The first step in dealing with this problem involves rescaling the space X. It is easy to check that scaling by $\alpha = (2,2)$ gives $X^\alpha = \mathrm{bd}\,[0,2]^2$. In Example 6.52 we discuss the effect of this scaling on f.

We begin by establishing that scalings are nice continuous maps in the sense that their minimal cubical representations are acyclic-valued.

6.4 Homology of Continuous Maps

Proposition 6.48 *Let X be a cubical set and let α be a scaling vector. The map $M_{\Lambda_X^\alpha}$ is acyclic-valued.*

Proof. By definition, for any $x \in X$

$$M_{\Lambda_X^\alpha}(x) = (\operatorname{ch}(\Lambda_X^\alpha(\operatorname{ch}(x)))).$$

Since $\operatorname{ch}(x)$ is a cube, it follows that $\Lambda_X^\alpha(\operatorname{ch}(x))$ is a cubical rectangle and consequently $\operatorname{ch}(\Lambda_X^\alpha(\operatorname{ch}(x))) = \Lambda_X^\alpha(\operatorname{ch}(x))$ is also a cubical rectangle. Therefore, by Proposition 2.81, the set $M_{\Lambda_X^\alpha}(x)$ is acyclic. □

In the sequel we will have to compose the homology of scalings. In order to be able to apply Lemma 6.41, we need to make the following observations. Let X, Y, and Z be cubical sets and let α and β be scaling vectors. First, note that

$$M_{\Lambda_Y^\beta} \circ M_{\Lambda_X^\alpha} = M_{\Lambda_Y^\beta \circ \Lambda_X^\alpha}. \tag{6.22}$$

Indeed, by an argument similar to that in the proof of Proposition 6.48

$$(M_{\Lambda_Y^\beta} \circ M_{\Lambda_X^\alpha})(x) = \operatorname{ch}(\Lambda_Y^\beta(\operatorname{ch}(\Lambda_X^\alpha(\operatorname{ch}(x))))) = \Lambda_X^{\beta\alpha}(\operatorname{ch}(x)) = M_{\Lambda_Y^\beta \circ \Lambda_X^\alpha}(x).$$

Next, since $\Lambda_Y^\beta \circ \Lambda_X^\alpha = \Lambda_X^{\alpha\beta}$, the map $M_{\Lambda_Y^\beta \circ \Lambda_X^\alpha}$ is acyclic-valued by Proposition 6.48. We have proved the following proposition.

Proposition 6.49 *Let X, Y, and Z be cubical sets and let α and β be scaling vectors. If $\Lambda^\alpha(X) \subset Y$ and $\Lambda^\beta(Y) \subset Z$, then $M_{\Lambda_Y^\beta \circ \Lambda_X^\alpha} = M_{\Lambda_Y^\beta} \circ M_{\Lambda_X^\alpha}$ and this map is acyclic-valued.*

Since the minimal representations of scalings are acyclic-valued, they induce maps on homology. Furthermore, since scalings just change the size of the space, one would expect that they induce isomorphisms on homology. The simplest way to check this is to show that their homology maps have inverses. Therefore, given a cubical set X and a scaling vector α, let $\Omega_X^\alpha : X^\alpha \to X$ be defined by

$$\Omega_X^\alpha(x) := (\alpha_1^{-1} x_1, \alpha_2^{-1} x_2, \ldots, \alpha_d^{-1} x_d).$$

Obviously, $\Omega_X^\alpha = (\Lambda_X^\alpha)^{-1}$. However, we need to know that it induces a map on homology.

Lemma 6.50 $M_{\Omega_X^\alpha} : X^\alpha \rightrightarrows X$ *is acyclic-valued.*

Proof. Let $x \in X^\alpha$. Then $\operatorname{ch}(x) = P$ for some $P \in \mathcal{K}_k(X^\alpha)$ and

$$M_{\Omega_X^\alpha}(x) = \operatorname{ch}(\Omega_X^\alpha(\operatorname{ch}(x)))$$
$$= \operatorname{ch}(\Omega_X^\alpha(P)).$$

Since obviously $\Omega_X^\alpha(P)$ is a Cartesian product of bounded intervals, the conclusion follows from Corollary 6.13 and Proposition 2.81. □

Proposition 6.51 *If X is a cubical set and α is a scaling vector, then*

$$(\Lambda_X^\alpha)_* : H_*(X) \to H_*(X^\alpha) \quad \text{and} \quad (\Omega_X^\alpha)_* : H_*(X^\alpha) \to H_*(X)$$

are isomorphisms. Furthermore,

$$(\Lambda_X^\alpha)_*^{-1} = (\Omega_X^\alpha)_*.$$

Proof. It follows from Proposition 6.48 and Lemma 6.50 that $M_{\Lambda_X^\alpha}$, $M_{\Omega_X^\alpha}$, $M_{\Omega_X^\alpha \circ \Lambda_X^\alpha} = M_{\text{id}_X}$, and $M_{\Lambda_X^\alpha \circ \Omega_X^\alpha} = M_{\text{id}_{X^\alpha}}$ are acyclic-valued. Moreover, for any $x' \in X^\alpha$,

$$M_{\Lambda_X^\alpha} \circ M_{\Omega_X^\alpha}(x') = \text{ch}\left(\Lambda_X^\alpha(\text{ch}\left(\Omega_X^\alpha(\text{ch}(x'))\right))\right)$$
$$= \Lambda_X^\alpha(\text{ch}\left(\Omega_X^\alpha(\text{ch}(x'))\right)).$$

By Corollary 6.13, $\text{ch}(\Omega_X^\alpha(\text{ch}(x')))$ is a rectangle. Hence, $M_{\Lambda_X^\alpha} \circ M_{\Omega_X^\alpha}(x)$ is a rectangle, since it is the rescaling of a rectangle. It follows from Proposition 2.81 that it is acyclic. Similarly, for any $x \in X$,

$$M_{\Omega_X^\alpha} \circ M_{\Lambda_X^\alpha}(x) = \text{ch}\left(\Omega_X^\alpha(\text{ch}\left(\Lambda_X^\alpha(\text{ch}(x))\right))\right)$$
$$= \text{ch}\left(\Omega_X^\alpha(\Lambda_X^\alpha(\text{ch}(x)))\right)$$
$$= \text{ch}(x)$$

is acyclic. Hence, by Lemma 6.41, and Proposition 6.40,

$$(\Lambda_X^\alpha)_* \circ (\Omega_X^\alpha)_* = (\Lambda_X^\alpha \circ \Omega_X^\alpha)_* = (\text{id}_{X^\alpha})_* = \text{id}_{H_*(X^\alpha)}$$

and

$$(\Omega_X^\alpha)_* \circ (\Lambda_X^\alpha)_* = (\Omega_X^\alpha \circ \Lambda_X^\alpha)_* = (\text{id}_X)_* = \text{id}_{H_*(X)}. \qquad \square$$

As indicated in the introduction, the purpose of scaling is to allow us to define the homology of an arbitrary continuous map between cubical sets. Thus, given a continuous map $f : X \to Y$ and a scaling vector α, define

$$f^\alpha := f \circ \Omega_X^\alpha.$$

Observe that $f^\alpha : X^\alpha \to Y$.

Example 6.52 To indicate the relationship between f and f^α we return to Example 6.37. Consider $\alpha = (2,2)$. Then X^α is the boundary of the square $[0,2]^2$, which can be parameterized by $t \in [0,8]$ with $0 \sim 8$ in the same manner as done in (5.10):

$$x(t) = (x_1(t), x_2(t)) := \begin{cases} (0,0) + t(1,0) & \text{if } t \in [0,2], \\ (2,0) + (t-2)(0,1) & \text{if } t \in [2,4], \\ (2,2) + (t-4)(-1,0)) & \text{if } t \in [4,6], \\ (0,1) + (t-6)(0,-1) & \text{if } t \in [6,8]. \end{cases} \quad (6.23)$$

It is easy to check that the formula relating f^α to f in terms of the parameter t is
$$f^\alpha(x(t)) = f(x(t/2)).$$
The graphs of f^α and M_{f^α} as functions of t are shown in Figure 6.5.

It is seen from the presented graph that the value of $M_{f^\alpha}(x)$ for each x is a vertex, if x is a vertex, and a union of two elementary intervals with a common vertex otherwise. To formally check this using the definition of f given in Example 6.37, consider any $(x_1, x_2) \in \overset{\circ}{Q}$ where, for example, $Q = [0,1] \times [2]$. Then

$$\begin{aligned} M_{f^\alpha}(x_1, x_2) &= \text{ch}\left(f^\alpha(\text{ch}(x_1, x_2))\right) \\ &= \text{ch}\left(f^\alpha(Q)\right) \\ &= \text{ch}\left(f([0, 1/2] \times [1])\right) \\ &= \text{ch}\left(\lambda([0, 1/2])\right) \\ &= \text{ch}\left([0] \times [0,1] \cup [0,1] \times [1]\right) \\ &= [0] \times [0,1] \cup [0,1] \times [1]. \end{aligned}$$

This set is star-shaped, hence it is acyclic by Proposition 2.84. Similar checks can be performed for all elementary edges. If x is a vertex in X^α, a similar check shows that $f^\alpha(x)$ is a vertex in X. In this case, $M_{f^\alpha}(x) = \{f^\alpha(x)\}$. It is an elementary vertex, hence it is acyclic. We conclude that M_{f^α} is acyclic-valued, and hence M_{f^α} is a cubical representation of f^α.

Theorem 6.53 *Let X and Y be cubical sets and $f : X \to Y$ be continuous. Then there exists a scaling vector α such that M_{f^α} is acyclic-valued. Moreover, if β is another scaling vector such that M_{f^β} is acyclic-valued, then*

$$f_*^\alpha(\Lambda_X^\alpha)_* = f_*^\beta(\Lambda_X^\beta)_*.$$

Proof. Since X is a cubical set, it is compact (see Definition 12.61 and Theorem 12.62). By Theorem 12.64, f is uniformly continuous (see Definition 12.63). Therefore, we may choose $\delta > 0$ such that for $x, y \in X$

$$\text{dist}(x, y) \le \delta \quad \Rightarrow \quad \text{dist}(f(x), f(y)) \le \frac{1}{2}. \tag{6.24}$$

Let α be a scaling vector such that $\min\{\alpha_i \mid i = 1, \ldots, n\} \ge 1/\delta$. Since $\text{diam}\,\text{ch}(x) \le 1$, we get from (6.24) that

$$\text{diam}\,f^\alpha(\text{ch}(x)) \le \frac{1}{2}.$$

Therefore, it follows from Proposition 6.39 that M_{f^α} is acyclic-valued.

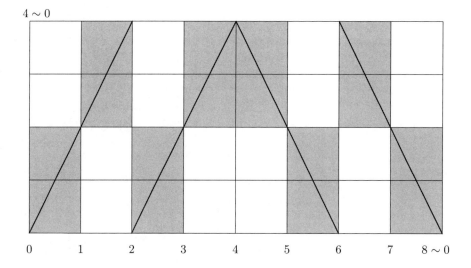

Fig. 6.5. The graph of f^α and of M_{f^α} as functions of the parameter $t \in [0,8]$ discussed in Example 6.52. Note that M_{f^α} sends points in open elementary intervals to unions of two intervals and any vertex to a vertex.

Now assume that the scaling vector β is such that M_{f^β} is also acyclic-valued. Also, assume for the moment that for each $i = 1, \ldots, n$, α_i divides β_i. Let $\gamma_i := \beta_i/\alpha_i$. Then $\gamma = (\gamma_1, \ldots, \gamma_d)$ is a scaling vector. Clearly, $\Lambda_X^\beta = \Lambda_{X^\alpha}^\gamma \circ \Lambda_X^\alpha$. Moreover, by Proposition 6.49, $M_{\Lambda_{X^\alpha}^\gamma} \circ M_{\Lambda_X^\alpha}$ is acyclic-valued. Therefore, it follows from Lemma 6.41 that

$$(\Lambda_X^\beta)_* = (\Lambda_{X^\alpha}^\gamma)_* \circ (\Lambda_X^\alpha)_*.$$

On the other hand, we also have

$$f^\alpha = f^\beta \circ \Lambda_{X^\alpha}^\gamma,$$

hence

$$M_{f^\beta}(M_{\Lambda_{X^\alpha}^\gamma}(x)) = \operatorname{ch}(f^\beta(\operatorname{ch}(\Lambda_{X^\alpha}^\gamma(\operatorname{ch}(x))))) = \operatorname{ch}(f^\alpha(\operatorname{ch}(x))) = M_{f^\alpha}(x).$$

Thus $M_{f^\beta} \circ M_{\Lambda_{X^\alpha}^\gamma}$ is acyclic-valued and we get from Lemma 6.41 that

$$f_*^\alpha = f_*^\beta \circ (\Lambda_{X^\alpha}^\gamma)_*.$$

Consequently,

$$f_*^\alpha \circ (\Lambda_X^\alpha)_* = f_*^\beta \circ (\Lambda_{X^\alpha}^\gamma)_* \circ (\Lambda_X^\alpha)_* = f_*^\beta \circ (\Lambda_X^\beta)_*.$$

Finally, if it is not true that α_i divides β_i, for each $i = 1, \ldots, n$, then let $\theta = \alpha\beta$. By what we have just proven

$$f_*^\alpha \circ (\Lambda_X^\alpha)_* = f_*^\theta \circ (\Lambda_X^\theta)_* = f_*^\beta \circ (\Lambda_X^\beta)_*,$$

which settles the general case. □

We can now give the general definition for the homology map of a continuous function.

Definition 6.54 Let X and Y be cubical sets and let $f : X \to Y$ be a continuous function. Let α be a scaling vector such that M_{f^α} is acyclic-valued. Then $f_* : H_*(X) \to H_*(Y)$ is defined by

$$f_* := f_*^\alpha \circ (\Lambda_X^\alpha)_*.$$

Given $k \in \mathbf{Z}$, the restriction of f_* to the kth dimensional homology is denoted by f_{*k}.

The definition of f_* can be illustrated by diagrams as follows. First, f^α is defined so as to complete the following commutative diagram.

$$\begin{array}{ccc} X^\alpha & \xrightarrow{f^\alpha} & Y \\ \downarrow{\scriptstyle \Omega^\alpha} & \nearrow{\scriptstyle f} & \\ X & & \end{array}$$

Next, f_* is defined so as to complete the diagram below.

$$\begin{array}{ccc} H_*(X^\alpha) & \xrightarrow{f_*^\alpha = (M_{f^\alpha})_*} & H_*(Y) \\ \uparrow{\scriptstyle \Lambda_*^\alpha} & \nearrow{\scriptstyle f_*} & \\ H_*(X) & & \end{array}$$

By Theorem 6.53, this definition is independent of the particular scaling vector used. However, we need to reconcile this definition of the homology map with that of (6.20). So assume that $f : X \to Y$ is such that M_f is acyclic-valued. Let α be the scaling vector where each $\alpha_i = 1$. Then $f^\alpha = f$, $\Lambda_X^\alpha = \text{id}_X$, and

$$f_* = f_* \circ (\text{id}_X)_* = (f^\alpha)_* \circ (\Lambda_X^\alpha)_*;$$

hence, the two definitions of f_* agree.

Several more natural questions can be posed. We know from Proposition 6.38 that if the minimal representation of a continuous map $f : X \to Y$ is acyclic-valued, then any of its acyclic-valued representations may be used to compute its homology. Given the examples presented so far, one may be lulled into assuming that if the minimal representation is not acyclic-valued, then no representation is acyclic-valued. As the following example demonstrates, this is false. This in turn raises the question: If one has an acyclic-valued representation, can it be used to compute the homology of f?

228   6 Homology of Maps

Example 6.55 Take X, f, and λ as in Example 6.37. Let $Y := [0,1] \times [0,1]$ and consider $g : X \to Y$ given by the same formula as f, namely by

$$g(x_1, x_2) := \begin{cases} \lambda(x_2) & \text{if } (x_1, x_2) \in K_1 \cup K_3, \\ \lambda(x_1) & \text{if } (x_1, x_2) \in K_2 \cup K_4. \end{cases}$$

Then $M_f = M_g$, that is, the minimal representation of g is not acyclic-valued. However, taking the cubical map $F : X \rightrightarrows Y$ given by

$$F(x_1, x_2) := [0,1] \times [0,1],$$

we easily see that F is an acyclic-valued representation of g. This gives an affirmative answer to the first question.

The answer to the other question is also positive, as the following proposition shows.

Proposition 6.56 *Assume $F : X \rightrightarrows Y$ is an acyclic-valued representation of a continuous map $f : X \to Y$. Then*

$$f_* = F_*.$$

Proof. Let α be a scaling such that M_{f^α} is acyclic-valued. Then for any $x \in X$ we have

$$\begin{aligned} M_{f^\alpha} \circ M_{\Lambda_X^\alpha}(x) &= \operatorname{ch}\left(f^\alpha(\operatorname{ch}\left(\Lambda_X^\alpha(\operatorname{ch}(x))\right))\right) \\ &= \operatorname{ch}\left(f^\alpha(\Lambda_X^\alpha(\operatorname{ch}(x)))\right) \\ &= \operatorname{ch}\left(f(\operatorname{ch}(x))\right) \\ &= M_f(x) \\ &\subset F(x). \end{aligned}$$

Therefore, by Proposition 6.30,

$$f_* = (M_{f^\alpha})_* \circ (M_{\Lambda_X^\alpha})_* = F_*.$$

The final issue we need to deal with involves the composition of continuous functions. We will use the following technical lemma.

Lemma 6.57 *Let X and Y be cubical sets and let $f : X \to Y$ be continuous. Let α be a scaling vector. If M_f is acyclic-valued, then $M_{\Lambda_Y^\alpha} \circ M_f$ is acyclic-valued.*

Proof. Let $x \in X$. Observe that

$$M_{\Lambda_Y^\alpha} \circ M_f(x) = \operatorname{ch}\left(\Lambda_Y^\alpha(\operatorname{ch}(M_f(x)))\right) = \Lambda_Y^\alpha(M_f(x)).$$

Since $M_f(x)$ is acyclic-valued, it follows from Proposition 6.51 that $\Lambda_Y^\alpha(M_f(x))$ is also acyclic-valued. □

6.4 Homology of Continuous Maps

Theorem 6.58 *Assume $f : X \to Y$ and $g : Y \to Z$ are continuous maps between cubical sets. Then*

$$(g \circ f)_* = g_* \circ f_*.$$

Proof. Let $h := g \circ f$. Select a scaling vector β such that M_{g^β} is acyclic-valued and for any $x, y \in Y^\beta$

$$\text{dist}(x, y) \leq 2 \quad \Rightarrow \quad \text{dist}(g^\beta(x), g^\beta(y)) \leq \frac{1}{2}. \tag{6.25}$$

Similarly, select a scaling vector α such that M_{f^α} and M_{h^α} are acyclic-valued, and for any x, y in X^α

$$\text{dist}(x, y) \leq 2 \quad \Rightarrow \quad \text{dist}(\Lambda^\beta \circ f^\alpha(x), \Lambda^\beta \circ f^\alpha(y)) \leq \frac{1}{2}. \tag{6.26}$$

Then, by Proposition 6.39 the maps $\Lambda^\beta \circ f^\alpha$ and $g^\beta \circ (\Lambda^\beta \circ f^\alpha) = h^\alpha$ have acyclic-valued minimal representations. In particular, we can apply Lemma 6.57 to $f^\alpha : X^\alpha \to Y$ and the scaling vector β, which allows us to conclude from Lemma 6.41 that

$$(\Lambda^\beta_{X^\alpha} \circ f^\alpha)_* = \Lambda^\beta_{X^\alpha *} \circ f^\alpha_*. \tag{6.27}$$

Moreover, by (6.25) and (6.26), for any $x \in X^\alpha$

$$\text{diam}(g^\beta \circ \text{ch} \circ \Lambda^\beta \circ f^\alpha \circ \text{ch}(x)) < 1.$$

Therefore, by Proposition 6.39,

$$M_{g^\beta} \circ M_{\Lambda^\beta \circ f^\alpha}(x) = (\text{ch} \circ g^\beta \circ \text{ch} \circ \Lambda^\beta \circ f^\alpha \circ \text{ch})(x)$$

is acyclic and consequently the composition $M_{g^\beta} \circ M_{\Lambda^\beta \circ f^\alpha}$ is acyclic-valued. This implies that

$$h^\alpha_* = (g^\beta \circ \Lambda^\beta_{X^\alpha} \circ f^\alpha)_* = g^\beta_* \circ (\Lambda^\beta_{X^\alpha} \circ f^\alpha)_*. \tag{6.28}$$

Finally, we get from (6.27) and (6.28)

$$\begin{aligned}
(g \circ f)_* &= h_* \\
&= h^\alpha_* \circ (\Lambda^\alpha_X)_* \\
&= g^\beta_* \circ (\Lambda^\beta_{X^\alpha} \circ f^\alpha)_* \circ (\Lambda^\alpha_X)_* \\
&= g^\beta_* \circ (\Lambda^\beta_{X^\alpha})_* \circ f^\alpha_* \circ (\Lambda^\alpha_X)_* \\
&= g_* \circ f_*.
\end{aligned}$$

$\square$

Theorem 6.58 and Proposition 6.40 indicate that there is a natural or *functorial* relationship between continuous maps and their induced maps on homology. This is explained in greater detail in Chapter 11. For the moment it is sufficient to recognize that these results allow us to conclude that the homology of two homeomorphic cubical sets is the same. This indicates that cubical homology is a *topological property*, in the sense that it is invariant under a homeomorphism of cubical sets.

Corollary 6.59 (Topological invariance) *Let X and Y be homeomorphic cubical sets. Then*
$$H_*(X) \cong H_*(Y).$$

Proof. Let $h : X \to Y$ be a homeomorphism. By Theorem 6.58 and Proposition 6.40,
$$(h^{-1})_* \circ h_* = (h^{-1} \circ h)_* = (\text{id}_X)_* = \text{id}_{H_*(X)},$$
and by the same argument
$$h_* \circ (h^{-1})_* = \text{id}_{H_*(Y)}.$$
Thus h_* is an isomorphism with the inverse $(h_*)^{-1} = (h^{-1})_*$. □

Because of the results of Section 2.3 we have a good understanding of the topological meaning of $H_0(X)$ for any cubical set X. Thus it is natural to try to understand how the homology map acts on this homology group.

Proposition 6.60 *If $f : X \to X$ is a continuous map on a connected cubical set, then $f_* : H_0(X) \to H_0(X)$ is the identity map.*

Proof. Consider first the case when M_f is acyclic-valued. Then the homology map $f_{*0} : H_0(X) \to H_0(X)$ is determined by an appropriate chain map $\varphi_0 : C_0(X) \to C_0(X)$, which in turn can be determined by M_f. So let $Q \in \mathcal{K}_0(X)$. By definition, $M_f(Q) = \text{ch}(f(Q))$, which is an elementary cube. Let $P \in \mathcal{K}_0(\text{ch}(f(Q)))$. Then we can define $\varphi_0(\widehat{Q}) = \widehat{P}$, in which case $f_*([\widehat{Q}]) = [\widehat{P}]$. By Theorem 2.59, $[\widehat{Q}] = [\widehat{P}] \in H_0(X)$, and hence we have the identity map on $H_0(X)$.

When M_f is not acyclic-valued, we have to go through the process of rescaling. We leave the details as an exercise. □

Exercises

6.10 Prove the conclusions announced in Example 6.43 about

(a) the immersion $j : X \to \{0\} \times X$ discussed in Example 4.6,
(b) the translation $t : X \to \mathbf{m} + X$ discussed in Example 4.7,
(b) the projection $p : Q \to Q$ discussed in Example 4.8.

6.11 Prove Proposition 6.45.

6.12 Let X, λ, and f be as in Example 6.37.

(a) Find a chain selector of M_{f^α} where $\alpha = (2, 2)$ (see Example 6.52).
(b) Compute the homology map of f. You may either compute it by hand or use the homology program for that.

6.13 Do the same as in Exercise 6.12 for the map given by

$$f(x_1, x_2) := \begin{cases} \lambda(x_2) & \text{if } (x_1, x_2) \in K_1, \\ \lambda(x_1) & \text{if } (x_1, x_2) \in K_2, \\ \lambda(1 - x_2) & \text{if } (x_1, x_2) \in K_3, \\ \lambda(1 - x_1) & \text{if } (x_1, x_2) \in K_4. \end{cases}$$

6.14 Complete the proof of Proposition 6.60 by examining the case when M_f is not acyclic-valued.

6.5 Homotopy Invariance

We now have a homology theory at our disposal. Given a cubical set X, we can compute its homology groups $H_*(X)$, and given a continuous map f between cubical sets, we can compute the induced map on homology f_*. What is missing is how these algebraic objects relate back to topology. Section 2.3 gives a partial answer in that we show that $H_0(X)$ counts the number of connected components of X. In this section we shall use homology to provide further classification of maps and spaces.

Definition 6.61 Let X and Y be cubical sets. Let $f, g : X \to Y$ be continuous functions. f is *homotopic* to g if there exists a continuous map $h : X \times [0, 1] \to Y$ such that

$$h(x, 0) = f(x) \quad \text{and} \quad h(x, 1) = g(x)$$

for each $x \in X$. The map h is called a *homotopy* between f and g. The relation f homotopic to g is denoted by $f \sim g$.

It is left as an exercise to check that homotopy is an equivalence relation. Homotopy of maps can be used to define an equivalence between spaces.

Definition 6.62 Two cubical sets X and Y have the same *homotopy type* or, for short, X and Y are *homotopic* if there exist continuous functions $f : X \to Y$ and $g : Y \to X$ such that

$$g \circ f \sim \text{id}_X \quad \text{and} \quad f \circ g \sim \text{id}_Y,$$

where id_X and id_Y denote the identity maps. The relation X homotopic to Y is denoted by $X \sim Y$.

Example 6.63 Two cubical sets can appear to be quite different and still be homotopic. For example, it is clear that the unit ball with respect to the supremum norm $B_0^d = [-1, 1]^d$ is not homeomorphic to the point $\{0\}$. On the other hand, these two sets are homotopic. To see this let $f : B_0^d \to \{0\}$ be defined by $f(x) = 0$ and let $g : \{0\} \to B_0^d$ be defined by $g(0) = 0$. Observe that $f \circ g = \mathrm{id}_{\{0\}}$ and hence $f \circ g \sim \mathrm{id}_{\{0\}}$. To show that $g \circ f \sim \mathrm{id}_{B_0^d}$, consider the map $h : B_0^d \times [0,1] \to B_0^d$ defined by

$$h(x,s) = (1-s)x.$$

Clearly, $h(x,0) = x = \mathrm{id}_{B_0^d}$ and $h(x,1) = 0$.

Notice that the homotopy constructed in the above example is also a deformation retraction (see Definition 2.70) of B_0^d onto $\{0\}$. We have introduced the concept of deformation retraction for the purpose of elementary collapses. What is the relation between that concept and the concept of homotopy? We start from the following definition.

Definition 6.64 Given a pair $A \subset X$, a continuous map $r : X \to A$ such that $r(a) = a$ for all $a \in A$ is called a *retraction*. If such a map exists, A is called a *retract* of X.

Note that, in particular, any retract of X must be closed in X. The notion of a retraction is the topological analog of a projection in linear algebra.

Definition 2.70 implies that a deformation retraction h is a homotopy between the identity map on X and the retraction $x \to h(x,1)$ of X onto A. More precisely, we have the following result.

Theorem 6.65 *If A is a deformation retract of X, then A and X have the same homotopy type.*

Proof. Let h be a deformation retraction of X to A. Define $f : X \to A$ by $f(x) := h(x,1)$ and let $g : A \to X$ be the inclusion map. It follows from the properties of F that $f \circ g = \mathrm{id}_A$ and $g \circ f \sim \mathrm{id}_X$. □

Definition 6.66 A cubical set X is *contractible* if the identity map on X is homotopic to a constant map; equivalently, if there exists a deformation retraction of X to a single point of X.

The relation between the contractibility and connectedness is illustrated in the following example.

Example 6.67 Any single vertex $\{v\}$ of any cubical set $X \subset \mathbf{R}^d$ is a retract of X with $r : X \to \{v\}$, $r(x) := v$. However, $\{0\}$ is not a deformation retract of the two-point set $X = \{0,1\}$. Indeed, if there was a deformation retraction h, the continuous map $s \to h(1,s)$ would send the connected interval $[0,1]$ onto the disconnected set X, which would contradict Theorem 12.46. By the same argument one may show that no disconnected space is contractible.

The next theorem provides the bridge from topology to algebra.

Theorem 6.68 *Let X and Y be cubical sets and let $f, g : X \to Y$ be homotopic maps. Then*

$$f_* = g_*.$$

Proof. First note that since X is cubical, $X \times [0,1]$ is also cubical and, for $k = 0, 1$ and $x \in X$, the maps $j_k : X \to X \times [0,1]$ given by

$$j_k(x) := (x, k)$$

are continuous. Let $h : X \times [0,1] \to Y$ be the homotopy joining f and g, that is, a continuous function such that $h(x,0) = f(x)$ and $h(x,1) = g(x)$, which may also be written as

$$f = h \circ j_0,$$
$$g = h \circ j_1. \qquad (6.29)$$

Let $\mathcal{F} : \mathcal{K}_{\max}(X) \rightrightarrows \mathcal{K}(X \times [0,1])$ be defined for $Q \in \mathcal{K}_{\max}(X)$ by

$$\mathcal{F}(Q) := \{Q \times [0,1]\}$$

and let $F := \lfloor \mathcal{F} \rfloor$. Then by Theorem 6.14 and Corollary 6.19, F is a lower semicontinuous cubical map. One easily verifies that if $x \in X$ and $P \in \mathcal{K}(X)$ is the unique cube such that $x \in \overset{\circ}{P}$, then

$$F(x) = P \times [0,1].$$

Therefore, F is acyclic-valued. Since obviously $j_k(x) \in F(x)$, we see that F is an acyclic-valued representation of j_0 and j_1. Therefore, we have, by (6.29), Theorem 6.58 and Proposition 6.56,

$$f_* = h_* \circ (j_0)_* = h_* \circ F_* = h_* \circ (j_1)_* = g_*. \qquad \square$$

From here we get the following result generalizing Corollary 6.59 on topological invariance of homology. The proof is similar to that of Corollary 6.59 and is left as exercise.

Corollary 6.69 (Homotopy invariance) *Let X and Y be cubical sets of the same homotopy type. Then*

$$H_*(X) \cong H_*(Y).$$

Exercises

6.15 Prove that homotopy is an equivalence relation.

6.16 Let X, Y be cubical sets and $f, g : X \to Y$ be continuous maps. Under the following assumptions on X and Y, construct a homotopy $f \sim g$.
- $X = Y = [0,1]$.
- $X = \Gamma^1$ and $Y = [0,1]$, where Γ^1 is the boundary of the unit square $Q = [0,1]^2$.
- X is any cubical set and $\{y\} \in Y$ is a deformation retract of Y.

6.17 Show that if A is a deformation retract of X and B is a deformation retract of A, then B is a deformation retract of X.

6.18 A subset $C \subset \mathbf{R}^d$ is called *star-shaped*[1] if there exists a point $x_0 \in C$ such that for any $y \in C$ the line segment $[x_0, y]$ is contained in C. A convex set (recall Definition 11.1) is a particular case of a star-shaped set. Prove that any star-shaped set is contractible.

6.19 Show that every tree is contractible to a vertex.

6.20 In Exercise 6.12 you should have obtained the trivial homology map. Prove this in a different way, by explicitly defining a homotopy from f to a constant map.

6.21 Prove Corollary 6.69.

6.22 Let $\Pi := \{f : \Gamma^1 \to \Gamma^1 \mid f \text{ is continuous}\}$. Give a lower bound on the number of homotopy classes in Π.

6.6 Bibliographical Remarks

The notion of representable sets discussed in Section 6.1 is introduced in [64, 65]. The study of homological properties of multivalued maps goes back to Vietoris and Begle [7] in the 1940s. Systematic approaches to the homology theory of upper semicontinuous maps have been developed by A. Granas, L. Górniewicz, H.W. Sieberg, and G. Skordev [33, 34, 77]. The idea of using multivalued maps as a tool for studying homological properties of continuous maps comes from [56, 44, 64]. The first attempts to provide a framework for computing the homology of a continuous map via multivalued representation are those due to M. Allili [1] and [2], in the setting of upper semicontinuous maps. The chain selector theorem 6.22 and Theorem 6.25 have been proved in a similar setting in [1, 2], but they are an explicit cubical analogy of a well-known acyclic carrier theorem (see, e.g., [69, Chapter 1, Theorem 13.3]). The concept of rescaling is used for the first time in this book except for [43] where results presented here are summarized. Proving the topological invariance and the homotopy invariance of homology is among the primary goals of any homology theory and can be found in classical algebraic topology textbooks in the context of simplicial, singular, cellular, or other theories. Corollary 6.59 and Corollary 6.69 are the first direct presentations of those properties of homology in the class of cubical sets.

---

[1] It can be easily verified that any cubical star-shaped set in the sense of Definition 2.82 is star-shaped in the sense of this more general definition.

7

Computing Homology of Maps

In Chapter 6 we have provided a theoretical construction for producing a homology map $f_* : H_*(X) \to H_*(Y)$ given an arbitrary continuous function $f : X \to Y$ between cubical sets $X \subset \mathbf{R}^d$ and $Y \subset \mathbf{R}^{d'}$. In this chapter we provide algorithms that allow us to use the computer to obtain f_*.

Four steps are involved in computing the homology of a map. The first is to construct a function $\mathcal{F} : \mathcal{K}_{\max}(X) \to \operatorname{CRect}(\mathbf{R}^{d'})$ that assigns to every $Q \in \mathcal{K}_{\max}(X)$ a collection of elementary cubes $\mathcal{F}(Q)$ such that

$$f(Q) \subset |\mathcal{F}(Q)|, \qquad (7.1)$$

where $|\mathcal{F}(Q)|$ is defined as in (6.6).

The second step is to use the map $\mathcal{F}$ to construct a multivalued lower semicontinuous representation $F : X \rightrightarrows Y$ of f by means of the formula

$$F(x) := \lfloor \mathcal{F} \rfloor(x) = \bigcap \{ |\mathcal{F}(Q)| \mid x \in Q \in \mathcal{K}_{\max}(X) \} \qquad (7.2)$$

(see (6.7), Theorem 6.14 and Corollary 6.19). The third step is to compute a chain selector of F, and the fourth and final step is to determine the homology map from the chain selector.

Steps one and two will be done in the next section. Observe that the first is purely an approximation problem, which has many classical numerical solutions and most of them can be adapted to our needs. However, given the goals of this book we do not want to divert too much from homology, so we only briefly describe one of the simplest approaches based on the class of rational maps already considered in Section 5.1. In Section 7.2 we provide the details of computing a chain selector of a multivalued lower semicontinuous representation. The fourth, and final, step, in which we determine the homology map from the chain selector is discussed in Section 7.3. We conclude this chapter with an optional section that describes a geometrical approach to finding a preboundary of a cycle in an acyclic cubical set.

7.1 Producing Multivalued Representation

As indicated in the introduction, the first step is essentially an approximation problem that we solve in the context of rational maps discussed in Section 5.1. Though rational functions are simple to represent (for example, one can take just the list of the coefficients as a computer representation), they encompass a reasonably large class of maps for which interval arithmetic provides a simple means of obtaining appropriate approximations. Using Eqs. (5.1–5.4) we can easily overload arithmetic operators so that they can operate on intervals (compare to Section 14.4 for operator overloading).

Since we identify a sequence of d intervals $I_1, I_2, \ldots, I_d$ with the rectangle $I_1 \times I_2 \times \cdots \times I_d$, the variables of data type

typedef rectangle := **array**[1 :] **of** interval

may be interpreted as storing either rectangles or just sequences of intervals. Without going into details we just assume that the coefficients of a rational function are stored in the data type

typedef rationalMap.

Now, using the concepts of interval arithmetic and depending on the particular definition of the data structure rationalMap, it is not too difficult to write an algorithm

function evaluate(rationalMap f, rectangle R)

which, given a rational map f and a rectangle R, returns a possibly small rectangle R' such that $f(R) \subset R'$.

For instance, if we restrict our attention to the simplest case of a polynomial of one variable with integer coefficients, then we can take

typedef polynomialMap = **array**[0 :] **of int**

and then the algorithm evaluate may look as follows:

```
function evaluate(polynomialMap f, rectangle R)
interval s = 0, xp = 1;
for i := 0 to lastIndex(f) do
    s := s + f[i] * xp;
    xp := xp * R[1];
endfor;
return s;
```

7.1 Producing Multivalued Representation   237

Extending this approach to a rational function of one variable is straightforward. Extending it to more variables is lengthy but not difficult. We leave it as an exercise.

Once we have an appropriate data structure rationalMap and a respective algorithm evaluate, we can define $\mathcal{F}(Q)$ by

$$\mathcal{F}(Q) := \left\{ P \in \mathcal{K} \mid P \subset \bar{f}(Q) \right\},$$

where $\bar{f}(Q)$ stands for the evaluation of f on Q in interval arithmetic, computed by the algorithm evaluate. However, since in this case $|\mathcal{F}(Q)|$ is exactly $\bar{f}(Q)$, we can eliminate $\mathcal{F}$ from Eq. (7.2) and obtain the multivalued lower semicontinuous representation of f directly by

$$F(x) := \bigcap \left\{ \bar{f}(Q) \mid x \in Q \in \mathcal{K}_{\max}(X) \right\}. \tag{7.3}$$

An elementary cube P is a *neighbor* of an elementary cell $\overset{\circ}{Q}$ if

$$P \cap \overset{\circ}{Q} \neq \emptyset.$$

The following algorithm is used to find all neighbors of a given elementary cell. It uses a simple algorithm degenerateDimensions which, given elementary cube $Q = I_1 \times I_2 \times \cdots \times I_d$, returns a list of integers i for which I_i is degenerate.

Algorithm 7.1 Neighbors of an elementary cell
    function neighbors(cube Q)
    L := degenerateDimensions(Q);
    N := ∅;
    for each K **in** subsets(L) **do**
        for each M **in** subsets(K) **do**
            R := Q;
            for each k **in** K **do**
                if k **in** M **then**
                    R[k]{left} := Q[k]{left} − 1;
                else
                    R[k]{right} := Q[k]{right} + 1;
                endif;
            endfor;
            N := **join**(N, R);
        endfor;
    endfor;
    return N;

Example 7.2 If $Q = [1, 2] \times [3] \subset \mathbf{R}^2$, then the output of Algorithm 7.1 is $\{[1, 2] \times [2, 3], [1, 2] \times [1, 2]\}$.

Proposition 7.3 *Assume Algorithm 7.1 is called with Q containing the representation of an elementary cube Q. Then it stops and returns a list of the neighbors of $\overset{\circ}{Q}$.*

Proof. Obviously, the algorithm always stops. Let

$$Q = I_1 \times I_2 \times \cdots \times I_d$$

be the decomposition of Q into the elementary intervals and put

$$L := \{i = 1, 2, \ldots, d \mid \dim I_i = 0\}.$$

Let N denote the list of cubes returned by the algorithm and let

$$R = J_1 \times J_2 \times \cdots \times J_d$$

be one of the cubes added to the list. It follows from the construction of R that $J_i = I_i$ whenever $i \notin L$ or J_i is degenerate and J_i is an elementary interval containing I_i in all other cases. Therefore, R is an elementary cube and $J_i \cap \overset{\circ}{I_i} \neq \emptyset$ for every $i = 1, 2, \ldots, d$. It follows that $R \cap \overset{\circ}{Q} \neq \emptyset$.

Assume in turn that R is an elementary cube satisfying $R \cap \overset{\circ}{Q} \neq \emptyset$. Then $J_i \cap \overset{\circ}{I_i} \neq \emptyset$ for every $i = 1, 2, \ldots, d$, which means that either I_i is nondegenerate and $I_i = J_i$ or I_i is degenerate and $I_i \subset J_i$. Let

$$K := \{i \in L \mid \dim J_i = 1\}$$

and let M be the set of such $i \in K$ that the left endpoint of J_i is less than the left endpoint of I_i. It is straightforward to verify that the elementary cube R is added to the list N when the variable K takes value K and the variable M takes value M. □

For the next algorithm we need to describe first three elementary algorithms, whose details are left as an exercise. The algorithms are

function rectangularCover(cubicalSet X)
function rectangleIntersection(rectangle M, N)
function elementaryCubes(rectangle M, cubicalSet X)

We assume that the first algorithm accepts a cubical set and returns the smallest cubical rectangle that contains the it, the second just returns the cubical rectangle that is the intersection of the two cubical rectangles given as arguments, and the third accepts a cubical rectangle and a cubical set on input and returns their intersection. Now we are ready to present an algorithm constructing a cubical representation of a rational map.

7.1 Producing Multivalued Representation

Algorithm 7.4 Multivalued representation
    function multivaluedRepresentation(
           cubicalSet X, Y, rationalMap f)
E := cubicalChainGroups(X);
for i := 0 **to** lastIndex(E) **do**
    for each Q **in** E[i] **do**
        M := rectangularCover(Y);
        L := neighbors(Q);
        for each R **in** L **do**
            if R **in** X **then**
                fR := evaluate(f, R);
                M := rectangleIntersection(M, fR);
            endif;
        endfor;
        F[i]{Q} := elementaryCubes(M, Y);
    endfor;
endfor;
return F;

Theorem 7.5 *Assume Algorithm 7.4 is called with* X, Y *containing the representation of two cubical sets* X, Y *and* f *is a representation of a rational function* $f : X \to Y$. *Then it stops and returns an array* F *containing a cubical representation of* f.

Proof. Let F be the multivalued map represented by F. It is straightforward to verify that F satisfies Eq. (7.3) and consequently also Eq. (7.2). Thus it follows from Theorem 6.14 and Corollary 6.19 that F is cubical and lower semicontinuous. It remains to prove (6.19), but this property follows from (5.6) and (7.3). □

Exercises

7.1 Count the number of times multiplication is performed in evaluating a polynomial of degree n in algorithm polynomialMap. Design an improved version of this algorithm, such that the number of required multiplications is reduced by half.

7.2 Write algorithms overloading the operators +, −, ∗, and / so that they can be used with variables of type interval in the sense of (5.1–5.4).

7.3 Design a respective data structure, and write an algorithm that evaluates a rational function of one variable on an interval.

7.4 Design a respective data structure, and write an algorithm that evaluates a polynomial of two variables on intervals.

7.5 Design a respective data structure, and write an algorithm that evaluates a rational function of an arbitrary finite number of variables on intervals.

7.6 To complete Algorithm 7.1, write an algorithm degenerateDimensions, which, given an elementary cube $Q = I_1 \times I_2 \times \cdots \times I_d$, returns the list of those i for which I_i is degenerate.

7.7 To complete Algorithm 7.4, write algorithms

1. **function** rectangularCover(cubicalSet X)
2. **function** rectangleIntersection(rectangle M, N)
3. **function** elementaryCubes(rectangle M, cubicalSet X)

7.2 Chain Selector Algorithm

Our next step in the computation of the map induced in homology is the construction of chain selectors from cubical representations of continuous maps.

The first issue to address is the question of acyclicity. There is no guarantee that Algorithm 7.4 will return an acyclic-valued representation. We only can test if the representation is acyclic and if it is not, undertake the process of rescaling discussed in Section 6.4.2. The following algorithm may be used to test if a multivalued map is acyclic-valued.

Algorithm 7.6 Acyclicity test
    function acyclityTest(multivaluedMap F)
    for i := 0 **to** lastIndex(F) **do**
        K := **keys**(F[i]);
        for each Q **in** K **do**
            H := **homology**(F[i]{Q});
            for j := 1 **to** lastIndex(H);
                if length(H[j]{generators}) > 0 **then return** false;
            endfor;
            if length(H[0]{generators}) > 1 **then return** false;
            if H[0]{orders}[0] $\neq$ "infinity" **then return** false;
        endfor;
    endfor;
    return true;

The following theorem may be easily obtained by verifying Definition 2.73 and Definition 6.21.

Theorem 7.7 *Assume Algorithm 7.6 is called with F containing the representation of a cubical map $F : X \rightrightarrows Y$. If F is acyclic-valued, then it returns* **true**. *Otherwise it returns* **false**.

7.2 Chain Selector Algorithm

We now turn to the question of finding a chain selector of an acyclic-valued map. We begin with the following simple algorithm evaluating a chain map at a chain. In this algorithm we assume that the operation of adding chains as well as multiplying an integer by a chain are overloaded. Also, we use the structure

typedef chainMap = **array**[0 :] **of** hash{cube} **of** chain;

to store chain maps. Note that formally the structure does not differ from the structure boundaryMap, which we use to store the boundary map.

Algorithm 7.8 Chain evaluation
    function evaluate(chainMap phi, int i, chain c)
    chain d := ();
    K := **keys**(c);
    for each Q **in** K **do**
        d := d + c{Q} * phi[i]{Q};
    endfor;
    return d;

The following proposition is straightforward.

Proposition 7.9 *Assume Algorithm 7.8 is called with* phi *containing a chainmap* ϕ, *i containing an integer i and* c *containing a chain c. Then it stops and returns the value* $\phi_i(c)$ *of the chain map* ϕ *on the chain* c.

Now we are ready to present the algorithm computing the chain selector. In this algorithm we will use algorithm preBoundary discussed in detail in Section 3.7. The input to the algorithm computing the chain selector will use the following data structure.

typedef multivaluedMap = **array**[0 :] **of** hash{cube} **of** cube;

We will also use an algorithm selectVertex (the construction is left to the reader), which, given an elementary cube, returns one of its vertices.

Algorithm 7.10 Chain selector
    function chainSelector(multivaluedMap F)
    if not acyclityTest(F) **then return** "Failure";
    chainMap phi := ();
    K := **keys**(F[0]);
    for each Q **in** K **do**
        phi[0]{Q} := selectVertex(F[0]{Q});
    endfor;
    for i := 1 **to** lastIndex(F) **do**

```
        K := keys(F[i]);
        for each Q in K do
            z := evaluate(phi, boundaryOperator(Q));
            phi[i]{Q} := preBoundary(z, F[i]{Q});
        endfor;
    endfor;
    return phi;
```

The reader may easily verify that the construction of the chain selector in Algorithm 7.10 follows directly the proof of Theorem 6.22. Using Theorem 3.81, we obtain the following result.

Proposition 7.11 *Assume Algorithm 7.10 is called with F containing an acyclic-valued, representable lower semicontinuous map F. Then it stops and returns a chain selector of F.*

7.3 Computing Homology of Maps

We begin with an algorithm, which, given a cycle and a basis of a group of cycles, will return the chain written as a linear combination of the elements of this basis. We assume the basis is stored in the data structure

```
    typedef basis = list of chain;
```

Algorithm 7.12 Coordinates in a given basis
```
    function coordinates(chain c, basis B)
    list of cube canonicalBasis := ();
    for each b in B do
        canonicalBasis := union(canonicalBasis, keys(b));
    endfor;
    matrix V;
    y := canonicalCoordinates(c, canonicalBasis);
    for j := 1 to length(B);
        V[j] := canonicalCoordinates(B[j], canonicalBasis);
    endfor;
    return Solve(V, y);
```

Proposition 7.13 *Assume Algorithm 7.12 is called with c containing a cycle c and B containing a list of basis vectors $B = \{b_1, b_2, \ldots, b_m\}$ of a group of cycles. Then it stops and returns the vector containing the coordinates of c in the basis B.*

Proof. Since B is a basis, it follows that there is an integer vector $\mathbf{x} = (x_1, x_2, \ldots, x_m)$, such that

$$c = \sum_{i=1}^{m} x_i b_i. \tag{7.4}$$

We need to show that the algorithm returns **x**. Observe that applying the coordinate isomorphism ξ_B to Eq. (7.4) we obtain

$$\xi_B c = \sum_{i=1}^{m} x_i \xi_B b_i.$$

Putting $\mathbf{y} := \xi_B c$, $\mathbf{v}_i := \xi_B b_i$ and $V := [\mathbf{v}_1 \mathbf{v}_2 \ldots \mathbf{v}_m]$, the last equation may be rewritten as

$$\mathbf{y} = V\mathbf{x}. \tag{7.5}$$

Thus finding **x** requires solving Eq. 7.5 and it is straightforward to verify that this is exactly what Algorithm 7.12 does. □

Now we are ready to present an algorithm computing the homology of a continuous map.

Algorithm 7.14 Homology of a continuous map
    function homologyOfMap(cubicalSet X, Y, rationalMap f)
    HX := homology(X);
    HY := homology(Y);
    F := multivaluedRepresentation(X, Y, f);
    phi := chainSelector(F);
    for k := 0 **to** lastIndex(HX[k]) **do**
        C := HX[k]{generators};
        D := HY[k]{generators};
        for each c **in** C **do**
            d := evaluate(phi, c);
            xi := coordinates(d, D);
            Hphi[k]{c} := xi;
        endfor;
    endfor;
    return (HX, HY, Hphi);

This algorithm produces the homology of a continuous map in the sense of the following theorem.

Theorem 7.15 *Assume Algorithm 7.14 is called with X, Y containing representations of two cubical sets X, Y and f containing a representation of a continuous map $f : X \to Y$. If chainSelector does not fail, then the algorithm stops and returns a triple (HX, HY, Hphi) such that HX, HY represent the homology of X and Y in the sense of Theorem 3.79. In particular HX[k]{generators} stores a basis $\{z_1, z_2, \ldots, z_m\}$ of $Z_k(X)$ such that $\{[z_1], [z_2], \ldots, [z_m]\}$ are generators of the homology group $H_k(X) = Z_k(X)/B_k(X)$ in the sense of Corollary 3.60. Similarly, HY[k]{generators}* stores a basis $\{\bar{z}_1, \bar{z}_2, \ldots, \bar{z}_n\}$ of $Z_k(Y)$ such that $\{[\bar{z}_1], [\bar{z}_2], \ldots, [\bar{z}_n]\}$ are generators of the homology group $H_k(Y)$. Moreover, Hphi is an array of hashes. The keys of the kth hash constitute the basis of $Z_k(X)$ stored in HX[k]{*generators*}. Given one such generator $z_i \in \mathbf{Z}_k(X)$, the hash element Hphi[k]{z_i} is the coordinate vector $\mathbf{x} = (x_1, x_2, \ldots x_n)$ such that

$$H_k(f)([z]) = \sum_{j=1}^{n} x_j [\bar{z}_j].$$

Proof. By Theorem 7.5 the variable F stores a cubical representation of f. Since algorithm chainSelector does not fail, the representation is acyclic-valued and by Proposition 7.11 phi stores a chain selector of the multivalued representation. Thus the conclusion follows from Proposition 6.56, Definition 6.26, and Definition 4.3. □

Algorithm 7.14 fails only in the case when multivaluedRepresentation fails to find an acyclic-valued representation. As we already have said, in this case one must undertake the process of rescaling. The details of an algorithm which can automatically perform this task are left as an exercise.

Exercise _____

7.8 Write an algorithm which performs the necessary rescaling when the construction of an acyclic-valued representation in Algorithm 7.4 fails.

7.4 Geometric Preboundary Algorithm (optional section)

In Section 3.7 we give an algebraic algorithm solving the equation

$$\partial c = z \tag{7.6}$$

for c having given a reduced cycle z in an acyclic cubical set X. A solution c is called a preboundary of z. That algorithm is, in turn, used for computing chain selectors in this chapter. As we mentioned earlier, the manipulation of large matrices in the algebraic preboundary algorithm due to a large number of cubes generating the groups $C_{m-1}(X)$ and $C_m(X)$ may lead to lengthy calculations. We are therefore interested in searching for algorithms with a more geometric flavor that may prove to be more efficient.

We shall present here a geometric algorithm that applies to reduced cycles z contained in a fixed cubical rectangle $R \subset \mathbf{R}^d$. For every such z the algorithm returns a chain $\operatorname{Pre}(z)$, which solves Eq. (7.6), that is

$$\partial \operatorname{Pre}(z) = z. \tag{7.7}$$

The algorithm uses recursion based on the dimension of R. More precisely, let R' be a $(d-1)$-dimensional face of R. The projection of R onto R' induces a

7.4 Geometric Preboundary Algorithm (optional section)

projection $\pi : \mathcal{C}(R) \to \mathcal{C}(R)$ with the image $\mathcal{C}(R')$ defined as it has been done for the unit cube in Example 4.8. We let $z' := \pi z$, and compute first $\operatorname{Pre}(z')$. Then $\operatorname{Pre}(z)$ is obtained by adding to $\operatorname{Pre}(z')$ a quantity, which we refer to as a "rib". It is made up of all $(k+1)$-dimensional unitary cubes through which z is projected, with appropriately chosen coefficients. We will show that our ribs are related to a chain homotopy between the projection chain map and identity constructed as in Example 4.11.

We start with a simple example.

Example 7.16 Consider the two-dimensional rectangle $R = [0, 100] \times [0, 1]$ and consider the reduced 0-cycle $z = \widehat{V_1} - \widehat{V_0}$, where $V_1 = [100] \times [1]$ and $V_0 = [0] \times [0]$. The reader is invited to count the number of subsequent elementary collapses needed to reduce R to its single vertex $[V_0]$. The calculation we propose below is based on a shortcut of those collapses.

When projecting R onto $R' = [0] \times [0, 1]$, we get a new reduced cycle $z' = \widehat{V_1'} - \widehat{V_0}$, where $V_1' = [0] \times [1]$. Moreover, V_1 is projected to V_1' through the interval $[0, 100] \times [1]$. We associate with this interval a "rib" defined as the sum of duals of all elementary intervals passed on the way:

$$\operatorname{Rib}(V_1) := (\sum_{i=1}^{100} \widehat{[i-1, i]}) \diamond \widehat{[1]}. \tag{7.8}$$

It is easy to check that $\partial_1 \operatorname{Rib}(V_1) = \widehat{V_1} - \widehat{V_1'}$. The new cycle z' lies in the rectangle R' of dimension 1, so we reduced the dimension. We will show that

$$\operatorname{Pre}(z) := \operatorname{Pre}(z') + \operatorname{Rib}(V_1)$$

gives the right formula. The next projection of R' onto V_0 (which is, in this case, an elementary collapse) induces the projection of V_1' to V_0 along the interval $[0] \times [0, 1]$. The corresponding rib is $\operatorname{Rib}(V_1') := \widehat{[0]} \diamond \widehat{[0, 1]}$. Since the projected cycle is $z'' = \widehat{V_0} - \widehat{V_0} = 0$, we put $\operatorname{Pre}(z') := \operatorname{Rib}(V_1')$. We get

$$\partial_1 \operatorname{Pre}(z) = \widehat{V_1'} - \widehat{V_0} + \widehat{V_1} - \widehat{V_1'} = z.$$

What permits us to write Eq. (7.8) without passing through those stages is the geometric intuition that the boundary of $[0, 100]$ is $\{0, 100\}$. We display an algebraic analogy of that simple fact by introducing the following notation for any pair of integers $k < l$:

$$\overrightarrow{[k, l]} := \sum_{i=k+1}^{l} \widehat{[i-1, i]}.$$

It is clear that $\overrightarrow{[k, l]} = \widehat{[k, l]}$ if $l = k + 1$ and an instant verification shows that

$$\partial \overrightarrow{[k, l]} = \widehat{[l]} - \widehat{[k]}.$$

With this notation, Eq. (7.8) becomes transparent:

$$\text{Rib}(V_1) := \overrightarrow{[0, 100]} \diamond \widehat{[1]}$$

and

$$\text{Pre}(z) = \widehat{[0]} \diamond \widehat{[0, 1]} + \overrightarrow{[0, 100]} \diamond \widehat{[1]}.$$

When using this notation we must be aware of the nonuniqueness of such expressions as

$$\overrightarrow{[0, 100]} - \overrightarrow{[50, 150]} = \overrightarrow{[0, 49]} - \overrightarrow{[101, 150]}.$$

Before we start the main construction, we need to generalize Examples 4.8 and 4.11 to arbitrary cubical rectangles. Let R be a cubical rectangle of the form

$$R = [m_1, M_1] \times [m_2, M_2] \times \ldots \times [m_d, M_d]$$

of dimension greater than 0 and let j_1 be the first nondegenerate coordinate of R, that is, the smallest integer j with the property $m_j \neq M_j$. Consider now the rectangle

$$R' = \{x \in R \mid x_{j_1} = m_{j_1}\} \tag{7.9}$$

perpendicular to the first nontrivial direction j_1 of R. Let $p : R \to R$ be the projection with the image R', defined coordinatewise by

$$p(x)_i := \begin{cases} x_i & \text{if } i \neq j_1, \\ m_{j_1} & \text{otherwise.} \end{cases} \tag{7.10}$$

We denote by $p' : R \to R'$ the associated projection given by the same formula but with the target space R'. The two maps are related by the identity $p = ip'$, where $i : R' \hookrightarrow R$ is the inclusion map. For integer $k = 0, 1, \ldots, \dim(R)$ consider an elementary cube $Q \in \mathcal{K}_k(R)$. It can be expressed as the product of elementary intervals $Q = Q_1 \times Q_2 \times \cdots \times Q_d$. We have two cases: Either Q_{j_1} is nondegenerate, say $Q_{j_1} = [a_{j_1}, a_{j_1} + 1]$, or it is degenerate, say $Q_{j_1} = [a_{j_1}]$. The projection p maps Q in both cases to an elementary cube

$$p(Q) = Q_1 \times \cdots \times Q_{j_1-1} \times [m_{j_1}] \times Q_{j_1+1} \times \cdots \times Q_d.$$

In the first case, this gives an elementary cube of dimension $k - 1$. But, in the second case, $p(Q)$ is an elementary cube of the same dimension k.

This permits us to associate with p homomorphisms $\pi_k : C_k(R) \to C_k(R)$ on the level of chain groups defined on generators as follows.

$$\pi_k(\widehat{Q}) := \begin{cases} \widehat{p(Q)} = \widehat{Q_1} \diamond \cdots \diamond \widehat{Q_{j_1-1}} \diamond \widehat{[m_{j_1}]} \diamond \widehat{Q_{j_1+1}} \cdots \diamond \widehat{Q_d} \\ \quad \text{if } Q_{j_1} = [a_{j_1}], \\ 0 \quad \text{otherwise.} \end{cases} \tag{7.11}$$

Note that $\pi_k(\widehat{Q}) = \widehat{Q}$ if $Q \subset R'$ hence π is a linear projection onto $\mathcal{C}(R')$. Here is an analogy of 4.19:

7.4 Geometric Preboundary Algorithm (optional section)

Lemma 7.17 *The collection of projections $\pi = \{\pi_k\} : \mathcal{C}(R) \to \mathcal{C}(R)$ is a chain map. In particular, π maps cycles to cycles.*

Proof. Let $Q \in \mathcal{K}_k(R)$. We need to show that $\partial \pi \widehat{Q} = \pi \partial \widehat{Q}$. Since j_1 is the first nontrivial coordinate of R, Q can be decomposed as $Q = V \times Q_{j_1} \times P$ where $\dim(V) = 0$ and either $Q_{j_1} = [a_{j_1}]$ or $Q_{j_1} = [a_{j_1}, a_{j_1} + 1]$. By using Eq. (2.7) we obtain

$$\partial \widehat{Q} = 0 + \widehat{V} \diamond \widehat{\partial Q_{j_1}} \diamond \widehat{P} + (-1)^{\dim(Q_{j_1})} \widehat{V} \diamond \widehat{Q_{j_1}} \diamond \partial \widehat{P}.$$

If $Q_{j_1} = [a_{j_1}]$, then $\widehat{\partial Q_{j_1}} = 0$, so we get

$$\pi \partial \widehat{Q} = \pi(\widehat{V} \diamond \widehat{Q_{j_1}} \diamond \partial \widehat{P}) = \widehat{V} \diamond \widehat{[m_{j_1}]} \diamond \partial \widehat{P}$$

and also

$$\partial \pi \widehat{Q} = \partial(\widehat{V} \diamond \widehat{[m_{j_1}]} \diamond \widehat{P}) = \widehat{V} \diamond \widehat{[m_{j_1}]} \diamond \partial \widehat{P} = \pi \partial \widehat{Q}.$$

If $Q_{j_1} = [a_{j_1}, a_{j_1} + 1]$, then $\pi \widehat{Q} = 0$ by definition, so $\partial \pi \widehat{Q} = 0$. On the other hand,

$$\pi \partial \widehat{Q} = \pi(\widehat{V} \diamond (\widehat{[a_{j_1} + 1]} - \widehat{[a_{j_1}]}) \diamond \widehat{P} - \widehat{V} \diamond \widehat{Q_{j_1}} \diamond \partial \widehat{P})$$
$$= \widehat{V} \diamond \widehat{[m_{j_1}]} \diamond \widehat{P} - \widehat{V} \diamond \widehat{[m_{j_1}]} \diamond \widehat{P} - 0$$
$$= 0. \qquad \square$$

We will later see that our chain map π is in fact induced by p in the sense given in Chapter 6. We denote by $\pi' : \mathcal{C}(R) \to \mathcal{C}(R')$ the associated projection given by the same formula but with the target space $\mathcal{C}(R')$. The two maps are related by the identity $\pi = \iota \pi'$, where $\iota : \mathcal{C}(R') \hookrightarrow \mathcal{C}(R)$ is the inclusion chain map.

Next, given any $Q \in \mathcal{K}_k(R)$ as discussed above, define

$$\text{Rib}(Q) := \begin{cases} \widehat{Q_1} \diamond \cdots \diamond \widehat{Q_{j_1-1}} \diamond \overrightarrow{[m_{j_1}, a_{j_1}]} \diamond \widehat{Q_{j_1+1}} \diamond \widehat{Q_d} \\ \qquad \text{if } Q_{j_1} = [a_{j_1}] \text{ and } a_{j_1} > m_{j_1}, \\ 0 \qquad \text{otherwise.} \end{cases} \quad (7.12)$$

Note that, by definition, $\text{Rib}(Q)$ is a $(k+1)$-chain in R. If it is not trivial, then its support is the $(k+1)$-dimensional rectangle through which Q is projected to $p(Q)$. With a little knowledge of anatomy we may observe that ribs add up to a chest. Let us define maps $\text{Chest}_k : C_k(R) \to \mathbf{C}_{k+1}(R)$ for each k by the formula

$$\text{Chest}_k(c) := \sum_{Q \in \mathcal{K}_k(R)} \langle \widehat{Q}, c \rangle \, \text{Rib}(Q). \quad (7.13)$$

Note that each $c \in C_k(R)$ can be presented as

248   7 Computing Homology of Maps

$$c = \sum_{Q \in \mathcal{K}_k(R)} \langle c, \widehat{Q} \rangle \, \widehat{Q},$$

so the above formula defines a linear map. The reader may now guess that this map is a chain homotopy: We postpone the proof of it until the end of our construction.

Now let $z \in Z_k(R)$ be a k-cycle (reduced cycle if $k = 0$) in a cubical rectangle R. We construct, by induction with respect to the dimension n of R, a $(k+1)$-chain $\mathrm{Pre}\,(z)$, satisfying Eq. (7.7). Obviously, the dimension of R is at least k, so the induction starts from $n = k$.

If $n = k$, we must have $z = 0$ so we may put

$$\mathrm{Pre}\,(z) := 0.$$

Indeed, in this case, $C_{k+1}(R) = 0$ so $B_k(R) = 0$ and $\tilde{Z}_k(R) = 0$ since $\tilde{H}_k(R) = 0$. Knowing that $z \in \tilde{Z}_k(R)$, it follows that $z = 0$.

Suppose the construction is done for dimensions up to a certain $n \geq k$ and now let $\dim(R) = n+1$.

By Lemma 7.17, $\pi(z)$ is a k-cycle contained in R', which has dimension n. Hence the $(k+1)$-chain $\mathrm{Pre}\,(\pi(z))$ is well defined and

$$\partial_{k+1}\mathrm{Pre}\,(\pi_k(z)) = \pi_k(z)$$

by the induction hypothesis.

We define

$$\mathrm{Pre}\,(z) := \mathrm{Pre}\,(\pi(z)) + \mathrm{Chest}\,_k(z). \tag{7.14}$$

Theorem 7.18 *For every cycle c the chain $\mathrm{Pre}\,(z)$ has the following properties*

$$\mathrm{Pre}\,(z) = \mathrm{Pre}\,(\pi(z)) + \mathrm{Chest}\,_k(z)$$
$$\partial \mathrm{Pre}\,(z) = z$$

Proof. The first property follows immediately from the recursive construction. The other is an immediate consequence of the following lemma. □

Lemma 7.19 *The collection of maps $\mathrm{Chest}\,_k : C_k(R) \to \mathbf{C}_{k+1}(R)$ is the chain homotopy between π and $\mathrm{id}\,_{C(R)}$. As a consequence, if $z \in Z_k(R)$, then*

$$\partial_{k+1}\,(\mathrm{Chest}\,_k(z)) = z - \pi_k(z). \tag{7.15}$$

Proof. We use the same arguments as those used in Example 4.11. We should verify that Eq. (4.8) holds for $\psi = \pi$ and $\varphi = \mathrm{id}$. Because it is enough to check this on the canonical basis, we need to verify that for any $Q \in \mathcal{K}_k(R)$

$$\partial_{k+1}\mathrm{Chest}\,_k\widehat{Q} + \mathrm{Chest}\,_{k-1}\partial_k\widehat{Q} = \widehat{Q} - \pi\widehat{Q}. \tag{7.16}$$

7.4 Geometric Preboundary Algorithm (optional section)   249

We decompose Q, as in the proof of Lemma 7.17, into $Q = V \times Q_{j_1} \times P$, where $\dim(V) = 0$ and either $Q_{j_1} = [a_{j_1}]$ or $Q_{j_1} = [a_{j_1}, a_{j_1}+1]$. Consider the case $Q_1 = [a_{j_1}]$, where $a_{j_1} > m_{j_1}$. Then

$$\partial_{k+1}\text{Chest}_k\widehat{Q} + \text{Chest}_{k-1}\partial_k\widehat{Q}$$
$$= \partial_{k+1}(\widehat{V} \diamond \overrightarrow{[m_{j_1}, a_{j_1}]} \diamond \widehat{P}) + \text{Chest}_{k-1}(\widehat{V} \diamond \widehat{[a_{j_1}]} \diamond \partial_k \widehat{P})$$
$$= \widehat{V} \diamond \widehat{[a_{j_1}]} \diamond \widehat{P} - \widehat{V} \diamond \widehat{[m_{j_1}]} \diamond \widehat{P}$$
$$\quad - \widehat{V} \diamond \overrightarrow{[m_{j_1}, a_{j_1}]} \diamond \partial_k \widehat{P} + \widehat{V} \diamond \overrightarrow{[m_{j_1}, a_{j_1}]} \diamond \partial_k \widehat{P}$$
$$= \widehat{Q} - \pi\widehat{Q}.$$

Next let $Q_1 = [m_{j_1}]$. Then each term in the left-hand side of Eq. (7.16) is zero by the definition of Chest_k and the right-hand side is zero because $\widehat{\pi}Q = \widehat{Q}$.

The last case to consider is $Q_1 = [a_{j_1}, a_{j_1}+1]$, where $m_{j_1} \leq a_{j_1}$ and $a_{j_1} + 1 \leq M_{j_1}$. Then $\text{Chest}_k \widehat{Q} = 0$ and $\pi\widehat{Q} = 0$ so we get

$$\partial_{k+1}\text{Chest}_k Q + \text{Chest}_{k-1}\partial_k\widehat{Q}$$
$$= 0 + \text{Chest}_{k-1}(\widehat{V} \diamond [a_{j_1}+1] \diamond \widehat{P} - \widehat{V} \diamond \widehat{[a_{j_1}]} \diamond \widehat{P} - \widehat{V} \diamond \overrightarrow{[a_{j_1}, a_{j_1}+1]} \diamond \partial_k \widehat{P})$$
$$= \widehat{V} \diamond \overrightarrow{[m_{j_1}, a_{j_1}+1]} \diamond \widehat{P} - \widehat{V} \diamond \overrightarrow{[m_{j_1}, a_{j_1}]} \diamond \widehat{P} - 0$$
$$= \widehat{V} \diamond \widehat{[a_{j_1}, a_{j_1}+1]} \diamond \widehat{P} = \widehat{Q} = \widehat{Q} - \pi\widehat{Q}.$$

This proves that Chest is a chain homotopy. Finally, if z is a cycle, then $\text{Chest}_{k-1}\partial_k z = \text{Chest}_{k-1} 0 = 0$ so we obtain Eq. (7.15). □

Example 7.20 Consider the cycle

$$z := \widehat{E_8} + \widehat{E_7} + \widehat{E_6} + \widehat{E_5} - \widehat{E_4} - \widehat{E_3} - \widehat{E_2} - \widehat{E_1}$$

with edges

$$E_1 := [0,1] \times [0], \quad E_2 := [0,2] \times [0], \quad E_3 := [2] \times [0,1],$$
$$E_4 := [2] \times [1,2], \quad E_5 := [1,2] \times [2], \quad E_6 := [1] \times [1,2],$$
$$E_7 := [0,1] \times [1], \quad E_8 := [0] \times [0,1].$$

The cycle z represents the clockwise-oriented contour in $\mathbf{R}^2$ indicated in Figure 7.1. The smallest rectangle containing $|z|$ is $R = [0,2]^2$ and the projection p sends it to $R' = [0] \times [0,2]$. We get

$$\pi(\widehat{E_1}) = \pi(\widehat{E_2}) = \pi(\widehat{E_7}) = \pi(\widehat{E_5}) = 0$$

because those edges project to vertices of R. Next

$$\pi(\widehat{E_8}) = \pi(\widehat{E_3}) = \widehat{E_8}, \quad \pi(\widehat{E_6}) = \pi(\widehat{E_4}) = \widehat{[0] \times [1,2]},$$

250   7 Computing Homology of Maps

so we get $z' := \pi(z) = 0$. From the above calculus we get

$$\operatorname{Rib}(E_1) = \operatorname{Rib}(E_2) = \operatorname{Rib}(E_7) = \operatorname{Rib}(E_5) = 0$$

and $\operatorname{Rib}(E_8) = 0$ because $E_8 \subset R'$. The only nontrivial ribs are

$$\operatorname{Rib}(E_3) = \widehat{Q_1} + \widehat{Q_2}, \quad \operatorname{Rib}(E_4) = \widehat{Q_3} + \widehat{Q_4}, \quad \operatorname{Rib}(E_6) = \widehat{Q_4},$$

where Q_1, Q_2, Q_3, and Q_4 are the squares indicated in Figure 7.1.

We get

$$\begin{aligned}\operatorname{Pre}(z) &= \operatorname{Pre}(z') - \operatorname{Rib}(E_3) - \operatorname{Rib}(E_4) + \operatorname{Rib}(E_6) \\ &= 0 - \widehat{Q_1} - \widehat{Q_2} - \widehat{Q_3} - \widehat{Q_4} + \widehat{Q_4} \\ &= -(\widehat{Q_1} + \widehat{Q_2} + \widehat{Q_3}).\end{aligned}$$

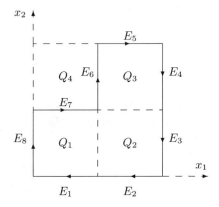

Fig. 7.1. Cycle $z = \widehat{E_8} + \widehat{E_7} + \widehat{E_6} + \widehat{E_5} - \widehat{E_4} - \widehat{E_3} - \widehat{E_2} - \widehat{E_1}$.

Example 7.21 Consider the 1-cycle $z = -\widehat{E_1} - \widehat{E_2} + \widehat{E_3} + \widehat{E_4} + \widehat{E_5} - \widehat{E_6}$ in $\mathbf{R}^3$ with edges

$$E_1 := [1] \times [0] \times [0,1], \quad E_2 := [0,1] \times [0] \times [0], \quad E_3 := [0] \times [0,1] \times [0],$$
$$E_4 := [0] \times [1] \times [0,1], \quad E_5 := [0,1] \times [1] \times [1], \quad E_6 := [1] \times [0,1] \times [1].$$

The cycle z represents the contour on the surface of the unit cube $[0,1]^3$ indicated in Figure 7.2. The smallest rectangle containing $|z|$ is $R = [0,1]^3$ and we have $R' = [0] \times [0,1] \times [0,1]$. We get

$$\pi(\widehat{E_2}) = \pi(\widehat{E_5}) = 0$$

7.4 Geometric Preboundary Algorithm (optional section)

because those edges project to vertices of R'. Next

$$\pi(\widehat{E_1}) = [0] \times \widehat{[0]} \times [0,1], \quad \pi(\widehat{E_3}) = \widehat{E_3},$$
$$\pi(\widehat{E_4}) = \widehat{E_4}, \quad \pi(\widehat{E_6}) = [0] \times \widehat{[0,1]} \times [1],$$

so we get

$$\pi(z) = -[0] \times \widehat{[0]} \times [0,1] + \widehat{E_3} + \widehat{E_4} - [0] \times \widehat{[0]} \times [0,1],$$

which represents an oriented contour of the square $Q' := [0] \times [0,1] \times [0,1]$. We leave the verification of the formula

$$\text{Pre}\,(z') = \widehat{Q'}$$

as an exercise.

By the same arguments as in the previous example we get

$$\text{Rib}\,(E_2) = \text{Rib}\,(E_3) = \text{Rib}\,(E_4) = \text{Rib}\,(E_5) = 0.$$

The only nontrivial ribs are

$$\text{Rib}\,(E_1) = \widehat{Q_1} := [0,1] \times \widehat{[0]} \times [0,1], \quad \text{Rib}\,(E_6) = \widehat{Q_6} := [0,1] \times \widehat{[0,1]} \times [1].$$

The formula now gives

$$\text{Pre}\,(z) = \text{Pre}\,(z') - \text{Rib}\,(E_1) - \text{Rib}\,(E_6)$$
$$= \widehat{Q'} - \widehat{Q_1} - \widehat{Q_6}.$$

We can see that the square Q' represents the left face of the cube in Figure 7.2, Q_1 represents its bottom face, and Q_6 represents its front face.

Remark 7.22 The preboundary construction is presented for one fixed rectangle R, but there may be many cubical rectangles R containing $|z|$ and it might be advantageous to vary them. In that case, we will add the variable R as an index to expressions for the previously defined functions, for example,

$$\pi_R(c), \text{ Rib}_R(Q), \text{ Chest}_R(z).$$

In particular, we may want to replace R in our previous discussion by $\text{Rect}\,(z)$ defined as the smallest rectangle containing the support of z. It is easily seen that

$$\text{Rect}\,(z) = [m_1, M_1] \times [m_2, M_2] \times \ldots \times [m_d, M_d], \qquad (7.17)$$

where $m_i = \min \bigcup \{Q_i \mid \langle z, \widehat{Q} \rangle \neq 0\}$ and $M_i = \max \bigcup \{Q_i \mid \langle z, \widehat{Q} \rangle \neq 0\}$. Note that if we put $R = \text{Rect}\,(z)$ and $R' = \pi_R(R)$, then R' might not be equal to $\text{Rect}\,(\pi_R(z))$. We only know that $\text{Rect}\,(\pi_R(z)) \subset R'$. In Example 7.20 we have $\pi(z) = 0$, so $\text{Rect}\,(z) = \emptyset$. By passing from R' to $\text{Rect}\,(\pi_R(z))$ after each projection, we may sometimes considerably speed up the construction of $\text{Pre}\,(z)$.

252   7 Computing Homology of Maps

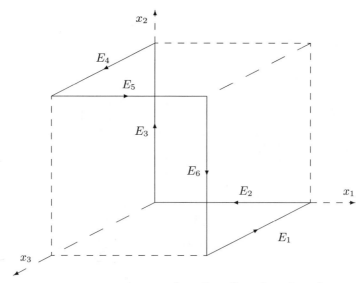

Fig. 7.2. Cycle $z = -\widehat{E_1} - \widehat{E_2} + \widehat{E_3} + \widehat{E_4} + \widehat{E_5} - \widehat{E_6}$.

Remark 7.23 It is visible from Figure 7.2 that different coboundaries of z may be obtained by projecting z to different faces of the cube. However, (7.11) and (7.14) are dependent on the choice of x_{j_1} as the first nondegenerate coordinate of R. If a different coordinate is chosen, then a delicate analysis of sign adjustment is required due to the alternating sign in the formula for the boundary map.

We end this chapter by presenting an algorithm based on the discussed preboundary construction. That algorithm uses several functions whose construction is left as an exercise. First, Eq. (7.17) may be used to define the function minRectangle, which, given a chain c stored in variable of type chain, returns the minimal rectangle containing $|c|$ stored in variable of type rectangle. Eq. (7.11) gives rise to the function projection and Eq. (7.13) give rise to the function chest. Both functions take two arguments: a chain c represented in the variable of type chain and a rectangle R represented in the variable of type rectangle. The first function returns the projection of the chain c onto the group of chains of R' given by (7.9). The other function returns the chest of c in R.

Algorithm 7.24 Cubical preboundary
    function cuPreBoundary(chain z)
    R := minRectangle(z);
    if $\dim(z) = \dim(R)$ **then**
        return 0;
    else

```
    z̄ := projection(z, R);
    return cuPreBoundary(z̄) + chest(z, R);
endif;
```

The properties of this algorithm are summarized in the following theorem.

Theorem 7.25 *Assume Algorithm 7.24 is called with* **z** *containing a hash representation of a reduced cycle z. Then it returns the hash representation of a preboundary of z, that is, a chain c such that $\partial c = z$.*

Proof. Let n_i denote the dimension of the rectangle stored in the variable R on the ith call of cuPreBoundary. Since every recursive call of cuPreBoundary is applied to the projection of the original chain, $\{n_i\}$ is a strictly decreasing sequence of nonnegative integers. Therefore the sequence is finite and the algorithm must stop. Thus the conclusion follows from Theorem 7.18 by an easy induction argument. □

Exercises

7.9 Consider a projection $p : R \to R'$ obtained by replacing the first nondegenerate coordinate j_1 in the preboundary construction by the last nondegenerate coordinate j_n, where $n = \dim(R)$. Obtain a version of Eq. (7.14) for that projection.

7.10 Let $X = A \cup B$, where $A = [0, 2] \times [0, 1]$ and $B = [0, 1] \times [0, 2]$ and let $Y = A \cap B = [0, 1]^2$ Construct, by analogy to the projection on the rectangle, a chain projection $\pi : \mathcal{C}(X) \to \mathcal{C}(X)$ with the image $\mathcal{C}(Y)$. Construct the chain homotopy between π and $\text{id}_{\mathcal{C}(X)}$, and use it to carry over the geometric preboundary construction to cycles in X. Note that X is a simple example of a star-shaped set.

7.11 Present algorithms for computing the functions used in Algorithm 7.24, namely

(a) minRectangle,
(b) projection,
(c) rib,
(d) chest.

7.5 Bibliographical Remarks

The constructions and algorithms presented in Sections 7.1 to 7.3 are new or recent contributions, and various programs derived from these constructions are in progress. The first steps toward computing the homology of maps via its multivalued representations are proposed in the work of Allili [1] and also presented in [2]. Algorithm 7.24 is due to [3]. Its generalization to certain classes of finitely generated chain complexes is presented in [42]. The first implementations of the material of this chapter in the Homology program are due to Mazur, Szybowski, and Pilarczyk [52, 72]. An alternative approach to computing the homology of maps can be found in [58] and [67].

Part II

Extensions

8
Prospects in Digital Image Processing

8.1 Images and Cubical Sets

In Chapter 1 we consider digital images to motivate the use of cubical complexes as the geometric building blocks of homology. As a concrete example we discuss the image of the Sea of Tranquillity (see Figure 1.5). However, such a complicated image and the discussion surrounding it can mask some of the even more elementary difficulties in the process of analog to digital conversion. Since the focus of this book is on computational homology, we will show that it is easy for pixel data to produce the wrong topological information.

Consider the two slanted line segments, $V \subset \mathbf{R}^2$, indicated in Figure 8.1(a). Let us think of it as a physical object; for example, two line segments drawn in black ink on a piece of white paper. Also, consider a highly idealized description of the information in a very coarse digital photo of V. In particular, assume that the digital photo consists of 25 pixels and that each pixel is either black or white depending on whether or not V intersects the area associated to each pixel. This is indicated in Figure 8.1. How V intersects each pixel is indicated in (a), and the corresponding assignment of black and white is indicated in (b).

Observe that the pixels in Figure 8.1(b) naturally define three sets: $\mathcal{Z}$, which consists of all 25 pixels in the 5×5 region; $\mathcal{X}$, the set of all black pixels; and $\mathcal{Y}$, the set of all white pixels. If we identify each pixel with a two-dimensional elementary cube, then $\mathcal{X}$, $\mathcal{Y}$, and $\mathcal{Z}$ give rise to cubical sets X, Y, and Z defined by (6.6):

$$X := |\mathcal{X}|, \quad Y := |\mathcal{Y}|, \quad Z := |\mathcal{Z}|.$$

Thus the combinatorial information is transformed to topological information. A word of caution is due here. Notice that $\mathcal{X}$ and $\mathcal{Y}$ are complementary sets of pixels. However, the corresponding cubical sets X and Y are not complementary sets of points because they are not disjoint: They have some common vertices and edges. Thus the correspondence between the combinatorial objects and the topological ones is not exact. It might seem that the

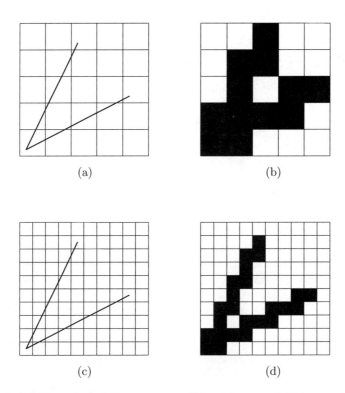

Fig. 8.1. (a) Two slanted line segments that define a set V in a cubical grid. (b) Two-dimensional cubes that intersect the set V are colored black. (c) The set $\Lambda^{(2,2)}(V)$ in a cubical grid. (d) Two-dimensional cubes that intersect $\Lambda^{(2,2)}(V)$ are colored black.

identification of pixels with two-dimensional open elementary cells could solve the problem of the lack of disjointness. But then $Z = X \cup Y$ would not give the full rectangle: All vertices and edges would be missing. This inconsistency comes from the fact that the information about zero- and one-dimensional faces, which is crucial for the computation of homology, cannot be exhibited by pixels themselves because those are the smallest distinguishable units in the numerical picture and often regarded as points.

Let us return to the set V. It is not cubical and, therefore, we cannot yet compute its homology (in Chapter 11 we generalize homology to a larger class of spaces). However, it is homotopic to a single point and thus should have the same homology as a point. On the other hand, we are thinking of X as a cubical approximation or representation of V, but it is clearly not acyclic. Furthermore, we cannot blame the resolution of our camera. Figures 8.1(c) and (d) indicate what happens if we apply the same procedure using a digital photo with 100 pixels. Observe that the homology of the cubical approximation remains unchanged.

The moral of this example is that producing discretized data and then interpreting it in the form of a cubical complex is not a fail-safe procedure. On the other hand, as we will explain in the remainder of this chapter, the reader should not assume that this example implies that digital approximations are doomed to failure.

Moving on to higher-dimensional examples, in biomedical imaging tomography is a standard tool. While each single tomographic image represents a two-dimensional slice of the object, a three-dimensional image can be reconstructed by using a set of parallel contiguous slices. Each single data point is now referred to as a voxel and interpreted as an elementary three-dimensional cube. Thus, to each collection of binary voxel data, we can associate a three-dimensional cubical complex.

As an example, consider the MRI shown in Figure 8.2. This is one out of 124 two-dimensional slices of volumetric image data of the brain. It is easily noted that in contrast to looking at a picture of the moon's surface, viewing three-dimensional images is a nontrivial task. There are, of course, nice software packages that allow one to view and rotate three-dimensional objects. Thus, with a sufficiently powerful computer, we can easily study the surface of a reasonably complex object. For example, as indicated in Figure 8.2, a smooth rendering of a brain's surface can be obtained.

However, if we want to be able to simultaneously study the interior of the object, then the problem of visualization becomes much more difficult.

Of course, this is precisely the point of introducing homology groups. Because we can compute them directly from a cubical complex, to determine the cavities of the brain in Figure 8.2, it is sufficient to be given a binary voxel representation of the organ. No visualization is necessary. Using thresholding techniques considerably more sophisticated than those presented earlier, a binary image was obtained (a single slice of this binary image is shown in Figure 8.2). The homology of the resulting three-dimensional complex was computed, resulting in Betti numbers $\beta_0 = 1$, $\beta_1 = 3$, and $\beta_2 = 1$.

8.2 Patterns from Cahn–Hilliard

As indicated in Section 1.1, the Cahn–Hilliard equation (1.1) is derived as a model of binary alloys and produces complicated patterns. The geometric structure of the solutions has a bearing on the physical properties of the material. In this section we indicate how cubical computational homology can be used to gain some insight into this structure.

We begin with a discussion concerning the appropriateness of using cubical complexes. Let us begin by recalling the level set S of Figure 1.7. As noted in the caption, this figure is actually a triangulated surface obtained by rendering the actual numerical data $\{u(i,j,k,\tau) \mid 1 \leq i,j,k \leq 128\}$. Figure 8.3

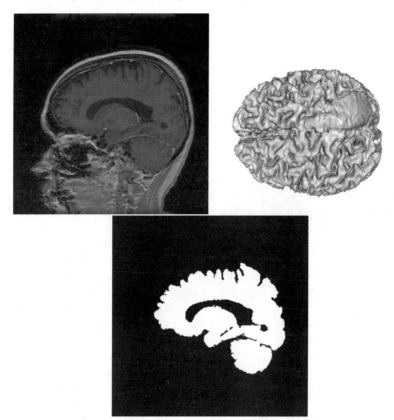

Fig. 8.2. (a) A single two-dimensional slice of an MRI image of a brain. The tissue of interest is indicated in light gray. The full data set consists of 256 by 256 by 124 voxels. (b) A view of the surface of the segmented brain. (c) A slice of the segmented binary image on which the homology program can operate. In this case the cubical complex is indicated in white. (From [4]. Copyright © 2001 IEEE. Reprinted with permission.)

represents a rendering obtained by graphing elementary cubes that intersect S. Computing the homology of the resulting cubical set X using CHomP leads to $H_0(X) = \mathbf{Z}$, $H_1(X) = \mathbf{Z}^{1705}$, and $H_2(X) = \mathbf{Z}$. Observe that this differs from the homology groups of S (see Figure 1.7).

The obvious question is: Which set of homology groups is correct? Unfortunately, the answer is not so obvious. The first difficulty arises from the fact that we are numerically computing a partial differential equation. Of course, this introduces errors that in turn lead to geometric errors. The second difficulty is that S, though it looks smoother than X, is essentially an interpolation of the data; thus small features can be overlooked. Thus it is possible that several generators of H_1 or H_2 were lost. On the other hand,

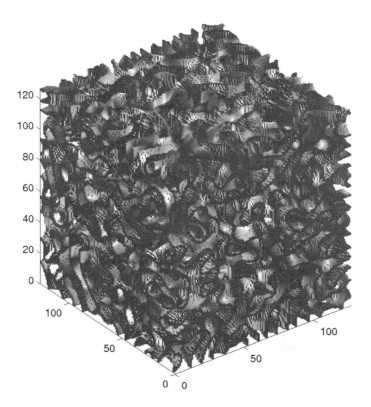

Fig. 8.3. A rendering of the elementary cubes that intersect the surface S of Figure 1.7. The homology groups for this cubical set X are $H_0(X) = \mathbf{Z}$, $H_1(X) = \mathbf{Z}^{1705}$, and $H_2(X) = \mathbf{Z}$.

as indicated in the first section of this chapter, it is possible that the cubical approximation has introduced extra generators. Resolving this issue is important, but it —requires a careful analysis of the numerical procedures involved in solving the differential equation and interpolating the data. As such they lie outside the scope of this book.

Having acknowledged the difficulties, let us also note that even though the homology groups are different, they differ by very small amounts—the relative difference between $\dim H_1(S)$ and $\dim H_1(X)$ is approximately 0.2%. Thus it appears that this simple calculation captures information about the geometry of solutions to (1.1), and we wish to exploit this further.

As indicated earlier, since the Cahn–Hilliard equation is a nonlinear partial differential equation, obtaining analytic results is extremely difficult. However, in the simpler case of a one-dimensional domain, there is a substantial body of a rigorous information about the dynamics. So consider

8 Prospects in Digital Image Processing

$$\frac{\partial u}{\partial t} = -(\epsilon^2 u_{xx} + u - u^3)_{xx}, \quad x \in [0,1], \tag{8.1}$$

$$u_x(0) = u_x(1) = u_{xxx}(0) = u_{xxx}(1) = 0, \tag{8.2}$$

where $\epsilon = 0.01$. Figure 8.4 indicates a numerical solution to (8.1) at various points in time on a grid consisting of 512 elements. This has been obtained by starting with a small, but random, initial condition $u(x,0)$ satisfying $\int_0^1 u(x,0)\,dx = 0$. It should be remarked that while $|u(x,0)|$ is small, it is not constant. This solution exhibits the typical behavior for the Cahn–Hilliard equation. Initially the magnitude of the solutions grows. This is referred to as *spinodal decomposition*. Observe the appearance and growth of many steep *interfaces* between regions where $u(x,t) \approx \pm 1$. Also observe that the number of interfaces decreases as time increases. In fact, it can be proven that for sufficiently large t, $u(x,t)$ will have a single interface; however, these interfaces disappear at an extremely slow rate.

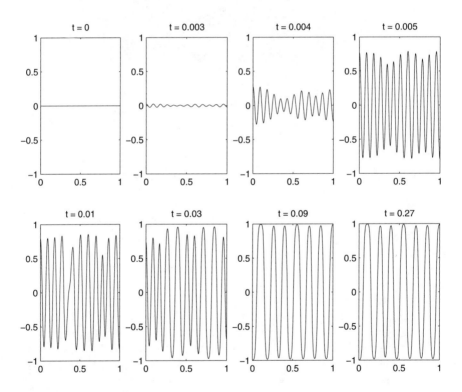

Fig. 8.4. Plots of a single solution to (8.1). The first graph is that of $u(x,0)$, a small randomly chosen initial condition satisfying $\int_0^1 u(x,0)\,dx = 0$. The remaining graphs represent $u(x,t)$ at the various values of t. The process of spinodal decomposition can be seen in the first few plots, while in the latter plot coarsening is taking place.

While in the case of a one-dimensional domain, quantitative statements can be associated with the above-mentioned observations; for higher-dimensional problems, our understanding is much more limited. To see how homology can be used to investigate these phenomena, observe that the patterns are related to the sets $U^+(t) := \{x \in \Omega \mid u(x,t) > 0\}$ and $U^-(t) := \{x \in \Omega \mid u(x,t) < 0\}$. The more components these sets have, the more complicated the pattern.

With this remark we now turn to the two-dimensional domain $\Omega = [0,1]^2$. There are two reasons for using $[0,1]^2$ as opposed to $[0,1]^3$. The first is that the memory requirements for three-dimensional numerical simulations are nontrivial. The second is that to help develop the reader's intuition we augment the computational results with plots. As should be clear from Figures 1.7 and 8.3, viewing the three-dimensional images is of limited value.

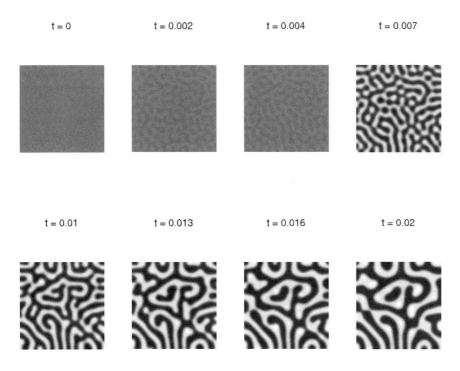

Fig. 8.5. Plots of a single solution to (1.1) with $\Omega = [0,1]^2$. The first graph is that of $u(x_1, x_2, 0)$, a small randomly chosen initial condition satisfying $\int_0^1 \int_0^1 u(x_1, x_2, 0)\, dx_1\, dx_2 = 0$. The remaining graphs represent $u(x,t)$ at the various values of t. The process of spinodal decomposition can be seen in the first few plots, while in the latter plots coarsening is taking place.

The plots of a numerical solution to (1.1) on a 512×512 grid with $\epsilon = 0.01$ are indicated in Figure 8.5. These plots are the two-dimensional analogs of those in Figure 8.4. However, the geometry is much more interesting. As above we define $U^+(t) := \{(x_1, x_2) \in \Omega \mid u(x_1, x_2, t) > 0\}$ and $U^-(t) := \{(x_1, x_2) \in \Omega \mid u(x_1, x_2, t) < 0\}$. One measure of the geometric structures of these sets is their homology. Since the data at our disposal are $\{u(i, j, t) \mid 1 \leq i, j \leq 512\}$, thresholding is easy: The square corresponding to $u(i, j, t)$ belongs to $\mathcal{K}_2(U^+(t))$ if $u(i, j, t) > 0$ and to $\mathcal{K}_2(U^-(t))$ if $u(i, j, t) < 0$. Figure 8.6 indicates the cubical sets obtained after thresholding.

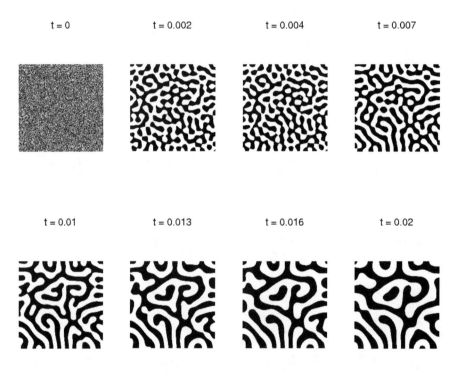

Fig. 8.6. Plots of the images in Figure 8.5 after thresholding. The black-and-white pixels identify $U^-(t)$ and $U^+(t)$, respectively.

We now use CHomP to compute $H_*(U^\pm(t))$ for a set of t-values. The results are presented in Figure 8.7. A few comments and observations need to be made.

1. The value of $\dim H_*(U^\pm(0))$ is not included in the plots because the numbers are extremely large since the initial condition is a random function and thus changes sign frequently.
2. Many of the components and holes of $U^\pm(0)$ disappear quickly, within the first few time steps. This property can be predicted analytically based on the form of the partial differential equation.

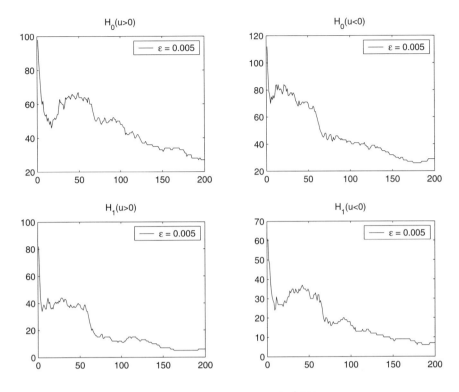

Fig. 8.7. Plots of $\dim H_*(U^\pm(t))$.

3. Once the extremely small-scale topological structures have been eliminated, there is a period of time during which the topology remains the same or perhaps becomes more complicated.
4. After 50 time steps the topological complexity rapidly decreases. This suggests that this is the time at which the coarsening process occurs.
5. The rate of decrease in the topological complexity decays rapidly.
6. Throughout the plot there are small-scale fluctuations in the Betti numbers. Though possibly true, other, less desirable factors might contribute to this phenomenon. The two prime suspects are numerical error and the topological errors that can arise from using cubical approximations to continuous objects.
7. These figures are derived from a single random initial condition, and it is not at all clear if they are representative of a typical solution.

These last two comments suggest that we should examine more solutions. Using CHomP this is easily done, and Figure 8.8 portrays the Betti numbers based on the average of 100 simulations. More precisely, the functions

$$\frac{1}{100}\sum_{m=1}^{100} \dim H_*(U_m^{\pm}(t))$$

are plotted, where the $U_m^{\pm}(t)$ are the cubical sets obtained during the mth simulation. Observe that these averaged plots reinforce comments 2, 3, 4, and 5.

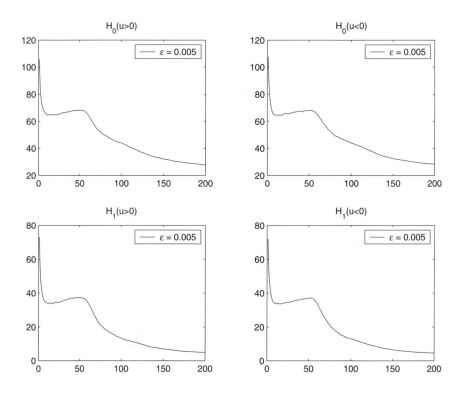

Fig. 8.8. Plots of $\dim H_*(U^{\pm}(t))$ averaged over 100 random initial conditions.

This book is not the appropriate venue in which to discuss the significance of these results as they relate either to metallurgy or to the geometry of partial differential equations. However, what should be clear is that this problem naturally generates a large amount of geometric data. The computational homology techniques developed in Part I allow us to reduce this data to simple graphs that still describe important properties of the system.

8.3 Complicated Time-Dependent Patterns

While the Cahn–Hilliard equation produces complicated patterns, these patterns change slowly with time and eventually disappear. We now turn to a problem in which the important phenomenon is the fact that the patterns are constantly changing and the structure at a single point in time is of limited interest.

Spiral waves occur in a wide variety of physical systems including the Belousov–Zhabotinskii chemical reaction, corrosion on metal surfaces, aggregation of slime molds, electrochemical waves in brain tissue, and contractions of muscle cells in cardiac tissue (see [70], especially Chapter 12 and Section 12.7 for an introduction to the biological and mathematical mechanisms behind these phenomena). In the last example, the breakup of these waves from fairly simple rotating spirals to highly complicated interactive patterns is associated with the onset of ventricular fibrillation, a potentially fatal cardiac rhythm. Due to their importance an enormous amount of work has been done in an attempt to understand their behavior, and much is known. However, as is indicated by the numerically generated images in Figures 8.9 and 8.10, these systems generate very complicated spatial patterns and the structure of these patterns change with time. Thus, obtaining a quantitative description or classification of these waves both spatially and temporally is a very challenging problem.

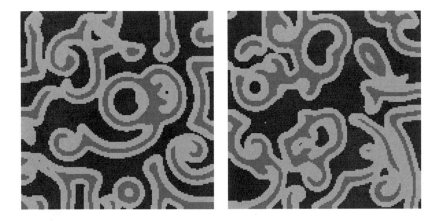

Fig. 8.9. Numerically generated spiral wave with parameter values $a = 0.75$, $b = 0.06$, and $1/\epsilon = 12$. The number of grid points in each direction is 141. The left figure occurs at time $t = 1000$ while the right figure occurs at $t = 2000$. The dark gray represents the excited media, that is $u(x, y) \geq 0.9$.

Figures 8.9 and 8.10 were generated by numerically solving the following system of partial differential equations, known as the FitzHugh-Nagumo equations, which provides an extremely simple model for electropotential waves in a two-dimensional slice of cardiac tissue

$$\frac{\partial u}{\partial t} = \frac{1}{80^2}\left(\frac{\partial^2 u}{\partial x^2} + \frac{\partial^2 u}{\partial y^2}\right) + \frac{u}{\epsilon}(1-u)\left(u - \frac{v+b}{a}\right),$$

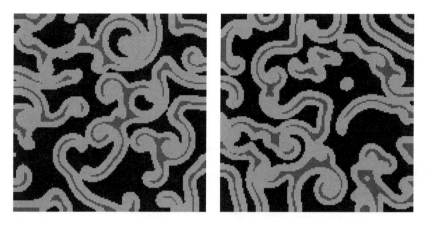

Fig. 8.10. Numerically generated spiral wave with parameter values $a = 0.65$, $b = 0.06$, and $1/\epsilon = 12$. The number of grid points in each direction is 141. The left figure occurs at time $t = 1000$ while the right figure occurs at $t = 2000$. The dark gray represents the excited media, that is $u(x, y) \geq 0.9$.

$$\frac{\partial v}{\partial t} = u^3 - v,$$

where $t \in [0, \infty)$ and $(x, y) \in [0, 1] \times [0, 1] \subset \mathbf{R}^2$ with Neumann boundary conditions, that is,

$$\frac{\partial u}{\partial \nu}(x, y, t) = 0 \quad \text{and} \quad \frac{\partial v}{\partial \nu}(x, y, t) = 0$$

if (x, y) is on the boundary of $[0, 1] \times [0, 1]$ and ν is the outward normal to the boundary.

We shall not attempt to explain this model here (see [70], for instance) but will only focus on using homology to investigate the complicated behavior this system exhibits. If the value of the function u exceeds 0.9 at a particular point in time, then we say that the material or media is excited. For a fixed time t the points in the domain $[0, 1] \times [0, 1]$, where $u(x, y, t) \geq 0.9$, are shaded dark gray and we are interested in how these gray regions change with time.

The images obtained in Figures 8.9 and 8.10 have been derived from the same initial conditions, but with only the parameter value a changed. Thus it is obvious that varying just a single parameter produces different wave patterns.

We shall attempt to address two questions. The first is: Can we find a characteristic that distinguishes the geometry of the waves at the two parameter values? For example, one could imagine counting the number of components corresponding to the excited media. The second has to do with the temporal variability of the complexity. Obviously, each image exhibits a complex spatial pattern. Is there a way to measure how complicated each pattern is and, if so, is this level of complexity constant in time or does it vary?

To answer these questions, the excited media was treated as a three-dimensional subset of the domain cross time, namely as a subset of $[0,1] \times [0,1] \times [0,\infty)$. The justification for this is that in many of the physical examples, for example, cardiac tissue, the movement of the waves through the media is of primary importance. The two-dimensional images shown in Figures 8.9 and 8.10 cannot capture this information.

Let us be more precise about the procedure that was employed. At each time step the partial differential equation was solved numerically on a grid consisting of 141×141 points. Thus, at time step k, there was a collection of values $(u(i,j,k), v(i,j,k)) \in \mathbf{R}^2$, where $i,j \in \{1,\ldots,141\}$, which was updated to a new set of values $(u(i,j,k+1), v(i,j,k+1))$ that represented the discretized solution at the next time step $k+1$. The excited elements of the media were defined to be those points at which $u(i,j,k) \geq 0.9$. Thus the cubical set $X \subset [0,141] \times [0,141] \times [0,\infty)$ was given by

$$[i-1,i] \times [j-1,j] \times [k,k+1] \in \mathcal{K}_3(X) \quad \Leftrightarrow \quad u(i,j,k) \geq 0.9.$$

In this way X represented a cubical approximation of the exited wave.

For each example associated with Figures 8.9 and 8.10, the following procedure was performed. Beginning with the same initial condition the system was numerically integrated for $50,000$ time steps. To better capture the behavior with respect to time, the cubical sets were constructed using the data from time sequences of length 1000. To be more precise, for the first example the cubical sets W_l^1, for $l = 1,2,\ldots,50$, were defined by using the data points $\{u(i,j,k) \mid l \times 1000 \leq k < (l+1) \times 1000\}$ generated when $a = 0.75$. The sets W_l^2 were defined similarly for $a = 0.65$.

Finally, the homology groups $H_*(W_l^i)$ were computed. Since most waves collide with other waves, H_0 provides very little information. Similarly, there are very few volumes actually enclosed by the waves, thus H_2 is not interesting. Plotted in Figure 8.11 are the first Betti numbers β_1 of W_l^1 and W_l^2 as a function of l, respectively.

With regard to the questions we posed earlier, two comments need to be made. The first is that since the scales on the vertical axis differ by an order of magnitude, it is clear that the wave structures at the two-parameter values $a = 0.75$ and $a = 0.65$ are different. The second is perhaps more interesting. Observe that β_1 varies both tremendously and erratically. This suggests that not only does the spatial complexity change with time, but that it changes in a chaotic or unpredictable manner.

8.4 Size Function

In the example of the previous section we gained information by computing the homology groups of a cubical set directly related to a numerical image. We now turn to a problem where the homology group by itself does not provide enough information.

270   8 Prospects in Digital Image Processing

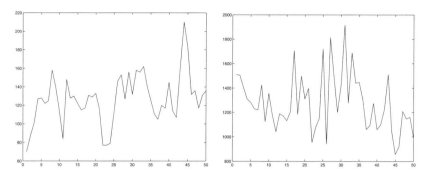

Fig. 8.11. Plots of β_1 of W_l^1 and W_l^2 as a function of l, respectively.

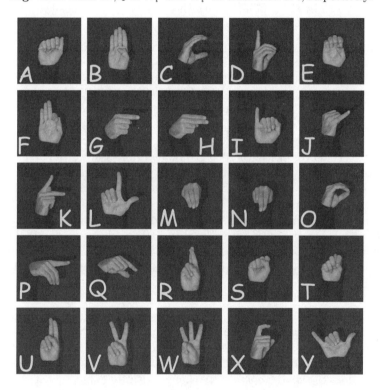

Fig. 8.12. A sample photographic record of the international sign language alphabet provided by courtesy of P. Frosini, A. Verri, and C. Uras. The letter Z is not recorded because it is realized by a dynamical gesture so it requires a video recording rather than a single photograph.

8.4 Size Function

Consider two sample photographic records of the *International Sign Language* alphabet presented in Figure 8.12. There is an extensive body of research devoted to developing an automated approach to recognizing hand gestures from images as meaningful signs. A common approach to pattern recognition is based on matching an image with a known model. Unfortunately, these methods have a high failure rate whenever the relation between the observed object and the model cannot be defined precisely. In the sign language, the shapes of hands showing the same sign may vary considerably depending on the hand of a particular person, the precision of the gesture, and the light or angle under which the photograph has been taken. Therefore, for this problem the tolerance in matching objects needs to be quite high. On the other hand, it cannot be too high because many signs are very similar in shape. Compare for example, the signs "K" and "V." In both, two fingers are pointed straight, though at different angles. A tool to compare these signs must be insensitive to small variations of angles under which fingers are pointed out, but it cannot be completely rotation-invariant, otherwise we cannot distinguish between the two signs.

Topology might seem to be a proper tool for comparing given shapes with model shapes due to its insensitivity to size. In the case of a sign langauge, we cannot advance much by directly applying homology to images of the hands in Figure 8.12 because all except for the letter "O" represent contractible sets. Hence their homology groups are all trivial. We shall present here a more subtle approach based on the concept of *size functions* introduced by Frosini [26] (see also [28] and the references therein) and applied to sign language by Uras and Verri [84].

We start from the following preliminary observation. Due to the imperfection of photography and variable levels of light, the information given by shades of gray in the image of a hand can be very misleading. The most reliable information provided by the techniques of numerical imaging is the information on the contour limiting a hand image. With this in mind we change our focus from the full images of the hands shown in Figure 8.12 to their contours as indicated in Figure 8.13.

The curves in the latter figure are still fairly complicated and so, for the sake of simplicity, we shall describe the basic concepts of sign functions using the two contours in $\mathbf{R}^2$ indicated in Figure 8.14.

We start with a continuous function $f : \mathbf{R}^2 \to [0, \infty)$, which is called a *measuring function*. In this example our choice is the function $f : \mathbf{R}^2 \to [0, \infty)$,

$$f(x) = \|x - a\|_0, \tag{8.3}$$

measuring the distance from a given *reference point* $a \in \mathbf{R}^2$ to x. In Figure 8.14, the reference points a_0 and a_1 are chosen to be within the contours E_0 and E_1, respectively.

We postpone the discussion on the pertinence of this particular choice of measuring function and limit ourselves to the remark that the difference in shape of E_0 and E_1 can be observed in the fact that the values of this function

272   8 Prospects in Digital Image Processing

Fig. 8.13. Hand contours corresponding to the sign language alphabet in Figure 8.12. (From [84]. Copyright © 1994 IEEE. Reprinted with permission.)

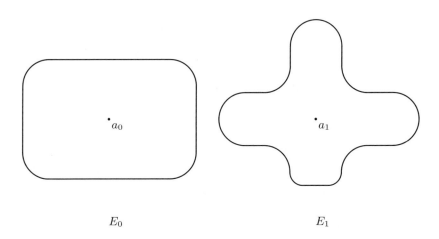

Fig. 8.14. Two distinct contours E_0 and E_1 along with reference points a_0 and a_1.

have greater oscillation on E_1 than on E_0. With this in mind we now present a method for measuring these levels of oscillation using topology.

Consider a bounded subset $E \subset \mathbf{R}^2$. For $\alpha \in [0, \infty)$, define *level sets* of f in E by
$$E^\alpha = \{x \in E \mid 0 \leq f(x) \leq \alpha\}.$$
We will study how the topology of the level sets change as the value of the parameter α changes. Since we will only apply the theory to a contour E in $\mathbf{R}^2$, the level sets will be pieces of that contour. We will concentrate on the number of connected components of the level sets.

For a fixed $\beta \geq \alpha$, define an equivalence relation in E^α as follows:
$$x \stackrel{\beta}{\sim} y \text{ if } x \text{ and } y \text{ belong to the same connected component of } E^\beta.$$
We will study this relation for the pairs (α, β) where $\beta \geq \alpha \geq 0$ are either reals or integers.

Definition 8.1 The function $n: \{(\alpha, \beta) \in \mathbf{R}^2 \mid \beta \geq \alpha \geq 0\} \to \mathbf{Z}$ given by
$$n(\alpha, \beta) = \text{number of equivalence classes of } E^\alpha \text{ under } \stackrel{\beta}{\sim}$$
is the *size function* associated to the set E and the measuring function f.

Since we want to be able to compute the values of the sign function, we must discretize the contours. The approach we have chosen in this chapter is approximating a given set by a cubical set. Given a bounded set $E \subset \mathbf{R}^2$, let
$$\tilde{E} := \operatorname{ch}(E).$$
Recall that this is the smallest cubical set containing E.

Since we restrict the study of size function to cubical sets, we will also restrict the values of α and β to nonnegative integers $\mathbf{Z}^+$. This assumption is reasonable if we take the point of view that the features we wish to extract are large with respect to the grid size. Restating this as an imaging problem, the assumption is that the objects of interest are represented by a significant number of pixels. Thus we shall study the equivalence relation $x \stackrel{\beta}{\sim} y$ on the sets $\tilde{E}^\alpha$, where we assume now that α and β are in $\mathbf{Z}^+$. We may also assume that the reference point a for our measuring function given in (8.3) has integer coordinates. It is left as an exercise to show that $\tilde{E}^\alpha$ is a cubical set for any integer α. Finally, let
$$\tilde{n}: \{(\alpha, \beta) \in \mathbf{Z}^2 \mid \beta \geq \alpha \geq 0\} \to \mathbf{Z}$$
be the restriction of the size function n associated with $\tilde{E}^\alpha$ and d to integer values of α and β.

In Figure 8.15 we have placed the curves E_0 and E_1 along with the points $a_0 = (0,0)$ and $a_1 = (0,0)$ into cubical grids and used gray to indicate the sets $\tilde{E}_0$ and $\tilde{E}_1$.

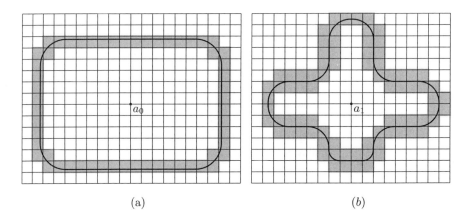

(a)   (b)

Fig. 8.15. The two distinct closed curves E_0 and E_1 placed in cubical grids along with reference points a_0 and a_1. For $i = 0, 1$, the cubical sets $\tilde{E}_i$ are indicated by the shaded squares.

To develop some intuition as to what the size function measures, set $\alpha = 4$ and $\beta = 7$. In Figure 8.16, the sets $\tilde{E}_i^\alpha$ are indicated by medium shading and the sets $\tilde{E}_i^\beta \setminus \tilde{E}_i^\alpha$ are indicated by dark shading. Observe that $\tilde{E}_i^\beta$ consists of those points in the original curve that lie in the medium- or darkly-shaded squares. Notice that there are no medium-shaded squares in Figure 8.16(a) and thus $\tilde{E}_0^4 = \emptyset$. Therefore,

$$\tilde{n}_0(4, 7) = 0.$$

On the other hand, as Figure 8.16(b) indicates, $\tilde{E}_1^4$ consists of four components, but two of the four components are connected to the same darkly shaded component, thus

$$\tilde{n}_1(4, 7) = 3.$$

Observe that having chosen the grid and measuring function, we can make tables of values as is indicated in Figure 8.17 of $\tilde{n}_i(\alpha, \beta)$ for $\alpha, \beta \in \mathbf{Z}^+$. The second table shows more variation of the values of our size function. This can be related to the fact that $\tilde{E}_1$ wiggles more away and toward the center than $\tilde{E}_0$.

Thus the size function may provide an automatic way of distinguishing curves of different shapes. Does it necessarily do so? In our example it did, but, in general, one particular choice of a measuring function might not be sufficient to distinguish properties we care about. Consider, for example, an image of a handwritten letter **p**. If our measuring function is invariant under the reflection in the x-axis, then the corresponding size function would not permit us to distinguish it from an image of a handwritten letter **b**. Thus, in practice, a collection of measuring functions is applied to compared images

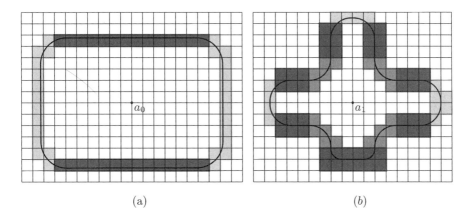

Fig. 8.16. The shaded squares indicate the cubical approximations $\tilde{E}_0$ and $\tilde{E}_1$ of the two simple curves E_0 and E_1. The reference points are $a_0 = (0,0)$ and $a_1 = (0,0)$. The medium-shaded cubes indicate $\tilde{E}_i^4$. The darkest shaded cubes indicate $\tilde{E}_i^7 \setminus \tilde{E}_i^4$.

$\beta \setminus \alpha$	6	7	8	9
6	2			
7	2	2		
8	2	2	2	
9	1	1	1	1

$\beta \setminus \alpha$	3	4	5	6	7	8	9
3	4						
4	4	4					
5	3	3	3				
6	3	3	3	3			
7	3	3	3	3	3		
8	1	1	1	1	1	1	
9	1	1	1	1	1	1	1

Fig. 8.17. Size function tables for $\tilde{E}_0$ (left) and $\tilde{E}_1$ (right). The values of α and β for which $\tilde{n}_i = 0$ have not been included.

and the data coming from the corresponding size functions are studied by statistical techniques.

For example, in [84] a measuring function f_0 is defined as the Euclidean distance from a point x to the lower half-space bounded by the horizontal line L passing through a reference point a. Then a family of 72 measuring functions f_θ was defined by rotating the line L around the reference point by angles $\theta = 0°, 5°, 10°, \ldots, 355°$ and statistical analysis of results was performed.

Our discussion of the size function began with a contour E. This was done in an attempt to provide the reader with an intuitive understanding of this tool. However, from the viewpoint of image processing, it is more realistic to think of the starting point as a digitalized image. With all the caveats of Section 8.1 still in mind, we shall go one more step and assume that we begin with binary images such as those indicated in Figure 8.18. Thus our goal is to distinguish or identify the associated cubical sets using size functions.

276   8 Prospects in Digital Image Processing

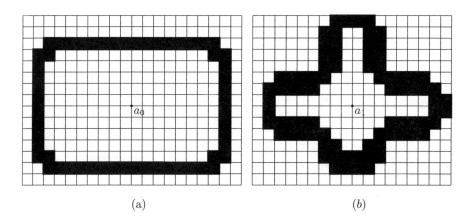

Fig. 8.18. The two images presented as cubical sets $\tilde{E}_0$ and $\tilde{E}_1$ along with reference points a_0 and a_1.

By now the reader may be asking, where does the homology come into play? Recall that we know how to count connected components using homology. In particular, by Theorem 2.59 the number of connected components equals the rank of $H_0(X)$. The following notation will prove useful. Given a pair of cubical sets $A \subset X$, let $q(X, A)$ be the number of connected components of X that are disjoint from A.

Proposition 8.2

$$\tilde{n}(\alpha, \beta) = \operatorname{rank} H_0(\tilde{E}^\beta) - q(\tilde{E}^\beta, \tilde{E}^\alpha).$$

Proof. Since we assume $\beta \geq \alpha$, it follows that $\tilde{E}^\alpha \subset \tilde{E}^\beta$. Thus any connected component of $\tilde{E}^\alpha$, being a connected set, must be contained in a connected component of $\tilde{E}^\beta$. Hence, it is contained in a certain equivalence class of the relation $x \stackrel{\beta}{\sim} y$. Of course, if two connected components of $\tilde{E}^\alpha$ lie in the same connected component of $\tilde{E}^\beta$, then they give rise to the same equivalence class. Thus $\tilde{n}$ is the number of those connected components of $\tilde{E}^\beta$ that contain at least one connected component of $\tilde{E}^\alpha$. Obviously, $\operatorname{rank} H_0(\tilde{E}^\beta)$ is the sum of the number of connected components of $\tilde{E}^\beta$ that contain a component of $\tilde{E}^\alpha$ and the number of connected components of $\tilde{E}^\beta$ that do not contain any. But a connected component B of $\tilde{E}^\beta$ that does not contain any connected component of $\tilde{E}^\alpha$ must be disjoint from $\tilde{E}^\alpha$. Indeed, suppose that B is not disjoint from $\tilde{E}^\alpha$, and let $a \in B \cap \tilde{E}^\alpha$. Let A be the connected component of $\tilde{E}^\alpha$ containing a. Then $A \cup B$ is a connected set containing B (see Theorem 12.53). By the maximality, $A \cup B = B$ so $A \subset B$, thus B contains a connected component of $\tilde{E}^\alpha$. Hence

$$\operatorname{rank} H_0(\tilde{E}^\beta) = \tilde{n}(\alpha, \beta) + q(\tilde{E}^\beta, \tilde{E}^\alpha),$$

and the conclusion follows. □

In the light of Proposition 8.2, in order to provide a purely homological formula for computing $\tilde{n}$, we need a homological method for computing $q(X, A)$ where $A \subset X$. This is provided in Chapter 9 where the concept of relative homology is introduced.

We finish this section with some remarks that are intended to put our discussion of size functions into proper perspective. As presented here size functions are related to one topological feature—the variation in the number of connected components of level sets of measuring functions. For this purpose, 0-dimensional homology is sufficient. Of course, from the viewpoint of computations there are more efficient algorithms for counting connected components than directly computing H_0. On the other hand, when one considers higher-dimensional images, a wider menagerie of topological features and higher-dimensional homologies comes into the play. Generalizing size functions in this direction leads to Morse theory, an extremely rich topic and one of the fundamental tools of differential topology.

Exercises

8.1 Let X be a cubical set in $\mathbf{R}^2$. Consider the following measuring functions on X:

(a) $f(x) := x_2$;
(b) $f(x) := ||x||_0$.

Show that the associated sets X^α are cubical sets for any nonnegative integer α.

8.2 Let X be a cubical set in $\mathbf{R}^2$ and let f be the size function considered in Exercise 8.1(a). Prove that the associated size function depends only on the integer parts floor α and floor β of, respectively, α and β.

8.5 Bibliographical Remarks

The background for the differential equations presented in Sections 8.3 can be found in Murray [70]. As has been repeatedly stressed, the use of homology to investigate complicated patterns arising from numerical simulations is extremely new in part because of the lack of efficient easily applicable computational tools. Thus the material in Sections 8.2 and 8.3 is merely indicative of current research projects. For a further discussion the reader can consult [30] and [29].

The concept of size function discussed in Section 8.4 was introduced by Frosini [26] and further developed by the group Vision Mathematics. We refer the reader to [27, 28] and references therein for further study. The discussed application of size functions to the sign language alphabet comes from [84]. The idea of presenting the homological formula for size functions comes from [5]. Generalizations of the method of size functions aimed to extract higher-dimensional features of images are discussed in [10, 11, 27, 32]. Those generalizations lead to Morse theory. The reader interested in pursuing this topic from a mathematical point of view is encouraged to consult [54, 11, 37]. On the computational side there are no standard references. The subject is too new and is still undergoing rapid advances both theoretically and in terms of its domain of application. However, the reader may find the following articles of interest [6, 22].

9
Homological Algebra

We finished the previous chapter with the observation that given a space X and a subset A it would be useful to have a means of computing the number of components of X that are disjoint from A. Using this as motivation, in Section 9.1 we introduce the concept of relative homology. Though motivated by a simple problem, it turns out that relative homology is an extremely powerful tool, both as a means of extracting topology from the homological algebra and as an abstract computational technique. The language of these computational methods takes the form of exact sequences, which are discussed in Section 9.2.

The reader may recall that by Theorem 2.78 if X, Y and $X \cap Y$ are acyclic sets, then $X \cup Y$ is acyclic. The Mayer–Vietoris sequence presented in Section 9.4 is the generalization of this theorem—it allows one to compute the homology of $X \cup Y$ in terms of X, Y, and $X \cap Y$ for arbitrary cubical sets. However, obtaining this sequence requires a fundamental algebraic result, the construction of the connecting homomorphism, described in Section 9.3.

9.1 Relative Homology

Given a pair of cubical sets $A \subset X$, we will define the relative homology groups $H_*(X, A)$. These groups can be used directly to measure how the topological structure of X and A differ. In Subsection 9.1.2 we consider the case of another pair of cubical sets $D \subset Y$ along with a continuous map $f : X \to Y$ with the property that $f(A) \subset D$ and show that there is an induced homomorphism on the relative homology groups.

9.1.1 Relative Homology Groups

Since we often want to talk about a pair of cubical sets, one of which is the subset of the other, the following definition is useful.

Definition 9.1 A pair of cubical sets X and A with the property that $A \subset X$ is called a *cubical pair* and denoted by (X, A).

Recall that we have ended Chapter 8 with the formula for the size function

$$\tilde{n}(\alpha, \beta) = \operatorname{rank} H_0(\tilde{E}^\beta) - q(\tilde{E}^\beta, \tilde{E}^\alpha), \tag{9.1}$$

where $q(\tilde{E}^\beta, \tilde{E}^\alpha)$ is the number of connected components of $\tilde{E}^\beta$ that are disjoint from $\tilde{E}^\alpha$. The first term on the right-hand side of the formula is expressed in the language of homology, and we would also like to find a way of expressing in this language the term $q(\tilde{E}^\beta, \tilde{E}^\alpha)$ or, more generally, the term $q(X, A)$, where (X, A) is a cubical pair.

In a sense, we want to ignore the set A and everything connected to it. Moving from topology to algebra requires us to consider the chain groups $C_0(A)$ and $C_0(X)$. At this level, ignoring a subset $A \in X$ can be viewed in terms of the quotient operation

$$C_0(X)/C_0(A).$$

In this quotient space, the equivalence class of any dual vertex of A is zero. However, recall that we also want to ignore everything connected to A. Of course, there may be a vertex P in X that is not in A, but is edge-connected to it. For such P, the equivalence class of $\widehat{P}$ in the quotient space $C_0(X)/C_0(A)$ is not trivial. However, recall Theorem 2.59: For such a P, there must be a vertex Q in A such that $\widehat{P}$ is homologous to $\widehat{Q}$.

Before proceeding to the next step, let us revisit this discussion in terms of a concrete example.

Example 9.2 Let $X = [0, 2] \cup \{3\}$ and let $A = [0, 1]$ (see Figure 9.1). Then the vertex $[3]$ is the only connected component disjoint from A. The quotient group $C_0(X)/C_0(A)$ is generated by equivalence classes of $\widehat{[2]}$ and $\widehat{[3]}$. But $\widehat{[2]}$ is homologous to $\widehat{[1]} \in C_0(A)$. In fact,

$$\partial \widehat{[1, 2]} = \widehat{[2]} - \widehat{[1]}.$$

This gives a suggestion that we should only count vertices of X that are not homologous to a vertex of A. Thus we want a homology theory in which dual vertices homologous to those of A are counted as those "homologous to 0 relative to A." This motivates the following definition.

Definition 9.3 Let (X, A) be a cubical pair. The *relative chains of X modulo A* are the elements of the quotient groups

$$C_k(X, A) := C_k(X)/C_k(A).$$

The equivalence class of a chain $c \in \mathcal{C}(X)$ relative to $\mathcal{C}(A)$ is denoted by $[c]_A$. By Exercise 13.16, the groups $C_k(X, A)$ are free abelian; hence it makes sense

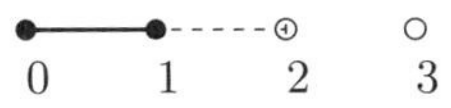

Fig. 9.1. Cubical pair (X, A) discussed in Example 9.2.

to introduce the following concept. The *relative chain complex of X modulo A* is given by

$$\{C_k(X, A), \partial_k^{(X,A)}\},$$

where $\partial_k^{(X,A)} : C_k(X, A) \to C_{k-1}(X, A)$ is defined by

$$\partial_k^{(X,A)}([c]_A) := [\partial_k c]_A. \tag{9.2}$$

It is left to the reader to verify that this map is well defined and it satisfies the fundamental property of the boundary map:

$$\partial_{k-1}^{(X,A)} \partial_k^{(X,A)} = 0.$$

The relative chain complex gives rise to the *relative k-cycles*,

$$Z_k(X, A) := \ker \partial_k^{(X,A)},$$

the *relative k-boundaries*,

$$B_k(X, A) := \operatorname{im} \partial_{k+1}^{(X,A)},$$

and finally the *relative homology groups*

$$H_k(X, A) := Z_k(X, A)/B_k(X, A).$$

Because $\partial^{(X,A)}$ is induced by the canonical boundary operator ∂, it is common to simplify the notation and let $\partial = \partial^{(X,A)}$. We will use that notation whenever it is not confusing.

Proposition 9.4 *Let X be a connected cubical set and let A be a nonempty cubical subset of X. Then*

$$H_0(X, A) = 0.$$

Proof. We need to show that every element of the group

$$Z_0(X, A) = C_0(X, A) = C_0(X)/C_0(A)$$

is a relative boundary. By Exercise 13.16, $Z_0(X, A)$ is generated by the set

$$\{[\hat{P}]_A \mid P \in \mathcal{K}_0(X) \setminus \mathcal{K}_0(A)\}.$$

Take any $Q \in \mathcal{K}_0(A)$. Then $[\widehat{Q}]_A = 0$. Take any $P \in \mathcal{K}_0(X) \setminus \mathcal{K}_0(A)$. Since X is connected, Theorem 2.59 implies that all dual vertices are homologous in X so there exists $c \in C_1(X)$ such that $\partial_1 c = \widehat{P} - \widehat{Q}$. By (9.2),

$$\partial_k^{(X,A)}([c]_A) := [\widehat{P} - \widehat{Q}]_A = [\widehat{P}]_A - [\widehat{Q}]_A = [\widehat{P}]_A,$$

hence $[\widehat{P}]_A \in B_0(X, A)$. □

We are now in position to answer the question addressed at the beginning of this section.

Proposition 9.5 *Let (X, A) be a cubical pair. Then the number of connected components of X that do not intersect A is the dimension of $H_0(X, A)$.*

Proof. Let $X = \cup_{i=1}^n X_i$, where the X_i are the connected components of X. Let $A_i = X_i \cap A$. Recall that $\widehat{\mathcal{K}}_k(X_i)$ and $\widehat{\mathcal{K}}_k(A_i)$ form bases of $C_k(X_i)$ and $C_k(A_i)$. Since $A_i \subset X_i$, $\widehat{\mathcal{K}}_k(A_i) \subset \widehat{\mathcal{K}}_k(X_i)$. This leads to three observations. The first is that

$$C_k(X) = \bigoplus_{i=1}^n C_k(X_i).$$

The second is that

$$C_k(X, A) = \frac{C_k(X)}{C_k(A)} = \bigoplus_{i=1}^n \frac{C_k(X_i)}{C_k(A_i)} = \bigoplus_{i=1}^n C_k(X_i, A_i),$$

and the third is that the set

$$\left\{ [c]_A \mid c \in \widehat{\mathcal{K}}_k(X_i) \setminus \widehat{\mathcal{K}}_k(A_i) \right\}$$

is a basis for $C_k(X_i, A_i)$ (see Exercise 13.16).

By definition,

$$\partial_k^{(X,A)}([c]_A) = [\partial_k c]_A.$$

It follows from Proposition 2.39 that if $c \in C_k(X_i, A_i)$, then $\partial_k^{(X,A)}([c]_A) \in C_{k-1}(X_i, A_i)$. Therefore,

$$\partial_k^{(X,A)} = \bigoplus_{i=1}^n \partial_k^{(X_i, A_i)},$$

from which the reader can check that

$$H_0(X, A) = \bigoplus_{i=1}^n H_0(X_i, A_i).$$

The result now follows from Proposition 9.4. □

As an immediate consequence of Propositions 8.2 and 9.5, we get the awaited characterization of size functions.

Corollary 9.6 *Let $\tilde{n}$ be the size function discussed in Section 8.4. Then*
$$\tilde{n}(\alpha,\beta) = \operatorname{rank} H_0(\tilde{E}^\beta) - \operatorname{rank} H_0(\tilde{E}^\beta, \tilde{E}^\alpha).$$

In Section 8.4 we consider the size function corresponding to a particular example of a measuring function, but the same formula would hold true for any measuring function on a cubical set that produces cubical level sets for integer parameter values.

As the following theorem indicates, relative homology also provides us with a new insight into the definition of reduced homology $\tilde{H}(X)$ studied in Section 2.7.

Theorem 9.7 *Let X be a cubical set and $P \in \mathcal{K}_0(X)$ a chosen vertex of X. Then*
$$H_*(X, P) \cong \tilde{H}_*(X).$$
More precisely, we have the identities
$$\tilde{H}_k(X) = H_k(X) = H_k(X, P) \text{ for all } k \geq 1. \tag{9.3}$$
The isomorphism
$$H_0(X, P) \cong \tilde{H}_0(X) \tag{9.4}$$
is induced by the isomorphism $\phi : \tilde{Z}_0(X) \to Z_0(X, P)$ given on the elements of the basis
$$\{\widehat{Q} - \widehat{P} \mid Q \in \mathcal{K}_0(X) \setminus \{P\}\} \tag{9.5}$$
of $\tilde{Z}_0(X)$ by
$$\phi(\widehat{Q} - \widehat{P}) := [Q]_P. \tag{9.6}$$

Proof. Since $C_k(P) = 0$ for all $k \neq 0$, we have
$$C_k(X, P) = \frac{C_k(X)}{\{0\}} = C_k(X) \text{ for all } k \geq 1,$$
and
$$\partial_k^{(X,P)} = \partial_k \text{ for all } k \geq 2.$$
By Definition 2.94 of the reduced cubical chain complex, we get
$$\tilde{C}_k(X) = C_k(X) = C_k(X, P) \text{ for all } k \geq 1$$
and
$$\tilde{\partial}_k = \partial_k = \partial_k^{(X,P)} \text{ for all } k \geq 2.$$
Hence the identities in (9.3) follow for all $k \geq 2$. The identity for $k = 1$ is a bit more delicate. We know that $\tilde{H}_1(X) = H_1(X)$, but the previous argument only shows that $B_1(X, P) = B_1(X)$. So we need to show that $Z_1(X, P) = Z_1(X)$. Let $c \in C_1(X) = C_1(X, P)$. The condition $\partial_1^{(X,P)}([c]_P) = 0$ is equivalent to $\partial_1(c) \in C_0(P)$. But $C_0(P) = \mathbf{Z}\widehat{P}$. Therefore, $\partial_1^{(X,P)}([c]_P) = 0$ if and only if

there exists $n \in \mathbf{Z}$ such that $\partial_1(c) = n\widehat{P}$. If this is the case, then the definition of $\tilde{\partial}$ implies that
$$0 = \tilde{\partial}_0(\tilde{\partial}_1(c)) = \epsilon(n\widehat{P}) = n.$$
Hence $\partial_1^{(X,P)}([c]_P) = 0$ if and only if $\partial_1(c) = 0$.

Finally, we show (9.4). By Exercise 13.16, the set
$$\{[\widehat{Q}]_P \mid Q \in \mathcal{K}_0(X) \setminus \{P\}\}$$
is a basis for $Z_0(X, P)$. We leave as an exercise the verification that the set in (9.5) is a basis for $\tilde{Z}_0(X) := \ker \epsilon$. Equation (9.6) defines a bijection between the two bases and so it extends by linearity to the isomorphism $\phi : \tilde{Z}_0(X) \to Z_0(X, P)$. It remains to show that $\phi(\tilde{B}_0(X)) = B_0(X, P)$. Indeed, any boundary in $\tilde{B}_0(X) = B_0(X)$ is a linear combination of boundaries of dual edges, and those are differences of dual vertices. So consider an edge E with $\partial_1(\widehat{E}) = \widehat{Q_1} - \widehat{Q_0}$. Then
$$\phi(\partial_1(\widehat{E})) = \phi\left((\widehat{Q_1} - \widehat{P}) - (\widehat{Q_0} - \widehat{P})\right) = [\widehat{Q_1}]_P - [\widehat{Q_0}]_P$$
$$= [\partial_1 \widehat{E}]_P = \partial_1^{(X,P)}[\widehat{E}]_P. \quad \square$$

Given a cubical pair (X, A), Proposition 9.5 provides us with a complete topological interpretation of the algebraic quantity $H_0(X, A)$. We would also like to have some intuition concerning $H_k(X, A)$ for $k \geq 1$ and A different from a point. With this in mind we consider the following two examples, the generalization of which is left as an exercise.

Example 9.8 Let $X = [0, 1]$ and $A = \{0, 1\}$. The set of elementary chains $\{\widehat{0}, \widehat{1}\}$ is a basis for both $C_0(X)$ and $C_0(A)$. Therefore, $C_0(X, A) = 0$. Clearly, $\widehat{[0, 1]}$ is a basis for $C_1(X)$ while $C_1(A) = 0$. Hence, $\left[\widehat{[0, 1]}\right]_A$ is a basis for $C_1(X, A)$. Since $C_0(X, A) = 0$, we have $Z_1(X, A) = C_1(X, A)$. Thus, $\left[\widehat{[0, 1]}\right]_A$ is a basis for $Z_1(X, A)$ and since obviously $B_1(X, A) = 0$, we get $H_1([0, 1], \{0, 1\}) = \mathbf{Z}$. Therefore
$$H_k([0, 1], \{0, 1\}) = \begin{cases} \mathbf{Z} & \text{if } k = 1, \\ 0 & \text{otherwise.} \end{cases}$$

Example 9.9 Let $X = [0, 1]^2$ and $A = \text{bd}\,([0, 1]^2)$. The reader can check that $C_k(X) = C_k(A)$ for all $k \neq 2$. Thus $C_k(X, A) = 0$ for all $k \neq 2$. On the other hand, $\widehat{[0, 1]^2}$ is a basis for $C_2(X)$ while $C_2(A) = 0$. Hence, $\left[\widehat{[0, 1]^2}\right]_A$ is a basis for $C_2(X, A)$. Since $C_1(X, A) = 0$, $\left[\widehat{[0, 1]^2}\right]_A$ is a basis for $Z_2(X, A)$ and thus $H_2([0, 1], \{0, 1\}) = \mathbf{Z}$. Therefore
$$H_k([0, 1], \{0, 1\}) = \begin{cases} \mathbf{Z} & \text{if } k = 2, \\ 0 & \text{otherwise.} \end{cases}$$

9.1 Relative Homology

From these two examples we see that relative homology can be used to capture important topological information about otherwise acyclic sets. In each case, by an appropriate choice of A, the nontriviality of $H_*(X, A)$ indicates the dimension of $\mathcal{K}_{\max}(X)$.

Example 9.10 Consider the cubical pair (Y, D) where $Y = [-3, 3] \subset \mathbf{R}$ and $D = [-3, -1] \cup [2, 3]$ (see Figure 9.2). Let us compute $H_*(Y, D)$. Since $\mathcal{C}_k(Y) = 0$ for $k \geq 2$, $H_k(Y, D) = 0$ for $k \geq 2$. By Proposition 9.4, $H_0(Y, D) = 0$. Thus only $H_1(Y, D)$ remains to be determined. However, since $C_2(Y, D) = 0$, $H_1(Y, D) = Z_1(Y, D)$.

Let $z := \widehat{[-1, 0]} + \widehat{[0, 1]} + \widehat{[1, 2]}$ and $\zeta := [z]_D$. It is easy to verify that $\partial z = \widehat{[2]} - \widehat{[-1]}$, which is in $C_0(D)$, so $\partial^{(Y,D)} \zeta = 0$. Thus $\zeta \in Z_1(Y, D)$. We want to show that ζ generates $Z_1(Y, D)$, which would permit us to conclude that $H_1(Y, D) = Z_1(Y, D) \cong \mathbf{Z}$. First, let $c \in C_1(Y)$ be a chain with $\partial^{(Y,D)}[c]_D = 0$. One can decompose c as

$$c = c_1 + c_2 + c_3,$$

where c_1 is supported in $[-3, -1]$, c_2 in $[-1, 2]$, and c_3 in $[2, 3]$. Then $[c_1]_D = [c_3]_D = 0$ so their boundaries are also zero. Consequently, $\partial^{(Y,D)}[c_1]_D = \partial^{(Y,D)}[c_2]_D = 0$. Next,

$$c_2 = \alpha_1 \widehat{[-1, 0]} + \alpha_2 \widehat{[0, 1]} + \alpha_3 \widehat{[1, 2]}.$$

The condition $\partial c_2 \in C_0(D)$ implies that $\alpha_1 = \alpha_2 = \alpha_3$, so forcingly $c_2 = \alpha_1 z$. Finally, $[c]_D = [c_2]_D = \alpha_1 \zeta$.

Let $X = [-1, 2]$ and $A = \{-1, 2\}$. Performing a computation similar to that of the previous example leads to the following observation:

$$H_*([-3, 3], [-3, -1] \cup [2, 3]) \approx H_*([-1, 2], \{-1, 2\}).$$

Fig. 9.2. The cubical set $Y = [-3, 3]$ discussed in Example 9.10. The set D is marked with solid lines and A is marked with solid circles. The part of D outside the circles can be "excised" without affecting relative homology.

One can ask if it is merely a coincidence that these homology groups are isomorphic or whether there is a deeper underlying relationship. Since the relative chains of the pair (X, A) are formed by quotienting out by those elementary chains that lie in the subspace A, it seems reasonable to conjecture that if one adds the same cubes to both X and A, then the group of relative chains does not change and hence the homology should not change. Theorem 9.14, presented shortly, confirms this, though at first glance its statement may appear somewhat different.

9.1.2 Maps in Relative Homology

Of course, to compare the relative homology groups of different cubical pairs, we need to be able to talk about maps. So let (X, A) and (Y, D) be cubical pairs, and let $f : X \to Y$ be a continuous map. The most basic question is whether f induces a map from $H_*(X, A)$ to $H_*(Y, D)$. In order to answer that question, we should first look at a chain map $\varphi : \mathcal{C}(X) \to \mathcal{C}(Y)$ and find out when it induces a chain map $\bar{\varphi} : \mathcal{C}(X, A) \to \mathcal{C}(Y, D)$. The relative chain groups are quotient groups so as the reader can check, the following holds true.

Proposition 9.11 *Let (X, A) and (Y, D) be cubical pairs, and let $\varphi : \mathcal{C}(X) \to \mathcal{C}(Y)$ be a chain map such that*

$$\varphi(\mathcal{C}(A)) \subset \mathcal{C}(D).$$

Then the induced map $\bar{\varphi} : \mathcal{C}(X, A) \to \mathcal{C}(Y, D)$ given by the formula

$$\bar{\varphi}([c]_A) := [\varphi(c)]_D$$

is well defined and is a chain map.

This leads to the following definition. The map $f : X \to Y$ is called a *continuous map between cubical pairs* (X, A) and (Y, D) if $f : X \to Y$ is continuous and $f(A) \subset D$. We then write

$$f : (X, A) \to (Y, D).$$

To generate a map on the level of relative homology, that is,

$$\bar{f}_* : H_*(X, A) \to H_*(Y, D),$$

we proceed as in the case of $f_* : H_*(X) \to H_*(Y)$. In order to understand the definition, we must initially distinguish $\bar{f}_*$ from f_*, but as we get used to this concept we will use f_* for either map and decide from the context if this is a map on relative homology.

Since $f : X \to Y$ is continuous, there exists an appropriate scaling vector α, such that $M_{f^\alpha} : X^\alpha \rightrightarrows Y$ is acyclic. Recall that the definition of $f_* : H_*(X) \to H_*(Y)$ involves scaling $\Lambda^\alpha : X \to X^\alpha$ and its inverse scaling $\Omega^\alpha : X^\alpha \to X$. In order to extend this definition to relative homology, we must first know that all involved maps send pairs to pairs. Indeed, since A is cubical, $A^\alpha = \Lambda^\alpha(A)$ is cubical, too, so we have a well-defined map on pairs

$$\Lambda^\alpha : (X, A) \to (X^\alpha, A^\alpha).$$

Next, the acyclic representation M_{Λ^α} maps any elementary cube $Q \in \mathcal{K}(A)$ to $Q^\alpha \subset A^\alpha$ so its chain selector $\theta : \mathcal{C}(X) \to \mathcal{C}(X^\alpha)$ must send $\mathcal{C}(A)$ to $\mathcal{C}(A^\alpha)$. Therefore, by Proposition 9.11, we have a well-defined map

$$\bar{\theta} : \mathcal{C}(X, A) \to \mathcal{C}(X^\alpha, A^\alpha)$$

and, consequently,

$$\bar{\Lambda}_*^\alpha := \bar{\theta}_* : H_*(X, A) \to H_*(X^\alpha, A^\alpha).$$

By the same reasoning we get the map

$$\bar{\Omega}_*^\alpha : H_*(X^\alpha, A^\alpha) \to H_*(X, A).$$

By the same arguments as those presented in the proof of Proposition 6.51, those two maps are isomorphisms and mutual inverses on relative homologies.

From the definition of f^α it follows that $f^\alpha(A^\alpha) \subset D$. The fact that D and A^α are cubical implies that $M_{f^\alpha}(A^\alpha) \subset D$. Now let $\varphi : \mathcal{C}(X^\alpha) \to \mathcal{C}(Y)$ be a chain selector for M_{f^α}. For any $Q \in \mathcal{K}(A^\alpha)$, $|\varphi(\widehat{Q})| \subset M_{f^\alpha}(Q) \subset D$, and hence $\varphi(\mathcal{C}(A^\alpha)) \subset \mathcal{C}(D)$. Thus, φ induces a chain map between the relative chain complexes

$$\bar{\varphi} : \mathcal{C}(X^\alpha, A^\alpha) \to \mathcal{C}(Y, D).$$

We can finally define $\bar{f}_*$ as follows.

Definition 9.12 Let $f : (X, A) \to (Y, D)$ be a continuous map between cubical pairs. Let the scaling α, the chain maps θ and φ be as in the preceding discussion. Then the map $\bar{f}_* : H_*(X, A) \to H_*(Y, D)$ induced by f in relative homology is given by the formula

$$\bar{f}_* := \bar{\theta}_* \bar{\varphi}_*.$$

The "bar" notation can be omitted for simplicity of presentation. The reader can check that this definition is independent of the particular choice of scaling.

All important properties of the map f_* discussed in Section 6.4.2 carry over to maps induced in relative homology. In particular, we get the following generalization of Theorem 6.58:

Theorem 9.13 Assume $f : (X, A) \to (Y, D)$ and $g : (Y, D) \to (Z, B)$ are continuous maps between cubical pairs. Then $(g \circ f) : (X, A) \to (Z, B)$ is a map between cubical pairs and

$$\overline{(g \circ f)}_* = \bar{g}_* \circ \bar{f}_*.$$

Theorem 9.14 (Excision isomorphism theorem) Let (X, A) be a cubical pair. Let $U \subset A$ be open in X and representable. Then $X \setminus U$ is a cubical set and the inclusion map $i : (X \setminus U, A \setminus U) \to (X, A)$ induces an isomorphism

$$i_* : H_*(X \setminus U, A \setminus U) \to H_*(X, A).$$

Proof. It follows from Proposition 6.3(ii) that $X \setminus U$ is representable and from Proposition 6.3(iii) that $X \setminus U$ is a cubical set.

First we will prove that

$$\mathcal{K}(X) \setminus \mathcal{K}(X \setminus U) \subset \mathcal{K}(A). \tag{9.7}$$

To this end take $Q \in \mathcal{K}(X) \setminus \mathcal{K}(X \setminus U)$. Then $Q \cap U \neq \emptyset$. By Proposition 2.15(iv), $\operatorname{cl} \overset{\circ}{Q} \cap U \neq \emptyset$. Since U is open in X, we get $\overset{\circ}{Q} \cap U \neq \emptyset$. Hence in particular, $\overset{\circ}{Q} \cap \operatorname{cl} U \neq \emptyset$. Therefore, by Proposition 2.15(vi), $Q \subset \operatorname{cl} U \subset \operatorname{cl} A \subset A$. This proves (9.7).

For the inclusion map $j : X \setminus U \to X$, we have $M_j(Q) = Q$ for every $Q \in \mathcal{K}(X \setminus U)$. Thus the inclusion map $\iota : \mathcal{C}(X \setminus U) \to \mathcal{C}(X)$ is a chain selector for M_j. Let $\pi : \mathcal{C}(X) \to \mathcal{C}(X, A)$ be the projection map. Take an element $[c]_A \in \mathcal{C}(X, A)$. Put

$$c' := \sum_{Q \in \mathcal{K}(X \setminus U)} c(Q)\widehat{Q}.$$

Then $\pi(\iota(c')) = [c']_A$. We have

$$c - c' = \sum_{Q \in \mathcal{K}(X) \setminus \mathcal{K}(X \setminus U)} c(Q)\widehat{Q},$$

and by (9.7) $|c - c'| \subset A$. This shows that $\pi(\iota(c')) = [c]_A$, that is $\pi \circ \iota : \mathcal{C}(X \setminus U) \to \mathcal{C}(X, A)$ is surjective.

We will show that

$$\ker \pi \circ \iota = \mathcal{C}(A \setminus U).$$

Take $c \in \ker \pi \circ \iota$. This is equivalent to $c \in \mathcal{C}(X \setminus U) \cap \mathcal{C}(A)$, which happens if and only if $|c| \subset A \setminus U$, that is, $c \in \mathcal{C}(A \setminus U)$. Therefore, by the fundamental epimorphism theorem (Corollary 13.44), the map ι induces an isomorphism of relative chain groups. It follows from Proposition 4.5 that also i_* is an isomorphism. $\square$

Exercises

9.1 Let $\mathcal{C} = \{C_k, \partial_k\}$ be an abstract chain complex. Let $\mathcal{C}'$ be a chain subcomplex. Define $H_*(\mathcal{C}, \mathcal{C}')$.

9.2 Let $Q \in \mathcal{K}_q$ be an elementary cube. Prove that

$$H_k(Q, \operatorname{bd}(Q)) \cong \begin{cases} \mathbf{Z} & \text{if } k = q, \\ 0 & \text{otherwise.} \end{cases}$$

9.3 Prove Proposition 9.11.

9.4 Let $X = [-4, 4] \times [-4, 4]$, $A = [-4, -2] \times [-4, 4] \cup [2, 4] \times [-4, 4]$, $Y = [-8, 8] \times [-2, 2]$, and $D = [-8, -2] \times [-2, 2] \cup [2, 8] \times [-2, 2]$. Let $f : (X, A) \to (Y, D)$ be the map given by the formula $f(x_1, x_2) = (2x_1, \frac{1}{2}x_2)$. Compute $H_*(X, A)$, $H_*(Y, D)$, and the map $f_* : H_*(X, A) \to H_*(Y, D)$.

9.2 Exact Sequences

Relative homology turns out to be a very powerful tool. However, if the reader solved Exercise 9.4, then it is clear that our ability to compute relative homology groups is rather limited. Thus we want to look for more efficient methods. Ideally, given a pair of cubical sets (X, A), we would have a theorem that relates $H_*(X, A)$ to $H_*(X)$ and $H_*(A)$. As we shall see in Section 9.3, such a theorem exists, but before it can be stated we need to develop some more tools in homological algebra.

From the algebraic point of view, homology begins with a chain complex $\{C_k, \partial_k\}$, which can be thought of as a sequence of abelian groups and maps

$$\ldots \to C_{k+1} \xrightarrow{\partial_{k+1}} C_k \xrightarrow{\partial_k} C_{k-1} \to \ldots$$

with the property that

$$\operatorname{im} \partial_{k+1} \subset \ker \partial_k.$$

A very special case of this is the following.

Definition 9.15 A sequence (finite or infinite) of groups and homomorphisms

$$\ldots \to G_3 \xrightarrow{\psi_3} G_2 \xrightarrow{\psi_2} G_1 \to \ldots$$

is *exact* at G_2 if

$$\operatorname{im} \psi_3 = \ker \psi_2.$$

It is an *exact sequence* if it is exact at every group. If the sequence has a first or last element, then it is automatically exact at that group.

Example 9.16 Let $\pi : \mathbf{Z} \to \mathbf{Z}_2$ be given by $\pi(n) := n \bmod 2$. Then the sequence

$$0 \longrightarrow \mathbf{Z} \xrightarrow{\times 2} \mathbf{Z} \xrightarrow{\pi} \mathbf{Z}_2 \longrightarrow 0,$$

where $\times 2$ is multiplication by 2, is exact. Notice that two of the homomorphisms are not marked. This is because they are uniquely determined. Let us consider the issue of exactness at

$$0 \longrightarrow \mathbf{Z} \xrightarrow{\times 2} \mathbf{Z}.$$

Observe that the image of 0 is 0; therefore, exactness is equivalent to the assertion that the kernel of multiplication by 2 is 0. This is true because $\times 2 : \mathbf{Z} \to \mathbf{Z}$ is a monomorphism. We now check for exactness at

$$\mathbf{Z} \xrightarrow{\times 2} \mathbf{Z} \xrightarrow{\pi} \mathbf{Z}_2.$$

Observe that the image of $\times 2$ is the set of even integers. This is precisely the kernel of π, since $\pi(n) = n \bmod 2 = 0$ if and only n is even. Hence the sequence is exact at $\mathbf{Z}$.

Finally, consider
$$\mathbf{Z} \xrightarrow{\pi} \mathbf{Z}_2 \longrightarrow 0.$$
Notice that $\pi(\mathbf{Z}) = \mathbf{Z}_2$, since $\pi(1) = 1$.

Since the sequence is exact at each group, it is exact.

To develop our intuition concerning exact sequences, we will prove a few simple lemmas.

Lemma 9.17 $G_1 \xrightarrow{\psi_1} G_0 \xrightarrow{\phi} 0$ is an exact sequence if and only if ψ_1 is an epimorphism.

Proof. ($\Rightarrow$) Assume that $G_1 \xrightarrow{\psi_1} G_0 \xrightarrow{\phi} 0$ is an exact sequence. Since $\phi : G_0 \to 0$, $\ker \phi = G_0$. By exactness, $\operatorname{im} \psi_1 = \ker \phi = G_0$, that is, ψ_1 is an epimorphism.

($\Leftarrow$) If ψ_1 is an epimorphism, then $\operatorname{im} \psi_1 = G_0$. Since $\phi : G_0 \to 0$, $\ker \phi = G_0$. Therefore, $\operatorname{im} \psi_1 = \ker \phi$. □

Lemma 9.18 $0 \xrightarrow{\phi} G_1 \xrightarrow{\psi_1} G_0$ is an exact sequence if and only if ψ_1 is a monomorphism.

Proof. ($\Rightarrow$) Assume that the sequence is exact. Clearly, $\operatorname{im} \phi = 0$, thus $\ker \psi_1 = 0$, which implies that ψ_1 is a monomorphism.

($\Leftarrow$) If ψ_1 is a monomorphism, then $\ker \psi_1 = 0$. Since $\phi : 0 \to G_1$, $\operatorname{im} \phi = 0$. Therefore, $\operatorname{im} \phi = \ker \psi_1$. □

Similar arguments lead to the following result.

Lemma 9.19 *Assume that*
$$G_3 \xrightarrow{\psi_3} G_2 \xrightarrow{\psi_2} G_1 \xrightarrow{\psi_1} G_0$$
is an exact sequence. Then the following are equivalent:

1. ψ_3 *is an epimorphism;*
2. ψ_2 *is the zero homomorphism;*
3. ψ_1 *is a monomorphism.*

Definition 9.20 A *short exact sequence* is an exact sequence of the form
$$0 \to G_3 \xrightarrow{\psi_3} G_2 \xrightarrow{\psi_2} G_1 \to 0.$$

Stated as a definition, it may appear that a short exact sequence is a rather obscure notion. However, it appears naturally in many situations.

Example 9.21 Consider a cubical pair (X, A) and for each k the following sequence:
$$0 \to C_k(A) \xrightarrow{\iota_k} C_k(X) \xrightarrow{\pi_k} C_k(X, A) \to 0, \tag{9.8}$$

where ι_k is the inclusion map and π_k is the quotient map. That this is a short exact sequence follows from simple applications of the previous lemmas. To begin with, ι_k is a monomorphism because it is an inclusion. Therefore, by Lemma 9.18,
$$0 \to C_k(A) \xrightarrow{\iota_k} C_k(X)$$
is exact. Similarly, by definition of relative chains, π_k is an epimorphism. Hence, Lemma 9.17 implies that
$$C_k(X) \xrightarrow{\pi_k} C_k(X, A) \to 0$$
is exact. So all that remains is to show that the sequence is exact at $C_k(X)$.

By definition, the kernel of π_k is $C_k(A)$. Moreover, since ι_k is an inclusion, $\operatorname{im} \iota_k = C_k(A)$, which means that $\operatorname{im} \iota_k = \ker \pi_k$.

The short exact sequence (9.8) is called the *short exact sequence of a pair*.

The short exact sequence (9.8) has a nice feature, defined below, which distinguishes it from the short exact sequence in Example 9.16.

Definition 9.22 A short exact sequence
$$0 \to G_3 \xrightarrow{\psi_3} G_2 \xrightarrow{\psi_2} G_1 \to 0 \tag{9.9}$$
splits if there exists a subgroup $H \subset G_2$ such that
$$G_2 = \operatorname{im} \psi_3 \oplus H.$$

Note that the above equation is also equivalent to $G_2 = \ker \psi_2 \oplus H$.

Example 9.23 The short exact sequence of a pair splits because
$$C_k(X) = C_k(A) \oplus H = \iota(C_k(A)) \oplus H,$$
where
$$H = \bigoplus_{Q \in \mathcal{K}_k(X) \setminus \mathcal{K}_k(A)} \mathbf{Z}\widehat{Q}.$$

The short exact sequence in Example 9.16 does not split, because the middle group $\mathbf{Z}$ is cyclic so it cannot be decomposed as a direct sum of two nontrivial subgroups.

The existence of a complementing subgroup H does not always seem evident. There are characterizations of splitting in terms of the left inverse of ψ_3 and the right inverse of ψ_2 which we leave as exercises.

Exercises

9.5 Prove Lemma 9.19.

9.6 Let X be an acyclic cubical set. The associated chain complex induces a sequence of maps between the chains

$$\ldots \to C_{k+1}(X) \xrightarrow{\partial_{k+1}} C_k(X) \xrightarrow{\partial_k} C_{k-1}(X) \to \ldots$$

Prove that this sequence is exact except at $C_0(X)$. Observe from this that homology measures how far a sequence induced by a chain complex is from being exact.

9.7 Consider the short exact sequence in (9.9). Show that the following conditions are equivalent.

(a) The sequence splits.
(b) ψ_3 has a left inverse, that is, there exists $\pi : G_2 \to G_3$ such that $\pi\psi_3 = \mathrm{id}_{G_3}$.
(c) ψ_2 has a right inverse, that is, there exists $\iota : G_1 \to G_2$ such that $\psi_3\iota = \mathrm{id}_{G_1}$.

9.3 The Connecting Homomorphism

In the previous section we have defined the notion of an exact sequence and proved some simple lemmas. In this section we shall prove the fundamental theorem of homological algebra. As a corollary we will answer the motivating question of how the relative homology groups are related to the homology groups of each space in the pair.

Definition 9.24 Let $\mathcal{A} = \{A_k, \partial_k^{\mathcal{A}}\}$, $\mathcal{B} = \{B_k, \partial_k^{\mathcal{B}}\}$, and $\mathcal{C} = \{C_k, \partial_k^{\mathcal{C}}\}$ be chain complexes. Let 0 denote the trivial chain complex, that is the chain complex in which each group is the trivial group. Let $\varphi : \mathcal{A} \to \mathcal{B}$ and $\psi : \mathcal{B} \to \mathcal{C}$ be chain maps. The sequence

$$0 \to \mathcal{A} \xrightarrow{\varphi} \mathcal{B} \xrightarrow{\psi} \mathcal{C} \to 0$$

is a *short exact sequence of chain complexes* if for every k

$$0 \to A_k \xrightarrow{\varphi_k} B_k \xrightarrow{\psi_k} C_k \to 0$$

is a short exact sequence.

Theorem 9.25 *("Zig-zag lemma") Let*

$$0 \to \mathcal{A} \xrightarrow{\varphi} \mathcal{B} \xrightarrow{\psi} \mathcal{C} \to 0$$

be a short exact sequence of chain complexes. Then for each k there exists a homomorphism

$$\partial_* : H_{k+1}(\mathcal{C}) \to H_k(\mathcal{A})$$

such that

$$\cdots \to H_{k+1}(\mathcal{A}) \xrightarrow{\varphi_*} H_{k+1}(\mathcal{B}) \xrightarrow{\psi_*} H_{k+1}(\mathcal{C}) \xrightarrow{\partial_*} H_k(\mathcal{A}) \to \cdots$$

is a long exact sequence.

The map ∂_* is called *connecting homomorphism*.

Proof. Before we begin the proof, observe that this is a purely algebraic result. Thus the only information that we have is that $\mathcal{A}$, $\mathcal{B}$, and $\mathcal{C}$ are chain complexes and that they are related by short exact sequences. The proof, therefore, can only consist of seeing how elements of the different groups are mapped around. To keep track of this it is useful to make use of the following commutative diagram.

$$\begin{array}{ccccccccc}
& & \downarrow \partial^{\mathcal{A}}_{k+2} & & \downarrow \partial^{\mathcal{B}}_{k+2} & & \downarrow \partial^{\mathcal{C}}_{k+2} & & \\
0 & \longrightarrow & A_{k+1} & \xrightarrow{\varphi_{k+1}} & B_{k+1} & \xrightarrow{\psi_{k+1}} & C_{k+1} & \longrightarrow & 0 \\
& & \downarrow \partial^{\mathcal{A}}_{k+1} & & \downarrow \partial^{\mathcal{B}}_{k+1} & & \downarrow \partial^{\mathcal{C}}_{k+1} & & \\
0 & \longrightarrow & A_k & \xrightarrow{\varphi_k} & B_k & \xrightarrow{\psi_k} & C_k & \longrightarrow & 0 \\
& & \downarrow \partial^{\mathcal{A}}_k & & \downarrow \partial^{\mathcal{B}}_k & & \downarrow \partial^{\mathcal{C}}_k & & \\
0 & \longrightarrow & A_{k-1} & \xrightarrow{\varphi_{k-1}} & B_{k-1} & \xrightarrow{\psi_{k-1}} & C_{k-1} & \longrightarrow & 0 \\
& & \downarrow \partial^{\mathcal{A}}_{k-1} & & \downarrow \partial^{\mathcal{B}}_{k-1} & & \downarrow \partial^{\mathcal{C}}_{k-1} & &
\end{array}$$

Step 1: The construction of ∂_*. The first step is to define the map

$$\partial_* : H_{k+1}(\mathcal{C}) \to H_k(\mathcal{A}).$$

Let $\gamma \in H_{k+1}(\mathcal{C})$. This means that there is some cycle $c \in C_{k+1}$ such that $\gamma = [c]$. Somehow, we need to map c to a cycle in A_k. Looking back at the commutative diagram there are two directions (down or left) we might try to go. The first is down, but $\partial^{\mathcal{C}}_{k+1} c = 0$ since c is a cycle. So the other option is to go left. Unfortunately, the map ψ_{k+1} goes in the wrong direction. However, by assumption,

$$0 \longrightarrow A_{k+1} \xrightarrow{\varphi_{k+1}} B_{k+1} \xrightarrow{\psi_{k+1}} C_{k+1} \longrightarrow 0$$

is exact. Thus $C_{k+1} = \operatorname{im} \psi_{k+1}$. Therefore, there exists $b \in B_{k+1}$ such that $\psi_{k+1}(b) = c$. In terms of the diagram we have the following:

$$\begin{array}{ccc}
b & & c \\
B_{k+1} & \xrightarrow{\psi_{k+1}} & C_{k+1} \\
& & \downarrow \partial^{\mathcal{C}}_{k+1} \\
& & C_k \\
& & 0
\end{array}$$

Of course, if $\ker \psi_{k+1} \neq 0$, then b is not uniquely defined. Let us ignore the nonuniqueness for the moment and continue our attempts to get to A_k. Given that $b \in B_{k+1}$, there are, again, two directions (left or down) we might try to go to get to A_k. Let's try to go down first, so we get $\partial_{k+1}^{\mathcal{B}} b \in B_k$. In order to get to A_k, we need to ask whether $\partial_{k+1}^{\mathcal{B}} b \in \operatorname{im} \varphi_k$. Recall that

$$0 \longrightarrow A_k \xrightarrow{\varphi_k} B_k \xrightarrow{\psi_k} C_k \longrightarrow 0$$

is exact, so $\operatorname{im} \varphi_k = \ker \psi_k$. Therefore, $\partial_{k+1}^{\mathcal{B}} b \in \operatorname{im} \varphi_k$ if and only if $\partial_{k+1}^{\mathcal{B}} b \in \ker \psi_k$. Now recall that the big diagram commutes, so in particular the following subdiagram commutes.

$$\begin{array}{ccc}
b & & c \\
B_{k+1} & \xrightarrow{\psi_{k+1}} & C_{k+1} \\
\downarrow \partial_{k+1}^{\mathcal{B}} & & \downarrow \partial_{k+1}^{\mathcal{C}} \\
B_k & \xrightarrow{\psi_k} & C_k \\
\partial b & & 0
\end{array}$$

But this means that

$$\psi_k \partial_{k+1}^{\mathcal{B}} b = \partial_{k+1}^{\mathcal{C}} \psi_{k+1} b = \partial_{k+1}^{\mathcal{C}} c = 0.$$

Thus $\partial_{k+1}^{\mathcal{B}} b \in \ker \psi_k = \operatorname{im} \varphi_k$. Therefore, there exists an $a \in A_k$ such that $\varphi_k a = \partial_{k+1}^{\mathcal{B}} b$. Moreover, this a is unique because the sequence is exact and hence φ_k is a monomorphism.

We have made it to A_k, but we have to verify that a is a cycle. Again we call on the commutativity of the big diagram to conclude that the following subdiagram commutes

$$\begin{array}{ccccc}
& & & & b \\
& & & & B_{k+1} \\
& & a & & \downarrow \partial_{k+1}^{\mathcal{B}} \\
0 & \longrightarrow & A_k & \xrightarrow{\varphi_k} & B_k \quad \partial b \\
& & \downarrow \partial_k^{\mathcal{A}} & & \downarrow \partial_k^{\mathcal{B}} \\
0 & \longrightarrow & A_{k-1} & \xrightarrow{\varphi_{k-1}} & B_{k-1} \\
& & \partial a & & 0
\end{array}$$

and hence we have

$$\varphi_{k-1} \partial_k^{\mathcal{A}} a = \partial_k^{\mathcal{B}} \varphi_k a = \partial_k^{\mathcal{B}} \partial_{k+1}^{\mathcal{B}} b = 0.$$

Again by exactness, φ_{k-1} is a monomorphism and, therefore, $\partial_k^{\mathcal{A}} a = 0$.

In summary, starting with a cycle $c \in C_{k+1}$ we have produced a cycle $a \in A_k$, which is summarized in the following diagram.

9.3 The Connecting Homomorphism

$$\begin{array}{ccc} & b & c \\ & B_{k+1} \xrightarrow{\psi_{k+1}} & C_{k+1} \\ & \downarrow \partial^B_{k+1} & \\ A_k \xrightarrow{\varphi_k} & B_k & \\ a & \partial b & \end{array} \quad (9.10)$$

So define
$$\partial_*([c]) := [a]. \quad (9.11)$$

In order to know that this map is well defined, we must verify two things: First, as we mentioned before, b is not uniquely determined, so we have to check that the equivalence class $[a]$ is independent of the choice of b. Second, we must check that $[a]$ is independent of the choice of the representant c of $\gamma = [c]$. We will do the first verification while leaving the second one as an exercise. So let $\bar{b} \in B_{k+1}$ be such that $\psi_{k+1}\bar{b} = c$. Before going further, observe that
$$\psi_{k+1}(b - \bar{b}) = \psi_{k+1}b - \psi_{k+1}\bar{b} = c - c = 0.$$
Therefore,
$$b - \bar{b} \in \ker \psi_{k+1} = \operatorname{im} \varphi_{k+1},$$
and there exists $q \in A_{k+1}$ such that $\varphi_{k+1}q = b - \bar{b}$.

Moreover, let $\bar{a}$ be the unique element in A_k such that $\varphi_k \bar{a} = \partial^B_{k+1}\bar{b}$. If we want the definition of ∂_* to be independent of the choice of b, we have to prove that $[a] = [\bar{a}]$. Thus we need to show that $a - \bar{a}$ is a boundary element. Again, we go back to the commutative diagram:

$$\begin{array}{ccc} & q & b - \bar{b} \\ & A_{k+1} \xrightarrow{\varphi_{k+1}} & B_{k+1} \\ & \downarrow \partial^A_{k+1} & \downarrow \partial^B_{k+1} \\ 0 \longrightarrow & A_k \xrightarrow{\varphi_k} & B_k \\ & \partial q & \partial(b - \bar{b}) \end{array}$$

Hence,
$$\varphi_k \partial^A_{k+1} q = \partial^B_{k+1}(b - \bar{b}) = \varphi_k(a - \bar{a}).$$

But, by exactness φ_k is a monomorphism; therefore, $a - \bar{a} = \partial^A_{k+1} q \in \operatorname{im} \partial^A_{k+1}$. Finally, we leave it as an exercise to verify that our map ∂_* is a homomorphism.

We now turn to the task of showing that the homology sequence is exact.

Step 2: Exactness at $H_k(\mathcal{B})$. We need to show that $\operatorname{im} \varphi_* = \ker \psi_*$, and we will do it in two steps.

($\operatorname{im} \varphi_* \subset \ker \psi_*$) Let $\beta \in \operatorname{im} \varphi_*$. Then there exists $\alpha \in H_k(\mathcal{A})$ such that $\varphi_* \alpha = \beta$. On the chain level, there is a cycle $a \in A_k$ such that $[a] = \alpha$. By definition, $\beta = [\varphi_k a]$ and
$$\psi_* \beta := [\psi_k(\varphi_k a)].$$

By exactness, $\psi_k \varphi_k = 0$, and hence $\psi_* \beta = 0$.

($\ker \psi_* \subset \operatorname{im} \varphi_*$) Let $\beta \in \ker \psi_*$ and let b be a cycle in B_k such that $\beta = [b]$. By assumption, $0 = [\psi_k b]$, and hence $\psi_k b = \partial^C_{k+1} c$ for some $c \in C_{k+1}$. By exactness, ψ_{k+1} is an epimorphism, so there exists $b' \in B_{k+1}$ such that $\psi_{k+1} b' = c$. Viewing this in terms of the commutative diagram, we have

$$
\begin{array}{ccccc}
b' & & c & & \\
B_{k+1} & \xrightarrow{\psi_{k+1}} & C_{k+1} & \longrightarrow & 0 \\
\downarrow \partial^B_{k+1} & & \downarrow \partial^C_{k+1} & & \\
0 \longrightarrow A_k \xrightarrow{\varphi_k} B_k & \xrightarrow{\psi_k} & C_k & \longrightarrow & 0 \\
b & & \partial c & &
\end{array}
$$

though we cannot claim that $\partial^B_{k+1} b' = b$.

However, we have

$$\psi_k(b - \partial^B_{k+1} b') = \psi_k b - \psi_k(\partial^B_{k+1} b') = 0.$$

This implies that $b - \partial^B_{k+1} b' \in \ker \psi_k = \operatorname{im} \varphi_k$, by exactness. Therefore, there exists a unique (φ_k is a monomorphism) $a \in A_k$ such that $\varphi_k a = b - \partial^B_{k+1} b'$. Moving down in the commutative diagram, we have

$$
\begin{array}{ccccc}
a & b - \partial b' & \partial^C_{k+1} c & & \\
0 \longrightarrow A_k & \xrightarrow{\varphi_k} B_k & \xrightarrow{\psi_k} C_k & \longrightarrow & 0 \\
\downarrow \partial^A_k & \downarrow \partial^B_k & \downarrow \partial^C_k & & \\
0 \longrightarrow A_{k-1} & \xrightarrow{\varphi_{k-1}} B_{k-1} & \xrightarrow{\psi_{k-1}} C_{k-1} & \longrightarrow & 0 \\
& 0 & & &
\end{array}
$$

Thus

$$\varphi_{k-1} \partial^A_k a = \partial^B_k (b - \partial^B_{k+1} b') = \partial^B_k b - 0 = 0,$$

because by assumption b is a cycle. Since φ_{k-1} is a monomorphism, $\partial^A_k a = 0$ and hence a is a cycle.

Finally,

$$\varphi_*[a] = [\varphi_k a] = [b - \partial^B_{k+1} b'] = [b].$$

Therefore, $\beta = [b] \in \operatorname{im} \varphi_*$.

Step 3: Exactness at $H_k(\mathcal{C})$. Again the proof is in two steps.

($\operatorname{im} \psi_* \subset \ker \partial_*$) Assume $\gamma = [c] \in \operatorname{im} \psi_*$ for some cycle $c \in C_k$. Then $\gamma = [\psi_k b]$, where b is a cycle in B_k. Let a cycle $a \in A_{k-1}$ be associated with c and b as in (9.10) and (9.11). We have $\varphi_{k-1} a = \partial^B_k b = 0$. Since φ_{k-1} is a monomorphism, $a = 0$. Thus

$$\partial_* \gamma = [a] = 0.$$

(ker $\partial_* \subset \operatorname{im}\psi_*$) Now assume that $\partial_*[c] = 0$ for some cycle $c \in C_k$. On the other hand, $\partial_*[c] = [a]$ for a cycle $a \in A_{k-1}$. Thus a is a boundary element, that is, there exists $\bar{a} \in A_k$ such that $\partial_k^{\mathcal{A}}\bar{a} = a$.

Expanding (9.10), one has

$$\begin{array}{ccccc} \bar{a} & & b & & \\ A_k & \xrightarrow{\varphi_k} & B_k & \xrightarrow{\psi_k} & C_k \\ \downarrow \partial_k^{\mathcal{A}} & & \downarrow \partial_k^{\mathcal{B}} & & \\ A_{k-1} & \xrightarrow{\varphi_{k-1}} & B_{k-1} & & \\ a & & \partial b & & \end{array}$$

Since the diagram commutes,

$$\partial_k^{\mathcal{B}} \varphi_k \bar{a} = \varphi_{k-1} \partial_k^{\mathcal{A}} a = \partial_k^{\mathcal{B}} b.$$

This implies that

$$\partial_k^{\mathcal{B}}(b - \varphi_k \bar{a}) = \partial_k^{\mathcal{B}} b - \partial_k^{\mathcal{B}} \varphi_k \bar{a} = 0.$$

Thus $b - \varphi_k \bar{a}$ is a cycle. Furthermore,

$$\psi_*([b - \varphi_k \bar{a}]) = [\psi_k(b - \varphi_k \bar{a})] = [\psi_k b - \psi_k \varphi_k \bar{a}] = [\psi_k b] = [c].$$

Therefore, $[c] \in \operatorname{im}\psi_*$.

Step 4: Exactness at $H_{k-1}(\mathcal{A})$. (im $\partial_* \subset \ker \varphi_*$) Let a be a cycle in A_{k-1} such that $\alpha = [a] \in \operatorname{im}\partial_*$. Then $\varphi_{k-1} a = \partial_k^{\mathcal{B}} b$ for some $b \in B_k$. Hence,

$$\varphi_* \alpha = [\varphi_{k-1} a] = [\partial_k^{\mathcal{B}} b] = 0.$$

(ker $\varphi_* \subset \operatorname{im}\partial_*$) Let $\alpha = [a] \in \ker \varphi_*$. Then $\varphi_{k-1} a = \partial_k^{\mathcal{B}} b$ for some $b \in B_k$. Let $c = \psi_k b$. By exactness on the chain level,

$$\partial_k^{\mathcal{C}} c = \partial_k^{\mathcal{C}} \psi_k b = \psi_{k-1} \partial_k^{\mathcal{B}} b = \psi_{k-1} \varphi_{k-1} a = 0,$$

so c is a cycle, hence, $\partial_*[c]$ is defined. Observe that $\partial_*[c] = [a]$. Thus $\alpha \in \operatorname{im}\partial_*$. □

While the proof may have seemed endless, this theorem is very powerful. For example, in the case of cubical pairs, we have the following corollary.

Corollary 9.26 (The exact homology sequence of a pair) *Let (X, A) be a cubical pair. Then there is a long exact sequence*

$$\ldots \to H_{k+1}(A) \xrightarrow{\iota_*} H_{k+1}(X) \xrightarrow{\pi_*} H_{k+1}(X, A) \xrightarrow{\partial_*} H_k(A) \to \ldots,$$

where $\iota : \mathcal{C}(A) \hookrightarrow \mathcal{C}(X)$ is the inclusion map and $\pi : \mathcal{C}(X) \to \mathcal{C}(X, A)$ is the quotient map.

Proof. The conclusion follows from Example 9.21 and Theorem 9.25. □

Example 9.27 To see how this result can be used, consider $X = [-4, 4] \times [-4, 4]$ and $A = [-4, -2] \times [-4, 4] \cup [2, 4] \times [-4, 4]$. We want to compute $H_*(X, A)$. The long exact sequence for this pair is

$$\cdots \to H_2(A) \xrightarrow{\iota_*} H_2(X) \xrightarrow{\pi_*} H_2(X, A) \xrightarrow{\partial_*}$$
$$H_1(A) \xrightarrow{\iota_*} H_1(X) \xrightarrow{\pi_*} H_1(X, A) \xrightarrow{\partial_*} \quad (9.12)$$
$$H_0(A) \xrightarrow{\iota_*} H_0(X) \xrightarrow{\pi_*} H_0(X, A) \xrightarrow{\partial_*} 0.$$

X is a rectangle and hence acyclic. A is the disjoint union of two rectangles and therefore the disjoint union of two acyclic sets. Thus $H_0(X) = \mathbf{Z}\eta$ and $H_0(A) = \mathbf{Z}\zeta_1 \oplus \mathbf{Z}\zeta_2$, where η, ζ_1, and ζ_2 are homology classes of arbitrarily chosen elementary vertices, respectively, in X, and in each component of A.

Furthermore, Proposition 9.4 implies that $H_0(X, A) = 0$. So we can rewrite (9.12) as

$$\cdots \to \quad 0 \quad \xrightarrow{\iota_*} 0 \xrightarrow{\pi_*} H_2(X, A) \xrightarrow{\partial_*}$$
$$0 \quad \xrightarrow{\iota_*} 0 \xrightarrow{\pi_*} H_1(X, A) \xrightarrow{\partial_*}$$
$$\mathbf{Z}\zeta_1 \oplus \mathbf{Z}\zeta_2 \xrightarrow{\iota_*} \mathbf{Z}\eta \xrightarrow{\pi_*} \quad 0 \quad \xrightarrow{\partial_*} 0.$$

By exactness, $H_k(X, A) = 0$ for $k \geq 2$. Since ι is the inclusion map, and all vertices of A are homologous in X, ι_* sends both ζ_1, and ζ_2 to η. Hence $\iota_* : \mathbf{Z}\zeta_1 \oplus \mathbf{Z}\zeta_2 \to \mathbf{Z}\eta$ is an epimorphism and $\ker \iota_* = \mathbf{Z}(\zeta_2 - \zeta_1) \cong \mathbf{Z}$. Again by exactness, $\partial_* : H_1(X, A) \to \mathbf{Z}\zeta_1 \oplus \mathbf{Z}\zeta_2$ is a monomorphism and $\operatorname{im} \partial_* = \ker \iota_* \cong \mathbf{Z}$. Therefore, $H_1(X, A) \cong \mathbf{Z}$.

Relative homology provides a necessary criterion for A to be a deformation retract of X.

Proposition 9.28 *If (X, A) is a cubical pair and A is a deformation retract of X, then $H_*(X, A) = 0$.*

Proof. Consider the following portion of the long exact sequence of a pair:

$$\cdots \to H_k(A) \xrightarrow{\iota_*} H_k(X) \xrightarrow{\pi_*} H_k(X, A) \xrightarrow{\partial_*} H_{k-1}(A) \xrightarrow{\iota_*} H_{k-1}(X) \to \cdots.$$

By Theorem 6.65, A and X have the same homotopy type. Thus by Corollary 6.69, $H_*(A) \cong H_*(X)$. Furthermore, by the proof of Theorem 6.65, the inclusion map $\iota_* : H_*(A) \to H_*(X)$ is an isomorphism.

Returning to the exact sequence, since ι_* is an isomorphism, $\pi_*(H_k(X)) = 0$ and hence ∂_* is a monomorphism. However, since ι_* is an isomorphism, $\partial_*(H_k(X, A)) = 0$ and thus $H_k(X, A) = 0$. □

Exercises

9.8 Let
$$0 \to \mathcal{A} \xrightarrow{\varphi} \mathcal{B} \xrightarrow{\psi} \mathcal{C} \to 0$$
be a short exact sequence of chain complexes. Show that if any two of them have trivial homology, then so has the third one.

9.9 Complete the proof of Theorem 9.25 by showing that

(a) the construction of $\partial_* \gamma$ is independent of the choice of a cycle c such that $[c] = \gamma$;
(b) the map ∂_* is a homomorphism.

9.10 Let $A \subset X$ be cubical sets. Let $Q \in \mathcal{K}(X) \setminus \mathcal{K}(A)$ be a free face in X. Furthermore, assume $Q \prec P \in \mathcal{K}(X) \setminus \mathcal{K}(A)$. Let X' be the cubical set obtained from X via an elementary collapse of P by Q. Prove that

$$H_*(X, A) \cong H_*(X', A).$$

9.11 Use Exercise 9.10 and excision to compute $H_*(X, A)$, where X and A consist of the shaded and darkly shaded squares.

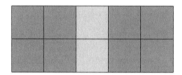

9.12 Let (X, A) be a cubical pair of acyclic sets. Show that $H_*(X, A) = 0$.

9.4 Mayer–Vietoris Sequence

Recall that Theorem 2.78 shows that two acyclic sets whose intersection is acyclic can be combined to form a larger acyclic set. In this section we want to generalize this result to show how the homology of a set can be computed from subsets.

Theorem 9.29 (Mayer–Vietoris) *Let X be a cubical space. Let A_0 and A_1 be cubical subsets of X such that $X = A_0 \cup A_1$ and let $B = A_0 \cap A_1$. Then there is a long exact sequence*

$$\ldots \to H_k(B) \to H_k(A_0) \oplus H_k(A_1) \to H_k(X) \to H_{k-1}(B) \to \ldots.$$

Before we start the proof, we need the following lemma on direct sums of abstract chain complexes, which is of interest by itself.

Lemma 9.30 Let $(\mathcal{C}, \partial)$ and $(\mathcal{C}', \partial')$ be two chain complexes. Define the structure $(\mathcal{C} \oplus \mathcal{C}', \partial \oplus \partial')$ as follows: $\mathcal{C} \oplus \mathcal{C}'$ is a sequence of groups $C_k \oplus C'_k$ and $\partial \oplus \partial'$ is a sequence of homomorphisms $\partial \oplus \partial' : C_k \oplus C'_k \to C_{k-1} \oplus C'_{k-1}$ defined by

$$(\partial \oplus \partial')(c, c') := (\partial c, \partial' c').$$

Then $(\mathcal{C} \oplus \mathcal{C}', \partial \oplus \partial')$ is a chain complex and

$$H_k(\mathcal{C} \oplus \mathcal{C}') \cong H_k(\mathcal{C}) \oplus H_k(\mathcal{C}').$$

Proof. For the simplicity of presentation we identify here the groups C_k and C'_k with the subgroups $C_k \times 0$ and, respectively, $0 \times C'_k$ of the direct sum $C_k \oplus C'_k$, defined via the product $C_k \times C'_k$ (see Chapter 13). This will make the proof a little sketchy, so some details are left as an exercise.

The verification that $\partial \oplus \partial'$ is a boundary map is obvious. Hence $\mathcal{C} \oplus \mathcal{C}'$ is a chain complex. It is also easy to verify that

$$\ker(\partial_k \oplus \partial'_k) = \ker \partial_k \oplus \ker \partial'_k$$

and

$$\mathrm{im}\,(\partial_k \oplus \partial'_k) = \mathrm{im}\, \partial_k \oplus \mathrm{im}\, \partial'_k.$$

From this it is easy to verify that

$$H_k(\mathcal{C} \oplus \mathcal{C}') = \frac{\ker \partial_k \oplus \ker \partial'_k}{\mathrm{im}\, \partial_k \oplus \mathrm{im}\, \partial'_k} \cong \frac{\ker \partial_k}{\mathrm{im}\, \partial_k} \oplus \frac{\ker \partial'_k}{\mathrm{im}\, \partial'_k} = H_k(\mathcal{C}) \oplus H_k(\mathcal{C}'). \quad \square$$

Proof of Theorem 9.29. The proof consists of finding an appropriate short exact sequence of chain complexes and then applying Theorem 9.25. Lemma 9.30 lets us write the short exact sequence we are looking for as

$$0 \to C_k(B) \xrightarrow{\varphi_k} C_k(A_0) \oplus C_k(A_1) \xrightarrow{\psi_k} C_k(X) \to 0 \qquad (9.13)$$

for all k, where the maps φ_k and ψ_k remain to be defined. Clearly, $B \subset A_0$ and $B \subset A_1$; therefore, $\mathcal{C}(B) \subset \mathcal{C}(A_0)$ and $\mathcal{C}(B) \subset \mathcal{C}(A_1)$. For any chain $c \in \mathcal{C}(B)$, define

$$\varphi_k c := (c, -c).$$

Similarly, $C_k(A_0) \subset C_k(X)$ and $C_k(A_1) \subset C_k(X)$, so we may define

$$\psi_k(d, e) := d + e.$$

The verification that φ and ψ are chain maps is obvious. We need to show that the sequence (9.13) is exact.

We begin by checking exactness at $C_k(B)$. This is equivalent to showing that φ_k is a monomorphism. But $\varphi_k c = (c, -c) = 0$ implies that $c = 0$.

Exactness at $C_k(A_0) \oplus C_k(A_1)$ is slightly more complicated. Observe that

$$\psi_k \varphi_k(c) = \psi_k(c, -c) = c + (-c) = 0.$$

This implies that $\operatorname{im}\varphi_k \subset \ker\psi_k$. To prove the other inclusion, assume that

$$0 = \psi_k(d, e) = d + e.$$

Thus $d = -e$. In particular, $d \in C_k(A_0) \cap C_k(A_1) = C_k(B)$. Then $\varphi_k(d)$ is well defined and $\varphi_k(d) = (d, -d) = (d, e)$, so $(d, e) \in \operatorname{im}\varphi_k$.

Finally, it needs to be checked that (9.13) is exact at $C_k(X)$. Let $c \in C_k(X)$. Then by Proposition 2.77, $c = c_0 + c_1$ for some $c_0 \in C_k(A_0)$ and $c_1 \in C_k(A_1)$. But

$$\psi_k(c_0, c_1) = c,$$

and, therefore, $c \in \operatorname{im}\psi_k$.

Since (9.13) is an exact sequence, the long exact sequence exists by Theorem 9.25. □

The Mayer–Vietoris theorem provides us with an excellent tool to compute homology groups of interesting spaces.

Recall that in examples and exercises of Chapter 2 a fair amount of effort is spent studying the set Γ^d, the boundary of the unit cube $[0,1]^{d+1}$, for $d=1$ and $d=2$. The direct computation of $H_k(\Gamma^d)$ increases in complexity as d increases. For $d=0$, the result is obvious because then $\Gamma^0 = \{0,1\}$ is of dimension 0, so the only nontrivial homology group is

$$H_0(\Gamma^0) = C_0(\Gamma^0) = \mathbf{Z}\widehat{[0]} \oplus \mathbf{Z}\widehat{[1]} \cong \mathbf{Z} \oplus \mathbf{Z}.$$

When $d=1$, we were able to compute the homology by hand in Example 2.46, but already for $d=2$ the direct computation seemed endless and we were tempted to apply the Homology program. Getting the formula for an arbitrary n seemed beyond our technical abilities. We can now derive it from the Mayer–Vietoris sequence.

Theorem 9.31 *Consider* $\Gamma^d = \operatorname{bd}[0,1]^{d+1}$, *where* $d \geq 1$. *Then*

$$H_k(\Gamma^d) \cong \begin{cases} \mathbf{Z} & \text{if } k=0 \text{ or } k=d, \\ 0 & \text{otherwise.} \end{cases}$$

Proof. Since Γ^d is connected for $d \geq 1$, we know that $H_0(\Gamma^d) \cong \mathbf{Z}$. The formula for $H_k(\Gamma^d)$ will be derived for $k > 0$ by induction on d. By Example 2.46, we know the result is true for $d=1$.

So assume that $d \geq 2$ and that the result it is true for Γ^k, $k=1,\ldots,d-1$. We will show it is true for Γ^d. Consider the $(d-1)$-dimensional face

$$P := [0] \times [0,1]^{d-1}$$

of Γ^d and let

$$X' := \Gamma^d \setminus \overset{\circ}{P}$$

be the union of complementary faces. By definition, $P \cup X = \Gamma^d$. By Theorem 2.76, P is acyclic. Note that X' is precisely the image of $Q = [0,1]^{d+1}$

under the elementary collapse of Q by P, so by Theorem 2.68 it is acyclic. Furthermore, the cubical set
$$Y := P \cap X' = [0] \times \Gamma^{d-1}$$
is homeomorphic to Γ^{d-1}, so we know its homology by the induction hypothesis.

We now use the Mayer–Vietoris sequence, with $X = \Gamma^d$, $A_0 = P$, $A_1 = X'$, and $B = Y$, to compute $H_k(\Gamma^d)$.

The case $k = 1$ is actually the hardest one so we start from the highest dimension $k = d$. The portion of the sequence of our interest is
$$\to H_d(P) \oplus H_d(X') \to H_d(\Gamma^d) \to H_{d-1}(Y) \to H_{d-1}(P) \oplus H_{d-1}(X') \to .$$
Since P and X' are acyclic and $d \geq 2$, this reduces to
$$\to 0 \oplus 0 \to H_d(\Gamma^d) \to H_{d-1}([0] \times \Gamma^{d-1}) \to 0 \oplus 0 \to 0 \to .$$
By exactness and by the induction hypothesis,
$$H_d(\Gamma^d) \cong H_{d-1}([0] \times \Gamma^{d-1}) \cong H_{d-1}(\Gamma^{d-1}) \cong \mathbf{Z}.$$
We leave it to the reader to check that $H_k(\Gamma^d) = 0$ for $1 \leq k < d$ and for $k > d$.

If $k = 1$, then the portion of the sequence that interests us is
$$\to H_1(P) \oplus H_1(X') \to H_1(\Gamma^d) \to H_0(Y) \to H_0(P) \oplus H_0(X') \to H_0(\Gamma^d) \to 0.$$
Again by the acyclicity, $H_1(P) = H_1(X') = 0$, so by exactness the map
$$H_1(\Gamma^d) \to H_0(Y)$$
is a monomorphism. On the other hand, the map
$$H_0(Y) \to H_0(P) \oplus H_0(X')$$
is induced by the map
$$\varphi_0 : C_0(Y) \to C_0(P) \oplus C_0(X')$$
given by $\varphi_0(c) = (c, -c)$. Since Y, P, and X' are connected, the homology group of each is generated by the homology class of any chosen elementary vertex. So let us choose $V \in Y \cap P \cap X'$. Then
$$\varphi_*([\widehat{V}]) = ([\widehat{V}], -[\widehat{V}]),$$
which means that φ_{0*} is a monomorphism. By exactness,
$$H_1(\Gamma^d) \to H_0(Y)$$
is the zero map, but we proved that it is a monomorphism. This forces
$$H_1(\Gamma^d) = 0. \qquad \square$$

Exercises

9.13 Consider the d-dimensional unit sphere

$$S_0^d := \{x \in \mathbf{R}^{d+1} \mid ||x||_0 = 1\},$$

where $||x||_0$ is the supremum norm (see Chapter 12). Note that S_0^d is a cubical set. By using the homeomorphism in Exercise 12.18 and Corollary 6.69, one can conclude that S_0^d has the same homology groups as Γ^d.

(a) Derive this result directly by applying the Mayer–Vietoris sequence to

$$X := S_0^d, \quad A_0 := \{x \in S_0^d \mid x_1 \leq 0\}, \quad A_1 := \{x \in S_0^d \mid x_1 \geq 0\},$$

and $B := A_0 \cap A_1 = [0] \times S_0^{d-1}$.

(b) Show that the reduced homology groups of spheres are

$$\tilde{H}_k(S_0^d) \cong \begin{cases} \mathbf{Z} & \text{if } k = d, \\ 0 & \text{otherwise.} \end{cases}$$

9.14 In Exercise 1.11 you are asked to make numerical experiments and give a conjecture about the homology of a graph G^n with n vertices where every pair of distinct vertices is connected by one edge. Now you are in position to prove your conjecture. More precisely, use the Mayer–Vietoris sequence to derive a formula for the dimension of $H_1(G^n)$ by induction on n.

9.15 Let $A \subset X$ and $B \subset Y$ be cubical sets. Prove the existence of the *relative Mayer–Vietoris sequence*

$$\ldots H_k(X \cap Y, A \cap B) \to H_k(X, A) \oplus H_k(Y, B) \to H_k(X \cup Y, A \cup B) \to \ldots$$

9.16 Let $A \subset X$ and $Q \in \mathcal{K}(X)$. Assume $Q \cap A$ is acyclic. Using Exercises 9.12 and 9.15, show that

$$H_*(X, A) \cong H_*(X, A \cup Q).$$

9.17 Use Exercise 9.16 and excision to compute $H_*(X, A)$, where $X = [-n, n]$ and $A = \{\pm n\}$ for any positive integer n.

9.5 Weak Boundaries

Recall that the algorithms computing homology of chain complexes and of cubical sets in Chapter 3 make use of Algorithm 3.58 Quotient group finder. The output of that algorithm consists of two matrices and a number

$$(U, B, s)$$

described in Theorem 3.59. In the context of computing homology of a chain complex, when the groups G and H discussed in Theorem 3.59 are $G := Z_k$ and $H := B_k$, the meaning of this output is the following:

1. The matrix B is the matrix of the inclusion map $B_k \hookrightarrow Z_k$ in the Smith normal form,
2. The columns of U are the canonical coordinates of the new basis
$$\{\mathbf{u}_1, \mathbf{u}_2, \ldots, \mathbf{u}_m\}$$
of Z_k.
3. The columns of UB are the canonical coordinates of the new basis
$$\{b_1\mathbf{u}_1, b_2\mathbf{u}_2, \ldots, b_n\mathbf{u}_n\}$$
of B_k, where $b_i := B[i,i]$ are the diagonal entries of B.
4. The number s tells us that $b_i = 1$ if and only if $i \leq s$.
5. The quotient group decomposition is

$$H_k(\mathcal{C}) = \bigoplus_{i=s+1}^{n} \langle [\mathbf{u}_i] \rangle \oplus \bigoplus_{i=n+1}^{m} \mathbf{Z}[\mathbf{u}_i]. \tag{9.14}$$

If we only care about the identification of $H_k(\mathcal{C})$ up to isomorphism, all necessary information is contained in the matrix B. Indeed,

$$H_k(\mathcal{C}) \cong \mathbf{Z}^r \oplus \bigoplus_{i=s+1}^{n} \mathbf{Z}_{b_i},$$

where $r := m - n$, b_i are the diagonal entries of B, $b_i > 1$ for $i = s+1, s+2, \ldots, n$, and $b_i | b_{i+1}$.

We shall push a little further the analysis of $H_k(\mathcal{C})$. The decomposition in (9.14) can be expressed as

$$H_k(\mathcal{C}) = T_k \oplus F_k,$$

where

$$T_k := \bigoplus_{i=s+1}^{n} \langle [\mathbf{u}_i] \rangle \cong \bigoplus_{i=s+1}^{n} \mathbf{Z}_{b_i} \tag{9.15}$$

is the torsion subgroup of $H_k(\mathcal{C})$ (see Exercise 9.18 and Definition 13.10), and the complement

$$F_k := \bigoplus_{i=n+1}^{m} \mathbf{Z}[\mathbf{u}_i]$$

is a maximal free subgroup of $H_k(\mathcal{C})$. Since T_k is the torsion subgroup of $H_k(\mathcal{C})$, it is independent of a choice of basis, but its complement F_k in the above decomposition is not (see Exercise 3.13). Notice that we have isomorphisms

$$H_k(\mathcal{C})/T_k \cong F_k \cong \bigoplus_{i=n+1}^{m} \mathbf{Z}\mathbf{u}_i \cong \mathbf{Z}^r. \tag{9.16}$$

On the other hand, if we only care about identification of our free component of the homology group, we can achieve the same goal by taking the quotient

$$Z_k/W_k,$$

where
$$W_k := \bigoplus_{i=1}^{n} \mathbf{Z}\mathbf{u}_i \subset Z_k. \tag{9.17}$$
Indeed, since W_k is generated by a subset of the basis $\{\mathbf{u}_i\}$ of Z_k, we get
$$Z_k/W_k \cong \bigoplus_{i=n+1}^{m} \mathbf{Z}\mathbf{u}_i \cong H_k(\mathcal{C})/T_k. \tag{9.18}$$

The group W_k is called the *group of weak boundaries* of C_k. This name is justified by the following definition and proposition.

Definition 9.32 A chain $c \in C_k$ is called a *weak boundary* if there exists a nonzero integer β such that $\beta c \in B_k$.

Proposition 9.33 *The group $W_k \subset Z_k$ given by Eq. (9.17) is the set of all weak boundaries of C_k. In particular, the definition of W_k given by Eq. (9.17) is independent of the choice of basis $\{\mathbf{u}_i\}$ of Z_k related to the Smith normal form B.*

Proof. If $c \in W_k$, then $c = \sum_{i=s}^{n} \alpha_i \mathbf{u}_i$, so $\beta c \in B_k$ for $\beta := b_s b_{s+1} \ldots b_n$, because $b_i \mathbf{u}_i \in B_k$. Hence c is a weak boundary.

Conversely, suppose that $c \in C_k$ is a weak boundary and let $\beta \neq 0$ be such that $\beta c \in B_k$. First, note that $c \in Z_k$. Indeed, $B_k \subset Z_k$, and hence $\beta \partial c = \partial(\beta c) = 0$. But this implies that $\partial c = 0$. Hence c can be expressed as a linear combination of the basis vectors $\{\mathbf{u}_1, \mathbf{u}_2, \ldots, \mathbf{u}_m\}$, $c = \sum_{i=1}^{m} \alpha_i \mathbf{u}_i$. Then
$$\sum_{i=1}^{m} \beta \alpha_i \mathbf{u}_i = \beta c \in B_k.$$
This implies that $\beta \alpha_i = 0$ for all $i > n$. Since $\beta \neq 0$, we get $\alpha_i = 0$ for all $i > n$ and
$$c = \sum_{i=1}^{n} \alpha_i \mathbf{u}_i \in W_k.$$
The last conclusion is obvious because the definition of a weak boundary is not related to any particular basis. □

Again a word of caution: All three groups in Eq. (9.18) are isomorphic, but, as pointed out before, the group in the middle is dependent on a choice of basis while the quotient groups on the left and right are independent.

Example 9.34 Consider the abstract chain complex studied in Example 3.63. The group $Z_0 = C_0$ is generated by all three columns $\{\mathbf{q}_1, \mathbf{q}_2, \mathbf{q}_3\}$ of the inverse change-of-coordinates transfer matrix Q while the group of weak boundaries W_0 is generated by $\{\mathbf{q}_1, \mathbf{q}_2\}$. Note that the second column $\mathbf{q}_2$ has the property $\mathbf{q}_2 \notin B_0$, but $2\mathbf{q}_2 \in B_0$. We get
$$H_0(\mathcal{C})/T_0 \cong Z_0/W_0 \cong \mathbf{Z}\mathbf{q}_3 \cong \mathbf{Z}.$$

The group $H_1(\mathcal{C})$ is free with $W_1 = B_1 = 0$, so its free component does not require any computation.

Exercise

9.18 Show that the group T_k defined by (9.15) is the torsion subgroup of $H_k(\mathcal{C})$.

9.6 Bibliographical Remarks

The contents of this chapter is an adaptation of the standard material on homological algebra to the needs of this book. For further reading see [36, 50, 69, 78].

10
Nonlinear Dynamics

This is perhaps the most challenging chapter of this book in that we attempt to show how homology can be used in nonlinear analysis and dynamical systems. As mentioned in the Preface, algebraic topology was developed to solve problems in these subjects. Thus, on one hand, some of the material (most notably Sections 10.4 and 10.5) is classical and has a rich history. On the other hand, the focus of Section 10.6, combining numerical analysis with homology via the computer to obtain mathematically rigorous results, is a cutting-edge topic. We assume that the background and interest of our readers are equally varied, ranging from an individual with no background in dynamics hoping to learn the mathematical theory to others whose primary goal is to understand how to apply these ideas to specific nonlinear systems.

In order to accommodate this variety, we have tried to make this chapter as modular as possible. For the reader with no background in dynamics, we recommend beginning with Sections 10.1 and 10.2, where, in as efficient a manner as possible, we introduce the essential ideas of the subject and indicate the potential complexity of nonlinear systems. Section 10.3 introduces the Ważewski principle, which is perhaps the most straightforward example of how nontrivial topology can be used to draw conclusions about dynamics. We recommend this section to all readers.

The most simply stated question (and for many applications the most important) about a continuous map $f : X \to X$ is whether it has a fixed point, that is a point $x \in X$ such that $f(x) = x$. This problem has received an enormous amount of attention. In Sections 10.4 and 10.5 we discuss two fundamental techniques, the Lefschetz fixed-point theorem and the degree theory.

Readers who are more interested in complex or chaotic dynamics can immediately turn to Sections 10.6 and 10.7 at the end of the chapter. The material of these two sections is written for the individual who wants to verify chaotic dynamics for specific problems.

10.1 Maps and Symbolic Dynamics

The simplest way to generate a dynamical system is to begin with a continuous map $f : X \to X$ defined on a metric space and iterate. In this setting X is referred to as the *phase space*. Given a point $x \in X$, its *forward orbit* is defined to be
$$\gamma^+(x, f) := \{f^n(x) \mid n = 0, 1, 2, \ldots\} \subset X.$$
Observe that the forward orbit of a point defines a subset of X. Unfortunately, forward orbits need not have particularly nice properties with respect to f. For example, while obviously $f(\gamma^+(x, f)) \subset \gamma^+(x, f)$, in general $f(\gamma^+(x, f)) \neq \gamma^+(x, f)$. To rectify this we would also like to consider the *backward orbit* of x_0, but since we have not assumed that f is invertible, it makes no sense to write f^{-n}. The following definition circumvents this technicality.

Definition 10.1 Let $f : X \to X$ be a continuous map and let $x \in X$. A *full solution* through x under f is a function $\gamma_x : \mathbf{Z} \to X$ satisfying the following two properties:

1. $\gamma_x(0) = x$;
2. $\gamma_x(n+1) = f(\gamma_x(n))$ for all $n \in \mathbf{Z}$.

A simple warning: As the following two examples indicate, full solutions need not exist and even if they do, they need not be unique. However, if f is invertible, then there exists a unique solution $\gamma_x(n) = f^n(x)$ for all $n \in \mathbf{Z}$. As these comments indicate, the invertibility or lack thereof of f has ramifications on the dynamics; thus it is useful to have the following definitions. The dynamical system generated by a continuous function f is *invertible* if f is a homeomorphism; otherwise, it is *noninvertible*.

Example 10.2 Consider the map $f : \mathbf{R} \to \mathbf{R}$ given by $f(x) = x^2$. Observe that any negative number fails to have a preimage. Therefore, if $x < 0$, then there does not exist a full solution γ_x.

Example 10.3 Consider the continuous map $f : [0, 1] \to [0, 1]$ given by
$$f(x) = \begin{cases} 2x & \text{if } 0 \leq x \leq \frac{1}{2}, \\ 2 - 2x & \text{if } \frac{1}{2} \leq x \leq 1. \end{cases}$$
Observe that $f(0) = 0$. Therefore, there is an obvious full solution $\gamma_0 : \mathbf{Z} \to [0, 1]$ through 0 defined by $\gamma_0(n) = 0$. However, we can also define another full solution $\gamma'_0 : \mathbf{Z} \to [0, 1]$ by
$$\gamma'_0(n) = \begin{cases} 0 & \text{if } n \geq 0, \\ 2^{n+1} & \text{if } n < 0. \end{cases}$$

Let $\gamma_x : \mathbf{Z} \to X$ be a full solution. The associated *orbit* of x is defined to be the set
$$\gamma_x(\mathbf{Z}) := \{\gamma_x(n) \mid n \in \mathbf{Z}\}.$$
Observe that $f(\gamma_x(\mathbf{Z})) = \gamma_x(\mathbf{Z})$. Thus orbits are invariant with respect to the dynamics. The following definition generalizes this observation.

Definition 10.4 Let $f : X \to X$ be a continuous map. A set $S \subset X$ is an *invariant set* under f if
$$f(S) = S.$$

One of the goals of dynamical systems is to understand the existence and structure of invariant sets. In general, this is an extremely difficult problem, in part because of the tremendous variety of types of orbits and combinations of orbits that can occur. To emphasize this point we begin our discussion with a very particular type of dynamical system known as symbolic dynamics. We will employ these systems in two ways. In this section, we will use them to rapidly introduce a variety of concepts from dynamical systems. Then, at the end of this chapter, we will use them as models for other, potentially more complicated, systems.

We begin by defining the phase space of a symbolic dynamical system.

Definition 10.5 The *symbol space* on n symbols is given by
$$\Sigma_n = \{1, \ldots, n\}^{\mathbf{Z}}$$
$$= \{\mathbf{a} = (\ldots, a_{-1}, a_0, a_1, \ldots) \mid a_j \in \{1, \ldots, n\} \text{ for all } j \in \mathbf{Z}\}.$$

The metric on Σ_n is defined to be
$$\text{dist}(\mathbf{a}, \mathbf{b}) := \sum_{j=-\infty}^{\infty} \frac{\delta(a_j, b_j)}{4^{|j|}},$$
where
$$\delta(t, s) = \begin{cases} 0 & \text{if } t = s, \\ 1 & \text{if } t \neq s. \end{cases}$$

Before continuing our development of symbolic dynamics, there is a useful observation to be made. Notice that if $\mathbf{a}, \mathbf{b} \in \Sigma_n$ and $a_k \neq b_k$, then
$$\text{dist}(\mathbf{a}, \mathbf{b}) \geq 4^{-|k|}. \tag{10.1}$$
On the other hand, if $a_j = b_j$ for all $|j| \leq k$, then
$$\text{dist}(\mathbf{a}, \mathbf{b}) \leq 2 \sum_{j=k+1}^{\infty} \frac{1}{4^j}.$$
Summing up the geometric series, we get
$$\text{dist}(\mathbf{a}, \mathbf{b}) \leq \frac{2}{3} 4^{-k}. \tag{10.2}$$

The *shift map* on n symbols, $\sigma : \Sigma_n \to \Sigma_n$, is defined by
$$(\sigma(\mathbf{a}))_k = a_{k+1}. \tag{10.3}$$

Before we can treat this as a dynamical system, we need to check that the shift map is continuous.

Proposition 10.6 $\sigma: \Sigma_n \to \Sigma_n$ is a uniformly continuous function.

Proof. We need to show that given $\epsilon > 0$ there exists $\delta > 0$ such that if dist $(\mathbf{a}, \mathbf{b}) < \delta$, then dist $(\sigma(\mathbf{a}), \sigma(\mathbf{b})) < \epsilon$. So fix $\epsilon > 0$. Then there exists $k > 0$ such that $\frac{2}{3} 4^{-k} < \epsilon$. Now take $\mathbf{a}, \mathbf{b} \in \Sigma_n$ such that dist $(\mathbf{a}, \mathbf{b}) < \delta := 4^{-(k+1)}$. By (10.1) we know that this implies that $a_j = b_j$ for all $|j| \le k+1$. Now observe that $\sigma(\mathbf{a})_j = \sigma(\mathbf{b})_j$ for all $|j| \le k$. Then, by (10.2), dist $(\sigma(\mathbf{a}), \sigma(\mathbf{b})) < \epsilon$. □

It is easy to check that σ is a homeomorphism, thus it is most appropriately used as a model for invertible systems. To construct an interesting noninvertible dynamical system, let

$$\Sigma_n^+ := \{\mathbf{a} = (a_0, a_1, \ldots) \mid a_j \in \{1, \ldots, n\} \text{ for all } j \in \mathbf{Z}^+\}$$

with a metric

$$\text{dist}(\mathbf{a}, \mathbf{b}) := \sum_{j=0}^{\infty} \frac{\delta(a_j, b_j)}{4^{|j|}}.$$

Observe that we can still use (10.3) to define the associated shift dynamics.

The importance of symbolic dynamics is that it is extremely easy to identify individual orbits with particular properties. As such, we can use it to provide concrete examples for a variety of important definitions. In what follows, $f: X \to X$ will always denote a continuous map.

Definition 10.7 An element $x \in X$ is a *fixed point* of f if $f(x) = x$.

Example 10.8 Consider $\sigma: \Sigma_2 \to \Sigma_2$. There are exactly two fixed points:

$$(\ldots, 1, 1, 1, 1, 1, \ldots) \quad \text{and} \quad (\ldots, 2, 2, 2, 2, 2, \ldots).$$

The corresponding fixed points for $\sigma: \Sigma_2^+ \to \Sigma_2^+$ are

$$(1, 1, 1, 1, 1, \ldots) \quad \text{and} \quad (2, 2, 2, 2, 2, \ldots).$$

Definition 10.9 Let $y, z \in X$ be fixed points for f. We say that x is a *heteroclinic point* from y to z if there exists a full solution γ_x such that

$$\lim_{n \to \infty} f^n(x) = z \quad \text{and} \quad \lim_{n \to -\infty} \gamma_x(n) = y.$$

The associated orbit is called a *heteroclinic orbit*. The point x is *homoclinic* if $y = z$.

Example 10.10 If $\mathbf{a} \in \Sigma_3$ is defined by

$$a_j = \begin{cases} 2 & \text{if } j < 0, \\ 3 & \text{if } j = 0, \\ 1 & \text{if } j > 0, \end{cases}$$

then $\mathbf{a}$ is a heteroclinic point from $(\ldots, 2, 2, 2, \ldots)$ to $(\ldots, 1, 1, 1, \ldots)$. To verify this statement, we need to show that $\lim_{n \to \infty} \sigma^n(\mathbf{a}) = (\ldots, 1, 1, 1, \ldots)$. Observe that by (10.2)

$$\text{dist}\,(\sigma^k(\mathbf{a}), (\ldots, 1, 1, 1, \ldots)) \leq \frac{2}{3} 4^{k-1}.$$

Therefore, as k tends to infinity, $\sigma^k(\mathbf{a})$ tends to $(\ldots, 1, 1, 1, \ldots)$. Since σ is invertible, there is a unique full solution $\gamma_\mathbf{a}$ given by $\gamma_\mathbf{a}(n) = \sigma^n(\mathbf{a})$. It is left to the reader to check that $\lim_{n \to -\infty} \sigma^n(\mathbf{a}) = (\ldots, 2, 2, 2, \ldots)$.

Definition 10.11 An element $x \in X$ is a *periodic point of period p* if there exists a positive integer p such that

$$f^p(x) = x.$$

The number p is the *minimal period* if $f^n(x) \neq x$ for $0 < n < p$. Observe that a fixed point is a periodic point with period 1. Given x_0, a periodic point of period p, its forward orbit $\{f^n(x_0) \mid n = 0, 1, \ldots, p-1\}$ is a *periodic orbit*.

Example 10.12 If $\mathbf{a} = (a_i)$ is a sequence satisfying $a_i = a_{i+p}$ for some positive p and all $i \in \mathbf{Z}$, then $\sigma^p(\mathbf{a}) = \mathbf{a}$, that is, $\mathbf{a}$ is a periodic point of period p. Observe that

$$(\ldots, 1, 2, 1, 2, 1, 2, 1, 2, \ldots)$$

is a periodic point of period 4 but has minimal period 2.

Periodic points play an important role in symbolic dynamics and so we need to be able to express them using a simpler notation. Thus if $\mathbf{a} = (a_i)$ is a periodic point of period p, we will write

$$\mathbf{a} = (\overline{a_0, a_1, \ldots, a_{p-1}}).$$

Using this notation, we can write the periodic orbit associated to the periodic point of Example 10.12 as

$$\{(\overline{1,2}), (\overline{2,1})\}.$$

Just as periodic points generalize fixed points, we want to extend the notion of heteroclinic orbits to connecting orbits. However, to do this we need to generalize the notion of limits of orbits.

Definition 10.13 Let $x \in X$. The *omega limit set* of x under f is

$$\omega(x, f) := \bigcap_{m=0}^{\infty} \text{cl}\,\{f^j(x) \mid j = m, m+1, \ldots\},$$

and the *alpha limit set* of a full solution γ_x of x under f is

$$\alpha(\gamma_x, f) := \bigcap_{m=0}^{\infty} \text{cl}\,(\gamma_x((-\infty, -m))).$$

Proposition 10.14 *If X is a compact set and $f : X \to X$ is continuous, then for any $x \in X$, $\omega(x, f)$ is a nonempty, compact, invariant set.*

Proof. Given any integer $m \geq 0$, define

$$S_m := \mathrm{cl}\, \{f^j(x) \mid j = m, m+1, \ldots\}.$$

Since S_m is a closed subset of the compact set X, it is compact and, by definition, it is nonempty. Furthermore, $S_{m+1} \subset S_m$ for each m. By Exercise 12.33,

$$\omega(x, f) = \bigcap_{m=0}^{\infty} S_m$$

is a nonempty compact set.

We still need to show that $\omega(x, f)$ is invariant. To do this it is sufficient to show that

$$f(\omega(x, f)) \subset \omega(x, f) \quad \text{and} \quad \omega(x, f) \subset f(\omega(x, f)).$$

To prove the first inclusion, take $y \in \omega(x, f)$. Then there exists an increasing sequence of positive integers k_i such that $\lim_{i\to\infty} \mathrm{dist}\,(f^{k_i}(x), y) = 0$. Since f is continuous,

$$\lim_{i\to\infty} \mathrm{dist}\,(f^{k_i+1}(x), f(y)) = 0.$$

Thus $f(y) \in \omega(x, f)$ and $f(\omega(x, f)) \subset \omega(x, f)$.

To prove the other inclusion, take $y \in \omega(x, f)$. Then $y = \lim_{i\to\infty} f^{k_i}(x)$ for some increasing sequence of integers k_i. Without loss of generality, assume that $k_i \geq 1$ for all i and consider the sequence $(f^{k_i-1}(x))_{i=1,2,3,\ldots}$. Since X is compact, this contains a convergent subsequence $(f^{k_{i_l}-1}(x))_{l=1,2,3,\ldots}$. Let $z = \lim_{l\to\infty} f^{k_{i_l}-1}(x)$. Then $z \in \omega(x, f)$. Furthermore, because f is continuous,

$$f(z) = \lim_{l\to\infty} f(f^{k_{i_l}-1}(x)) = \lim_{l\to\infty} f^{k_{i_l}}(x) = y.$$

Thus $y \in f(\omega(x, f))$ and, more generally, $\omega(x, f) \subset f(\omega(x, f))$. □

As will be demonstrated shortly, even though the symbol spaces Σ_n and Σ_n^+ are complicated sets, they are compact. Rather than proving this result twice, we state the following lemma, which is left to the reader to prove (see Exercise 10.2).

Lemma 10.15 *The space Σ_n is homeomorphic to Σ_n^+.*

Proposition 10.16 *The spaces Σ_n and Σ_n^+ are compact.*

Proof. Lemma 10.15 implies that Σ_n is compact if and only if Σ_n^+ is compact. Thus it is sufficient to prove the result for Σ_n^+.

10.1 Maps and Symbolic Dynamics   313

By definition, Σ_n^+ is compact if every sequence in Σ_n^+ contains a convergent subsequence. Thus, given an arbitrary infinite sequence $(\mathbf{a}^0, \mathbf{a}^1, \mathbf{a}^2, \dots)$ of elements of Σ_n^+, we need to find a convergent subsequence. Let

$$\mathbf{S} := \{\mathbf{a}^k \in \Sigma_n^+ \mid k = 0, 1, 2, \dots\},$$

that is, $\mathbf{S}$ is the set of elements of the original sequence. Observe that if $\mathbf{S}$ consists of a finite set of elements, then there must be at least one element $\mathbf{b} \in \Sigma_n^+$ that appears infinitely often in the sequence $(\mathbf{a}^0, \mathbf{a}^1, \mathbf{a}^2, \dots)$. Choose the subsequence consisting of those terms $\mathbf{a}^k$ that are equal to $\mathbf{b}$. Since this is a constant sequence, it converges to $\mathbf{b} \in \Sigma_n^+$. Having eliminated that case, we may assume now that $\mathbf{S}$ is an infinite set and that each term of the sequence $(\mathbf{a}^0, \mathbf{a}^1, \mathbf{a}^2, \dots)$ is repeated at most finitely many times. By eliminating the repetitions, we obtain an infinite subsequence in which no term is repeated. Therefore, we may assume, without a loss of generality, that the terms of $(\mathbf{a}^0, \mathbf{a}^1, \mathbf{a}^2, \dots)$ are pairwise distinct.

For the remaining case we will construct an infinite collection of sets and positive integers which will be used to define a convergent subsequence.

Each element $\mathbf{a}^k \in \mathbf{S}$ takes the form

$$\mathbf{a}^k = (a_0^k, a_1^k, \dots).$$

By definition, for each $k \in \mathbb{Z}$, $a_0^k \in \{1, \dots, n\}$. In particular, there are only a finite number of choices for the a_0^k. Therefore, in the infinite sequence of integers a_0^k, at least one value must be repeated infinitely often. Let us call this value b_0.

Define

$$\mathbf{S}^{(0)} := \{\mathbf{a}^k \in \mathbf{S} \mid a_0^k = b_0\}.$$

Clearly, $\mathbf{S}^{(0)}$ is an infinite set. Let us also define $m(0) := \min\{k \mid a_0^k = b_0\}$.

With this construction as a model we will now inductively define sets $\mathbf{S}^{(p)}$, positive integers $m(p)$, and integers $b_p \in \{1, \dots, n\}$ that satisfy the following properties for every $p = 1, 2, \dots$:

1. $\mathbf{S}^{(p)} \subset \mathbf{S}^{(p-1)}$ and contains an infinite number of elements.
2. If $\mathbf{a}^k \in \mathbf{S}^{(p)}$, then $a_i^k = b_i$ for $i = 0, \dots, p$.
3. $m(p) \geq \min\{k \mid \mathbf{a}^k \in \mathbf{S}^{(p)}, a_p^k = b_p\}$. Furthermore, $m(p) > m(p-1)$.

The $m(p)$ will be used to select an appropriate subsequence and the b_i will be used to define the limit point of the subsequence.

Let us assume that the sets $\mathbf{S}^{(p)}$, positive integers $m(p)$, and integers $b_p \in \{1, \dots, n\}$ have been defined for $p = \{0, 1, \dots, q\}$. We will now construct $\mathbf{S}^{(q+1)}$, $m(q+1)$, and b_{q+1}.

The set $\mathbf{S}^{(q)}$ contains an infinite number of elements; however,

$$\{a_{q+1}^k \mid \mathbf{a}^k \in \mathbf{S}^{(q)}\} \subset \{1, \dots, n\},$$

which is finite. Thus we can choose an integer $b_{q+1} \in \{1, \dots, n\}$ such that set

$$\mathbf{S}^{(q+1)} := \left\{ \mathbf{a}^k \in \mathbf{S}^{(q)} \mid a_{q+1}^k = b_{q+1} \right\}$$

has an infinite number of elements. Observe that $\mathbf{S}^{(q+1)} \subset \mathbf{S}^{(q)}$. By construction and the induction hypothesis, if $\mathbf{a}^k \in \mathbf{S}^{(q+1)}$, then $a_i^k = b_i$ for $i = 0, \ldots, q+1$. Finally, define

$$m(q+1) := \min \left\{ k \mid k > m(q),\ \mathbf{a}^k \in \mathbf{S}^{(q+1)},\ a_{q+1}^k = b_{q+1} \right\}.$$

The reader can check that $m(q+1)$ satisfies the desired properties.

We are finally in the position to prove the existence of a convergent subsequence. Let $\mathbf{b} := (b_i) \in \Sigma_n^+$. Consider the sequence $(\mathbf{a}^{m(p)})_{p=0,1,2,\ldots}$. This is clearly a subsequence of the original sequence. We will now argue that this sequence converges to $\mathbf{b}$. Since $\mathbf{a}^{m(p)} \in \mathbf{S}^{m(p)}$, $a_i^{m(p)} = b_i$ for $i = 0, \ldots, p$. Thus by (10.2),

$$\operatorname{dist}(\mathbf{a}^{m(p)}, \mathbf{b}) < \frac{2}{3} 4^{-p+1},$$

which shows that $\lim_{p \to \infty} \mathbf{a}^{m(p)} = \mathbf{b}$. $\square$

Corollary 10.17 *If $\mathbf{a} \in \Sigma_n$, then $\alpha(\mathbf{a}, \sigma) \neq \emptyset$ and $\omega(\mathbf{a}, \sigma) \neq \emptyset$.*

Proof. By Proposition 10.16, Σ_n is compact. The conclusion for the omega limit set now follows from Proposition 10.14. Since σ is invertible, the same argument applies to alpha limit sets (see Exercise 10.3). $\square$

Clearly, the same result holds for omega limit sets in Σ_n^+.

Example 10.18 To see that alpha and omega limit sets are appropriate generalizations of the limits of a heteroclinic orbit, consider $\mathbf{a} \in \Sigma_3$ from Example 10.10. Let us check that

$$\omega(\mathbf{a}, \sigma) = (\bar{1}) \quad \text{and} \quad \alpha(\mathbf{a}, \sigma) = (\bar{2}).$$

Since $\lim_{k \to \infty} \sigma^k(\mathbf{a}) = (\bar{1})$, $(\bar{1}) \in \operatorname{cl}\{\sigma^j(x) \mid j = m, m+1, \ldots\}$ for all m. Thus $(\bar{1}) \in \omega(\mathbf{a}, \sigma)$. Let $\mathbf{b} = (b_i) \in \Sigma_n$ and assume $\mathbf{b} \neq (\bar{1})$. Then there exists q such that $b_q \neq 1$. Inequality (10.1) allows us to conclude that $\operatorname{dist}(\mathbf{b}, \sigma^j(\mathbf{a})) \geq 4^{|q|}$ for every $j > |q|$. In particular,

$$\mathbf{b} \notin \operatorname{cl}\{\sigma^j(\mathbf{a}) \mid j = |q|+1, |q|+2, \ldots\},$$

and hence $\mathbf{b} \notin \omega(\mathbf{a}, \sigma)$.

A similar argument shows that $\alpha(\mathbf{a}, \sigma) = (\bar{2})$.

Example 10.19 Alpha and omega limit sets can be quite complicated. Consider the point $\mathbf{a} = (a_i) \in \Sigma_2$ such that $\{a_i \mid i \geq 0\}$ defines the following sequence:

1 2 11 12 21 111 112 121 211 212 1111 1112 1121 1211 1212 2111 . . . .

In words, the sequence is built of all possible blocks of 1 and 2 that do not contain two consecutive 2s. The sequence begins with blocks of length 1, followed by all blocks of length 2, then blocks of length 3, and so on.

By Corollary 10.17, $\omega(\mathbf{a}, \sigma)$ exists, but we can provide no simple description of all the orbits it contains.

Up to this point we have focused on orbits and invariant sets with a given structure. In this chapter we provide theorems that allow one to prove the existence of particular orbits. However, for many nonlinear problems, demonstrating the existence of particular orbits is extremely difficult and one is willing to settle for proving the existence of some invariant set within a prescribed region of phase space. This more modest goal leads to the following concept.

Definition 10.20 Let $f : X \to X$ be continuous and let $N \subset X$. The *maximal invariant set* in N under f is

$$\operatorname{Inv}(N, f) := \{x \in N \mid \text{there exists a full solution } \gamma_x : \mathbf{Z} \to N\}.$$

As we indicated in the introduction, we will also use symbolic dynamics to understand the dynamics of other systems. However, to do this we will need a broader collection of examples than just the shift dynamics on n symbols. For this reason we need to introduce the notion of a *subshift of finite type*. Let $A = [a_{ij}]$ be a *transition matrix* on Σ_n—that is, an $n \times n$ matrix with entries of the form $a_{ij} = 0, 1$. Define

$$\Sigma_A := \{\mathbf{a} \in \Sigma_n \mid a_{s_k s_{k+1}} = 1 \text{ for } k \in \mathbf{Z}\}.$$

It is easy to check that $\sigma(\Sigma_A) = \Sigma_A$. Therefore, we can define $\sigma_A : \Sigma_A \to \Sigma_A$ by restricting σ to Σ_A. The dynamical system $\sigma_A : \Sigma_A \to \Sigma_A$ is referred to as the *subshift of finite type for the transition matrix A*.

Example 10.21 Let

$$A = \begin{bmatrix} 1 & 1 & 0 & 0 & 0 \\ 1 & 0 & 0 & 1 & 0 \\ 1 & 1 & 0 & 0 & 1 \\ 0 & 1 & 0 & 0 & 0 \\ 0 & 0 & 0 & 0 & 0 \end{bmatrix}$$

be a transition matrix on Σ_5. Let us consider the subshift of finite type $\sigma_A : \Sigma_A \to \Sigma_A$. Several simple observations can be made. The first is that there is only one fixed point:

$$(\ldots, 1, 1, 1, 1, 1, \ldots).$$

The second is that no orbits contain the symbols 3 or 5. This is easily seen from the fact that $a_{5j} = 0$, for all j, and $a_{i3} = 0$, for all i. The third is that locally not many patterns of symbols are permitted. To be more precise, given $\mathbf{s} \in \Sigma_A$,

- If $s_0 = 1$, then $s_{\pm 1} \in \{1, 2\}$.
- If $s_0 = 2$, then $s_{\pm 1} \in \{1, 4\}$.
- If $s_0 = 4$, then $s_{-1} = s_1 = 2$.

On the other hand, even with all these restrictions σ_A still exhibits very rich dynamics. Observe, for example, that there are infinitely many periodic points, including:
$$\overline{(2, 1)}, \quad \overline{(2, 4)}, \quad \overline{(2, 1, 2, 4)}.$$

From the previous example it should be clear that the form of the transition matrix A determines how rich the dynamics of the corresponding subshift will be. With this in mind, we make the following definition.

Definition 10.22 An $n \times n$ transition matrix A is *irreducible* if for every pair $1 \leq i, j \leq n$, there exists an integer $k = k(i, j)$ such that $\left(A^k\right)_{ij} \neq 0$.

Example 10.23 The matrix
$$A = \begin{bmatrix} 0 & 0 & 1 & 0 & 0 \\ 1 & 0 & 0 & 0 & 0 \\ 1 & 1 & 0 & 0 & 0 \\ 1 & 1 & 1 & 1 & 1 \\ 0 & 0 & 1 & 1 & 0 \end{bmatrix}$$

is not irreducible since, for any $k \geq 1$, $\left(A^k\right)_{1,5} = 0$. On the other hand, A contains two submatrices that are irreducible. In particular, let $B = [A_{ij}]$ for $1 \leq i, j \leq 3$ and let $C = [A_{ij}]$ for $4 \leq i, j \leq 5$. Observe that
$$B = \begin{bmatrix} 0 & 0 & 1 \\ 1 & 0 & 0 \\ 1 & 1 & 0 \end{bmatrix}, \quad B^2 = \begin{bmatrix} 1 & 1 & 0 \\ 0 & 0 & 1 \\ 1 & 0 & 1 \end{bmatrix}, \quad B^3 = \begin{bmatrix} 1 & 0 & 1 \\ 1 & 1 & 0 \\ 1 & 1 & 1 \end{bmatrix}.$$

Every entry is nonzero in B, B^2, or B^3. Therefore, B is irreducible. It is left to the reader to check that C is irreducible.

Example 10.24 Let A be as defined in Example 10.21. Observe that A is not irreducible since $\left(A^k\right)_{5j} = 0$ for all $k \geq 0$ and all $1 \leq j \leq 5$. On the other hand, if we consider the submatrix of A determined by the first, second, and fourth columns and rows, we obtain
$$A' = \begin{bmatrix} 1 & 1 & 0 \\ 1 & 0 & 1 \\ 0 & 1 & 0 \end{bmatrix},$$

which is irreducible.

We leave the proof of the following proposition as an exercise.

Proposition 10.25 *Let A be an irreducible transition matrix. Then*
1. *The set of periodic orbits is dense in Σ_A.*
2. *There exists a dense orbit in Σ_A.*

Remark 10.26 Observe that this proposition implies that if A is irreducible and there exists more than one periodic orbit in Σ_A, then the dynamics must be fairly complicated and is typically referred to as *chaotic*. A detailed discussion of chaotic dynamics, its properties, and how one can identify it is beyond the scope of this book. However, the interested reader is referred to [73] for an introduction to the subject.

Exercises

10.1 Prove that
$$\text{dist}(\mathbf{a}, \mathbf{b}) := \sum_{j=-\infty}^{\infty} \frac{\delta(a_j, b_j)}{4^{|j|}}$$
defines a metric on Σ_n.

10.2 Let $h : \Sigma_n^+ \to \Sigma_n$ be defined as follows. Given $\mathbf{a} \in \Sigma_n^+$, let $\mathbf{b} = h(\mathbf{a})$ be given by
$$b_0 = a_0, \quad b_{-i} = a_{2i-1}, \quad b_i = a_{2i}$$
for all $i \geq 1$. Show that h is a homeomorphism.

10.3 Suppose that $f : X \to X$ is a homeomorphism.
(a) Show that every $x \in X$ admits a unique full solution given by
$$\gamma_x(n) := f^n(x), \quad n \in \mathbf{Z},$$
where $f^n := (f^{-1})^{|n|}$ for $n < 0$.
(b) Show that the solutions of f^{-1} are given by $\tilde{\gamma}_x(n) = \gamma_x(-n)$, where γ_x is a solution of f.
(c) For $N \subset X$, show that $\text{Inv}(N, f^{-1}) = \text{Inv}(N, f)$.
(d) By the uniqueness of full solutions established in (a), it makes sense to write $\alpha(x, f)$ rather than $\alpha(\gamma_x, f)$. Show that
$$\alpha(x, f^{-1}) = \omega(x, f) \quad \text{and} \quad \omega(x, f^{-1}) = \alpha(x, f).$$

10.4 Construct an example of a continuous map in which some elements have multiple full solutions.

10.5 Let S be an invariant set under the continuous map $f : X \to X$. Prove that if $x \in S$, then a full solution $\gamma_x : \mathbf{Z} \to S$ exists.

10.6 List all periodic orbits of period 3, 4, and 5 for $\sigma : \Sigma_2 \to \Sigma_2$, the full shift on two symbols. How many periodic orbits of period n are there?

10.7 Prove Proposition 10.25.

10.2 Differential Equations and Flows

Historically, the theory of dynamical systems has its roots in the qualitative theory of ordinary differential equations. Thus we now turn our attention to this topic and briefly review some standard results whose proofs can be found in [35, 38].

Let $f : \mathbf{R}^d \to \mathbf{R}^d$. Consider an open interval $I \subset \mathbf{R}$ such that $0 \in I$. A differentiable function

$$x : I \to \mathbf{R}^d,$$
$$t \mapsto x(t)$$

is a *solution* to the ordinary differential equation

$$\dot{x} = f(x),$$

where the dot stands for the derivative with respect to t, if $(dx/dt)(t) = f(x(t))$ for all $t \in I$. Usually the t variable in this context is referred to as *time*. An initial-value problem is formulated as follows: Given the initial time $t = 0$ and an initial condition $x_0 \in \mathbf{R}^d$, find a solution $x : I \to \mathbf{R}^d$ with the property that $x(0) = x_0$. The following theorem gives conditions under which this problem has a unique solution.

Theorem 10.27 (Existence and uniqueness theorem) *If $f \in C^1(\mathbf{R}^d, \mathbf{R}^d)$, then for any $x_0 \in \mathbf{R}^d$, there exist an open interval I (possibly unbounded) containing 0 and a unique solution $\varphi(\cdot, x_0) : I \to \mathbf{R}^d$ of $\dot{x} = f(x)$ with the property that $\varphi(0, x_0) = x$.*

To simplify the discussion, we shall usually consider functions f such that for each $x \in \mathbf{R}^d$ the solution is defined for all $t \in \mathbf{R}$. Observe that this implies that $\varphi : \mathbf{R} \times \mathbf{R}^d \to \mathbf{R}^d$. Furthermore, as shown in [38, Chapter 8.7], $\varphi(t+s, x) = \varphi(t, \varphi(s, x))$ and $\varphi(0, x) = x$. The fact that, for fixed x, $\varphi(\cdot, x)$ is a solution implies that φ is differentiable in t. That φ is a continuous function of both variables follows from a classical result known as continuity with respect to initial conditions ([35, Theorem 3.4]). Abstracting this leads to the following definitions.

Definition 10.28 Let X be a topological space. A continuous function $\varphi : \mathbf{R} \times X \to X$ is a *flow* if

1. $\varphi(0, x) = x$ for all $x \in X$, and
2. $\varphi(t+s, x) = \varphi(t, \varphi(s, x))$ for all $t, s \in \mathbf{R}$, and $x \subset X$.

Definition 10.29 The *orbit* of the point $x \in X$ under the flow φ is

$$\varphi(\mathbf{R}, x) := \{\varphi(t, x) \mid t \in \mathbf{R}\}.$$

10.2 Differential Equations and Flows

Observe that an orbit in a flow is analogous to the solution curve of a differential equation.

Since the goal of dynamical systems is to understand the existence and structure of orbits, we are once again led to the following concepts.

Definition 10.30 A subset $S \subset \mathbf{R}^d$ is *invariant* under the flow φ if, for every element $x \in S$, its orbit $\varphi(\mathbf{R}, x) \subset S$, or equivalently $\varphi(t, S) = S$ for every $t \in \mathbf{R}$.

Definition 10.31 The *maximal invariant set* in $N \subset X$ under the flow φ is

$$\operatorname{Inv}(N, \varphi) := \{x \in N \mid \varphi(\mathbf{R}, x) \subset N\}.$$

Definition 10.32 Let $x \in X$. The *alpha* and *omega limit sets* of x under φ are

$$\alpha(x, \varphi) := \bigcap_{t \leq 0} \operatorname{cl}(\varphi((-\infty, t), x)) \quad \text{and} \quad \omega(x, \varphi) := \bigcap_{t \geq 0} \operatorname{cl}(\varphi([t, \infty), x)).$$

A proof similar to the one used to demonstrate Proposition 10.14 leads to the following result.

Proposition 10.33 *If X is a compact set, then for any $x \in X$ the sets $\alpha(x, \varphi)$ and $\omega(x, \varphi)$ are nonempty, compact, and invariant.*

We can make definitions concerning the types of orbits of flows that are similar to those for maps.

Definition 10.34 An element $x \in X$ is an *equilibrium point* of φ if $\varphi(t, x) = x$ for all $t \in \mathbf{R}$. It is a *periodic point* of period $\tau > 0$ if $\varphi(\tau, x) = x$.

We can go a step further and observe the very strong relationship between flows and maps. In particular, given a flow $\varphi : \mathbf{R} \times X \to X$, we can always define the *time-τ map* $\varphi_\tau : X \to X$ by fixing a particular time τ and letting

$$\varphi_\tau(x) := \varphi(\tau, x). \tag{10.4}$$

This relationship demonstrates that the problem of finding periodic points for flows is closely related to finding fixed points of maps. Clearly, x is a periodic point for φ with period τ if and only if x is a fixed point of φ_τ.

The reader should also recognize that time-τ maps represent a very special class of maps.

Proposition 10.35 *Let $\varphi : \mathbf{R} \times X \to X$ be a flow. Then φ_τ is homotopic to the identity map, namely*

$$\varphi_\tau \sim \operatorname{id}_X.$$

Proof. The flow itself provides the homotopy. To be more precise, define $h : X \times [0, 1] \to X$ by $h(x, s) = \varphi(s\tau, x)$. Since φ is continuous, h is continuous. Furthermore, $h(0, x) = \varphi(0, x) = x$ and $h(1, x) = \varphi(\tau, x) = \varphi_\tau(x)$. □

10.3 Ważewski Principle

In the previous two sections we have introduced basic concepts from dynamical systems. We now turn to the question of using topology to prove the existence of an invariant set within a specified region. In the case of flows there is an elementary but deep method called the Ważewski principle.

Let X be a topological space and let $\varphi : \mathbf{R} \times X \to X$ be a flow. Consider $W \subset X$. The set of points in W that eventually leave in forward time is

$$W^0 := \{x \in W \mid \text{there exists } t > 0 \text{ such that } \varphi(t, x) \notin W\}.$$

The set of points that immediately leave W is denoted by

$$W^- := \{x \in W \mid \varphi([0, t), x) \not\subset W \text{ for all } t > 0\}.$$

Clearly, $W^- \subset W^0 \subset W$. Observe that it is possible that $W^0 = \emptyset$, in which case $W^- = \emptyset$.

Definition 10.36 A set W is a *Ważewski set* if the following conditions are satisfied:

1. $x \in W$ and $\varphi([0, t], x) \subset \operatorname{cl} W$ imply that $\varphi([0, t], x) \subset W$.
2. W^- is closed in W^0.

Theorem 10.37 *Let W be a Ważewski set. Then W^- is a strong deformation retract of W^0 and W^0 is open in W.*

Proof. Observe that if $W^0 = \emptyset$, then the result is trivially true. Therefore, assume that $W^0 \neq \emptyset$. To construct the strong deformation retraction

$$h : W^0 \times [0, 1] \to W^0$$

of W^0 onto W^-, we first define $\tau : W^0 \to [0, \infty)$ by

$$\tau(x) = \sup\{t \geq 0 \mid \varphi([0, t], x) \subset W\}.$$

By the definition of W^0, $\tau(x)$ is finite. Continuity of the flow implies that $\varphi([0, \tau(x)], x) \subset \operatorname{cl}(W)$.

Since W is a Ważewski set, $\varphi(\tau(x), x) \in W$ and, in fact, the definition of τ implies that $\varphi(\tau(x), x) \in W^-$. Observe that $\tau(x) = 0$ if and only if $x \in W^-$.

Assume for the moment that τ is continuous, and define h by

$$h(x, \sigma) = \varphi((\sigma\tau(x)), x).$$

Obviously,

$$h(x, 0) = \varphi(0, x) = x,$$
$$h(x, 1) = \varphi(\tau(x), x) \in W^-,$$

and, for $y \in W^-$,

$$h(y, \sigma) = \varphi(\sigma\tau(y), y) = \varphi(0, y) = y.$$

Therefore, h is a strong deformation retraction of W^0 onto W^-.

It remains to show that the function τ is continuous. By Proposition 12.37, this is equivalent to showing that it is both upper and lower semicontinuous in the sense of Definition 12.36.

We first show that it is upper semicontinuous, that is, that the set $\{x \in W^0 \mid \tau(x) < \alpha\}$ is open for any $\alpha \in \mathbf{R}$. Let α be given and let $x \in W^0$ be such that $\tau(x) < \alpha$. We must show the existence of a neighborhood U_x of x in W^0 such that $\tau(y) < \alpha$ for all $y \in U_x$. By the definition of τ, we have $\varphi([\tau(x), \alpha], x) \not\subset W$. By Definition 10.36(1), there exists $t_0 \in [\tau(x), \alpha]$ such that $\varphi(t_0, x) \notin \operatorname{cl} W$. Thus we can choose a neighborhood V of $\varphi(t_0, x)$ in X such that $V \cap \operatorname{cl} W = \emptyset$. By the continuity of φ, there exists an open neighborhood U of x in X such that $\varphi(t_0, U) \subset V$. Let $U_x := U \cap W$. For all $y \in U_x$, $\varphi(t_0, y) \notin W$. This implies, first, that $U_x \subset W^0$ and, second, that $\tau(y) < t_0 \leq \alpha$. Thus τ is upper semicontinuous.

By the same arguments applied to any $x \in W^0$ and α chosen so that $\tau(x) < \alpha$, the conclusion $U_x \subset W^0$ implies that W^0 is open in W.

It remains to prove that τ is lower semicontinuous; that is, for any given $\alpha \in \mathbf{R}$, the set $\{x \in W^0 \mid \tau(x) > \alpha\}$ is open in W^0. This is equivalent to showing that the set $A := \{x \in W^0 \mid \tau(x) \leq \alpha\}$ is closed in W^0. Let (x_n) be a sequence of points in A convergent to a point $x \in W^0$. Put $t_n := \tau(x_n)$. Then $t_n \in [0, \alpha]$, which is compact, so there is a convergent subsequence $t_{n_k} \to t_0$ with $t_0 \leq \alpha$. Since φ is continuous, $\varphi(t_{n_k}, x_{n_k}) \to \varphi(t_0, x)$.

We first prove that $\varphi(t_0, x) \in W^0$. Indeed, suppose the contrary. This means that $\varphi(t, x) \in W$ for all $t \geq t_0$. We show that this is also true for all $t \in [0, t_0]$. Choose any such t. Since $t_{n_k} \to t_0$, there exists a sequence of $s_{n_k} \in [0, t_{n_k}]$ converging to t. By the choice of t_{n_k}, $\varphi(s_{n_k}, x_{n_k}) \in W$. Hence, $\varphi(t, x) \in \operatorname{cl} W$. Since this is true for all $t \in [0, t_0]$, condition (1) in Definition 10.36 implies that $\varphi(t, x) \in W$. This together with the preceding observation shows that $\varphi(t, x) \in W$ for all $t \geq 0$ in contrary to the hypothesis that $x \in W^0$. Thus, $\varphi(t_0, x) \in W^0$.

Next, we have $\varphi(t_{n_k}, x_{n_k}) \in W^-$ for all k. Since W^- is closed in W^0, it follows that $\varphi(t_0, x) \in W^-$. Hence $\tau(x) \leq t_0 \leq \alpha$, so $x \in A$. This shows that A is closed in W^0. □

Corollary 10.38 (Ważewski principle) *If W is a Ważewski set and W^- is not a strong deformation retract of W, then $W \backslash W^0 \neq \emptyset$; namely there exist solutions that stay in W for all positive time.*

Proof. Suppose that, on the contrary, all solutions eventually leave W. Then $W^0 = W$, so we get a contradiction with Theorem 10.37. □

Corollary 10.39 *If W is a compact Ważewski set and W^- is not a strong deformation retract of W, then $\operatorname{Inv}(W, \varphi) \neq \emptyset$.*

322   10 Nonlinear Dynamics

Proof. By Corollary 10.38, there exists a point $x \in W$ such that $\varphi([0, \infty), x) \subset W$. By Proposition 10.33, $\omega(x, \varphi)$ is a nonempty, compact, invariant subset of W. □

Hopefully, the reader is impressed by the elegant relation between topology and dynamics expressed in the Ważewski principle. However, this is a book on algebraic topology, and so it is reasonable to wonder how homology fits into the picture.

Proposition 10.40 *Let W be a Ważewski set. Assume W and W^- are cubical. If $H_*(W, W^-) \neq 0$, then $\mathrm{Inv}\,(W, \varphi) \neq \emptyset$.*

Proof. If $H_*(W, W^-) \neq 0$, then, by Proposition 9.28, W^- is not a deformation retract of W. The conclusion now follows from Corollary 10.39. □

Let us now consider two examples. The first shows the power of Ważewski's principle, and the second shows an inherent weakness.

Example 10.41 Consider the differential equation

$$\dot{x} = Ax, \tag{10.5}$$

where

$$A = \begin{bmatrix} 1 & 0 \\ 0 & -1 \end{bmatrix}.$$

The reader can check that

$$\varphi\left(t, \begin{bmatrix} x_1 \\ x_2 \end{bmatrix}\right) = \begin{bmatrix} e^t x_1 \\ e^{-t} x_2 \end{bmatrix}$$

is a solution to (10.5). Observe that the only bounded invariant set is the equilibrium solution; every other orbit is unbounded in either forward or backward time.

Now consider the following differential equation:

$$\dot{x} = Ax + f(x), \tag{10.6}$$

where we do not know f explicitly, but assume that (10.6) generates a flow ψ. Instead, all the information we are given is that $\|f(x)\| < \kappa$ for all $x \in \mathbf{R}^2$. We would like to know if (10.6) has a bounded solution. Let $m = \mathrm{ceil}\,(\kappa)$ and let $W = [-m, m]^2 \subset \mathbf{R}^2$ (see Figure 10.1). We claim that W is a Ważewski set and

$$W^- = \left\{ \begin{bmatrix} x_1 \\ x_2 \end{bmatrix} \in W \,\middle|\, |x_1| = m \right\}. \tag{10.7}$$

Observe that if (10.7) is correct, then W is a Ważewski set since both W and W^- are closed.

Clearly, W^- is a subset of the boundary of W (otherwise, the point could not leave W immediately in forward time). We shall verify (10.7) by evaluating the vector field $Ax + f(x)$ at points on the boundary of W. Let

$$f(x) = \begin{bmatrix} f_1(x) \\ f_2(x) \end{bmatrix}.$$

The assumption that $\|f(x)\| < \kappa$ implies that $\|f_i(x)\| < \kappa$ for $i = 1, 2$. Consider

$$x = \begin{bmatrix} m \\ x_2 \end{bmatrix},$$

where $|x_2| \leq m$. Observe that at x,

$$\dot{x}_1 = m + f_1(x) > m - \kappa \geq 0.$$

Thus the second coordinate of the solution through x increases as time increases. Therefore, the solution leaves W immediately in forward time, that is, $x \in W^-$. We leave it to the reader to check the other cases as well as

$$H_1(W, W^-) \cong \mathbf{Z}.$$

Therefore, by Proposition 10.40, Inv $(W, \psi) \neq \emptyset$.

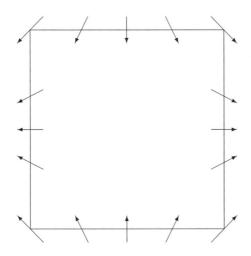

Fig. 10.1. A Ważewski set W, where W^- consists of the vertical edges.

Example 10.42 Consider the following one-parameter family of scalar ordinary differential equations

$$\dot{x} = x^2 + \lambda^2,$$

where $\lambda \in [-1, 1]$. Let $W_\lambda = [-1, 0]$. Then W_λ is a Ważewski set. Notice now that $W_0^- = \emptyset$, and hence

$$H_*(W_0, W_0^-) = H_*([-1,0], \emptyset) = H_*([-1,0]) \neq 0.$$

Therefore, Ważewski's principle detects the existence of a nontrivial invariant set, that is, the equilibrium $\{0\}$, for $\lambda = 0$, since W_0^- is not a strong deformation retract of W_0. However, for all other values of λ the invariant set is trivial. This is an important point. The use of Ważewski's principle to prove the existence of an invariant set at one parameter value does not allow one to conclude the existence of an invariant set at any other parameter value. In other words, Ważewski's principle is not robust with respect to perturbations.

10.4 Fixed-Point Theorems

While the Ważewski principle and index pairs provide a method to prove the existence of a nonempty invariant set, the former gives no information concerning the structure of the invariant set. With this in mind, we change topics slightly and consider some classical fixed-point theorems that are outgrowths of homotopy invariance introduced in Section 6.5. As will become clear, these theorems allow one to conclude the existence of an equilibrium or fixed point by means of an algebraic topological computation.

We start with the presentation of theorems relating homotopical properties of the sphere to fixed points of a continuous map on the corresponding disk (i.e., a closed ball). In order to be able to employ cubical homology, we use the supremum norm, which implies that spheres and balls are cubical sets. However, once the notion of homology is generalized to topological polyhedra (see Chapter 11), all the results of this section extend to spheres and balls with respect to any norm in $\mathbf{R}^d$.

10.4.1 Fixed Points in the Unit Ball

Recall that $\bar{B}_0^{d+1} := [-1,1]^{d+1}$, the unit ball with radius 1, and $S_0^d = \mathrm{bd}\, \bar{B}_0^{d+1}$ is the corresponding unitary sphere. Let $f : \bar{B}_0^{d+1} \to \bar{B}_0^{d+1}$ be a continuous map. If one wishes to remain in the setting of flows, consider $\varphi : \mathbf{R} \times \bar{B}_0^{d+1} \to \bar{B}_0^{d+1}$, and let $f = \varphi_\tau$ for some $\tau > 0$.

The following theorem due to K. Borsuk brings together three famous theorems: the noncontractibility theorem, the nonretraction theorem, and the Brouwer fixed-point theorem (1912).

Theorem 10.43 *Let $d \geq 0$. The following statements are equivalent:*

(a) S_0^d is not contractible.
(b) Every continuous map $f : \bar{B}_0^{d+1} \to \bar{B}_0^{d+1}$ has a fixed point.
(c) There is no retraction from $\bar{B}_0^{d+1}$ onto S_0^d.

10.4 Fixed-Point Theorems

Proof. All implications will be proved by contradiction.

We first show that (a) implies (b). Suppose that $f(x) \neq x$ for all $x \in \bar{B}_0^{d+1}$. Then $y - tf(y) \neq 0$ for all $y \in S_0^d$ and all $t \in [0,1]$. Indeed, if $t = 1$, that is immediate from what we just supposed, and if $0 \leq t < 1$, it follows from the inequality
$$\|tf(y)\|_0 < \|f(y)\|_0 \leq 1 = \|y\|_0.$$
Let $n : \mathbf{R}^{d+1} \setminus \{0\} \to S_0^d$ be given by $n(x) = x/\|x\|_0$. The map $h_1 : S_0^n \times [0,1] \to S_0^n$ given by
$$h_1(y,t) = n(y - tf(y))$$
is a homotopy from $\mathrm{id}_{S_0^d}$ to the map $g(y) = n(y - f(y))$, and the map h_0 given by
$$h_0(y,t) = n((1-t)y - f((1-t)y))$$
is a homotopy from g to the constant map $k(y) = y_0 := n(-f(0))$. By the transitivity property (see Exercise 6.15), $\mathrm{id}_{S_0^d} \sim k$, a contradiction.

We next show that (b) implies (c). Suppose that there is a retraction $r : \bar{B}_0^{d+1} \to S_0^d$. Then $f : \bar{B}_0^{d+1} \to \bar{B}_0^{d+1}$, $f(x) = -r(x)$ is fixed-point-free.

Finally, we show that (c) implies (a). Suppose that there is a homotopy $h : S_0^n \times [0,1] \to S_0^n$ joining $\mathrm{id}_{S_0^n}$ at $t = 0$ to $y_0 \in S_0^n$ at $t = 1$. Then the formula
$$r(x) := \begin{cases} y_0 & \text{if } \|x\| \leq \frac{1}{2}, \\ h(n(x), 2(1 - \|x\|_0)) & \text{if } \|x\| \geq \frac{1}{2} \end{cases}$$
defines a retraction r of $\bar{B}_0^{d+1}$ onto S_0^d. $\square$

Observe that this theorem merely proves the equivalence of these statements, but it does not address their truthfulness. This is resolved in the following corollary.

Corollary 10.44 *All three statements of Theorem 10.43 are true.*

Proof. By Exercise 9.13 for $d \geq 1$,
$$H_k(S_0^d) \cong \begin{cases} \mathbf{Z} & \text{if } k = 0, d, \\ 0 & \text{otherwise,} \end{cases}$$
while for $d = 0$,
$$H_k(S_0^d) \cong \begin{cases} \mathbf{Z}^2 & \text{if } k = 0, \\ 0 & \text{otherwise.} \end{cases}$$
In either case, it is clear that S_0^d is not acyclic and hence, by Corollary 6.69, is not contractible to a point. Thus condition (a) of Theorem 10.43 holds, and therefore the remaining two do also. $\square$

10.4.2 The Lefschetz Fixed-Point Theorem

The Brouwer fixed-point theorem of the previous section has the advantage that it is easy to state and has an elementary proof in the sense that it only requires knowledge of continuous functions. However, it is very limited since it only applies to disks in $\mathbf{R}^d$. The Lefschetz fixed-point theorem, which will be presented in this section, is one of the most important results in algebraic topology. In particular, it allows one to conclude the existence of a fixed point of a continuous map f based on the associated homology map f_*. Of course, one must pay a price for such a gem; the proof is no longer elementary—it involves a delicate mixture of algebra and topology. We begin, therefore, with a few necessary algebraic preliminaries.

Let $A = [a_{ij}]$ be an $n \times n$ matrix. The *trace* of A is defined to be the sum of the diagonal entries, that is,

$$\operatorname{tr} A := \sum_{i=1}^{n} a_{ii}.$$

If A and B are $n \times n$ matrices, then

$$\operatorname{tr} AB = \sum_{i,j=1}^{n} a_{ij} b_{ji} = \sum_{i,j=1}^{n} b_{ji} a_{ij} = \operatorname{tr} BA.$$

Let G be a finitely generated free abelian group, and let $\phi : G \to G$ be a group homomorphism. Since G is free abelian, it has a basis and, for a particular choice of basis, ϕ can be written as a matrix A. So in this case define

$$\operatorname{tr} \phi := \operatorname{tr} A.$$

To check that this is a well-defined concept, let $\{b_1, \ldots, b_n\}$ be a different basis for G.

Let $B : G \to G$ be an isomorphism corresponding to the change of basis. By Corollary 3.6, in the new basis the matrix representation of ϕ is given by $B^{-1}AB$. Observe that

$$\operatorname{tr}(B^{-1}AB) = \operatorname{tr}(B^{-1}(AB)) = \operatorname{tr}((AB)B^{-1}) = \operatorname{tr} A.$$

Given a homomorphism $f : G \to G$, where G is a finitely generated abelian group, we define the $\operatorname{tr} f$ as the $\operatorname{tr} \tilde{f}$, where $\tilde{f} : G/T(G) \to G/T(G)$ is the induced homomorphism of free abelian groups (see Corollary 3.62).

Definition 10.45 Let X be a cubical set and let $f : X \to X$ be a continuous map. The *Lefschetz number* of f is

$$L(f) := \sum_{k} (-1)^k \operatorname{tr} f_{*k}.$$

Theorem 10.46 (Lefschetz fixed-point theorem) *Let X be a cubical set and let $f : X \to X$ be a continuous map. If $L(f) \neq 0$, then f has a fixed point.*

This theorem is an amazing example of how closely the algebra is tied to the topology. To prove it, we need to understand how to relate the topology in the form of the map on the chain complexes to the algebra in the form of the induced homology maps on the free part of the homology groups.

We begin with a technical lemma.

Lemma 10.47 *Consider a finitely generated free abelian group G and a subgroup H such that G/H is free abelian. Let $\phi : G \to G$ be a group homomorphism such that $\phi(H) \subset H$. Then*

$$\operatorname{tr} \phi = \operatorname{tr} \phi' + \operatorname{tr} \phi \,|_H,$$

where $\phi' : G/H \to G/H$ is the induced homomorphism.

Proof. By Lemma 13.68, the facts that G is a finitely generated free abelian group and that H is a subgroup imply that H is also a finitely generated free abelian group. Let $\{v_1, \ldots, v_k\}$ be a basis for H and let $\{u_1 + H, \ldots, u_n + H\}$ be a basis for G/H. Since ϕ and ϕ' are group homomorphisms, there exist integers α_{ij} and β_{ij} such that

$$\phi'(u_j + H) = \sum_{i=1}^{n} \beta_{ij}(u_i + H),$$

$$\phi\,|_H (v_j) = \sum_{i=1}^{k} \alpha_{ij} v_i,$$

where the latter equation follows from the fact that $\phi(H) \subset H$. Observe that, using these bases, the matrix representations of ϕ and ϕ' are given by $A = [\alpha_{ij}]$ and $B = [\beta_{ij}]$, respectively.

By Exercise 13.17, $G \cong G/H \oplus H$; thus $\{u_1, \ldots, u_n, v_1, \ldots, v_k\}$ is a basis for G. Furthermore,

$$\phi(u_j) = \sum_{i=1}^{n} \beta_{ij} u_i + h_j \quad \text{for some } h_j \in H,$$

$$\phi(v_j) = \sum_{i=1}^{k} \alpha_{ij} v_i.$$

This means that, using this basis, the matrix representation of ϕ has the form

$$\begin{bmatrix} B & * \\ 0 & A \end{bmatrix}.$$

Clearly, $tr\phi = \operatorname{tr} \phi' + \operatorname{tr} \phi \,|_H$. □

The following theorem plays a crucial role in that it allows us to relate the trace of the homology map to the chain map that induced it.

Theorem 10.48 (Hopf trace theorem) *Let $\mathcal{C} := \{C_k, \partial_k\}$ be a finitely generated free chain complex and $\varphi : \mathcal{C} \to \mathcal{C}$ a chain map. Let $H_k := H_k(\mathcal{C})$ denote the corresponding homology groups with torsion subgroups T_k. Let $\phi_k : H_k/T_k \to H_k/T_k$ be the induced homomorphism. Then*

$$\sum_k (-1)^k \operatorname{tr} \varphi_k = \sum_k (-1)^k \operatorname{tr} \phi_k.$$

Proof. Let $Z_k := \ker \partial_k$ and $B_k := \operatorname{im} \partial_{k+1}$. We will also use the notation from Section 9.5, where W_k denotes the group of weak boundaries. Recall that

$$B_k \subset W_k \subset Z_k \subset C_k.$$

Furthermore, since φ is a chain map, each of these subgroups is invariant under φ_k, that is, $\varphi_k(B_k) \subset B_k$, $\varphi_k(W_k) \subset W_k$, and so forth. The map φ_k induces maps

$$\varphi_k |_{W_k} : W_k \to W_k,$$
$$\varphi'_k : Z_k/W_k \to Z_k/W_k,$$
$$\varphi''_k : C_k/Z_k \to C_k/Z_k.$$

From (9.18) we have that Z_k/W_k is free abelian for each k. Using Exercise 13.14 it follows that C_k/Z_k is free abelian for each k. Therefore, applying Lemma 10.47 twice gives

$$\operatorname{tr} \varphi_k = \operatorname{tr} \varphi''_k + \operatorname{tr} \varphi'_k + \operatorname{tr} \varphi_k |_{W_k}. \tag{10.8}$$

Again, from (9.18), $Z_k/W_k \cong H_k/T_k$ and furthermore under this isomorphism, φ'_k becomes ϕ_k. Therefore, (10.8) becomes

$$\operatorname{tr} \varphi_k = \operatorname{tr} \varphi''_k + \operatorname{tr} \phi_k + \operatorname{tr} \varphi_k |_{W_k}. \tag{10.9}$$

Similarly, C_k/Z_k is isomorphic to B_{k-1} and under this isomorphism φ''_k becomes $\varphi_{k-1} |_{B_{k-1}}$. Hence, (10.9) can be written as

$$\operatorname{tr} \varphi_k = \operatorname{tr} \varphi_{k-1} |_{B_{k-1}} + \operatorname{tr} \phi_k + \operatorname{tr} \varphi_k |_{W_k}. \tag{10.10}$$

We now show that $\operatorname{tr} \varphi_k |_{W_k} = \operatorname{tr} \varphi_k |_{B_k}$. As done explicitly in Section 9.5, there exist a basis $\{u_1, \ldots, u_l\}$ for W_k and integers $\alpha_1, \ldots, \alpha_l$, such that $\{\alpha_1 u_1, \ldots, \alpha_l u_l\}$ is a basis for B_k.

Observe that

$$\varphi_k |_{W_k} (u_j) = \sum_{i=1}^{l} a_{ij} u_i \tag{10.11}$$

and

$$\varphi_k |_{B_k} (\alpha_j u_j) = \sum_{i=1}^{l} b_{ij} \alpha_i u_i \tag{10.12}$$

for appropriate constants a_{ij} and b_{ij}. Both these maps are just restrictions of φ_k to the appropriate subspaces. So multiplying (10.11) by α_j must give rise to (10.12) and hence $\alpha_j a_{ij} = b_{ij}\alpha_i$ and, in particular, $\alpha_i a_{ii} = b_{ii}\alpha_i$. Therefore, $\operatorname{tr} \varphi_k |_{W_k} = \operatorname{tr} \varphi_k |_{B_k}$. Applying this to (10.10) gives

$$\operatorname{tr} \varphi_k = \operatorname{tr} \varphi_{k-1} |_{B_{k-1}} + \operatorname{tr} \phi_k + \operatorname{tr} \varphi_k |_{B_k}. \tag{10.13}$$

The proof is finished by multiplying (10.13) by $(-1)^k$ and summing. □

The Hopf trace formula is the key step in the proof of the Lefschetz fixed-point theorem. However, before beginning the proof, let us discuss the basic argument that will be used. Observe that an equivalent statement to the Lefschetz fixed-point theorem is the following: *If f has no fixed points, then $L(f) = 0$.* This is what we will prove. The Hopf trace formula provides us with a means of relating a chain map $\varphi : C(X) \to C(X)$ for f with $L(f)$. In particular, if we could show that $\operatorname{tr} \varphi = 0$, then it would be clear that $L(f) = 0$. Of course, the easiest way to check that $\operatorname{tr} \varphi = 0$ is for all the diagonal entries of φ to be zero. However, the diagonal entries of φ indicate how the duals of elementary cubes are mapped to themselves. If f has no fixed points, then the image of a small cube will not intersect itself and so the diagonal entries are zero. With this argument in mind we turn to the proof, which, as is often the case in mathematics, is presented in the reverse order.

Proof of Lefschetz fixed-point theorem. We argue by contradiction. Assume f has no fixed points. We want to show that $L(f) = 0$. Let

$$\epsilon := \inf_{x \in X} ||x - f(x)||,$$

where $|| \cdot || = || \cdot ||_0$ as defined by (12.3). Observe that $\epsilon > 0$. Indeed, since cubical sets are compact and the function $x \mapsto ||x - f(x)||$ is continuous, it has a minimum at some $x_0 \in X$ (see Theorem 12.66). But we are assuming that f has no fixed points, so $\epsilon = ||x_0 - f(x_0)|| > 0$. Take any integer $m > 2/\epsilon$ and a scaling vector α with $\alpha_i = m$ for all i. We will study the map

$$g := \Lambda_X^\alpha \circ f \circ \Omega_X^\alpha : X^\alpha \to X^\alpha.$$

Note that for any $y \in X^\alpha$,

$$||y - g(y)|| > 2. \tag{10.14}$$

Indeed, let $x = \Omega^\alpha(y)$. Then

$$\begin{aligned}
||y - g(y)|| &= ||\Lambda^\alpha(\Omega^\alpha(y) - f(\Omega^\alpha(y)))|| \\
&= ||\Lambda^\alpha(x - f(x))|| = m||x - f(x)|| \\
&> \frac{2}{\epsilon} \cdot \epsilon = 2.
\end{aligned}$$

We next prove a combinatorial version of the inequality $x \neq f(x)$ for g, that is,

$$Q \cap \operatorname{ch}(g(Q)) = \emptyset \qquad (10.15)$$

for any $Q \in \mathcal{K}(X^\alpha)$. Suppose the contrary and let $x \in Q \cap \operatorname{ch}(g(Q))$. By the definition of a closed hull, $\operatorname{dist}(x, g(Q)) \leq 1$, so there exists $y \in Q$ such that $||x - g(y)|| \leq 1$. Since $x, y \in Q$, it follows that $||x - y|| \leq 1$. Therefore,

$$||y - g(y)|| \leq ||y - x|| + ||x - g(y)|| \leq 2,$$

which contradicts (10.14).

Now let Λ^β be a rescaling of X^α such that

$$M_{g^\beta} : X^{\alpha\beta} \to X^\alpha$$

is acyclic-valued. Let θ be a chain selector of $M_{\Lambda^\beta} : X^\alpha \rightrightarrows X^{\alpha\beta}$ and let ψ be a chain selector of M_{g^β}. Then $g_* := \psi_* \circ \theta_* = (\psi \circ \theta)_*$. We next show that

$$Q \cap \left|\psi_k \circ \theta_k(\widehat{Q})\right| = \emptyset \qquad (10.16)$$

for all $Q \in \mathcal{K}_k(X^\alpha)$ and all $k = 0, 1, 2, \ldots$. Indeed, suppose the contrary and let $x \in Q \cap \left|\psi_k \circ \theta_k(\widehat{Q})\right|$. Since

$$\left|\psi_k \circ \theta_k(\widehat{Q})\right| \subset \bigcup \{M_{g^\beta}(\overset{\circ}{P}) \mid P \in \mathcal{K}(\left|\theta(\widehat{Q})\right|)\},$$

there exists an elementary cube $P \subset \left|\theta_k(\widehat{Q})\right|$ such that $x \in M_{g^\beta}(\overset{\circ}{P})$. We have

$$\left|\theta_k(\widehat{Q})\right| \subset M_{\Lambda^\beta}(\overset{\circ}{Q}) = \operatorname{ch}(\Lambda^\beta(\operatorname{ch}(\overset{\circ}{Q}))) = \operatorname{ch}(\Lambda^\beta(Q)) = \Lambda^\beta(Q),$$

so $P \subset \Lambda^\beta(Q)$. Hence, we get

$$\begin{aligned} x \in M_{g^\beta}(\overset{\circ}{P}) &= \operatorname{ch}(g^\beta(\operatorname{ch}(\overset{\circ}{P}))) = \operatorname{ch}(g \circ \Omega^\beta(P)) \\ &\subset \operatorname{ch}(g \circ \Omega^\beta(\Lambda^\beta(Q))) = \operatorname{ch}(g(Q)), \end{aligned}$$

so $x \in \operatorname{ch}(g(Q))$, which contradicts (10.15). Equation (10.16) implies that the diagonal entries in the matrix of the chain map $\psi_k \circ \theta_k$ with respect to the basis $\widehat{\mathcal{K}}_k(X^\alpha)$ are all zero; hence, $\operatorname{tr} \psi_k \circ \theta_k = 0$ for every k. By the Hopf trace formula, $L(g_*) = 0$. Finally, by Proposition 6.51,

$$L(f_*) = L(g_*) = 0. \qquad \square$$

Theorem 10.49 *Let X be an acyclic cubical set. Let $f : X \to X$ be continuous. Then f has a fixed point.*

Proof. Since X is acyclic, the only nonzero homology group is $H_0(X) \cong \mathbf{Z}$. But, by Proposition 6.60, $f_* : H_0(X) \to H_0(X)$ is the identity map. Therefore, $L(f) = 1$. $\square$

Corollary 10.50 *Let $\varphi : \mathbf{R} \times X \to X$ be a flow. If X is an acyclic cubical set, then φ has an equilibrium.*

Proof. Observe that for each $\tau > 0$, $\varphi_\tau : X \to X$ has a fixed point. In particular, let (t_n) be a sequence of positive times such that $\lim_{n\to\infty} t_n = 0$. For each t_n there exists $x_n \in X$ such that $\varphi_{t_n}(x_n) = x_n$. Since X is compact, (x_n) contains a convergent subsequence (x_{n_m}). Let $y = \lim_{m\to\infty} x_{n_m}$. Finally, let s_m be the minimal period of x_{n_m}.

We prove that y is an equilibrium, that is, that $\varphi(t, y) = y$ for any $t \in \mathbf{R}$. Assume not. Then there exists $\tau > 0$ such that $\varphi(\tau, y) = z \neq y$. Since $z \neq y$, there exists $\epsilon > 0$ such that $\|z - y\| > \epsilon$. By continuity,

$$\lim_{m\to\infty} \varphi(\tau, x_{n_m}) = z.$$

Now consider the set $\varphi([0, s_m), x_{n_m})$. Since x_{n_m} is periodic with period s_m, $\varphi(\mathbf{R}, x_{n_m}) = \varphi([0, s_m), x_{n_m})$. Since $\lim_{m\to\infty} s_m = 0$ and φ is continuous,

$$\lim_{m\to\infty} \operatorname{diam}(\varphi(\mathbf{R}, x_{n_m})) = 0.$$

However, this contradicts the fact that $\|z - y\| > \epsilon$. □

Both Theorem 10.49 and Corollary 10.50 follow from the fact that the Lefschetz number of the identity map on an acyclic space is nonzero because the only nontrivial homology group is H_0. The following concept generalizes this idea.

Definition 10.51 *Let X be a cubical set. The* Euler number *of X is*

$$E(X) := \sum_{i=0}^{\infty} (-1)^i \beta_i(X),$$

where $\beta_i(X)$ is the Betti number of $H_i(X)$.

Since φ_τ is homotopic to the identity for all $\tau \geq 0$, the proof of the following result is almost identical to that of Corollary 10.50.

Theorem 10.52 *Let $\varphi : \mathbf{R} \times X \to X$ be a flow on a cubical set. If $E(X) \neq 0$, then φ has an equilibrium.*

Exercises

10.8 Let X be a cubical set and let $\varphi : \mathcal{C}(X) \to \mathcal{C}(X)$ be a chain map. Let $T_k(X)$ denote the torsion subgroup of $H_k(X)$. Prove that $\varphi_k(T_k(X)) \subset T_k(X)$.

10.9 Let $X := \Gamma^1 = \operatorname{bd}[0, 1]^2$. Compute the Lefschetz number of the map $f : X \to X$ given by

(a) $f(x_1, x_2) = (1 - x_1, x_2)$,

(b) $f(x_1, x_2) = (1 - x_1, 1 - x_2)$.

If $L(f) \neq 0$, explicitly find fixed points of f. If $L(f) = 0$, show that f has no fixed points. One should be aware of the fact that the reverse implication of Theorem 10.46 is not true in general.

10.10 Let X be a cubical set, and let $F : X \rightrightarrows X$ be a lower semicontinuous cubical acyclic-valued map. Then F admits a chain selector φ, so $F_* = \varphi_*$, and the Lefschetz number $L(F)$ is defined for F by the same formula as it is done for a continuous map. A *fixed point* of F is a point x such that

$$x \in F(x).$$

Prove that if $L(F) \neq 0$, then F has a fixed point.
Hint: Extract arguments from the proof of Theorem 10.46.

10.5 Degree Theory

We finish the previous section with a rather remarkable result: If X is a cubical set and $E(X) \neq 0$, then *any* flow on X must have at least one equilibrium. Furthermore, this information is captured by a single number even though the homology groups $H_*(X)$ could be far from trivial. One might try to argue that this is not surprising since at the heart of the proof of Theorem 10.52 is a single number, the Lefschetz number. However, at this level it should appear even more impressive since given a continuous map $f : X \to X$, if $H_*(X)$ is high-dimensional, then $f_* : H_*(X) \to H_*(X)$ is represented by large matrices.

As important as the Lefschetz number is, it does have a serious limitation. Consider the linear differential equation (10.5) of Example 10.41. The associated flow φ is defined on all of $\mathbf{R}^2$, which is not a cubical set. At the same time, we cannot try to mimic the approach of restricting our attention to a cubical set of the form $[-m, m]^2$, since for any $t > 0$, $\varphi_t([-m, m]^2) \not\subset [-m, m]^2$.

Obviously, we would like to be able to circumvent this problem while still maintaining the simplicity of having a single number that can be used to guarantee the existence of equilibria. The proof of Theorem 10.52 is indirect in the sense that equilibria are obtained by analyzing the fixed points of the time τ maps of the flow for all positive τ. If we are working with a differential equation

$$\dot{x} = f(x), \quad x \in \mathbf{R}^d,$$

then the direct question is

Does the equation $f(x) = 0$ have a solution in a prescribed set $U \subset \mathbf{R}^d$?

Of course, there is an obvious relationship between fixed points of maps and zeros of a function. In particular, if $g : X \to X$, then

$$g(x) = x \quad \Leftrightarrow \quad f(x) := g(x) - x = 0.$$

With these observations in mind, we turn to the topic of this section, which is the degree of a map. In particular, it provides a single number that, if nonzero, guarantees the existence of a solution to $f(x) = 0$ in a prescribed region U. However, a necessary first step is to understand the degree of maps on spheres.

10.5.1 Degree on Spheres

Consider a continuous map $f : S_0^d \to S_0^d$. Recall that the reduced homology groups of spheres (see Exercise 9.13) are

$$\tilde{H}_k(S_0^d) \cong \begin{cases} \mathbf{Z} & \text{if } k = d, \\ 0 & \text{otherwise.} \end{cases}$$

In particular,
$$\tilde{H}_d(S_0^d) = \mathbf{Z}e,$$

where e is a generator that is the equivalence class of a certain d-cycle z in $Z_d(S_0^d)$. Obviously, there are only two choices of generators of $\mathbf{Z}e$: e or $-e$. Now the dth homology map of f,

$$f_{*d} : \tilde{H}_d(S_0^d) \to \tilde{H}_d(S_0^d),$$

acting on the generator e must take the form

$$f_{*d}(e) = ne, \tag{10.17}$$

where n is some integer. Moreover, that integer would be the same if we replace e by $-e$ because

$$f_{*d}(-e) = -f_{*d}(e) = -ne = n(-e).$$

Therefore, the number n completely characterizes f_{*d} and is independent of the choice of generator of $\tilde{H}_d(S_0^d)$.

Definition 10.53 *Given a continuous map $f : S_0^d \to S_0^d$, the degree of f denoted by $\deg(f)$ is the number n given by (10.17).*

Proposition 10.54 *Degree has the following properties:*

(a) $\deg(id_{S_0^d}) = 1$;
(b) $\deg(c) = 0$, where c is a constant map;
(c) $\deg(f \circ g) = \deg(f)\deg(g)$, where $f, g : S_0^d \to S_0^d$;
(d) if $f \sim g$, then $\deg(f) = \deg(g)$;
(e) if f can be extended to a continuous map $\tilde{f} : \bar{B}_0^{d+1} \to S_0^d$ such that $\tilde{f}(x) = f(x)$ for $x \in S_0^d$, then $\deg(f) = 0$.

Proof. The conclusions are immediate from the material presented in Chapter 6:
(a) follows from Proposition 6.40;
(b) follows from the definition of the homology of a map;
(c) follows from Theorem 6.58;
(d) follows from Corollary 6.68;
(e) we have $f = \tilde{f} \circ j$, where $j : S_0^d \to \bar{B}_0^{d+1}$ is the inclusion map. Then $f_* = \tilde{f}_* \circ j_*$ by Theorem 6.58. But $\bar{B}_0^{d+1}$ is acyclic, so $j_{*d} = 0$, hence $f_{*d} = 0$. For $d = 0$, we use this argument for maps on reduced homology groups: $j_{*0} : \tilde{H}_0^0(S_0^0) \to \tilde{H}_0^0(\bar{B}_0^1) = 0$. □

While our derivation of the definition of degree is extremely efficient, it provides no insight into the topological meaning of degree. The next example is meant to remedy this.

Example 10.55 Consider a parameterization of the unit circle

$$S_0^1 = \mathrm{bd}\,[-1,1]^2$$

by $t \in [0,8]$, which represents counterclockwise winding of the interval $[0,8]$ around the circle, starting from $(-1,1)$ as the image of $t = 0$ and ending at the same point at $t = 8$. This construction is analogous to what has been done for the boundary of the unit square in Section 5.4 and in Example 6.37. The explicit formula is

$$x(t) := \begin{cases} (-1,-1) + t(1,0) & \text{if } t \in [0,2], \\ (1,-1) + (t-2)(0,1) & \text{if } t \in [2,4], \\ (1,1) + (t-4)(-1,0)) & \text{if } t \in [4,6], \\ (-1,1) + (t-6)(0,-1) & \text{if } t \in [6,8]. \end{cases}$$

The above formula provides a continuous bijection of the interval $[0,8]$, with its endpoints identified together, onto S_0^1. Now let $f : S_0^1 \to S_0^1$ be given by

$$f(x(t)) := \begin{cases} x(2t) & \text{if } t \in [0,4], \\ x(2t-4) & \text{if } t \in [4,8]. \end{cases}$$

By the discussion in Section 5.4, f can be expressed in terms of parameter t as $t \mapsto 2t$ provided we make identification $2t \sim 2t - 4$. Hence, f is a doubling map on S_0^1. We interpret this geometrically as wrapping the circle twice in the counterclockwise direction. The calculation of M_f and its chain selector ϕ is left as Exercise 10.13. Note that rescaling is not necessary here. Once those steps are completed, it is easy to verify that $\phi_1(z) = 2z$ for the generating cycle

$$z = \left(\widehat{[-1,0]} + \widehat{[0,1]}\right) \diamond \widehat{[-1]} + \widehat{[1]} \diamond \left(\widehat{[-1,0]} + \widehat{[0,1]}\right)$$
$$- \left(\widehat{[-1,0]} + \widehat{[0,1]}\right) \diamond \widehat{[1]} - \widehat{[1]} \diamond \left(\widehat{[-1,0]} + \widehat{[0,1]}\right).$$

It follows that $\deg(f) = 2$.

It is now easy to guess that if we define, in a similar manner, a map expressed by $t \mapsto nt$, where n is any integer, then the induced degree should be n. For $|n| > 2$, however, one needs a rescaling in order to compute the homology map.

Observe that the degree is closely related to the Lefschetz number (see Definition 10.45):

Proposition 10.56 Let $f : S_0^d \to S_0^d$ be a continuous function and $d \geq 1$. Then
$$L(f) = 1 + (-1)^d \deg(f). \tag{10.18}$$

Proof. Since $H_k(S_0^d) = 0$ unless $k = 0$ or $k = d$, Definition 10.45 gives
$$L(f) = (-1)^0 \text{tr } f_{*0} + (-1)^d \text{tr } f_{*d} = \text{tr } f_{*0} + (-1)^d \deg(f).$$

By Example 6.27, f induces the identity in $H_0(S_0^d)$, so the first term in the sum is 1. □

Example 10.57 Consider the *antipodal map* $-\text{id} : S_0^d \to S_0^d$, which sends any x to its *antipode* $-x$. Clearly, the antipodal map has no fixed points on the sphere, so by the Lefschetz fixed-point theorem 10.46, $L(-\text{id}) = 0$. Thus, by Proposition 10.56,
$$\deg(-\text{id}_{S_0^d}) = (-1)^{d+1}.$$

The material presented above can serve to derive many interesting conclusions on fixed points of maps on odd- and even-dimensional spheres. By extending these results to Euclidean spheres S_2^d with the tools presented in Chapter 11, one can also derive the famous *Poincaré–Brouwer theorem* saying that there is no continuous nonvanishing tangent field on S_2^{2n}. Such fields exist on S_2^{2n-1}. We refer the reader to [21, 68].

Exercises

10.11 (a) Let $f, g : X \to S_0^d$ be two continuous maps. If $f(x) \neq -g(x)$ for all $x \in X$, show that $f \sim g$.
(b) Assume next that $X = S_0^d$. Deduce from (a) that if $f(x) \neq -x$ for all $x \in X$, then $f \sim 1_{S_0^d}$, and if $f(x) \neq x$ for all $x \in X$, then $f \sim -\text{id}$, where $-\text{id}$ is the antipodal map given by $-\text{id}(x) = -x$.

10.12 Here is an alternative way of proving that (c) implies (b) in Theorem 10.43. If $f : \bar{B}_0^{d+1} \to \bar{B}_0^{d+1}$ is fixed-point-free, the retraction $r : \bar{B}_0^{d+1} \to S_0^d$ can be defined as follows. For any $x \in \bar{B}_0^{d+1}$, since $f(x) \neq x$, there is a unique half-line emanating from $f(x)$ and passing through x. That open half-line intersects S_0^d at exactly one point $r(x)$. If $x \in S_0^d$, then obviously $r(x) = x$. Complete the proof by showing that r is continuous.

10.13 Complete the arguments in Example 10.55 by computing the multi-valued representation of f and its chain selector.

10.14 Consider the symmetry $f_i : S_0^d \to S_0^d$ sending $x = (x_1, x_2, \ldots, x_{d+1})$ to the point y with coordinates $y_j = x_j$ if $j \neq i$ and $y_i = -x_i$. Prove that $\deg(f_i) = -1$,

(a) in the case $d = 1$;
(b) in the general case.

10.15 Consider the coordinate exchange map $f_{i,i+1} : S_0^d \to S_0^d$ sending $x = (x_1, x_2, \ldots, x_{d+1})$ to the point y with coordinates $y_i = x_{i+1}$, $y_{i+1} = x_i$, and $y_j = x_j$ if $j \neq i, i+1$. Prove that $\deg(f_{i,i+1}) = -1$,

(a) in the case $d = 1$;
(b) in the general case.

10.16 Let $f : S_0^d \to S_0^d$ be a given continuous map.
(a) If $\deg(f) \neq (-1)^{d+1}$, show that f must have a fixed point.
(b) If $\deg(f) \neq 1$, show that f must send a point x to its antipode $-x$.

10.5.2 Topological Degree

The reader may be somewhat bewildered at this point. In the introduction to this section we motivated to study degree by stating that it would be used to prove the existence of solutions to $f(x) = 0$. Obviously, if $f : S_0^d \to S_0^d$, then for any $x \in S_0^d$, $f(x) \neq 0$. We will now resolve this apparent non sequitur.

Consider the open subset U of R^d given by

$$U := B_0^d(0, m) = (-m, m)^d,$$

where $m > 0$ is a given positive integer. Then U is a representable set whose closure

$$\operatorname{cl} U = \bar{B}^d(0, m) = [-m, m]^d$$

and whose boundary in $\mathbf{R}^d$ (see Definition 12.27)

$$\operatorname{bd} U = S_0^{d-1}(0, m) = \{x \in \operatorname{cl} U \mid x_i \in \{-m, m\} \text{ for some } i = 1, 2, \ldots, d\}$$

are cubical sets.

We are finally in a position to return to the original problem of finding solutions to

$$f(x) = 0 \tag{10.19}$$

for a continuous map $f : \operatorname{cl} U :\to R^d$ such that

$$f(x) \neq 0 \quad \text{for all } x \in \operatorname{bd} U. \tag{10.20}$$

Solutions x of (10.19) are called *roots* of f. The reason for imposing the condition (10.20) is that roots on the boundary of the domain are sensitive to small perturbations of f and cannot be detected by topological methods, as the following example illustrates.

Example 10.58 Consider $U = (-1, 1) \subset \mathbf{R}$ and a continuous function $f : [-1, 1] \to \mathbf{R}$. Put $a = f(-1)$ and $b = f(1)$. We examine the following cases.

Case 1. $a, b \neq 0$ and are of opposite signs, say $a < 0$ and $b > 0$. By the intermediate-value theorem, f must have a root in U. Moreover, any sufficiently small perturbation f_ϵ of f has the same property. Indeed, it is enough to choose
$$\epsilon \leq \frac{1}{2} \min\{|a|, |b|\}.$$
If $|f_\epsilon(x) - f(x)| \leq \epsilon$, we see that $f_\epsilon(-1) < 0$ and $f_\epsilon(1) > 0$, so the intermediate-value theorem can be applied to f_ϵ. We will later see that this is a very simple case of a map with nonzero degree.

Case 2. $a, b \neq 0$ have the same sign, say, $a, b > 0$. Then f may have no roots in U. For example, if $f(x) := 1$, f has no roots. If $f(x) := x^2$, it has a root $x = 0$, but it would disappear if we consider a small perturbation $f_\epsilon = x^2 + \epsilon$. We will later see that this is a very simple case of a map of degree zero.

Case 3. $a = 0$ or $b = 0$, so assume $a = 0$ and $b > 0$. Consider $f(x) = x + 1$. Then f has a root $x = -1$. A small perturbation $f_\epsilon = x + 1 + \epsilon$ of f admits no roots. Another small perturbation $f_{-\epsilon} = x + 1 - \epsilon$ falls into Case 1, so not only does it have a root but it also has the stability property described there.

We will reduce the problem of detecting roots of f to the study of maps in the unitary sphere S_0^{d-1} of the previous section. Define $\bar{f} : S_0^{d-1} \to S_0^{d-1}$ as follows:
$$\bar{f}(x) := \frac{f(mx)}{\|f(mx)\|_0}. \tag{10.21}$$

Note that $mx \in \mathrm{bd}\, U$ if and only if $x \in S_0^{d-1}$, so (10.20) implies that $\bar{f}$ is well defined.

Definition 10.59 Let $f : \mathrm{cl}\, U \to \mathbf{R}^d$ be a continuous map satisfying (10.20). Then the *topological degree* or, shortly, the *degree* of f on U is defined by the formula
$$\deg(f, U) := \deg(\bar{f}), \tag{10.22}$$
where $\deg(\bar{f})$ is given by Definition 10.53.

Here is the theorem relating the degree to the detection of roots.

Theorem 10.60 Let $f : \mathrm{cl}\, U \to \mathbf{R}^d$ be a continuous map satisfying (10.20). If $\deg(f, U) \neq 0$, then (10.19) has a solution in U.

Proof. We argue by contradiction. Suppose that
$$f(x) \neq 0$$
for all $x \in \mathrm{cl}\, U = \bar{B}_0^d(0, m)$. Then
$$f(mx) \neq 0$$

for all x in the closed unitary ball $\bar{B}_0^d$, so Eq. (10.21) can be extended to all $x \in \bar{B}_0^d$. By Proposition 10.54, $\deg(f, U) = \deg(\bar{f}) = 0$. □

We shall now discuss the homotopy invariance of degree analogous to the one established for the degree on spheres. Since we demand that f has no roots on bd U, it is natural to assume this property is maintained by a homotopy $h : \operatorname{cl} U \times [0,1] \to R^d$. More precisely, we require that $h(\cdot, t)$ have no roots on bd U for each $t \in [0,1]$. To make this more transparent, we will rephrase (10.20) in terms of maps on cubical pairs, that is we consider continuous maps

$$f : (\operatorname{cl} U, \operatorname{bd} U) \to (\mathbf{R}^d, \mathbf{R}^d \setminus \{0\}).$$

This notation means that $f : \operatorname{cl} U \to \mathbf{R}^d$ has the property $f(\operatorname{bd} U) \subset \mathbf{R}^d \setminus \{0\}$, which is equivalent to (10.20).

Definition 10.61 *Two maps $f, g : (\operatorname{cl} U, \operatorname{bd} U) \to (\mathbf{R}^d, \mathbf{R}^d - \{0\})$ are homotopic on pairs if there exists a continuous map*

$$h : (\operatorname{cl} U \times [0,1], \operatorname{bd} U \times [0,1]) \to (\mathbf{R}^d, \mathbf{R}^d \setminus \{0\})$$

such that $h(x, 0) = f(x)$ and $h(x, 1) = g(x)$ for all $x \in \operatorname{cl} U$.

Proposition 10.62 *Let $f, g : (\operatorname{cl} U, \operatorname{bd} U) \to (\mathbf{R}^d, \mathbf{R}^d \setminus \{0\})$ be homotopic on pairs. Then $\deg(f, U) = \deg(g, U)$.*

Proof. If $h : (\operatorname{cl} U \times [0,1], \operatorname{bd} U \times [0,1]) \to (\mathbf{R}^d, \mathbf{R}^d \setminus \{0\})$ is a homotopy on pairs joining f to g, then the map $\bar{h} : S_0^d \to S_0^d$ defined by

$$\bar{h}(x, t) := \frac{h(mx, t)}{\|h(mx, t)\|_0}$$

is a continuous homotopy on S_0^d. The conclusion now follows from Proposition 10.54(d). □

The following proposition links homotopy to stability under perturbation.

Proposition 10.63 *Let $f : (\operatorname{cl} U, \operatorname{bd} U) \to (\mathbf{R}^d, \mathbf{R}^d \setminus \{0\})$ be a continuous map. Then there exists an $\epsilon > 0$ such that if $f_\epsilon : \operatorname{cl} U \to \mathbf{R}^d$ satisfies*

$$\|f_\epsilon(x) - f(x)\|_0 \leq \epsilon \quad \text{for all } x \in U,$$

then $f_\epsilon : (\operatorname{cl} U, \operatorname{bd} U) \to (\mathbf{R}^d, \mathbf{R}^d \setminus \{0\})$ and $\deg(f_\epsilon) = \deg(f)$.

Proof. Since bd U is compact, and the continuous function $x \mapsto \|f(x)\|_0$ has no zero on bd U, it follows that

$$c := \inf\{\|f(x)\|_0 \mid x \in \operatorname{bd} U\} > 0.$$

It is enough to take any $\epsilon < c$ and the homotopy given by

$$h(x, t) := (1 - t)f(x) + tf_\epsilon(x).$$

We get

$$\|h(x,t) - f(x)\|_0 = \|-tf(x) + tf_\epsilon(x)\|_0 \leq t\|f(x) - f_\epsilon(x)\|_0 \leq \epsilon.$$

If $h(x,t) = 0$ for some $(x_0, t_0) \in \operatorname{bd} U \times [0,1]$, then we get $\|f(x_0)\|_0 \leq \epsilon$, which contradicts the choice of ϵ. Thus h is a homotopy on pairs and the conclusion follows from Proposition 10.62. □

We finish with a series of examples heading toward various applications of degree.

Example 10.64 Consider the map $f : \mathbf{R}^2 \to \mathbf{R}^2$ given by

$$f(x_1, x_2) := (x_1^2 - x_2^2, 2x_1 x_2). \tag{10.23}$$

We will show that $\deg(f, B_0^2) = 2$. We pass to polar coordinates

$$\begin{cases} x_1 = r\cos\theta, \\ x_2 = r\sin\theta, \end{cases}$$

where $r \geq 0$ and $\theta \in [0, 2\pi]$ are referred to, respectively, as *the radius* and *the argument* of $x = (x_1, x_2)$. By double angle formulas we get

$$f(r\cos\theta, r\sin\theta) = r^2(\cos 2\theta, \sin 2\theta).$$

The related map $\bar{f} : S_0^1 \to S_0^1$ is given by

$$\bar{f}(r\cos\theta, r\sin\theta) = \frac{(\cos 2\theta, \sin 2\theta)}{\max\{|\cos 2\theta|, |\sin 2\theta|\}},$$

so in terms of the argument θ it is a doubling map $\theta \mapsto 2\theta$ upon identification $\theta + 2\pi \sim \theta$. Note that the argument values

$$\theta = k\frac{\pi}{4},\ k = 0, 1, 2, \ldots, 8,$$

correspond to elementary vertices in S_0^1 and $\bar{f}$ maps them to vertices corresponding, respectively, to $2k\frac{\pi}{4}$. Intervals

$$\theta \in [k\frac{\pi}{4}, (k+1)\frac{\pi}{4}]$$

correspond to elementary intervals in S_0^1 and $\bar{f}$ maps them to pairs of two adjacent elementary intervals corresponding to

$$[2k\frac{\pi}{4}, (2k+1)\frac{\pi}{4}] \cup [(2k+1)\frac{\pi}{4}, (2k+2)\frac{\pi}{4}].$$

The map $\bar{f}$ is very similar to the map discussed in Example 10.55 except for two points: One, the loop there starts at the vertex $(-1, -1)$ and here at $(1, 0)$. Two, the map in Example 10.55 is linear on each elementary interval and here is not. But in the light of the above discussion, the multivalued representation and its chain selector can be calculated in the same way. After some calculation we get

$$\deg(f, B_0^2) = \deg(\bar{f}) = 2.$$

Equation (10.23) defining f looks complicated, but it can be demystified by the use of complex numbers. Any $x = (x_1, x_2) \in \mathbf{R}$ is identified with the complex number $z = x_1 + \mathbf{i}x_2$, where $\mathbf{i} := \sqrt{-1}$. The complex multiplication gives

$$z^2 = x_1^2 - x_2^2 + \mathbf{i}2x_1 x_2 = f(x_1, x_2),$$

so our result reflects the fact that 0 is the root of z^2 of multiplicity two. We may now be tempted to guess that

$$\deg(z^n, B_0^2) = n.$$

This can be proved in a similar way, but one needs to consider a homotopy joining the map on a circle given by $\theta \mapsto n\theta$ with the map given by $t \mapsto nt$ discussed in Example 10.55 and consider an appropriate rescaling.

As the reader may already have observed, it is easy to check that $f(0,0) = (0,0)$, and therefore using degree theory to prove the existence of a zero appears to be overkill. On the other hand, the same type of analysis used here leads to a proof of the fundamental theorem of algebra (see Exercise 10.19).

Example 10.65 Let A be any $d \times d$ real matrix and let $f_A : \mathbf{R}^d \to \mathbf{R}^d$ be given by $f_A(x) := Ax$. Assume that A is nonsingular, that is, $\det A \neq 0$. Then 0 is the only root of f_A in $\mathbf{R}^d$, so the degree of f_A is well defined on $U = B_0(0, m)$ for any value of m and

$$\deg(f_A, U) = \operatorname{sgn}(\det A) = (-1)^{\det A}. \tag{10.24}$$

Indeed, by linearity of f_A and homogeneity of the norm, the related map on the unit sphere is

$$\bar{f}(x) = \frac{f_A(mx)}{||f_A(mx)||_0} = \frac{Ax}{||Ax||_0}.$$

Any real matrix can be expressed as a product of real elementary matrices analogous to those studied in Chapter 3 in the integer case:

- $E_{i,i+1}$ obtained by exchange of rows i and $i+1$ in the identity matrix;
- $E_i(c)$ obtained by multiplication of row i by $c \in \mathbf{R} \setminus \{0\}$; and
- $E_{i,j,q}$ obtained by adding row j times q to row i, $i \neq j$, $q \in \mathbf{R}$.

By the product formula in Proposition 10.54 and the analogous product formula for determinants, the problem is reduced to finding the degree of an elementary matrix.

From Exercise 10.15 we get

$$\deg(f_{E_{i,i+1}}, U) = -1.$$

From Exercise 10.14 we get

$$\deg(f_{E_i(c)}, U) = -\operatorname{sgn} c.$$

Finally, the map $h : (\operatorname{cl} U \times [0,1], \operatorname{bd} U \times [0,1]) \to (\mathbf{R}^d, \mathbf{R}^d \setminus \{0\})$ given by

$$h(x,t) := E_{i,j,(1-t)q} x$$

is a homotopy joining $f_{E_{i,j,q}}$ to the identity map, hence

$$\deg(f_{E_{i,j,q}}, U) = 1.$$

Now the formula follows from Proposition 10.54(c) and the analogous product formula for determinants.

Example 10.66 Consider the differential equation

$$\dot{x} = f(x), \quad x \in \mathbf{R}^d, \qquad (10.25)$$

where $f : \mathbf{R}^d \to \mathbf{R}^d$ is a continuous map such that the associated flow is well defined on $\mathbf{R}^d \times \mathbf{R}$.

Let A be a nonsingular real $d \times d$ matrix and suppose that f satisfies the condition

$$\|f(x) - Ax\|_0 \le c \quad \text{for all } x \in \mathbf{R}^d.$$

for some $c > 0$. We show that Eq. (10.25) has an *equilibrium solution* given by $x(t) := \bar{x}$ for all t. This is equivalent to showing that there exist $\bar{x} \in \mathbf{R}^d$ such that $f(\bar{x}) = 0$. By the assumption on A, the number

$$M = \min\{\|Ax\|_0 \mid x \in S_0^{d-1}\}$$

is strictly positive. Choose an integer $m > M^{-1}c$. It is easy to show that the equation $f(x) = 0$ cannot have solutions on $\operatorname{bd} U$ for $U = (-m, m)^d$ so $\deg(f, U)$ is well defined. Moreover, by the same arguments as in the proof of Proposition 10.63, f is homotopic to f_A given by the matrix A, so by Example 10.65,

$$\deg(f, U) = \deg(f_A) = (-1)^{\det A}$$

and the conclusion follows. Note that we also found a bound for the norm of a possible stationary point, $\|\bar{x}\|_0 < M^{-1} c$.

Consider a continuous map

$$f : (\operatorname{cl} U, \operatorname{bd} U) \to (\mathbf{R}^d, \mathbf{R}^d \setminus \{0\}),$$

where $U = B_0(0, m)$. As we mention in the introduction to this section, the equation $f(x) = 0$ is equivalent to the fixed-point equation $g(x) = x$, where $g(x) = x - f(x)$. The condition $f(x) \neq 0$ on $\operatorname{bd} U$ is equivalent to saying that g does not have fixed points on $\operatorname{bd} U$. The presence of fixed points of g in U can be detected by the inequality

$$\deg(\operatorname{id}_{\operatorname{cl} U} - g, U) \neq 0.$$

Many interesting applications of the degree theory to nonlinear ordinary and partial differential equations require generalizing the degree to infinite-dimensional spaces of functions. The reader interested in pursuing this topic is referred to [21, 49].

Exercises

10.17 Let $U := (-1, 1)$, and let $f : [-1, 1] \to \mathbf{R}$ be any continuous map such that $f(t) \neq 0$ for $t = -1, 1$. Show that $\deg(f, U)$ is either 0, 1, or -1. Interpret your answer graphically.

10.18 Let $U = (-m, m)^2$. By taking for granted that $\deg(z^n, U) = n$, where z^n is the complex polynomial discussed in Example 10.64, construct a map $f : [-m, m]^2 \to \mathbf{R}^2$ with precisely n distinct roots in U and $\deg(f, U) = n$.

10.19 Consider $U = (-m, m)^2$ as in the previous exercise and the complex polynomial
$$f(z) := z^n + \alpha_{n-1} z^{n-1} + \cdots + \alpha_1 z + \alpha_0,$$
where α_i are real or complex coefficients. Again, by taking for granted that $\deg(z^n, U) = n$, show that $\deg(f, U) = n$ for sufficiently large m. Conclude that f must have a root in the plane. This conclusion is known as the *fundamental theorem of algebra*.

10.6 Complicated Dynamics

As we have mentioned in the introduction, one of the goals of this chapter is to show that the algebraic topological tools being developed can be combined with standard numerical techniques to provide rigorous proofs about the dynamics of nonlinear systems. In Section 10.3 it is shown that homology can be used to prove the existence of invariant sets. However, the mathematical justification relies on the Ważewski principle, which is not the ideal tool to combine with rigorous numerical computations. This can be seen from Example 10.42, where the existence of an invariant set for a particular system does not imply the existence of an invariant set for a perturbed system. Of course, because of floating-point errors and discretization errors, we can never expect numerical computations to exactly reproduce the nonlinear system being simulated. Therefore, the tools needed for computer-assisted proofs must have the feature that they guarantee the existence of invariant sets not only for a particular flow but also for the nearby flows. Obviously, this property by itself will not prove anything about the particular system from the study of the nearby system obtained by means of numerical computations. Nevertheless, without such a feature there is no hope for any computer-assisted proof. This means that we need to replace the Ważewski principle by something that persists under perturbations.

Another limitation of the Ważewski principle as described in the previous section is that it is not applicable to maps. Recall that the proof of Theorem 10.37 makes essential use of the flow to construct the deformation retraction.

An alternative approach, the Conley index, overcomes both these problems. This theory is applicable both to flows and to maps, and, furthermore, it has the important property that it can be used for rigorous computation. As such it provides a powerful method for computer-assisted proofs in dynamics. While a full development of the Conley index is beyond the scope of this book, our goal is to introduce some of the ideas that are essential to its effective computation. We will use the Ważewski principle to motivate the small part of the theory that we will present here. Since most numerical schemes result in maps (time discretization is a typical method for studying ordinary differential equations), we will present the ideas in the context of continuous functions rather than flows. However, it should be emphasized that there is a rigorous procedure for converting computations on such maps into results for the original flow.

10.6.1 Index Pairs and Index Map

As mentioned in the introduction, one of the goals of this section is to demonstrate that the computer can be used to rigorously study the dynamics of nonlinear systems. This leads us to adopt the following philosophy. We are interested in the dynamics of a continuous map $f : \mathbf{R}^d \to \mathbf{R}^d$, but because we use the computer to perform calculations, we restrict our attention to the information available in the form of combinatorial multivalued maps

$$\mathcal{F} : \mathcal{K}_{\max} \rightrightarrows \mathcal{K}_{\max},$$

where $\mathcal{K}_{\max} := \mathcal{K}_{\max}(\mathbf{R}^d)$ is the set of all maximal (i.e., d-dimensional) elementary cubes in $\mathbf{R}^d$. Such maps are introduced in Definition 5.3 in the context of approximating continuous functions. In Chapter 6 the related lower semicontinuous maps $F = \lfloor \mathcal{F} \rfloor$ [see (6.7)] are used for defining homomorphisms induced by maps in homology.

At the risk of being redundant, observe that we will be working with three objects: a continuous function f that generates the dynamical system that we are interested in studying; a combinatorial map $\mathcal{F}$ that is generated by and manipulated by means of a computer; and a multivalued topological map F

that allows us to pass from the combinatorics of the computer to statements about the dynamics of f via homology. How this last step is performed is discussed in this section.

Clearly, to make use of such a strategy, there needs to be an appropriate relationship between f and $\mathcal{F}$. To explain it, we begin with a definition.

Definition 10.67 A set $X \subset \mathbf{R}^d$ is a *full cubical set* if there exists a finite subset $\mathcal{X} \in \mathcal{K}_{\max}(\mathbf{R}^d)$ such that

$$X = |\mathcal{X}| := \bigcup \mathcal{X}.$$

Obviously, any full cubical set is cubical (though, of course, not all cubical sets of $\mathbf{R}^d$ are of this form; they may be composed of lower-dimensional elementary cubes). The set $\mathcal{X}$ can be recovered from X by the formula $\mathcal{X} = \mathcal{K}_{\max}(X)$.

Given any closed bounded subset A of $\mathbf{R}^d$, let

$$\mathrm{wrap}(A) := \{Q \in \mathcal{K}_{\max} \mid Q \cap A \neq \emptyset\}.$$

The set $\mathrm{wrap}(A) \subset \mathcal{K}_{\max}$ is the *cubical wrap* of A. This notation and terminology are extended to cubical wraps of points $x \in \mathbf{R}^d$ by setting

$$\mathrm{wrap}(x) := \mathrm{wrap}(\{x\})$$

and to cubical wraps of finite collections of elementary cubes $\mathcal{X}$ in $\mathbf{R}^d$ by setting

$$\mathrm{wrap}(\mathcal{X}) := \mathrm{wrap}(|\mathcal{X}|).$$

In the latter case we will also use the *collar* of $\mathcal{X}$ defined by

$$\mathrm{col}(\mathcal{X}) := \mathrm{wrap}(\mathcal{X}) \setminus \mathcal{X}.$$

It is left as an exercise to verify that, given a cubical set X, $|\mathrm{wrap}(X)|$ is equal to $\bar{B}_1(X)$, the closed cubical ball of radius 1 about X, defined in Exercise 6.3.

Definition 10.68 A combinatorial multivalued map $\mathcal{F} : \mathcal{K}_{\max} \rightrightarrows \mathcal{K}_{\max}$ is a *combinatorial enclosure* of $f : \mathbf{R}^d \to \mathbf{R}^d$ if, for every $Q \in \mathcal{K}_{\max}$,

$$\mathrm{wrap}(f(Q)) \subset \mathcal{F}(Q).$$

We shall need the following characterization of topological interiors of sets in terms of surroundings of points:

Proposition 10.69 *Assume X is a full cubical set. Then*

$$\mathrm{int}\, X = \{x \in \mathbf{R}^d \mid \mathrm{wrap}(x) \subset \mathcal{K}_{\max}(X)\}.$$

Proof. Set $\mathcal{X} := \mathcal{K}_{\max}(X)$. To prove that the left-hand side is contained in the right-hand side, take $x \in \operatorname{int} X$ and assume that there is a $Q \in \operatorname{wrap}(x) \setminus \mathcal{X}$. By Proposition 2.15(iv), we can choose a sequence (x_n) in $\overset{\circ}{Q}$ such that $x_n \to x$. Using Proposition 2.15(vi), one can show that $\overset{\circ}{Q} \cap X = \emptyset$. Therefore, $x_n \notin X$ and consequently $x \notin \operatorname{int} X$.

To show the opposite inclusion, assume $x \in \mathbf{R}^d$ is such that $\operatorname{wrap}(x) \subset \mathcal{X}$. Then by Proposition 6.10(i),

$$x \in \operatorname{oh}(x) \subset |\operatorname{wrap}(x)| \subset X,$$

and since $\operatorname{oh}(x)$ is open by Proposition 6.10(ii), it follows that $x \in \operatorname{int} X$. □

As an easy consequence we get the following result.

Proposition 10.70 *If* $\mathcal{A}, \mathcal{B} \subset \mathcal{K}_{\max}$, *then*

$$|\mathcal{A}| \cap \operatorname{int} |\mathcal{B}| \subset |\mathcal{A} \cap \mathcal{B}|.$$

Proof. Assume $x \in |\mathcal{A}| \cap \operatorname{int} |\mathcal{B}|$. Then there exists a $Q \in \mathcal{A}$ such that $x \in Q$. By Proposition 10.69, $Q \in \mathcal{B}$. Therefore, $x \in |\mathcal{A} \cap \mathcal{B}|$. □

An important immediate corollary of Proposition 10.69 is the following alternative condition for a combinatorial enclosure.

Proposition 10.71 *A combinatorial multivalued map* $\mathcal{F} : \mathcal{K}_{\max} \rightrightarrows \mathcal{K}_{\max}$ *is a combinatorial enclosure of* $f : \mathbf{R}^d \to \mathbf{R}^d$ *if, for every* $x \in \mathbf{R}^d$ *and* $Q \in \operatorname{wrap}(x)$,

$$f(Q) \subset \operatorname{int}(|\mathcal{F}(Q)|).$$

Proposition 10.71 is helpful in understanding the topology behind Definition 10.68, but the definition itself leads directly to an algorithm constructing a combinatorial enclosure of a given map. As in Chapter 7, we restrict the statement of the algorithm to rational maps.

Algorithm 10.72 Combinatorial enclosure
```
function combinatorialEnclosure(cubicalSet X, rationalMap f)
    for each Q in X do
        A := evaluate(f, Q);
        U := cubicalWrap(A);
        F{Q} := U;
    endfor;
    return F;
```

The presentation of an algorithm `cubicalWrap` computing $\operatorname{wrap}(X)$, where X is a rectangle, is left as an exercise.

Notice that the data structure `cubicalSet` makes no distinction between a cubical set X and the finite set $\mathcal{X} = \mathcal{K}_{\max}(X)$, so the algorithm `cubicalWrap` will equally apply to sets $\mathcal{X} \in \mathcal{K}_{\max}$.

Having established the relationship between f and $\mathcal{F}$, we now turn to the question of how homology can be used to understand the dynamics of f. As discussed in the introduction, our goal is to introduce a topological object that is analogous to the Ważewski principle and robust with respect to perturbations. We have dealt with the question of sensitivity to perturbations when we introduced topological degree (see Example 10.58). In that setting we imposed the condition that the function could not be zero on the boundary of the set of interest. The following definition captures the same idea, but in the more general context of invariant sets.

Definition 10.73 A compact set $N \subset X$ is an *isolating neighborhood* for a continuous map f if
$$\operatorname{Inv}(N, f) \subset \operatorname{int} N.$$
An invariant set S is *isolated* under f if there exists an isolating neighborhood N such that
$$S = \operatorname{Inv}(N, f).$$

The following proposition, which is essentially a restatement of the definition, provides another characterization of the isolating neighborhood.

Proposition 10.74 *A compact set N is an isolating neighborhood for f if and only if, for every x in the boundary N, there exists no full solution $\sigma_x : \mathbf{Z} \to N$.*

Returning to the Ważewski principle for motivation, recall that given a flow, one begins with a pair of spaces (W, W^-) where W is a Ważewski set and W^- is the set of points that leave W immediately. We shall now introduce a similar set of conditions for the case of maps.

Let $f : X \to X$ be a continuous map. By an *f-boundary* of a set $A \subset X$, we mean the set $\operatorname{bd}_f(A) := \operatorname{cl}(f(A) \setminus A) \cap A$.

Example 10.75 Consider, for example, the linear map $f : \mathbf{R}^2 \to \mathbf{R}^2$ given by the matrix
$$f = \begin{bmatrix} 2 & 0 \\ 0 & 1/2 \end{bmatrix}$$
and let $A := \{x \in \mathbf{R}^2 \mid \|x\| \leq 4\}$. Obviously, $f(A) = [-8, 8] \times [-2, 2]$. Thus
$$\operatorname{bd}_f(A) = \operatorname{cl}\left(([-8, 8] \times [-2, 2]) \setminus [-4, 4]^2\right) \cap [-4, 4]^2 = \{\pm 4\} \times [-2, 2].$$

Definition 10.76 An *index pair* for f is a pair of compact sets $P = (P_1, P_0)$, where $P_0 \subset P_1$, satisfying the following three properties:

1. *(isolation)* $\operatorname{Inv}(\operatorname{cl}(P_1 \setminus P_0), f) \subset \operatorname{int}(\operatorname{cl}(P_1 \setminus P_0))$;
2. *(positive invariance)* $f(P_0) \cap P_1 \subset P_0$;
3. *(exit set)* $\operatorname{bd}_f(P_1) \subset P_0$.

Example 10.77 Returning to Example 10.75, observe that the pair

$$(P_1, P_0) = (A, \operatorname{bd}_f(A))$$

is an index pair for f.

The following example may appear to be too trivial to be worth presenting; however, it will reappear in a crucial manner at the end of this subsection.

Example 10.78 Observe that $\emptyset$ is a cubical set. Furthermore, given any continuous function f,

$$\emptyset = \operatorname{Inv}(\emptyset, f) \subset \operatorname{int} \emptyset = \emptyset.$$

Therefore, $\emptyset$ is an isolated invariant set.

In line with the previous remarks, index pairs are intended to be analogs of Ważewski sets where P_0 plays a role similar to that of W^-. Therefore, one might hope that if $H_*(P_1, P_0) \not\cong 0$, then $\operatorname{Inv}(\operatorname{cl}(P_1 \setminus P_0), f) \neq \emptyset$. Unfortunately, as the following example indicates, this is wrong.

Example 10.79 Consider $f : \mathbf{R} \to \mathbf{R}$ given by $f(x) = x + 1$. Let $N = \{0\} \cup \{1\}$. Clearly, $\operatorname{Inv}(N, f) = \emptyset$. Let $(P_1, P_0) = (N, \{1\})$. We leave it to the reader to check that this is an index pair. Observe that

$$H_k(P_1, P_0) = H_*(\{0\} \cup \{1\}, \{1\}) \cong \begin{cases} \mathbf{Z} & \text{if } k = 0, \\ 0 & \text{otherwise}. \end{cases}$$

This example suggests that we need a more refined approach if we wish to use algebraic topology to capture the dynamics. It turns out that the reason the index pairs themselves do not work as a counterpart of the Ważewski theorem in the discrete case is that nonhomotopic maps can still lead to the same index pairs. As will be shown later, what is missing is information available in the map f. However, since our philosophy is that it is the combinatorial map $\mathcal{F}$ and not f that is known, we need to think about dynamics on the combinatorial level.

The basic definitions related to the dynamics of a map can be carried over to combinatorial maps, but we must remember that the variables are now elementary cubes and not points in $\mathbf{R}^d$.

Definition 10.80 A *full solution* through $Q \in \mathcal{K}_{\max}$ under $\mathcal{F}$ is a function $\Gamma_Q : \mathbf{Z} \to \mathcal{K}_{\max}$ satisfying the following two properties:

1. $\Gamma_Q(0) = Q$;
2. $\Gamma_Q(n+1) \in \mathcal{F}(\Gamma_Q(n))$ for all $n \in \mathbf{Z}$.

On occasion we shall make use of solutions that satisfy these two properties on subsets of $\mathbf{Z}$. In particular, if $\Gamma_Q : \mathbf{Z}^+ \to \mathcal{K}_{\max}$, then Γ_Q is a *positive* solution through Q, and if $\Gamma_Q : \mathbf{Z}^- \to \mathcal{K}_{\max}$, then Γ_Q is a *negative* solution through Q.

Definition 10.81 Let $\mathcal{N}$ be a finite subset of $\mathcal{K}_{\max}$. The *maximal invariant set* in $\mathcal{N}$ under $\mathcal{F}$ is

$$\text{Inv}\,(\mathcal{N}, \mathcal{F}) := \{Q \in \mathcal{N} \mid \text{there exists a full solution } \Gamma_Q : \mathbf{Z} \to \mathcal{N}\}.$$

We also define the *maximal positively invariant set* and the *maximal negatively invariant set* in $\mathcal{N}$ under $\mathcal{F}$, respectively, by

$$\text{Inv}^+(\mathcal{N}, \mathcal{F}) := \{Q \in \mathcal{N} \mid \text{there exists a positive solution } \Gamma_Q : \mathbf{Z}^+ \to \mathcal{N}\},$$
$$\text{Inv}^-(\mathcal{N}, \mathcal{F}) := \{Q \in \mathcal{N} \mid \text{there exists a negative solution } \Gamma_Q : \mathbf{Z}^- \to \mathcal{N}\}.$$

Although maximal invariant sets are the subject of our investigation, they are usually not explicitly known. However, their combinatorial counterparts can easily be determined. With this in mind we present an alternative characterization of $\text{Inv}\,(\mathcal{N}, \mathcal{F})$ that leads to an algorithm for computing it. First, given a combinatorial map $\mathcal{F} : \mathcal{K}_{\max} \rightrightarrows \mathcal{K}_{\max}$, we would like to define its inverse $\mathcal{F}^{-1} : \mathcal{K}_{\max} \rightrightarrows \mathcal{K}_{\max}$ by the formula:

$$\mathcal{F}^{-1}(R) := \{Q \in \mathcal{K}_{\max} \mid R \in \mathcal{F}(Q)\}.$$

There is a potential problem with this formula. On this level of generality it is possible that $\mathcal{F}^{-1}$ consists of an infinite number of cubes. The simplest way to avoid this is by restricting the domain of $\mathcal{F}$ to a finite subset $\mathcal{X} \subset \mathcal{K}_{\max}$.

Thus we define $\mathcal{F}_{\mathcal{X}} : \mathcal{X} \rightrightarrows \mathcal{K}_{\max}$ to be the restriction of $\mathcal{F}$ to $\mathcal{X}$ given by $\mathcal{F}_{\mathcal{X}}(Q) := \mathcal{F}(Q)$ for all $Q \in \mathcal{X}$. The inverse of $\mathcal{F}_{\mathcal{X}}$ is given by

$$\mathcal{F}_{\mathcal{X}}^{-1}(R) := \{Q \in \mathcal{X} \mid R \in \mathcal{F}(Q)\}$$

for all $R \in \mathcal{K}_{\max}$ and has values that consist of finite sets of cubes.

The following proposition is straightforward.

Proposition 10.82 *Assume $\mathcal{N} \subset \mathcal{K}_{\max}$ and $\mathcal{F} : \mathcal{K}_{\max} \rightrightarrows \mathcal{K}_{\max}$. Then*

$$\text{Inv}\,(\mathcal{N}, \mathcal{F}) = \text{Inv}^-(\mathcal{N}, \mathcal{F}) \cap \text{Inv}^+(\mathcal{N}, \mathcal{F}), \tag{10.26}$$
$$\mathcal{F}(\text{Inv}^-(\mathcal{N}, \mathcal{F})) \cap \mathcal{N} \subset \text{Inv}^-(\mathcal{N}, \mathcal{F}), \tag{10.27}$$
$$\mathcal{F}^{-1}(\text{Inv}^+(\mathcal{N}, \mathcal{F})) \cap \mathcal{N} \subset \text{Inv}^+(\mathcal{N}, \mathcal{F}). \tag{10.28}$$

The following theorem is a technical result needed in the algorithmic construction of index pairs.

Theorem 10.83 *Let $\mathcal{F} : \mathcal{K}_{\max} \rightrightarrows \mathcal{K}_{\max}$ be a combinatorial multivalued map and let $\mathcal{N}$ be a finite subset of $\mathcal{K}_{\max}$. Construct the sequence of subsets $\mathcal{S}_j$, $j = 0, 1, 2, \ldots$, of $\mathcal{K}_{\max}$ by induction as follows:*

$$\mathcal{S}_0 := \mathcal{N};$$
$$\mathcal{S}_{j+1} := \mathcal{F}(\mathcal{S}_j) \cap \mathcal{S}_j \cap \mathcal{F}_{\mathcal{N}}^{-1}(\mathcal{S}_j).$$

Then the sequence eventually becomes constant, that is, there exists an index j_0 such that $\mathcal{S}_j = \mathcal{S}_{j_0}$ for all $j \geq j_0$. Moreover, for such j_0, we have

$$\text{Inv}\,(\mathcal{N}, \mathcal{F}) = \mathcal{S}_{j_0}.$$

10.6 Complicated Dynamics   349

Proof. It is clear that the sequence of sets $\mathcal{S}_j$ is decreasing, namely $\mathcal{S}_{j+1} \subset \mathcal{S}_j$ for all $j = 0, 1, 2, \ldots$. Since $\mathcal{S}_0$ is finite, there must be an integer j_0 such that $\mathcal{S}_j = \mathcal{S}_{j_0}$ for all $j \geq j_0$, so the first statement is proved. To prove that $\text{Inv}(\mathcal{N}, \mathcal{F}) = \mathcal{S}_{j_0}$, first notice that we have

$$\mathcal{S}_{j_0} = \mathcal{S}_{j_0+1} = \mathcal{F}(\mathcal{S}_{j_0}) \cap \mathcal{S}_{j_0} \cap \mathcal{F}_\mathcal{N}^{-1}(\mathcal{S}_{j_0}).$$

This implies that for every element $Q \in \mathcal{S}_{j_0}$ there are elements $K \in \mathcal{S}_{j_0}$ and $L \in \mathcal{S}_{j_0}$ such that $Q \in \mathcal{F}(K)$ and $L \in \mathcal{F}(Q)$. The same arguments can be again applied to K and L. Thus a solution $\Gamma_Q : \mathbf{Z} \to \mathcal{K}_{\max}$ through Q under $\mathcal{F}$ can be defined by putting

$$\Gamma_Q(0) := Q \ , \ \Gamma_Q(-1) := K \ , \ \Gamma_Q(1) := L,$$

and applying induction twice, for $n = 1, 2, 3, \ldots$ and for $n = -1, -2, -3, \ldots$. Since $\mathcal{S}_{j_0} \subset \mathcal{N}$, the solution obviously has values in $\mathcal{N}$. Since Q is arbitrary in $\mathcal{S}_{j_0}$, this proves that $\mathcal{S}_{j_0} = \text{Inv}(\mathcal{S}_{j_0}, \mathcal{F}) \subset \text{Inv}(\mathcal{N}, \mathcal{F})$.

To prove the reverse inclusion, notice that

$$\mathcal{S}_{j_0} = \bigcap_{k=0}^{j_0} \mathcal{S}_k = \bigcap_{k=0}^{\infty} \mathcal{S}_k,$$

so it is enough to prove that $\text{Inv}(\mathcal{N}, \mathcal{F}) \subset \mathcal{S}_k$ for all $k = 0, 1, 2, \ldots$. We will prove this by induction on k. By definition, $\text{Inv}(\mathcal{N}, \mathcal{F}) \subset \mathcal{N} = \mathcal{S}_0$ so the hypothesis is true for $k = 0$. Now let $k \geq 0$ and suppose that $\text{Inv}(\mathcal{N}, \mathcal{F}) \subset \mathcal{S}_k$. Then any $Q \in \text{Inv}(\mathcal{N}, \mathcal{F})$ admits a solution $\Gamma_Q : \mathbf{Z} \to \text{Inv}(\mathcal{N}, \mathcal{F})$ through Q under $\mathcal{F}$. Put $Q_n := \Gamma_Q(n)$ for $n \in \mathbf{Z}$. By the induction hypothesis, $Q_n \in \mathcal{S}_k$ for all n. We have $Q = Q_0 \in \mathcal{F}(Q_{-1})$ and $Q_1 \in \mathcal{F}(Q)$. The last identity is equivalent to $Q \in \mathcal{F}_\mathcal{N}^{-1}(Q_1)$. We get

$$Q \in \mathcal{S}_k \cap \mathcal{F}(Q_{-1}) \cap \mathcal{F}_\mathcal{N}^{-1}(Q_1) \subset \mathcal{S}_k \cap \mathcal{F}(\mathcal{S}_k) \cap \mathcal{F}_\mathcal{N}^{-1}(\mathcal{S}_k) = \mathcal{S}_{k+1}.$$

Thus $Q \in \mathcal{S}_{k+1}$ and we have proved that $\text{Inv}(\mathcal{N}, \mathcal{F}) \subset \mathcal{S}_{k+1}$. □

The above theorem guarantees that the following algorithm always stops and returns the set $\text{Inv}(\mathcal{N}, \mathcal{F})$.

Algorithm 10.84 Combinatorial invariant set
    function invariantPart(set N, combinatorialMap F)
    F := restrictedMap(F, N);
    Finv := evaluateInverse(F);
    S := N;
    repeat
        S' := S;
        S'' := **intersection**(S, evaluate(F, S));
        S := **intersection**(S'', evaluate(Finv, S));
    until (S = S');
    return S;

The construction of algorithms restrictedMap and evaluateInverse evaluating, respectively, the restricted map $\mathcal{F}_N$ and the inverse map $\mathcal{F}^{-1}$ are left as exercises.

Definition 10.85 Let $\mathcal{N}$ be a finite subset of $\mathcal{K}_{\max}$. The set $\mathcal{N}$ is an *isolating neighborhood* if
$$\mathrm{wrap}(\mathrm{Inv}\,(\mathcal{N},\mathcal{F})) \subset \mathcal{N}. \tag{10.29}$$

Now, given a cubical multivalued map $\mathcal{F}$ and a candidate $\mathcal{N}$ for an isolating neighborhood in the sense of Definition 10.85, we can construct an index pair for f using the following algorithm.

Algorithm 10.86 Combinatorial index pair
```
    function indexPair(cubicalSet N, combinatorialMap F)
    S := invariantPart(N, F);
    M := cubicalWrap(S);
    if subset(M, N) then
        F := restrictedMap(F, M);
        C := collar(S);
        P0 := intersection(evaluate(F, S), C);
        repeat
            lastP0 := P0;
            P0 := intersection(evaluate(F, P0), C);
            P0 := union(P0, lastP0);
        until (P0 = lastP0);
        P1 := union(S, P0);
        Pbar1 := evaluate(F, P1);
        Pbar0 := setminus(Pbar1, S);
        return (P1, P0, Pbar1, Pbar0);
    else
        return "Failure";
    endif
```

Theorem 10.87 *Assume Algorithm 10.86 is called with* F *containing a combinatorial enclosure* $\mathcal{F}$ *of a single-valued continuous map* $f : \mathbf{R}^d \to \mathbf{R}^d$ *and* N *a representation* $\mathcal{N}$ *of a cubical set* N. *Furthermore, assume that it does not return* "Failure". *Then it returns representations* $\mathcal{P}_1, \mathcal{P}_0, \bar{\mathcal{P}}_1, \bar{\mathcal{P}}_0$ *of cubical sets* $P_1, P_0, \bar{P}_1, \bar{P}_0$, *such that if* $\mathcal{S} := \mathrm{Inv}\,(\mathcal{N}, \mathcal{F})$, *then*

(i) $|\mathcal{S}|$ *is an isolating neighborhood for* f,
(ii) (P_1, P_0) *is an index pair for* f *which isolates* $\mathrm{Inv}\,(|\mathcal{S}|, f)$,
(iii) $P_i \subset \bar{P}_i$ *for* $i = 0, 1$,
(iv) $f(P_i) \subset \bar{P}_i$ *for* $i = 0, 1$,

Proof. Let $\mathcal{P}_0^{(0)}$ denote the first value assigned to variable P0 and let $\mathcal{P}_0^{(i)}$ for $i = 1, 2, \ldots$ denote the value of this variable when leaving the ith pass of the repeat loop. Then

$$\mathcal{P}_0^{(0)} = \mathcal{F}(\mathcal{S}) \cap \operatorname{col}(\mathcal{S}), \tag{10.30}$$

$$\mathcal{P}_0^{(i+1)} = \mathcal{P}_0^{(i)} \cup \left(\mathcal{F}(\mathcal{P}_0^{(i)}) \cap \operatorname{col}(\mathcal{S}) \right) \text{ for } i = 0, 1, 2, \ldots. \tag{10.31}$$

In particular, for $i = 0, 1, 2, \ldots$,

$$\mathcal{P}_0^{(i)} \subset \operatorname{col}(\mathcal{S}), \tag{10.32}$$

$$\mathcal{P}_0^{(i)} \subset \mathcal{P}_0^{(i+1)}. \tag{10.33}$$

Moreover, there must be an index i_0 such that $\mathcal{P}_0^{(i_0)} = \mathcal{P}_0^{(i_0+1)}$, since otherwise $\{\mathcal{P}_0^{(i)}\}_{i \in \mathbf{N}}$ would be an infinite, strictly increasing sequence of finite subsets of $\operatorname{col}(\mathcal{S})$, which is obviously impossible. Therefore, the **repeat** loop always terminates, and consequently the algorithm always stops.

From the preceding discussion we see, in particular, that

$$\mathcal{P}_0 = \mathcal{P}_0^{(i_0)},$$

$$\mathcal{P}_0 \subset \operatorname{col}(\mathcal{S}), \tag{10.34}$$

$$\mathcal{P}_0 \cap \mathcal{S} = \emptyset. \tag{10.35}$$

By the assumption that the algorithm does not return `"Failure"`, we have $\operatorname{wrap}(\mathcal{S}) \subset \mathcal{N}$. Therefore, $\mathcal{N}$ is an isolating neighborhood for $\mathcal{F}$.

Suppose that $|\mathcal{S}|$ is not an isolating neighborhood for f. This means that

$$\operatorname{Inv}(|\mathcal{S}|, f) \cap \operatorname{bd} |\mathcal{S}| \neq \emptyset,$$

so there exists a full solution $\gamma_x : \mathbf{Z} \to |\mathcal{S}|$ through a boundary point $x \in \operatorname{bd} |\mathcal{S}|$ under f. Then $x \notin \operatorname{int} |\mathcal{S}|$. Thus by Proposition 10.69 we can choose $Q \in \operatorname{wrap}(x) \setminus \mathcal{S}$. We define $\Gamma_Q : \mathbf{Z} \to \mathcal{K}_{\max}$ as follows. We put $\Gamma_Q(0) := Q$, and for $n \neq 0$, we let $\Gamma_Q(n)$ be any element of $\operatorname{wrap}(\gamma_x(n))$. Then by Definition 10.68,

$$\Gamma_Q(n+1) \in \operatorname{wrap}(\gamma_x(n+1)) = \operatorname{wrap}(f(\gamma_x(n))) \subset \mathcal{F}(\Gamma_Q(n)),$$

which shows that Γ_Q is a full solution through Q under $\mathcal{F}$. Moreover, since $\gamma_x(n) \in |\mathcal{S}| \cap \Gamma_Q(n)$, it follows from (10.29) that $\Gamma_Q(n) \in \operatorname{wrap}(\mathcal{S}) \subset \mathcal{N}$. Therefore, $Q = \Gamma_Q(0) \in \operatorname{Inv}(\mathcal{N}, \mathcal{F}) = \mathcal{S}$, which contradicts the choice of Q.

To prove (ii), we must verify the three properties of index pairs in Definition 10.86. First observe that by the only assignment to the P1 variable,

$$\mathcal{P}_1 = \mathcal{S} \cup \mathcal{P}_0. \tag{10.36}$$

Thus, by (10.35),

$$\mathcal{P}_1 \setminus \mathcal{P}_0 = \mathcal{S}. \tag{10.37}$$

We have

$$P_1 \setminus P_0 = |\mathcal{P}_1| \setminus |\mathcal{P}_0| = \bigcup_{Q \in \mathcal{P}_1} Q \setminus |\mathcal{P}_0| = \bigcup_{Q \in \mathcal{S}} Q \setminus |\mathcal{P}_0|.$$

Therefore,
$$\operatorname{cl}(P_1 \setminus P_0) = \bigcup_{Q \in \mathcal{S}} \operatorname{cl}(Q \setminus |\mathcal{P}_0|) = \bigcup_{Q \in \mathcal{S}} Q = |\mathcal{S}|.$$

Thus
$$\operatorname{cl}(P_1 \setminus P_0) = |\mathcal{S}|, \tag{10.38}$$

which shows that the isolation property (1) is already proved in (i). To prove the positive invariance property (2), we first prove that

$$\mathcal{F}(\mathcal{P}_0) \cap \mathcal{S} = \emptyset. \tag{10.39}$$

We do this by contradiction. Suppose that there is a cube $Q \in \mathcal{P}_0$ such that $\mathcal{F}(Q) \cap \mathcal{S} \neq \emptyset$. Then $Q \in \mathcal{F}^{-1}(\mathcal{S})$. An easy induction argument based on Eqs. (10.27), (10.30) and (10.31) proves that

$$\mathcal{P}_0^{(i)} \subset \operatorname{Inv}^-(\mathcal{N}, \mathcal{F}). \tag{10.40}$$

Since $Q \in \mathcal{P}_0 = \mathcal{P}_0^{(i_0)} \subset \mathcal{N}$, we conclude by Eq. (10.26) and (10.28) that $Q \in \mathcal{F}^{-1}(\mathcal{S}) \cap \mathcal{N} \subset \operatorname{Inv}^+(\mathcal{N}, \mathcal{F})$ and by (10.40) that $Q \in \operatorname{Inv}^-(\mathcal{N}, \mathcal{F})$. Thus $Q \in \operatorname{Inv}(\mathcal{N}, \mathcal{F})$. But also $Q \in \mathcal{P}_0 \subset \operatorname{col}(\mathcal{S})$, a contradiction. Therefore, (10.39) is proved.

Since $\mathcal{S} = \mathcal{P}_1 \setminus \mathcal{P}_0$, Eq. (10.39) implies the following combinatorial counterpart of positive invariance:

$$\mathcal{F}(\mathcal{P}_0) \cap \mathcal{P}_1 \subset \mathcal{P}_0. \tag{10.41}$$

We are now ready to prove the positive invariance property (2). Let $x \in P_0$ and $f(x) \in P_1$. Then $x \in Q$ for some $Q \in \mathcal{P}_0$ and $f(x) \in R$ for some $R \in \mathcal{P}_1$. Thus, by Definition 10.68,

$$R \in \operatorname{wrap}(f(x)) \subset \mathcal{F}(Q) \subset \mathcal{F}(\mathcal{P}_0),$$

and by (10.41), $R \in \mathcal{P}_0$, which proves that $f(x) \in P_0$.

For proving the exit set property (3), we will show first that

$$\mathcal{F}(\mathcal{P}_1) \cap \operatorname{col}(\mathcal{S}) \subset \mathcal{P}_0, \tag{10.42}$$
$$\operatorname{bd} P_1 \subset |\operatorname{col}(\mathcal{S})|. \tag{10.43}$$

Indeed, we get from (10.36), (10.30), (10.31), and (10.33)

$$\begin{aligned}
\mathcal{F}(\mathcal{P}_1) \cap \operatorname{col}(\mathcal{S}) &\subset \mathcal{F}(\mathcal{S} \cup \mathcal{P}_0) \cap \operatorname{col}(\mathcal{S}) \\
&\subset \mathcal{F}(\mathcal{S}) \cap \operatorname{col}(\mathcal{S}) \cup \mathcal{F}(\mathcal{P}_0) \cap \operatorname{col}(\mathcal{S}) \\
&= \mathcal{P}_0^{(0)} \cup \mathcal{F}(\mathcal{P}_0^{(i_0)}) \cap \operatorname{col}(\mathcal{S}) \\
&= \mathcal{P}_0^{(i_0)} \cup \mathcal{F}(\mathcal{P}_0^{(i_0)}) \cap \operatorname{col}(\mathcal{S}) \\
&= \mathcal{P}_0^{(i_0+1)} = \mathcal{P}_0,
\end{aligned}$$

which proves (10.42). In order to prove (10.43), take $x \in \operatorname{bd} P_1$. Then $x \in Q$ for some $Q \in \mathcal{P}_1$ and, by Proposition 10.69, $x \in R$ for some $R \in \operatorname{wrap}(x) \setminus \mathcal{P}_1$. Since $\mathcal{P}_1 = \mathcal{S} \cup \mathcal{P}_0$, either $Q \in \mathcal{S}$ or $Q \in \mathcal{P}_0$. In the latter case $Q \in \operatorname{col}(\mathcal{S})$ by (10.34) and, consequently, $x \in |\operatorname{col}(\mathcal{S})|$. If $Q \in \mathcal{S}$, then $x \in R \cap |\mathcal{S}|$, that is, $R \in \operatorname{col}(\mathcal{S})$ and again $x \in \operatorname{col}(\mathcal{S})$. Thus (10.43) is proved. Now, by Proposition 10.71, Eq. (10.43), Proposition 10.70, and Eq. (10.42),

$$\begin{aligned}\operatorname{bd}_f(P_1) &= \operatorname{cl}(f(P_1) \setminus P_1) \cap P_1 \\ &\subset f(P_1) \cap \operatorname{bd} P_1 \\ &\subset \operatorname{int} |\mathcal{F}(\mathcal{P}_1)| \cap |\operatorname{col}(\mathcal{S})| \\ &\subset |\mathcal{F}(\mathcal{P}_1) \cap \operatorname{col}(\mathcal{S})| \subset |\mathcal{P}_0| = P_0.\end{aligned}$$

Thus also the exit property (3) is proved, which means that we are done with (ii).

Before showing (iii) we will prove that

$$\mathcal{P}_0 \subset \mathcal{F}(\mathcal{P}_1) \setminus \mathcal{S}. \tag{10.44}$$

For this end it is sufficient to show that

$$\mathcal{P}_0^{(i)} \subset \mathcal{F}(\mathcal{P}_1) \setminus \mathcal{S} \text{ for all } i = 0, 1, 2, \ldots, i_0. \tag{10.45}$$

Obviously,

$$\mathcal{P}_0^{(0)} = \mathcal{F}(\mathcal{S}) \cap \operatorname{col}(\mathcal{S}) \subset \mathcal{F}(\mathcal{P}_1) \setminus \mathcal{S}.$$

Arguing by induction, we see that for $i < i_0$ satisfying (10.45)

$$\mathcal{P}_0^{(i+1)} = \mathcal{P}_0^{(i)} \cup \left(\mathcal{F}(\mathcal{P}_0^{(i)}) \cap \operatorname{col}(\mathcal{S})\right) \subset \mathcal{F}(\mathcal{P}_1) \setminus \mathcal{S},$$

which proves (10.44).

Since $\bar{\mathcal{P}}_0 = \bar{\mathcal{P}}_1 \setminus \mathcal{S} = \mathcal{F}(\mathcal{P}_1) \setminus \mathcal{S}$, we see from (10.44) that $\mathcal{P}_0 \subset \bar{\mathcal{P}}_0$ and, consequently, $P_0 \subset \bar{P}_0$. We also have

$$\mathcal{P}_1 = \mathcal{S} \cup \mathcal{P}_0 \subset \mathcal{F}(\mathcal{S}) \cup (\mathcal{F}(\mathcal{P}_1) \setminus \mathcal{S}) \subset \mathcal{F}(\mathcal{P}_1) = \bar{\mathcal{P}}_1.$$

Therefore, $P_1 \subset \bar{P}_1$ and (iii) is proved.

To prove (iv), observe that for $i = 0, 1$

$$f(P_i) = \bigcup_{Q \in \mathcal{P}_i} f(Q) \subset \bigcup_{Q \in \mathcal{P}_i} |\mathcal{F}(Q)| = |\mathcal{F}(\mathcal{P}_i)|.$$

Therefore,

$$f(P_1) \subset |\mathcal{F}(\mathcal{P}_1)| = |\bar{\mathcal{P}}_1| = \bar{P}_1,$$

and by (10.39),

$$f(P_0) \subset |\mathcal{F}(\mathcal{P}_0)| \subset |\mathcal{F}(\mathcal{P}_0) \setminus \mathcal{S}| \subset |\mathcal{F}(\mathcal{P}_1) \setminus \mathcal{S}| = |\bar{\mathcal{P}}_0| = \bar{P}_0,$$

and (iv) is done. □

Here is the property of the constructed index pair that shall enable us to pass from dynamics to algebraic topology.

354   10 Nonlinear Dynamics

Lemma 10.88 *Let the map f and cubical sets $P_1, P_0, \bar{P}_1, \bar{P}_0$ be as discussed in Theorem 10.87. Then, the inclusion map $\iota : (P_1, P_0) \to (\bar{P}_1, \bar{P}_0)$ induces an isomorphism*
$$\iota_* : H_*(P_1, P_0) \to H_*(\bar{P}_1, \bar{P}_0)$$

Proof. Let $\mathcal{F}, \mathcal{S}, \mathcal{P}_1, \mathcal{P}_0, \bar{\mathcal{P}}_1$ and $\bar{\mathcal{P}}_0$ be as in the proof of Theorem 10.87. Set $\mathcal{R} := (\mathcal{F}(\mathcal{P}_1) \setminus \mathcal{S}) \setminus \mathcal{P}_0$. It is straightforward to verify that
$$\bar{\mathcal{P}}_1 \setminus \mathcal{P}_1 = \mathcal{R} = \bar{\mathcal{P}}_0 \setminus \mathcal{P}_0. \tag{10.46}$$

We will first prove that
$$Q \setminus P_0 = Q \setminus P_1 \text{ for } Q \in \mathcal{R}. \tag{10.47}$$

Since $P_0 \subset P_1$, we have $Q \setminus P_1 \subset Q \setminus P_0$. Assume the opposite inclusion does not hold. Then we can take an $x \in Q \setminus P_0$ such that $x \in P_1$. From (10.38) we get $x \in |\mathcal{S}|$ and consequently $Q \in \text{wrap}(\mathcal{S})$. But $Q \in \mathcal{R} = (\mathcal{F}(\mathcal{P}_1) \setminus \mathcal{S}) \setminus \mathcal{P}_0$. Therefore, $Q \in \mathcal{F}(\mathcal{P}_1) \cap \text{col}(\mathcal{S})$, and by (10.42), $Q \in \mathcal{P}_0$, which contradicts $Q \in \mathcal{R}$. This proves (10.47).

Therefore, we get from (10.46)
$$\bar{P}_1 \setminus P_1 = \bigcup_{Q \in \bar{\mathcal{P}}_1} Q \setminus P_1 = \bigcup_{Q \in \bar{\mathcal{P}}_1 \setminus \mathcal{P}_1} Q \setminus P_1 = \bigcup_{Q \in \mathcal{R}} Q \setminus P_1$$

and
$$\bar{P}_0 \setminus P_0 = \bigcup_{Q \in \bar{\mathcal{P}}_0} Q \setminus P_0 = \bigcup_{Q \in \bar{\mathcal{P}}_0 \setminus \mathcal{P}_0} Q \setminus P_0 = \bigcup_{Q \in \mathcal{R}} Q \setminus P_0.$$

Now from (10.47) we conclude that
$$\bar{P}_1 \setminus P_1 = \bar{P}_0 \setminus P_0. \tag{10.48}$$

Set $U := \bar{P}_1 \setminus P_1$. Obviously, U is a representable set as a difference of representable sets. Moreover, it follows from (10.48) that
$$P_i = \bar{P}_i \setminus U.$$

Therefore, by Theorem 9.14, the inclusion map $\iota : (P_1, P_0) \to (\bar{P}_1, \bar{P}_0)$ is an isomorphism. □

By virtue of Lemma 10.88 the map
$$\iota_*^{-1} : H_*(\bar{P}_1, \bar{P}_0) \to H_*(P_1, P_0)$$

is well defined. Since by Theorem 10.87(iv) we have a well defined map of pairs
$$f : (P_1, P_0) \to (\bar{P}_1, \bar{P}_0),$$

which also induces a map in homology, we are ready to give the following definition.

10.6 Complicated Dynamics

Definition 10.89 Let $P = (P_1, P_0)$ be a cubical index pair for f constructed by Algorithm 10.86. The associated *index map* is defined by

$$f_{P*} := \iota_*^{-1} \circ f_* : H_*(P_1, P_0) \to H_*(P_1, P_0).$$

As the following theorem indicates, the index map can be used to prove the existence of a nontrivial isolated invariant set. However, to state the theorem we need to introduce the following notion.

Definition 10.90 Let G be an abelian group. A homomorphism $L : G \to G$ is *nilpotent* if there exists a positive integer n such that $L^n = 0$.

Theorem 10.91 Let $P = (P_1, P_0)$ be a cubical index pair for f constructed by Algorithm 10.86. If the associated index map f_{P*} is not nilpotent, then $\mathrm{Inv}\,(\mathrm{cl}\,(P_1 \setminus P_0), f) \neq \emptyset$.

This theorem, which can be interpreted as a homological version of the Ważewski principle for maps, lies at the heart of the Conley index theory. Unfortunately, unlike the setting of flows where the proof follows from the construction of a deformation retraction, the proof of this theorem is not straightforward. We will return to this problem at the end of this section.

One of the themes that we have been pursuing is that we can use a combinatorial acyclic valued enclosure $\mathcal{F}$ and homology to make conclusions about the dynamics of a continuous map f. For example, assume that using $\mathcal{F}$ we obtain an index pair $P = (P_1, P_0)$ and that the resulting index map f_{P*} is not nilpotent, then by Theorem 10.91,

$$S := \mathrm{Inv}\,(\mathrm{cl}\,(P_1 \setminus P_0), f) \neq \emptyset.$$

Thus, we are able to use $\mathcal{F}$ to conclude the existence of a nontrivial invariant set. Of course, there is nothing unique about the choice of $\mathcal{F}$. So assume we use a different combinatorial acyclic valued enclosure $\mathcal{F}'$ and obtain a different index pair (Q_1, Q_0) for f, but that

$$S = \mathrm{Inv}\,(\mathrm{cl}\,(Q_1 \setminus Q_0), f).$$

Since the index pairs are different, it is not at all obvious that f_{Q*} must also be nilpotent. The Conley index theory assures us that it is. More succinctly, it tells us that f_{P*} is nilpotent if and only if f_{Q*} is nilpotent. Thus we can apply Theorem 10.91 independently of the particular index pair used. Unfortunately, a complete discussion of the Conley index is far beyond the scope of this book; therefore, we will only give a brief description and discuss some important consequences.

However, before doing so we wish to make an important observation. Since $\mathcal{F}$ is a combinatorial enclosure of f, it is also a combinatorial enclosure for any other continuous function g that is a sufficiently small perturbation of f. Therefore, if $P = (P_1, P_0)$ is an index pair for f derived from $\mathcal{F}$, then

356   10 Nonlinear Dynamics

$P = (P_1, P_0)$ is an index pair for g. Similarly, if f_{P*} is not nilpotent, then g_{P*} is not nilpotent. Thus this method not only proves that $\mathrm{Inv}\,(\mathrm{cl}\,(P_1 \setminus P_0), f) \neq \emptyset$, but in fact it demonstrates that $\mathrm{Inv}\,(\mathrm{cl}\,(P_1 \setminus P_0), g) \neq \emptyset$ for all g sufficiently close to f. Moreover, the enclosure $\mathcal{F}$ tells us what "sufficiently close" means.

Returning to the Conley index, consider two index pairs (P_1, P_0) and (Q_1, Q_0) for a continuous map f. The induced index maps are f_{P*} and f_{Q*}. If we add the additional assumption that

$$S = \mathrm{Inv}\,(\mathrm{cl}\,(P_1 \setminus P_0), f) = \mathrm{Inv}\,(\mathrm{cl}\,(Q_1 \setminus Q_0), f),$$

then the Conley index theory guarantees that the two index maps are equivalent in the following sense.

Definition 10.92 *Two group homomorphisms between abelian groups, $f : G \to G$ and $g : G' \to G'$, are* shift equivalent *if there exist group homomorphisms $r : G \to G'$ and $s : G' \to G$ and a natural number m such that*

$$r \circ f = g \circ r, \quad s \circ g = f \circ s, \quad r \circ s = g^m, \quad s \circ r = f^m.$$

It is left as an exercise to show that shift equivalence defines an equivalence relation. From the computational point of view distinguishing these equivalence classes is not obvious. On the other hand, the following proposition shows that shift equivalence preserves nilpotency, which is easily verifiable.

Proposition 10.93 *Let $f : G \to G$ and $g : G' \to G'$ be group homomorphisms that are shift equivalent. Then f is nilpotent if and only if g is nilpotent.*

Proof. Observe that it is sufficient to prove that if f is not nilpotent, then g is not nilpotent. Since f is not nilpotent and $s \circ r = f^m$, neither r nor s can be trivial maps. Assume that g is nilpotent or more specifically that $g^k = 0$. Then

$$r \circ f = g \circ r,$$
$$s \circ r \circ f \circ (s \circ r)^k = s \circ g \circ r \circ (s \circ r)^k,$$
$$f^m \circ f \circ (f^m)^k = s \circ g \circ (r \circ s)^k \circ r,$$
$$f^{m(k+1)+1} = s \circ g \circ (g^m)^k \circ r,$$
$$f^{m(k+1)+1} = 0.$$

This contradicts the assumption that f is not nilpotent. □

As the next example shows, shift equivalence provides a finer classification than nilpotency.

Example 10.94 Let $f, g : \mathbf{Z} \to \mathbf{Z}$ be the group homomorphisms given by

$$f(x) = 2x \quad \text{and} \quad g(x) = x.$$

Clearly, both f and g are not nilpotent. However, as the following argument shows, they are not shift equivalent. If they were, then there would exist group homomorphisms $r, s : \mathbf{Z} \to \mathbf{Z}$ such that

$$r \circ f = g \circ r \quad \text{and} \quad r \circ s = f^m$$

for some positive integer m. The first equation can only be solved by setting $r = 0$, which means that the second equation has no solution.

Shift equivalence can be used to define a form of the Conley index.

Definition 10.95 Let $f : X \to X$ be a continuous map. Let S be an isolated invariant set and let (P_1, P_0) be an index pair for f such that $S = \text{Inv}\,(\text{cl}\,(P_1 \setminus P_0), f)$. The *homology Conley index* for S is the shift equivalence class of the index map f_{P*}.

The deep result, the proof of which we shall *not* present, is that the Conley index of an isolated invariant set S is well defined. In other words, up to shift equivalence the index map does not depend on the index pair chosen.

We conclude this section by using the fact that the Conley index is well defined to prove Theorem 10.91.

Proof of Theorem 10.91: We shall prove the contrapositive; if $\text{Inv}\,(\text{cl}\,(P_1 \setminus P_0), f) = \emptyset$, then f_{P*} is nilpotent.

Recall Example 10.78, where it is demonstrated that $\emptyset$ is an isolated invariant set for any continuous function f. Observe that $(\emptyset, \emptyset)$ is an index pair for $\emptyset$. Let $g_* : H_*(\emptyset, \emptyset) \to H_*(\emptyset, \emptyset)$ be the induced index map. Clearly, $g_* = 0$ and hence g_* is nilpotent.

Since the Conley index is well defined, g_* and f_{P*} are shift equivalent. By Example 10.94, the fact that g_* is nilpotent implies that f_{P*} is nilpotent. □

10.6.2 Topological Conjugacy

Up to this point we have concentrated on finding relatively simple dynamical structures such as fixed points and periodic orbits. However, nonlinear systems can exhibit extremely complicated structures often referred to as chaotic dynamics or simply chaos. While there is no universally accepted definition of chaos, there are particular examples that everyone agrees are chaotic. The best understood are those that can be represented in terms of symbolic dynamics. In this section we shall indicate how homology can be used to verify the existence of this type of chaos, but for the moment we need to deal with the issues of what it means to represent one dynamical system in terms of another.

The following definition provides a topological means of concluding that the two dynamical systems generated by different continuous maps are equivalent.

Definition 10.96 Let $f : X \to X$ and $g : Y \to Y$ be continuous maps. A homeomorphism $\rho : X \to Y$ is a *topological conjugacy* from f to g if $\rho \circ f = g \circ \rho$ or equivalently if

$$\begin{array}{ccc} X & \xrightarrow{f} & X \\ \downarrow{\rho} & & \downarrow{\rho} \\ Y & \xrightarrow{g} & Y \end{array}$$

commutes. We can weaken the relationship between f and g by only assuming that ρ is a continuous surjective map. In this case it is called a *topological semiconjugacy*.

Observe that if $\gamma_x : \mathbf{Z} \to X$ is a solution of f through x, then $\sigma_{\rho(x)} : \mathbf{Z} \to Y$ given by

$$\sigma_{\rho(x)} := \rho \circ \gamma_x$$

is a solution of g through $\rho(x)$. If ρ is a homeomorphism, then ρ^{-1} exists and hence given a solution $\gamma_y : \mathbf{Z} \to Y$ of g through y, we can define a solution $\sigma_x := \rho^{-1} \circ \gamma_y$ of f through $x := \rho^{-1}(y)$. Therefore, if two dynamical systems are related by a topological conjugacy, then there is an exact correspondence among all the orbits of each system.

As a first application of topological conjugacy, let us consider the question of how much information a particular combinatorial enclosure $\mathcal{F}$ can provide about the continuous map f. As the following example illustrates, there is no guarantee that $\mathcal{F}$ provides us with any knowledge about the dynamics of f.

Example 10.97 Consider the family of logistic maps $f_\lambda : [0,1] \to [0,1]$ given by $f_\lambda(x) = \lambda x(1-x)$. For all $\lambda \in [0,4]$ a cubical enclosure is given by the unique map $\mathcal{F} : \mathcal{K}_{\max}([0,1]) \to \mathcal{K}_{\max}([0,1])$, which sends the elementary interval $[0,1]$ to itself. Observe that $f_0(x) = 0$. Thus, the homological information that we can extract from $\mathcal{F}$ cannot indicate any dynamics more complicated than the constant map f_0. However, as we shall see in Example 10.99, it can be shown that f_4 generates many interesting orbits.

A possible remedy to the problem described in the above example is to pass to smaller units, that is to rescale the space. The decision on how large the rescaling should be can only be made experimentally. However, it should be clear that the larger the scaling the easier it should be to separate individual orbits. The following example indicates that dynamics generated by the rescaled system is conjugate to the dynamics of the original system.

Example 10.98 Let X be a cubical set and $f : X \to X$ a continuous map. Recall that scaling $\Lambda^\alpha : X \to X^\alpha$ is a homeomorphism with the inverse $\Omega^\alpha : X^\alpha \to X$. Recall that $f^\alpha : X^\alpha \to X$ has been defined as $X^\alpha = f \circ \Omega^\alpha$. Define

10.6 Complicated Dynamics   359

$$\tilde{f}^\alpha := \Lambda^\alpha \circ f^\alpha : X^\alpha \to X^\alpha.$$

It is instantly verified that Λ^α is a conjugacy from f to $\tilde{f}^\alpha$.

We are finally ready to turn our attention to topological conjugacy and complicated dynamics. The following example involving the logistic map is a useful starting point.

Example 10.99 Let $f : [0,1] \to [0,1]$ be the *logistic map* given by $f(x) = 4x(1-x)$. Let $g : [0,1] \to [0,1]$ be the *tent map* given by

$$g(y) = \begin{cases} 2y & \text{if } 0 \le y \le \frac{1}{2}, \\ 2(1-y) & \text{if } \frac{1}{2} < y \le 1. \end{cases}$$

Since arcsin is the inverse of $\sin : [0, \frac{\pi}{2}] \to [0,1]$, the map $\rho : [0,1] \to [0,1]$ given by

$$\rho(x) = \frac{2}{\pi} \arcsin \sqrt{x}$$

is a homeomorphism with the inverse given by $\rho^{-1}(y) = \sin^2(\pi y/2)$. We shall check that this is a topological conjugacy. The condition $\rho \circ f = g \circ \rho$ is equivalent to checking that $\rho \circ f \circ \rho^{-1} = g$. By elementary trigonometric identities,

$$f(\rho^{-1}(y)) = 4\sin^2 \frac{\pi y}{2} \left(1 - \sin^2 \frac{\pi y}{2}\right) = 4\sin^2 \frac{\pi y}{2} \cos^2 \frac{\pi y}{2} = \sin^2(\pi y),$$

for all $y \in [0,1]$. Hence for $y \in [0, \frac{1}{2}]$ we get

$$\rho(f(\rho^{-1}(y))) = \frac{2}{\pi} \arcsin(\sin(\pi y)) = 2y.$$

and for $y \in [\frac{1}{2}, 1]$ we get

$$\rho(f(\rho^{-1}(y))) = \frac{2}{\pi} \arcsin(\sin(\pi y)) = \frac{2}{\pi} \arcsin(\sin(\pi(1-y))) = 2(1-y).$$

Therefore $\rho(f(\rho^{-1}(y))) = g(y)$.

In Example 10.99 g is a piecewise linear map and hence it is relatively easy to study individual trajectories. Since we have a topological conjugacy, any trajectory of g corresponds to a trajectory in f, and vice versa; thus we can use the dynamics of g to describe the dynamics of f. For example, it is fairly easy to discover that $\{\frac{2}{5}, \frac{4}{5}\}$ is a period 2 orbit for g. Therefore, f contains a period two orbit. Of course, even though g is a reasonably simple map, a complete description of all its orbits is not obvious. For example, answering the question of whether or not g has a periodic orbit of period 17 requires some rather tedious work. With this in mind we would like to have a family of interesting but well-understood dynamical systems. This was one of the reasons for introducing symbolic dynamics in Section 10.1.

What may appear puzzling at the moment is how one can relate the dynamics of a continuous map $f : X \to X$ to a particular subshift $\sigma_A : \Sigma_A \to \Sigma_A$. To avoid technical complications, let us assume that f is a homeomorphism. Now consider a collection of closed mutually disjoint sets $N_1, \ldots, N_n \subset X$. Let

$$S = \text{Inv}\left(\bigcup_{i=1}^n N_i, f\right).$$

Our goal is to understand the structure of the dynamics of S. Since S is an invariant set, $f(S) = S$. In particular, if $x \in S$, then

$$f^k(x) \in S \subset \bigcup_{i=1}^n N_i$$

for every $k \in \mathbf{Z}$. Furthermore, since the N_i are mutually disjoint, for each k there exists a unique i such that $f^k(x) \in N_i$. Define $\rho : S \to \Sigma_n$ by

$$\rho(x)_k = i \quad \text{if } f^k(x) \in N_i. \tag{10.49}$$

Proposition 10.100 *If the N_i are disjoint closed sets, then $\rho : S \to \Sigma_n$ as defined by (10.49) is continuous.*

The proof follows from the definition of continuity and is left as an exercise.

The reader should check that the following is a commutative diagram.

$$\begin{array}{ccc} S & \xrightarrow{f} & S \\ \downarrow{\rho} & & \downarrow{\rho} \\ \Sigma_n & \xrightarrow{\sigma} & \Sigma_n \end{array} \tag{10.50}$$

Notice, however, that while this diagram is correct, it does not provide us with any information. Consider, for example, the extreme case where S consists of a single fixed point. Then $\rho(S)$ is a single point and so there is nothing to be gained by knowing that $\rho : S \to \Sigma_n$.

On the other hand, assume for the moment that we can find a transition matrix A and an invariant set $S' \subset S$ such that

$$\begin{array}{ccc} S' & \xrightarrow{f} & S' \\ \downarrow{\rho} & & \downarrow{\rho} \\ \Sigma_A & \xrightarrow{\sigma_A} & \Sigma_A \end{array} \tag{10.51}$$

commutes. Moreover, assume that $\rho(S') = \Sigma_A$ that is ρ is surjective. In this case we can conclude that for every orbit in Σ_A there is a similar orbit in S.

To be more precise, assume, for example, that the periodic point $(\overline{1,2}) \in \Sigma_A$, then we can conclude that there is an orbit

$$\{\ldots, x_{-1}, x_0, x_1, \ldots\} \subset S$$

with the property that $x_i \in N_1$ if i is even and $x_i \in N_2$ if i is odd.

Of course, since we have not assumed that ρ is a homeomorphism, we cannot claim that there is a unique orbit in S with this property. Thus, this approach of finding a transition matrix A such that $\rho : S' \to \Sigma_A$ is a surjective map, namely producing a topological semiconjugacy, can only provide a lower bound on the complexity of the dynamics in S. On the other hand, as we shall now indicate, homology can be used to determine A and prove that we do have a semiconjugacy.

Exercises

10.20 Prove that shift equivalence defines an equivalence relation on the set of group homomorphisms between abelian groups.

10.21 Prove that ρ in Example 10.99 defines a topological conjugacy.

10.22 Prove that $\rho : S \to \Sigma_n$ defined by (10.49) is continuous and that the diagram (10.50) commutes.

10.7 Computing Chaotic Dynamics

The sheer complexity of chaotic dynamics makes it not only an interesting mathematical subject, but also a computational challenge. It is the second point that guides our presentation in this subsection. The goal of this book has been to explain homology from a computational point of view with an emphasis on what we feel are the essential ideas. In Chapter 8 we have indicated through simple examples our belief that an efficient computational homology theory based on cubes has a broad range of potential applications to problems where it is important to be able to resolve geometric structures. The material of this chapter has a slightly different flavor—the theoretical results are more sophisticated and there is a short, but significant, history of applications using these computational techniques. It seems natural to conclude with a nontrivial example that demonstrates the power of these methods.

However, being nontrivial also implies the need for efficient code that computes combinatorial enclosures, index pairs, the associated homology groups, and maps. Furthermore, we need to be able to compute relative homology groups and maps. While we defined these concepts, for reasons of space we decided not to include a discussion of the associated algorithms. These can be found in the literature (see bibliographical remarks at the end of the chapter) and the reader who has mastered the material up to this point should be able to follow the arguments in these articles.

Similarly, we have not discussed any algorithms for computing combinatorial enclosures—this is the topic for a different book. There is, however, general-purpose software $GAIO$ [17] that is capable of performing these computations along with that of computing isolating neighborhoods and index pairs. The essential input for this program is a nonlinear map $f : \mathbf{R}^d \to \mathbf{R}^d$, a rectangle $[a_1, b_1] \times \cdots \times [a_d, b_d] \subset \mathbf{R}^d$ that defines the region of phase space that is to be studied, and the number of subdivisions that determines the size of the cubes to be used in the approximation.

To put this discussion into perspective, we will show using a two-dimensional map how one can use these algorithms to rigorously demonstrate the existence of chaotic symbolic dynamics. We have chosen a two-dimensional example in order to include pictures to aid in the exposition. It is worth mentioning that these techniques are in principle dimension-independent, though the computational cost increases rapidly as a function of dimension.

Example 10.101 The Hénon map $h : \mathbf{R}^2 \to \mathbf{R}^2$ is a two-dimensional function

$$\begin{pmatrix} x \\ y \end{pmatrix} \mapsto \begin{pmatrix} 1 + y/5 - ax^2 \\ 5bx \end{pmatrix}, \tag{10.52}$$

which we shall study numerically at the parameter values $a = 1.3$ and $b = 0.2$.

The index pair $P = (P_1, P_0)$ indicated in Figure 10.2 was computed using GAIO. The initial rectangle was $[-0.9311, 1.2845]^2$. During the process of the computation each direction was subdivided 2^7 times, thus the individual cubes on which the acyclic-valued combinatorial enclosure $\mathcal{F}$ of h is defined have the form

$$\left[-0.9311 + i2^7, -0.9311 + (i+1)2^7\right] \times \left[-0.9311 + j2^7, -0.9311 + (j+1)2^7\right]$$

where $0 \leq i, j < 2^7$.[1] Finally, the index pair was constructed by finding a period two orbit for $\mathcal{F}$, a candidate for an isolating neighborhood and applying Algorithm 10.86.

Clearly, $P = (P_1, P_0)$ is not cubical, since it is not composed of unitary cubes. This is because GAIO subdivides as opposed to rescaling. However, it is equally clear that multiplying each square by an appropriate factor results in a cubical complex. Thus from now on we shall refer to the GAIO output as cubical.

By inspection (using Exercises 9.10 and 9.16) or using the homology program of CHomP, one can check that

$$H_k(P_1, P_0) \cong \begin{cases} \mathbf{Z}^2 & \text{if } k = 1, \\ 0 & \text{otherwise.} \end{cases}$$

The associated index map was computed by means of the homology map program of CHomP with the result

---

[1] It should be noted that an important feature of GAIO is that it is designed in such a way that not all these cubes are used in the calculation.

$$h_{P*1} = \begin{bmatrix} 0 & 1 \\ -1 & 0 \end{bmatrix}. \tag{10.53}$$

Since h_{P*1} is not nilpotent,

$$S_2 := \mathrm{Inv}\,(\mathrm{cl}\,(P_1 \setminus P_0), h) \neq \emptyset.$$

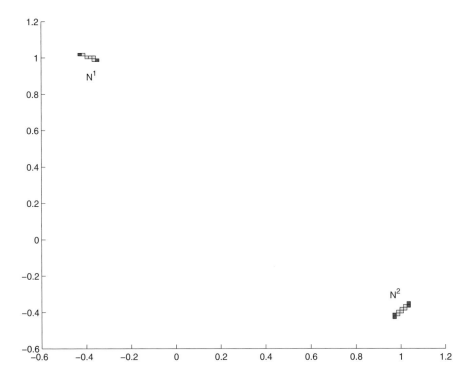

Fig. 10.2. An index pair $P = (P_1, P_0)$ for the Hénon map. All the squares form P_1, and the darker squares indicate P_0.

As the previous example shows, the Conley index can be used to conclude the existence of an invariant set. However, as will be demonstrated, it is possible to obtain much more detailed information concerning the structure of the dynamics of the invariant set.

The following result is a generalization of Theorem 10.46 to the setting of index pairs. It can be used to prove the existence of fixed points and periodic orbits.

Theorem 10.102 (Lefschetz fixed-point theorem for index pairs) *Let f_{P*} : $H_*(P_1, P_0) \to H_*(P_1, P_0)$ be the index map defined on a cubical index pair (P_1, P_0). If*

$$L(f_{P*}) := \sum_k (-1)^k \operatorname{tr} f_{P*k} \neq 0,$$

then $\operatorname{Inv}(\operatorname{cl}(P_1 \setminus P_0), f)$ contains a fixed point.

Furthermore, if for a fixed positive integer n,

$$\sum_k (-1)^k \operatorname{tr} f_{P*k}^n \neq 0,$$

then $\operatorname{Inv}(\operatorname{cl}(P_1 \setminus P_0), f)$ contains a periodic orbit of period n (this need not be the minimal period).

Example 10.103 Returning to the index pair (P_1, P_0) of Example 10.101, observe that

$$L(h_{P*}) = -\operatorname{tr}\begin{bmatrix} 0 & 1 \\ -1 & 0 \end{bmatrix} = 0 \quad \text{and} \quad L(h_{P*}^2) = -\operatorname{tr}\begin{bmatrix} -1 & 0 \\ 0 & -1 \end{bmatrix} = 2.$$

By Theorem 10.102, S_2 contains a periodic orbit of period 2. The converse of Theorem 10.102 is not true. Thus we cannot at this point conclude that S_2 does not contain any fixed points and that 2 is the minimal period of the periodic orbit.

In order to show that S_2 has no fixed points, we need to make use of the fact that the index pair is made up of two components. This leads to the following notation.

Definition 10.104 A *disjoint decomposition* of an index pair $P = (P_1, P_0)$ is a collection of disjoint cubical sets $\{\mathcal{N}^i \mid i = 1, ..., n\}$ such that

$$\operatorname{cl}(P_1 \setminus P_0) = \bigcup_{i=1}^n \mathcal{N}^i. \tag{10.54}$$

Let

$$S = \operatorname{Inv}\left(\bigcup_{i=1}^n \mathcal{N}^i, f\right).$$

Later in this section, when we turn to chaotic dynamics, the $\mathcal{N}^i$ will become the sets upon which the symbolic dynamics is defined (see the discussion at the end of 10.6.2). In particular, in what follows we define $\rho : S \to \Sigma_n$ by

$$\rho(x)_k = i \quad \text{if } f^k(x) \in \mathcal{N}^i. \tag{10.55}$$

Example 10.105 Returning, yet again, to Example 10.101, observe that we have a disjoint decomposition $\{\mathcal{N}^i \mid i = 1, 2\}$. Furthermore, from GAIO we obtain the fact that

$$\mathcal{F}(\mathcal{N}^i) \cap \mathcal{N}^i = \emptyset \quad \text{and} \quad \mathcal{F}(\mathcal{N}^i) \cap \mathcal{N}^j \neq \emptyset \text{ for } i \neq j,$$

where $\mathcal{N}^i := \mathcal{K}_{\max}(\mathcal{N}^i)$. This implies that there are no fixed points in S_2 and therefore S_2 must contain a periodic orbit with minimal period 2.

10.7 Computing Chaotic Dynamics   365

As this example shows, we can use the combinatorial enclosure $\mathcal{F}$ to argue that specific dynamics cannot occur for a particular example. To systematically make use of this information, we introduce the following definition.

Definition 10.106 Given a disjoint decomposition $\{N^i \mid i = 1,\ldots,n\}$, the associated *cubical transition graph* is a directed graph with vertices $\{N^i \mid i = 1,\ldots,n\}$ and directed edges $N^i \to N^j$ if and only if $\mathcal{F}(N^i) \cap N^j \neq \emptyset$.

Let us note that GAIO can provide the associated cubical transition graph for a disjoint decomposition.

Example 10.107 The analysis done in Examples 10.103 and 10.105 to find a period two orbit can be repeated to find periodic orbits of a higher period. The index pair $Q = (Q_1, Q_0)$ indicated in Figure 10.3 was found using GAIO. Clearly, there exists a disjoint decomposition of the form $\{N^i \mid i = 1,\ldots,4\}$. The associated cubical transition graph is

$$\begin{array}{ccc} N^1 & \longrightarrow & N^2 \\ \uparrow & & \downarrow \\ N^4 & \longleftarrow & N^3 \end{array} \qquad (10.56)$$

From this we can conclude that there are no periodic orbits of period less than or equal to 3 in $S_4 := \operatorname{Inv}(\operatorname{cl}(Q_1 \setminus Q_0), h)$.

The index information was computed using CHomP, resulting in

$$H_k(Q_1, Q_0) \cong \begin{cases} \mathbf{Z}^4 & \text{if } k = 1, \\ 0 & \text{otherwise} \end{cases}$$

and the associated index map in dimension one

$$h_{Q*1} = \begin{bmatrix} 0 & 0 & -1 & 0 \\ 0 & 0 & 0 & 1 \\ 0 & -1 & 0 & 0 \\ -1 & 0 & 0 & 0 \end{bmatrix}. \qquad (10.57)$$

Observe that

$$L(h^4_{Q*}) = -\operatorname{tr} \begin{bmatrix} -1 & 0 & 0 & 0 \\ 0 & -1 & 0 & 0 \\ 0 & 0 & -1 & 0 \\ 0 & 0 & 0 & -1 \end{bmatrix} = 4.$$

Thus, by Theorem 10.102, S_4 contains a periodic orbit of minimal period four.

Though the title of this subsection is Computing Chaotic Dynamics, up to this point we have only uncovered rather tame objects (e.g., periodic orbits). We will remedy this by exhibiting a semiconjugacy between an isolated

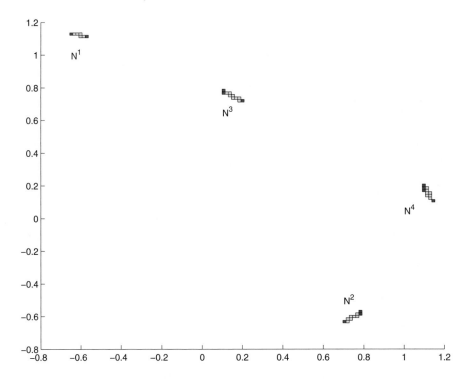

Fig. 10.3. An index pair $Q = (Q_1, Q_0)$ for the Hénon map. All the squares form Q_1 and the darker squares indicate Q_0.

invariant set for Hénon and nontrivial subshift dynamics. However, determining the desired subshift dynamics is somewhat subtle. Therefore, to introduce the ideas, we return to Example 10.107. In this case we have an index pair $Q = (Q_1, Q_0)$ with disjoint decomposition $\{N^i \mid i = 1, 2, 3, 4\}$. By (10.55) the symbol space should consist of four symbols, but since we are only proving the existence of a single orbit of period four, the subshift dynamics is almost trivial; there are exactly four elements

$$\overline{(1,2,3,4)},\ \overline{(4,1,2,3)},\ \overline{(3,4,1,2)},\ \overline{(2,3,4,1)}.$$

The transition matrix corresponding to this trivial subshift is

$$A = \begin{bmatrix} 0 & 0 & 0 & 1 \\ 1 & 0 & 0 & 0 \\ 0 & 1 & 0 & 0 \\ 0 & 0 & 1 & 0 \end{bmatrix}.$$

Our goal is to show that we have a topological semiconjugacy given by

10.7 Computing Chaotic Dynamics

$$S_4 \xrightarrow{h} S_4$$
$$\downarrow \rho \qquad \downarrow \rho \quad . \tag{10.58}$$
$$\Sigma_A \xrightarrow{\sigma} \Sigma_A$$

We will do this by indicating how the index map $h_{Q*} : H_*(Q_1, Q_0) \to H_*(Q_1, Q_0)$ can be used to track orbits as they pass through the individual sets N^i. The first step is to identify $H_*(Q_1, Q_0)$ with generators defined in terms of the sets N^i.

Proposition 10.108 *Let $\{N^i \mid i = 1, ..., n\}$ be a disjoint decomposition of an index pair $P = (P_1, P_0)$. Then*

$$H_*(P_1, P_0) \cong \bigoplus_{i=1}^{n} H_*(N^i \cup P_0, P_0).$$

Proof. The conclusion follows from the relative Mayer–Vietoris sequence (see Exercise 9.15) and induction on the number of sets n in the decomposition. We will present the first step and leave the rest of the proof as an exercise.

Assume $n = 2$ and consider the cubical pairs $(N^1 \cup P_0, P_0)$ and $(N^2 \cup P_0, P_0)$. The associated relative Mayer–Vietoris sequence is

$$\ldots H_k(P_0, P_0) \to H_k(N^1 \cup P_0, P_0) \oplus H_k(N^1 \cup P_0, P_0) \to H_k(P_1, P_0) \to \ldots .$$

Since $H_*(P_0, P_0) = 0$, we obtain the desired isomorphism. □

Proposition 10.108 implies that the index map can be viewed as a homomorphism from $\bigoplus_{i=1}^{n} H_*(N^i \cup P_0, P_0)$ to $\bigoplus_{i=1}^{n} H_*(N^i \cup P_0, P_0)$, in which case the index map is providing information about how points in the N^i are being mapped by f. The following discussion makes this precise.

Let

$$R^i := \bigcup_{j \neq i} N^j.$$

and let $r^i : (P_1, P_0) \hookrightarrow (P_1, P_0 \cup R^i)$ be the inclusion map. The map r^i induces a chain map, which in turn defines the homology map

$$r^i_* : H_*(P_1, P_0) \to H_*(P_1, P_0 \cup R^i).$$

There is also the inclusion map $\iota^i : (N^i \cup P_0, P_0) \hookrightarrow (P_1, P_0 \cup R^i)$. By excision, Theorem 9.14, ι^i_* is an isomorphism and, in particular, $(\iota^i_*)^{-1} : H_*(P_1, P_0 \cup R^i) \to H_*(N^i \cup P_0, P_0)$ is a well defined homomorphism.

Recall that Algorithm 10.86 returns not only a representation of the index pair P but also a representation of a pair $\bar{P} = (\bar{P}_1, \bar{P}_0)$, such that $f(P_j) \subset \bar{P}_j$ for $j = 0, 1$ and the inclusion $\iota : (P_1, P_0) \to (\bar{P}_1, \bar{P}_0)$ induces an isomorphism in homology (see Theorem 10.87 and Lemma 10.88). This gives rise to a map on pairs $f^i : (N^i \cup P_0, P_0) \to (\bar{P}_1, \bar{P}_0)$, which induces the group homomorphism

$$f^i_* : H_*(N^i \cup P_0, P_0) \to H_*(\bar{P}_1, \bar{P}_0).$$

Combining these maps, we can define $g^i_* : H_*(P_1, P_0) \to H_*(P_1, P_0)$ by

$$g^i_* := \iota^{-1}_* \circ f^i_* \circ (\iota^i_*)^{-1} \circ r^i_*. \tag{10.59}$$

Example 10.109 Let us return to Example 10.107. By Proposition 10.108,

$$\mathbf{Z}^4 \cong \bigoplus_{i=1}^{4} H_1(N^i \cup Q_0, Q_0)$$

and, in fact, as is easily checked, $H_1(N^i \cup Q_0, Q_0) \cong \mathbf{Z}$ for each $i = 1, \ldots, 4$. Let ξ^i be a generator for $H_1(N^i \cup Q_0, Q_0) \cong \mathbf{Z}$. Thus we can use $\{\xi^i \mid i = 1, \ldots, 4\}$ as a basis for $H_1(Q_1, Q_0)$. With an appropriate labelling of the N^i we can write

$$h_{Q*1} = \begin{bmatrix} 0 & 0 & 0 & -1 \\ -1 & 0 & 0 & 0 \\ 0 & 1 & 0 & 0 \\ 0 & 0 & -1 & 0 \end{bmatrix}. \tag{10.60}$$

Since $H_i(Q_1, Q_0) = 0$ for $i \neq 1$, also $h_{Q*i} = 0$ for $i \neq 0$.

Using (10.59), we obtain

$$g^1_{*1} = \begin{bmatrix} 0 & 0 & 0 & 0 \\ -1 & 0 & 0 & 0 \\ 0 & 0 & 0 & 0 \\ 0 & 0 & 0 & 0 \end{bmatrix}, \quad g^2_{*1} = \begin{bmatrix} 0 & 0 & 0 & 0 \\ 0 & 0 & 0 & 0 \\ 0 & 1 & 0 & 0 \\ 0 & 0 & 0 & 0 \end{bmatrix}, \quad g^3_{*1} = \begin{bmatrix} 0 & 0 & 0 & 0 \\ 0 & 0 & 0 & 0 \\ 0 & 0 & 0 & 0 \\ 0 & 0 & -1 & 0 \end{bmatrix}, \quad g^4_{*1} = \begin{bmatrix} 0 & 0 & 0 & -1 \\ 0 & 0 & 0 & 0 \\ 0 & 0 & 0 & 0 \\ 0 & 0 & 0 & 0 \end{bmatrix}$$

and $g^j_{*i} = 0$ for $j = 1, 2, 3, 4$ and $i \neq 0$. Finally, recall that for this example the elements of Σ_A are precisely the orbits

$$\left\{ \overline{(4,3,2,1)}, \overline{(3,2,1,4)}, \overline{(2,1,4,3)}, \overline{(1,4,3,2)} \right\}.$$

Let us now make a few suggestive observations concerning relationships between the g^i_* and the elements of Σ_A. First, consider the composition

$$g^{i_4}_* \circ g^{i_3}_* \circ g^{i_2}_* \circ g^{i_1}_*,$$

where $(i_4, i_3, i_2, i_1) \in \{1, 2, 3, 4\}^4$. Notice that

$$g^{i_4}_* \circ g^{i_3}_* \circ g^{i_2}_* \circ g^{i_1}_* \text{ is not nilpotent} \quad \Leftrightarrow \quad \overline{(i_4, i_3, i_2, i_1)} \in \Sigma_A.$$

On the level of dynamics this reflects the fact that these are the only paths in the cubical transition graph. Thus orbits containing any other sequence of digits can be excluded from the symbolic dynamics.

On the positive side, not only does $\overline{(i_4, i_3, i_2, i_1)} \in \Sigma_A$ imply that $g^{i_4}_{*1} \circ g^{i_3}_{*1} \circ g^{i_2}_{*1} \circ g^{i_1}_{*1}$ is not nilpotent, but in fact $L(g^{i_4}_* \circ g^{i_3}_* \circ g^{i_2}_* \circ g^{i_1}_*) \neq 0$.

10.7 Computing Chaotic Dynamics   369

As the following theorem indicates, the observations at the end of Example 10.109 are not coincidences.

Theorem 10.110 *Let $P = (P_1, P_0)$ be an index pair with disjoint decomposition $\{N^i \mid i = 1, \ldots, n\}$. Let $S = \text{Inv}(\text{cl}(P_1 \setminus P_0), f)$. If $g_*^{i_m} \circ g_*^{i_{m-1}} \circ \cdots \circ g_*^{i_1}$ is not nilpotent, then*

$$\overline{(i_1, i_2, \ldots, i_{m-1}, i_m)} \in \rho(S).$$

Furthermore, if

$$L(g_*^{i_m} \circ g_*^{i_{m-1}} \circ \cdots \circ g_*^{i_1}) \neq 0,$$

then $\rho^{-1}\left(\overline{(i_1, i_2, \ldots, i_{m-1}, i_m)}\right) \subset S$ contains a periodic orbit.

The proof of this theorem follows from [83, Corollaries 1.1 and 1.2]. Let us finally apply this theorem to obtain chaotic dynamics.

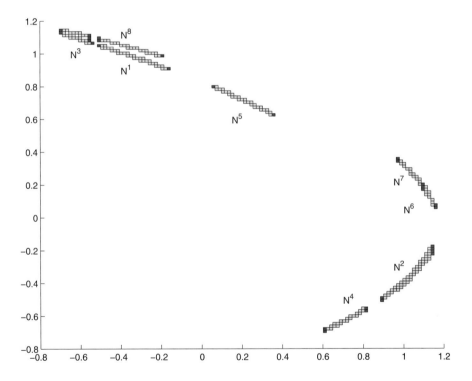

Fig. 10.4. An index pair $T = (T_1, T_0)$ for the Hénon map. All the squares form T_1 and the darker squares indicate T_0. $\text{cl}(T_1 \setminus T_0)$ contains $\text{cl}(P_1 \setminus P_0)$ and $\text{cl}(Q_1 \setminus Q_0)$ from Figures 10.2 and 10.3.

Example 10.111 Consider the index pair $T = (T_1, T_0)$ obtained using GAIO and shown in Figure 10.4. Observe that $\operatorname{cl}(T_1 \setminus T_0)$ contains $\operatorname{cl}(P_1 \setminus P_0)$ and $\operatorname{cl}(Q_1 \setminus Q_0)$ from Figures 10.2 and 10.3, and thus $S = \operatorname{Inv}(\operatorname{cl}(T_1 \setminus T_0))$ must contain the period 2 and period 4 orbits found earlier. We will show that there are many more orbits contained in S.

Observe that we have a disjoint decomposition of T consisting of eight regions $\{N^i \mid i = 1, \ldots, 8\}$. The corresponding cubical transition graph provided by GAIO is

$$
\begin{array}{ccc}
N^1 & N^3 \longrightarrow N^4 \\
\updownarrow \nearrow \uparrow & & \downarrow \\
N^2 & N^6 \longleftarrow N^5 \\
\uparrow & \swarrow \\
N^8 \longleftarrow N^7 &
\end{array}
\qquad (10.61)
$$

Using CHomP and Proposition 10.108, we obtain

$$\mathbf{Z}^{11} \cong H_*(T_1, T_0) \cong \bigoplus_{i=1}^{8} H_1(N^i \cup T_0, T_0).$$

Careful inspection of Figure 10.4 or using CHomP shows that

$$H_1(N^i \cup T_0, T_0) \cong \begin{cases} \mathbf{Z}^4 & \text{if } i = 3, \\ \mathbf{Z} & \text{otherwise.} \end{cases}$$

Using the generators of these groups as a basis, the output of CHomP implies that the matrix form of the index map in dimension one is

$$
h_{T*1} = \begin{bmatrix}
0 & -1 & 0 & 0 & 0 & 0 & 0 & 0 & 0 & 0 & 0 \\
1 & 0 & 0 & 0 & 0 & 0 & 0 & 0 & 0 & 0 & 1 \\
0 & 1 & 0 & 0 & 0 & 0 & 0 & 0 & 0 & 0 & 0 \\
0 & -1 & 0 & 0 & 0 & 0 & 0 & 0 & 0 & 0 & 0 \\
0 & 0 & 0 & 0 & 0 & 0 & 0 & 0 & -1 & 0 & 0 \\
0 & 0 & 0 & 0 & 0 & 0 & 0 & 0 & -1 & 0 & 0 \\
0 & 0 & 1 & 0 & 1 & 0 & 0 & 0 & 0 & 0 & 0 \\
0 & 0 & 0 & 0 & 0 & -1 & 0 & 0 & 0 & 0 & 0 \\
0 & 0 & 0 & 0 & 0 & 0 & -1 & 0 & 0 & 0 & 0 \\
0 & 0 & 0 & 0 & 0 & 0 & 1 & 0 & 0 & 0 & 0 \\
0 & 0 & 0 & 0 & 0 & 0 & 0 & 0 & 1 & 0 & 0
\end{bmatrix}. \qquad (10.62)
$$

Using (10.59) we have that

10.7 Computing Chaotic Dynamics

$$g_{*1}^1 = \begin{bmatrix} 0 & 0 & 0 & 0 & 0 & 0 & 0 & 0 & 0 & 0 & 0 \\ 1 & 0 & 0 & 0 & 0 & 0 & 0 & 0 & 0 & 0 & 0 \\ 0 & 0 & 0 & 0 & 0 & 0 & 0 & 0 & 0 & 0 & 0 \\ 0 & 0 & 0 & 0 & 0 & 0 & 0 & 0 & 0 & 0 & 0 \\ 0 & 0 & 0 & 0 & 0 & 0 & 0 & 0 & 0 & 0 & 0 \\ 0 & 0 & 0 & 0 & 0 & 0 & 0 & 0 & 0 & 0 & 0 \\ 0 & 0 & 0 & 0 & 0 & 0 & 0 & 0 & 0 & 0 & 0 \\ 0 & 0 & 0 & 0 & 0 & 0 & 0 & 0 & 0 & 0 & 0 \\ 0 & 0 & 0 & 0 & 0 & 0 & 0 & 0 & 0 & 0 & 0 \\ 0 & 0 & 0 & 0 & 0 & 0 & 0 & 0 & 0 & 0 & 0 \\ 0 & 0 & 0 & 0 & 0 & 0 & 0 & 0 & 0 & 0 & 0 \end{bmatrix}, \quad g_{*1}^2 = \begin{bmatrix} 0 & -1 & 0 & 0 & 0 & 0 & 0 & 0 & 0 & 0 & 0 \\ 0 & 0 & 0 & 0 & 0 & 0 & 0 & 0 & 0 & 0 & 0 \\ 0 & 1 & 0 & 0 & 0 & 0 & 0 & 0 & 0 & 0 & 0 \\ 0 & -1 & 0 & 0 & 0 & 0 & 0 & 0 & 0 & 0 & 0 \\ 0 & 0 & 0 & 0 & 0 & 0 & 0 & 0 & 0 & 0 & 0 \\ 0 & 0 & 0 & 0 & 0 & 0 & 0 & 0 & 0 & 0 & 0 \\ 0 & 0 & 0 & 0 & 0 & 0 & 0 & 0 & 0 & 0 & 0 \\ 0 & 0 & 0 & 0 & 0 & 0 & 0 & 0 & 0 & 0 & 0 \\ 0 & 0 & 0 & 0 & 0 & 0 & 0 & 0 & 0 & 0 & 0 \\ 0 & 0 & 0 & 0 & 0 & 0 & 0 & 0 & 0 & 0 & 0 \\ 0 & 0 & 0 & 0 & 0 & 0 & 0 & 0 & 0 & 0 & 0 \end{bmatrix},$$

$$g_{*1}^3 = \begin{bmatrix} 0 & 0 & 0 & 0 & 0 & 0 & 0 & 0 & 0 & 0 & 0 \\ 0 & 0 & 0 & 0 & 0 & 0 & 0 & 0 & 0 & 0 & 0 \\ 0 & 0 & 0 & 0 & 0 & 0 & 0 & 0 & 0 & 0 & 0 \\ 0 & 0 & 0 & 0 & 0 & 0 & 0 & 0 & 0 & 0 & 0 \\ 0 & 0 & 0 & 0 & 0 & 0 & 0 & 0 & 0 & 0 & 0 \\ 0 & 0 & 0 & 0 & 0 & 0 & 0 & 0 & 0 & 0 & 0 \\ 0 & 0 & 0 & 1 & 0 & 1 & 0 & 0 & 0 & 0 & 0 \\ 0 & 0 & 0 & 0 & 0 & 0 & 0 & 0 & 0 & 0 & 0 \\ 0 & 0 & 0 & 0 & 0 & 0 & 0 & 0 & 0 & 0 & 0 \\ 0 & 0 & 0 & 0 & 0 & 0 & 0 & 0 & 0 & 0 & 0 \\ 0 & 0 & 0 & 0 & 0 & 0 & 0 & 0 & 0 & 0 & 0 \end{bmatrix}, \quad g_{*1}^4 = \begin{bmatrix} 0 & 0 & 0 & 0 & 0 & 0 & 0 & 0 & 0 & 0 & 0 \\ 0 & 0 & 0 & 0 & 0 & 0 & 0 & 0 & 0 & 0 & 0 \\ 0 & 0 & 0 & 0 & 0 & 0 & 0 & 0 & 0 & 0 & 0 \\ 0 & 0 & 0 & 0 & 0 & 0 & 0 & 0 & 0 & 0 & 0 \\ 0 & 0 & 0 & 0 & 0 & 0 & 0 & 0 & 0 & 0 & 0 \\ 0 & 0 & 0 & 0 & 0 & 0 & 0 & 0 & 0 & 0 & 0 \\ 0 & 0 & 0 & 0 & 0 & 0 & 0 & 0 & 0 & 0 & 0 \\ 0 & 0 & 0 & 0 & 0 & 0 & -1 & 0 & 0 & 0 & 0 \\ 0 & 0 & 0 & 0 & 0 & 0 & 0 & 0 & 0 & 0 & 0 \\ 0 & 0 & 0 & 0 & 0 & 0 & 0 & 0 & 0 & 0 & 0 \\ 0 & 0 & 0 & 0 & 0 & 0 & 0 & 0 & 0 & 0 & 0 \end{bmatrix},$$

$$g_{*1}^5 = \begin{bmatrix} 0 & 0 & 0 & 0 & 0 & 0 & 0 & 0 & 0 & 0 & 0 \\ 0 & 0 & 0 & 0 & 0 & 0 & 0 & 0 & 0 & 0 & 0 \\ 0 & 0 & 0 & 0 & 0 & 0 & 0 & 0 & 0 & 0 & 0 \\ 0 & 0 & 0 & 0 & 0 & 0 & 0 & 0 & 0 & 0 & 0 \\ 0 & 0 & 0 & 0 & 0 & 0 & 0 & 0 & 0 & 0 & 0 \\ 0 & 0 & 0 & 0 & 0 & 0 & 0 & 0 & 0 & 0 & 0 \\ 0 & 0 & 0 & 0 & 0 & 0 & 0 & 0 & 0 & 0 & 0 \\ 0 & 0 & 0 & 0 & 0 & 0 & 0 & 0 & 0 & 0 & 0 \\ 0 & 0 & 0 & 0 & 0 & 0 & 0 & -1 & 0 & 0 & 0 \\ 0 & 0 & 0 & 0 & 0 & 0 & 0 & 1 & 0 & 0 & 0 \\ 0 & 0 & 0 & 0 & 0 & 0 & 0 & 0 & 0 & 0 & 0 \end{bmatrix}, \quad g_{*1}^6 = \begin{bmatrix} 0 & 0 & 0 & 0 & 0 & 0 & 0 & 0 & 0 & 0 & 0 \\ 0 & 0 & 0 & 0 & 0 & 0 & 0 & 0 & 0 & 0 & 0 \\ 0 & 0 & 0 & 0 & 0 & 0 & 0 & 0 & 0 & 0 & 0 \\ 0 & 0 & 0 & 0 & 0 & 0 & 0 & 0 & 0 & 0 & 0 \\ 0 & 0 & 0 & 0 & 0 & 0 & 0 & 0 & -1 & 0 & 0 \\ 0 & 0 & 0 & 0 & 0 & 0 & 0 & 0 & -1 & 0 & 0 \\ 0 & 0 & 0 & 0 & 0 & 0 & 0 & 0 & 0 & 0 & 0 \\ 0 & 0 & 0 & 0 & 0 & 0 & 0 & 0 & 0 & 0 & 0 \\ 0 & 0 & 0 & 0 & 0 & 0 & 0 & 0 & 0 & 0 & 0 \\ 0 & 0 & 0 & 0 & 0 & 0 & 0 & 0 & 0 & 0 & 0 \\ 0 & 0 & 0 & 0 & 0 & 0 & 0 & 0 & 0 & 0 & 0 \end{bmatrix},$$

$$g_{*1}^7 = \begin{bmatrix} 0 & 0 & 0 & 0 & 0 & 0 & 0 & 0 & 0 & 0 & 0 \\ 0 & 0 & 0 & 0 & 0 & 0 & 0 & 0 & 0 & 0 & 0 \\ 0 & 0 & 0 & 0 & 0 & 0 & 0 & 0 & 0 & 0 & 0 \\ 0 & 0 & 0 & 0 & 0 & 0 & 0 & 0 & 0 & 0 & 0 \\ 0 & 0 & 0 & 0 & 0 & 0 & 0 & 0 & 0 & 0 & 0 \\ 0 & 0 & 0 & 0 & 0 & 0 & 0 & 0 & 0 & 0 & 0 \\ 0 & 0 & 0 & 0 & 0 & 0 & 0 & 0 & 0 & 0 & 0 \\ 0 & 0 & 0 & 0 & 0 & 0 & 0 & 0 & 0 & 0 & 0 \\ 0 & 0 & 0 & 0 & 0 & 0 & 0 & 0 & 0 & 0 & 0 \\ 0 & 0 & 0 & 0 & 0 & 0 & 0 & 0 & 0 & 0 & 0 \\ 0 & 0 & 0 & 0 & 0 & 0 & 0 & 0 & 0 & 1 & 0 \end{bmatrix}, \quad g_{*1}^8 = \begin{bmatrix} 0 & 0 & 0 & 0 & 0 & 0 & 0 & 0 & 0 & 0 & 0 \\ 0 & 0 & 0 & 0 & 0 & 0 & 0 & 0 & 0 & 0 & 1 \\ 0 & 0 & 0 & 0 & 0 & 0 & 0 & 0 & 0 & 0 & 0 \\ 0 & 0 & 0 & 0 & 0 & 0 & 0 & 0 & 0 & 0 & 0 \\ 0 & 0 & 0 & 0 & 0 & 0 & 0 & 0 & 0 & 0 & 0 \\ 0 & 0 & 0 & 0 & 0 & 0 & 0 & 0 & 0 & 0 & 0 \\ 0 & 0 & 0 & 0 & 0 & 0 & 0 & 0 & 0 & 0 & 0 \\ 0 & 0 & 0 & 0 & 0 & 0 & 0 & 0 & 0 & 0 & 0 \\ 0 & 0 & 0 & 0 & 0 & 0 & 0 & 0 & 0 & 0 & 0 \\ 0 & 0 & 0 & 0 & 0 & 0 & 0 & 0 & 0 & 0 & 0 \\ 0 & 0 & 0 & 0 & 0 & 0 & 0 & 0 & 0 & 0 & 0 \end{bmatrix}.$$

Having computed this topological information, we can now prove the following theorem.

Theorem 10.112 *Let*

$$S = \text{Inv}\left(\bigcup_{i=1}^{8} N^i, h\right) \tag{10.63}$$

be the invariant set under the Hénon map at parameter values $a = 1.3$ *and* $b = 0.2$ *with the sets* $\{N^i \mid i = 1, \ldots, 8\}$ *indicated in Figure 10.4. Let*

$$A = \begin{bmatrix} 0 & 1 & 0 & 0 & 0 & 0 & 0 & 0 \\ 1 & 0 & 0 & 0 & 0 & 0 & 0 & 1 \\ 0 & 1 & 0 & 0 & 0 & 1 & 0 & 0 \\ 0 & 0 & 1 & 0 & 0 & 0 & 0 & 0 \\ 0 & 0 & 0 & 1 & 0 & 0 & 0 & 0 \\ 0 & 0 & 0 & 0 & 1 & 0 & 0 & 0 \\ 0 & 0 & 0 & 0 & 1 & 0 & 0 & 0 \\ 0 & 0 & 0 & 0 & 0 & 0 & 1 & 0 \end{bmatrix}.$$

Then there exists a semiconjugacy

$$\begin{array}{ccc} S & \xrightarrow{h} & S \\ \downarrow\rho & & \downarrow\rho \\ \Sigma_A & \xrightarrow{\sigma} & \Sigma_A \end{array} \tag{10.64}$$

Furthermore, for each periodic sequence $\mathbf{s} \in \Sigma_A$ *with period* p, $\rho^{-1}(\mathbf{s})$ *contains a periodic orbit with period* p.

Proof. To begin, observe that the cubical transition graph (10.61) implies that $\rho(S) \subset \Sigma_A$. Thus, we only have to show that $\Sigma_A \subset \rho(S)$. We begin with the observation that S is an isolated invariant set and therefore is compact. Thus $\rho(S)$ is a compact set and in particular, $\text{cl}(\rho(S)) = \rho(S)$. Let Π_A denote the set of periodic orbits in Σ_A. We shall prove that

$$\Pi_A \subset \rho(S). \tag{10.65}$$

Since A is irreducible, Proposition 10.25 implies that $\text{cl}(\Pi_A) = \Sigma_A$. Thus,

$$\Sigma_A = \text{cl}(\Pi_A) \subset \text{cl}(\rho(S)) = \rho(S)$$

and the conclusion is proved, if we complete the proof of (10.65).

For this end take $\mathbf{s} = (s_1, s_2, \ldots, s_p) \in \Pi_A$. By Theorem 10.110 it is enough to show that

$$L(g_*^{s_p} \circ g_*^{s_{p-1}} \circ \cdots \circ g_*^{s_1}) \neq 0. \tag{10.66}$$

10.7 Computing Chaotic Dynamics

Let $G(\mathbf{s})$ denote the matrix of $g_{*1}^{s_p} \circ g_{*1}^{s_{p-1}} \circ \cdots \circ g_{*1}^{s_1}$ in the selected basis. Since evidently $H_i(T_1, T_0) = 0$ for $i \neq 0$, all what we need to prove (10.66) is to verify that
$$\operatorname{tr} G(\mathbf{s}) \neq 0. \tag{10.67}$$

Given two finite sequences $\mathbf{t} = (t_1, t_2, \ldots, t_k)$ and $\mathbf{t}' = (t'_1, t'_2, \ldots, t'_{k'})$, by their product we mean the sequence
$$\mathbf{tt}' := (t_1, t_2, \ldots, t_k, t'_1, t'_2, \ldots, t'_{k'}).$$

It is straightforward to verify that in this case
$$G(\mathbf{tt}') = G(\mathbf{t}')G(\mathbf{t}). \tag{10.68}$$

Let us identify a simpler description of the periodic orbits in Σ_A. Let
$$\theta_1 := (1,2), \quad \theta_2 := (3,4,5,6), \quad \theta_3 := (3,4,5,7,8,2).$$

Consider the symbol space $\{1,2,3\}^{\mathbb{Z}}$ and transition matrix
$$B := \begin{bmatrix} 1 & 0 & 1 \\ 1 & 1 & 1 \\ 1 & 1 & 1 \end{bmatrix}. \tag{10.69}$$

Let Σ_B denote the associate subshift space. It is left to the reader to check that if $\mathbf{s}$ is a periodic orbit in Σ_A, then $\mathbf{s} = \overline{\theta_{t_1} \theta_{t_2} \cdots \theta_{t_q}}$, where $\mathbf{t} := \overline{(t_1, t_2 \ldots, t_q)}$ is a periodic orbit in Σ_B. Put $\Theta_i := G(\theta_i)$ for $i = 1, 2, 3$. Then, by (10.68),
$$G(\mathbf{s}) = \Theta_{t_q} \Theta_{t_{q-1}} \cdots \Theta_{t_1}. \tag{10.70}$$

Using the formulas for the matrices of g_{*1}^j one can compute that

$$\Theta_1 = \begin{bmatrix} -1 & 0 & 0 & 0 & 0 & 0 & 0 & 0 & 0 & 0 & 0 \\ 0 & 0 & 0 & 0 & 0 & 0 & 0 & 0 & 0 & 0 & 0 \\ 1 & 0 & 0 & 0 & 0 & 0 & 0 & 0 & 0 & 0 & 0 \\ -1 & 0 & 0 & 0 & 0 & 0 & 0 & 0 & 0 & 0 & 0 \\ 0 & 0 & 0 & 0 & 0 & 0 & 0 & 0 & 0 & 0 & 0 \\ 0 & 0 & 0 & 0 & 0 & 0 & 0 & 0 & 0 & 0 & 0 \\ 0 & 0 & 0 & 0 & 0 & 0 & 0 & 0 & 0 & 0 & 0 \\ 0 & 0 & 0 & 0 & 0 & 0 & 0 & 0 & 0 & 0 & 0 \\ 0 & 0 & 0 & 0 & 0 & 0 & 0 & 0 & 0 & 0 & 0 \\ 0 & 0 & 0 & 0 & 0 & 0 & 0 & 0 & 0 & 0 & 0 \\ 0 & 0 & 0 & 0 & 0 & 0 & 0 & 0 & 0 & 0 & 0 \end{bmatrix}, \quad \Theta_2 = \begin{bmatrix} 0 & 0 & 0 & 0 & 0 & 0 & 0 & 0 & 0 & 0 & 0 \\ 0 & 0 & 0 & 0 & 0 & 0 & 0 & 0 & 0 & 0 & 0 \\ 0 & 0 & 0 & 0 & 0 & 0 & 0 & 0 & 0 & 0 & 0 \\ 0 & 0 & 0 & 0 & 0 & 0 & 0 & 0 & 0 & 0 & 0 \\ 0 & 0 & 0 & -1 & 0 & -1 & 0 & 0 & 0 & 0 & 0 \\ 0 & 0 & 0 & -1 & 0 & -1 & 0 & 0 & 0 & 0 & 0 \\ 0 & 0 & 0 & 0 & 0 & 0 & 0 & 0 & 0 & 0 & 0 \\ 0 & 0 & 0 & 0 & 0 & 0 & 0 & 0 & 0 & 0 & 0 \\ 0 & 0 & 0 & 0 & 0 & 0 & 0 & 0 & 0 & 0 & 0 \\ 0 & 0 & 0 & 0 & 0 & 0 & 0 & 0 & 0 & 0 & 0 \\ 0 & 0 & 0 & 0 & 0 & 0 & 0 & 0 & 0 & 0 & 0 \end{bmatrix},$$

$$\Theta_3 = \begin{bmatrix} 0 & 0 & 0 & 1 & 0 & 1 & 0 & 0 & 0 & 0 & 0 \\ 0 & 0 & 0 & 0 & 0 & 0 & 0 & 0 & 0 & 0 & 0 \\ 0 & 0 & 0 & -1 & 0 & -1 & 0 & 0 & 0 & 0 & 0 \\ 0 & 0 & 0 & 1 & 0 & 1 & 0 & 0 & 0 & 0 & 0 \\ 0 & 0 & 0 & 0 & 0 & 0 & 0 & 0 & 0 & 0 & 0 \\ 0 & 0 & 0 & 0 & 0 & 0 & 0 & 0 & 0 & 0 & 0 \\ 0 & 0 & 0 & 0 & 0 & 0 & 0 & 0 & 0 & 0 & 0 \\ 0 & 0 & 0 & 0 & 0 & 0 & 0 & 0 & 0 & 0 & 0 \\ 0 & 0 & 0 & 0 & 0 & 0 & 0 & 0 & 0 & 0 & 0 \\ 0 & 0 & 0 & 0 & 0 & 0 & 0 & 0 & 0 & 0 & 0 \\ 0 & 0 & 0 & 0 & 0 & 0 & 0 & 0 & 0 & 0 & 0 \end{bmatrix}.$$

A straightforward but lengthy computation shows that we have the following multiplication table.

	Θ_1	Θ_2	Θ_3
Θ_1	$-\Theta_1$	0	$-\Theta_3$
Θ_2	$\Theta_2\Theta_1$	$-\Theta_2$	$\Theta_2\Theta_3$
Θ_3	Θ_1	$-\Theta_3$	Θ_3

We will show by induction in q that for every periodic orbit $\mathbf{t} = \overline{(t_1, t_2, \ldots, t_q)}$ in Π_B

$$\Theta_{t_q}\Theta_{t_{q-1}}\cdots\Theta_{t_1} = \begin{cases} \pm\Theta_{t_1^*} & \text{if } t_q \neq 2, \\ \pm\Theta_2 & \text{if } t_i = 2 \text{ for all } i = 1, 2, \ldots, q, \\ \pm\Theta_2\Theta_3 & \text{otherwise,} \end{cases} \quad (10.71)$$

where

$$i^* := \begin{cases} i & \text{if } i \neq 2, \\ 3 & \text{otherwise} \end{cases}$$

for $i \in \{1, 2, 3\}$.

If $q = 1$ the conclusion is obvious. Thus assume that $q > 1$. Let $\Theta_{\mathbf{t}} := \Theta_{t_q}\Theta_{t_{q-1}}\cdots\Theta_{t_1}$. It is straightforward to verify from the multiplication table that if $t_q = 3$, then

$$\Theta_{t_q}\Theta_{t_{q-1}} = \pm\Theta_{t_{q-1}^*}. \quad (10.72)$$

If $t_q = 1$, then $t_{q-1} \neq 2$, because otherwise $\mathbf{t} \notin \Pi_B$. Therefore in this case Eq. (10.72) is also satisfied. Since $t_{q-1}^* \in \{1, 3\}$, we get from the induction assumption that

$$\Theta_{\mathbf{t}} = \Theta_{t_{q-1}^*}\Theta_{t_{q-2}}\cdots\Theta_{t_1} = \pm\Theta_{t_1^*}, \quad (10.73)$$

which proves Eq. (10.71) for $t_q \neq 2$.

It remains to consider the case $t_q = 2$. In this case $\mathbf{t} \in \Pi_B$ implies that $t_{q-1} \neq 1$ and $t_1 \neq 1$ and consequently also $t_1^* = 3$.

If $t_{q-1} = 2$, we have $\Theta_{t_q}\Theta_{t_{q-1}} = \pm\Theta_2$, therefore

$$\Theta_{\mathbf{t}} = \pm\Theta_{t_{q-1}}\Theta_{t_{q-2}}\cdots\Theta_{t_1}$$

and we get the conclusion from the induction assumption. If $t_{q-1} = 3$, then

$$\Theta_{t_{q-1}}\Theta_{t_{q-2}}\cdots\Theta_{t_1} = \pm\Theta_{t_1^*},$$

which implies that
$$\Theta_{\mathbf{t}} = \pm\Theta_2\Theta_{t_1^*} = \pm\Theta_2\Theta_3,$$

and finishes the proof of Eq. (10.71).

It is straightforward to verify that $\operatorname{tr}\Theta_1 = \operatorname{tr}\Theta_2 = \operatorname{tr}\Theta_2\Theta_3 = -1$ and $\operatorname{tr}\Theta_3 = 1$. It follows from (10.71) that for every periodic orbit $\mathbf{t} = (t_1, t_2, \ldots, t_q)$ in Π_B
$$\operatorname{tr}\left(\Theta_{t_q}\Theta_{t_{q-1}}\cdots\Theta_{t_1}\right) = \pm 1.$$

Combining this result with (10.70) we obtain (10.67) and the theorem is proved. □

As a concluding remark, observe that A is irreducible and Π_A contains more than one orbit. Thus as indicated in Remark 10.26 the subshift dynamics $\sigma : \Sigma_A \to \Sigma_A$ exhibits chaotic dynamics. Therefore we can conclude that the invariant set S of the Hénon map given by (10.63) contains chaotic dynamics.

Exercises

10.23 Complete the proof of Proposition 10.108.

10.8 Bibliographical Remarks

For an introduction to the qualitative theory of differential equations, the reader is referred to [35, 38], for dynamical systems and chaos to [73], for fixed-point theory to [21], and for degree theory to [49].

The Ważewski principle discovered by Ważewski is presented in [85]. The first appearance of the Lefschetz fixed-point theorem is in the Lefschetz paper [48]. The Borsuk Theorem 10.43 on the equivalence of the Brouwer theorem, nonretraction and noncontractibility is presented in [21, Chapter II, Theorem 5.2].

Index pairs were introduced by Ch. Conley [13] in order to define the Conley index for flows. The first generalization of Conley index in the discrete setting was presented in [75] by means of the shape theory. Index maps, as a tool for studying the dynamics of discrete dynamical systems, were proposed in [61]. The language of shift equivalence was introduced to the index theory by Franks and Richeson [24] (it is equivalent to the categorical construction of Szymczak [80]).

Theorem 10.102 in a more general setting of the fixed-point index theory is proved in [60, 62]. A related earlier result is the relative Lefschetz fixed-point theorem proved by Bowszyc [9].

The idea of using multivalued maps to study the dynamics of a continuous map originated from [56] and it led to [44], where the Conley index theory is extended to upper semicontinuous acyclic-valued maps. The first algorithms concerning automated computation of index pairs were introduced in [64, 65]. Theorem 10.83 is due to A. Szymczak [81]. The definition of index pair used in this book comes from [66]. It was motivated by work of Szymczak [83] and Day [15] and is dictated by the efficiency of algorithms. Details on how it can be tied to classical index pairs and the general Conley index theory as well as some other algorithms computing index pairs may be found in [66]. For a general reference to the computational aspects of the Conley index theory, see [57, 55]. Theorem 10.91 is a special case of a general statement from the Conley index theory for discrete dynamical systems as developed in [61, 63, 80]. The methods used in Section 10.7 are based on work of Szymczak [82, 83], Day and Junge [16], and [58].

11
Homology of Topological Polyhedra

We have presented in this book a theory of cubical homology. Our justification for this approach lies in the applications described in Chapters 1, 8, and 10, where we are required to work with large sets of data and for which we need a computationally effective means of computing homology. In all these examples the data itself naturally generates cubical sets. However, this cubical homology theory is unconventional, and furthermore, there is a wide variety of other homology theories available.

This last sentence needs some explanation. Even within this book we have presented several alternative homologies such as reduced homology and relative homology. However, for each of these theories we have used cubes in $\mathbf{R}^d$ as our basic building blocks. Of course, we have also discussed and used the homology of abstract complexes, which should suggest that it is possible to define a homology theory that is based on other geometric objects. For example, a cube is just a special form of a polyhedron, and thus one might ask if it is possible to define a homology theory based on a different polyhedron. The answer is yes, and in fact, by far the most common theory is simplicial homology based on simplices or triangulations. We provide a very brief description of this homology in Section 11.1.

An obvious question is: What is the relationship between the cubical and simplicial theories? In Section 11.2 we discuss in some detail the advantages and disadvantages of using cubes versus simplices. For the moment let us simply make the remark that every cubical set can be represented in terms of simplices, but there are sets (e.g., a simple triangle) that can be represented in terms of simplices but are not cubical sets. However, even though there is an asymmetry on the geometric level, in Section 11.3 we show that cubical theory can be extended to the same class of spaces as the simplicial theory. This in turn implies that the simplicial and cubical theories are equivalent.[1]

---

[1] The proof of this last statement goes beyond the scope of this book. The interested reader is referred to [78, Chapter 4.8].

However, there is a deeper reason for showing that simplicial homology and cubical homology agree. Many of the standard references on algebraic topology [69, 78] include not only the simplicial theory, but also other homology theories, such as singular, CW, Čech, and Alexander Spanier. Furthermore, these texts also describe the relationships between these different theories. Thus by extending the cubical theory to polyhedra, we are placing it in its proper context in the larger field of algebraic topology.

The above-mentioned theories have their own strengths and weaknesses. We believe that the cubical theory is ideal for a wide variety of concrete computational problems. However, some of the other theories are much easier to work with on an abstract level, and, in particular, this allows one to more easily interpret the topological content of the homology groups and maps. We already have used this in Section 10.7. Observe that the computations are all done using the cubical theory. The mathematical justification (crucial portions of which were not provided) for the results claimed about the dynamics can be easily proved using other homology theories. Thus, knowing the relationship between the cubical and the other homology theories is essential.

In summary, this chapter is meant to provide a bridge between the introductory material of this book and the rich, deep, and widely developed field of algebraic topology. With this in mind, our presentation also changes its tone. Many of the technical details in Section 11.1 are left to the reader, and our discussion in Section 11.3 is much more abstract in nature.

11.1 Simplicial Homology

The cubical theory that has been developed throughout this book is built upon the geometry of cubes in $\mathbf{R}^d$. The simplicial theory, which is by far the most common, is based on simplices. Therefore, to present it we need to begin with some geometric preliminaries.

Definition 11.1 A subset K of $\mathbf{R}^d$ is called *convex* if, given any two points $x, y \in K$, the line segment

$$[x, y] := \{\lambda x + (1 - \lambda)y \mid 0 \leq \lambda \leq 1\}$$

joining x to y is contained in K.

Definition 11.2 The *convex hull* conv A of a subset A of $\mathbf{R}^d$ is the intersection of all closed and convex sets containing A.

Since $\mathbf{R}^d$ is a closed convex set containing A, conv $A \neq \emptyset$. Furthermore, the intersection of closed sets is closed and it is easily checked that the intersection of convex sets is convex. Thus conv A is the smallest closed convex set containing A.

It is intuitively clear that the convex hull of two points is a line segment joining those points, a convex hull of three noncolinear points is a triangle, and a convex hull of four noncoplanar points is a tetrahedron. We shall generalize those geometric figures to an arbitrary dimension under the name simplex.

Theorem 11.3 ([74, Chapter 2, Theorem 2.2]) *Let* $\mathcal{V} = \{v_0, v_1, \ldots, v_n\} \in \mathbf{R}^d$ *be a finite set. Then* $\operatorname{conv} \mathcal{V}$ *is the set of those* $x \in \mathbf{R}^d$ *that can be written as*

$$x = \sum_{i=0}^n \lambda_i v_i, \ 0 \leq \lambda_i \leq 1, \ \sum_{i=0}^n \lambda_i = 1. \tag{11.1}$$

In general, the coefficients λ_i need not be unique. For example, consider $v_1 = (0,0)$, $v_2 = (1,0)$, $v_3 = (1,1)$, and $v_4 = (0,1)$, the four vertices of the unit square $[0,1]^2$. Observe that

$$\left(\frac{1}{2}, \frac{1}{2}\right) = \frac{1}{2}v_1 + 0v_2 + \frac{1}{2}v_3 + 0v_4 = 0v_1 + \frac{1}{2}v_2 + 0v_3 + \frac{1}{2}v_4.$$

Definition 11.4 A finite set $\mathcal{V} = \{v_0, v_1, \ldots, v_n\}$ in $\mathbf{R}^d$ is *geometrically independent* if, for any $x \in \operatorname{conv} \mathcal{V}$, the coefficients λ_i in (11.1) are unique. If this is the case, the λ_i are called *barycentric coordinates* of x.

The proof of the following statement is left as an exercise.

Proposition 11.5 *Let* $\mathcal{V} = \{v_0, v_1, \ldots, v_n\} \in \mathbf{R}^d$. *Then* $\mathcal{V}$ *is geometrically independent if and only if the set of vectors* $\{v_1 - v_0, v_2 - v_0, \ldots, v_n - v_0\}$ *is linearly independent.*

Definition 11.6 Let $\mathcal{V} = \{v_0, v_1, \ldots, v_n\}$ be geometrically independent. The set $S = \operatorname{conv} \mathcal{V}$ is called a *simplex* or, more specifically, an *n-simplex spanned by vertices* $v_0, v_1, \ldots, v_n$. In certain situations we might want to use the term *geometric simplex* for S in order to distinguish it from an abstract oriented simplex, which will be defined later. The number n is called *the dimension of* S. If $\mathcal{V}'$ is a subset of $\mathcal{V}$ of $k \leq n$ vertices, the set $S' = \operatorname{conv} \mathcal{V}'$ is called a *k-face* of S.

Theorem 11.7 ([21, Chapter II, Proposition 1.2]) *Any two n-simplices are homeomorphic. Moreover, for any simplex* $S = \operatorname{conv}\{v_0, v_1, \ldots, v_n\}$, *the corresponding barycentric coordinate functions* $\lambda_i : S \to [0,1]$ *are continuous.*

Definition 11.8 A *simplicial complex* $\mathcal{S}$ is a finite collection of simplices such that

1. every face of a simplex in $\mathcal{S}$ is in $\mathcal{S}$,
2. the intersection of any two simplices in $\mathcal{S}$ is a face of each of them.

Definition 11.9 Given a simplicial complex $\mathcal{S}$ in $\mathbf{R}^d$, the union of all simplices of $\mathcal{S}$ is called the *polytope* of $\mathcal{S}$ and is denoted by $|\mathcal{S}|$. A subset $P \subset \mathbf{R}^d$ is a *polyhedron* if P is the polytope of some simplicial complex $\mathcal{S}$. In this case $\mathcal{S}$ is called a *triangulation* of P.

A polyhedron may have different triangulations. Figure 11.1 shows examples of subdivisions of a square to triangles. The first two are triangulations, but the last one is not since the intersection of a triangle in the lower left corner with the triangle in the upper right corner is not an edge of the latter one but a part of it. Observe that this is different from the cubical theory where a cubical subset of $\mathbf{R}^d$ has a unique cubical complex associated to it. Furthermore, any cubical set can be triangulated (see Exercise 11.6). However, as shown in Exercise 2.6, not every polyhedron is cubical. This means that the simplicial theory is more flexible.

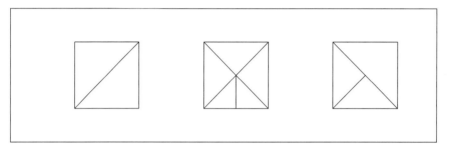

Fig. 11.1. Subdivisions of a square to triangles: The first two are triangulations, the last one is not.

Example 11.10 By a *torus* we mean any space homeomorphic to the product $S^1 \times S^1$ of two circles. By means of polar coordinates one can show that this space is homeomorphic to the surface in $\mathbf{R}^3$ obtained by rotation of the circle $(x-2)^2 + z^2 = 1$, $y = 0$ about the y-axis. This set can be described as the surface of a donut. Neither of the above surfaces is a polyhedron, but we shall construct one that is. Let G be the boundary of any triangle in $\mathbf{R}^2$. Then G is a simple closed curve; hence it is homeomorphic to the unit circle. Thus $T = G \times G \in \mathbf{R}^4$ is a torus. In order to construct a triangulation of T, we may visualize T as a square in Figure 11.2 with pairs of parallel sides glued together. More precisely, consider the square $[0,3]^2 = \text{conv}\{v_1, v_2, v_3, v_4\}$, where $v_1 = (0,0), v_2 = (0,3), v_3 = (3,3), v_4 = (0,3)$. Bend the square along the lines $x=1$ and $x=2$ and glue the directed edge $[v_1, v_4]$ to $[v_2, v_3]$ so that the vertex v_1 is identified with v_2 and v_4 with v_3. We obtain a cylinder in $\mathbf{R}^3$ with a boundary of a unilateral triangle in the plane $y=0$ as the base. We bend the cylinder along the lines $y=1$ and $y=2$ (this cannot be done in $\mathbf{R}^3$ without stretching, though it can be done in $\mathbf{R}^4$) and glue the edge $[v_1, v_2]$ to $[v_4, v_3]$. Note that the four vertices v_1, v_2, v_3, v_4 of the square became one. The bent lines divide the square into nine unitary squares. Each of them can be divided into two triangles as shown in Figure 11.2. Let $\mathcal{S}$ be the collection of all vertices, edges, and triangles of T obtained in this way. Although some vertices and edges are identified by gluing, the reader may verify that $\mathcal{S}$ satisfies the definition of a simplicial complex.

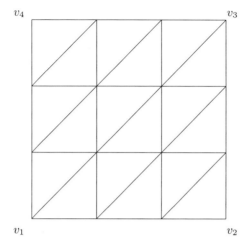

Fig. 11.2. Triangulation of a torus.

Having finished with the geometric preliminaries, we now turn to the problem of developing the associated algebra. The term "simplicial complex" suggests that there should be some natural structure of chain complex associated with it. That is not so easy to define due to problems with orientation that do not appear when we study cubical sets. We shall therefore proceed as we did with graphs in Chapter 1, that is, we shall start from chain complexes with coefficients in $\mathbf{Z}_2$. This will make definitions much more simple and, historically, this is the way homology was first introduced.

Let $C_n(\mathcal{S}; \mathbf{Z}_2)$ be the vector space generated by the set $\mathcal{S}^n$ of n-dimensional simplices of $\mathcal{S}$ as the canonical basis. More precisely, the basic vectors are duals $\widehat{S}$ of simplices S defined as in Definition 13.65. We put $C_n(\mathcal{S}; \mathbf{Z}_2) := 0$ if $\mathcal{S}$ has no simplices of dimension n. The *boundary map* $\partial_n : C_n(\mathcal{S}; \mathbf{Z}_2) \to C_{n-1}(\mathcal{S}; \mathbf{Z}_2)$ is defined on the dual of any basic element $S = \operatorname{conv}\{v_0, v_1, \ldots, v_n\}$ by the formula

$$\partial_n(\widehat{S}) = \sum_{i=0}^{n} \operatorname{conv} \widehat{(\mathcal{V} \setminus \{v_i\})}.$$

Thus, in the modulo 2 case, the algebraic boundary of a simplex corresponds precisely to its geometric boundary. We have the following.

Proposition 11.11 $\partial_{n-1}\partial_n = 0$ *for all* n.

Proof. For any simplex $S = \operatorname{conv}\{v_0, v_1, \ldots, v_n\}$,

$$\partial_{n-1}\partial_n(\widehat{S}) = \sum_{j \neq i} \sum_{i=0}^{n} \operatorname{conv} \widehat{(\mathcal{V} \setminus \{v_i, v_j\})}.$$

Each $(n-1)$-face of S appears in the above sum twice; therefore, the sum modulo 2 is equal to zero. □

Thus $\mathcal{C}(\mathcal{S}; \mathbf{Z}_2) := \{C_n(\mathcal{S}; \mathbf{Z}_2), \partial_n\}_{n \in \mathbf{Z}}$ has the structure of a chain complex with coefficients in $\mathbf{Z}_2$. The homology of that chain complex is the sequence of vector spaces

$$H_*(\mathcal{S}; \mathbf{Z}_2) = \{H_n(\mathcal{S}; Z_2)\} = \{\ker \partial_n / \operatorname{im} \partial_{n+1}\}.$$

The modulo 2 homology of graphs discussed in Section 1.6 is a special case of what we did above. The real goal, however, is to construct a chain complex corresponding to $\mathcal{S}$ with coefficients in $\mathbf{Z}$ as defined in Chapter 2. As we did it with graphs, we want to impose an orientation of vertices $v_0, v_1, \ldots, v_n$ spanning a simplex. In the case of graphs that was easy, since each edge joining vertices v_1, v_2 could be written in two ways, as $[v_1, v_2]$ or $[v_2, v_1]$ and it was sufficient to tell which vertex we wanted to write as the first and which as the last. In the case of simplices of dimension higher than one, there are many different ways of ordering the set of vertices.

Definition 11.12 Two orderings $(v_0, v_1, \ldots, v_n)$ and $(v_{p_0}, v_{p_1}, \ldots, v_{p_n})$ of vertices of an n-simplex S are said to have the same *orientation* if one can get one from another by an even number of permutations of neighboring terms

$$(v_{i-1}, v_i) \to (v_i, v_{i-1}).$$

This defines an equivalence relation on the set of all orderings of vertices of S. An *oriented simplex* $\sigma = [v_0, v_1, \ldots, v_n]$ is an equivalence class of the ordering $(v_0, v_1, \ldots, v_n)$ of vertices of a simplex $S = \operatorname{conv}\{v_0, v_1, \ldots, v_n\}$. If S and T are geometric simplices, the corresponding oriented simplices are denoted by σ and, respectively, τ.

It is easy to see that for $n > 0$ the above equivalence relation divides the set of all orderings into two equivalence classes. Hence we may say that the orderings that are not in the same equivalence class have the *opposite orientation*. We shall denote the pairs of opposite oriented simplices by σ, σ' or τ, τ'. An oriented simplicial complex is a simplicial complex $\mathcal{S}$ with one of the two equivalence classes chosen for each simplex of $\mathcal{S}$. The orientations of a simplex and its faces may be done arbitrarily; they do not need to be related.

Example 11.13 Let S be a triangle in $\mathbf{R}^2$ spanned by vertices v_1, v_2, v_3. Then the orientation equivalence class $\sigma = [v_1, v_2, v_3]$ contains the orderings (v_1, v_2, v_3), (v_2, v_3, v_1), (v_3, v_1, v_2) and the opposite orientation σ' contains (v_1, v_3, v_2), (v_2, v_1, v_3), (v_3, v_2, v_1). One may graphically distinguish the two orientations by tracing a closed path around the boundary of the triangle s following the order of vertices. The first equivalence class gives the counter-clockwise direction and the second one the clockwise direction. However, the meaning of clockwise or counterclockwise orientation is lost when we consider a triangle in a space of higher dimension. Let $\mathcal{S}$ be the complex consisting of s and all of its edges and vertices. Here are some among possible choices of orientations and their graphical representations in Figure 11.3:

1. $[v_1, v_2, v_3], [v_1, v_2], [v_2, v_3], [v_3, v_1]$;
2. $[v_1, v_2, v_3], [v_1, v_2], [v_2, v_3], [v_1, v_3]$;
3. $[v_1, v_3, v_2], [v_1, v_2], [v_2, v_3], [v_1, v_3]$.

On the first sight the second and third orientations seem wrong since the arrows on the edges of the triangle do not close a cycle, but do not worry: When we get to algebra, the "wrong" direction of the arrows will be corrected by the minus sign in the formula for the boundary operator.

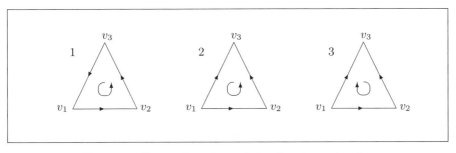

Fig. 11.3. Some orientations of simplices in a triangle.

Now let S^n be the set of all oriented n-simplices of $\mathcal{S}$. Recall that the free abelian group $\mathbf{Z}(S^n)$ generated by a finite set S^n is the set of all functions $c : S^n \to \mathbf{Z}$, generated by basic elements $\hat{\sigma}$ that can be identified with $\sigma \in S^n$. We would like to call this the group of n-chains, but there is a complication: If $n \geq 0$, each n-simplex of $\mathcal{S}$ corresponds to two elements of S^n. We therefore adopt the following definition.

Definition 11.14 The group of n-chains denoted by $C_n(\mathcal{S})$ is the subgroup of $\mathbf{Z}(S^n)$ consisting of those functions c that satisfy the identity

$$c(\sigma) = -c(\sigma')$$

if σ and σ' are opposite orientations of the same n-simplex s. We put $C_n(\mathcal{S}) := 0$ if $\mathcal{S}$ contains no n-simplices.

Proposition 11.15 The group $C_n(\mathcal{S})$ is a free abelian group generated by functions $\tilde{\sigma} = \hat{\sigma} - \hat{\sigma'}$ given by the formula

$$\tilde{\sigma}(\tau) := \begin{cases} 1 & \text{if } \tau = \sigma, \\ -1 & \text{if } \tau = \sigma', \\ 0 & \text{otherwise,} \end{cases}$$

where $\sigma, \sigma', \tau \in S^n$ and σ, σ' are opposite orientations of the same simplex. This set of generators is not a basis since $\tilde{\sigma'} = -\tilde{\sigma}$ for any pair σ, σ'. A basis is obtained by selecting one $\tilde{\sigma}$ from each pair of oriented simplices with mutually opposite orientations.

The choice of a basis in Proposition 11.15 is related to the choice of an orientation in $\mathcal{S}$. Note that our notation of generators which correspond to geometric simplices became very complicated: We already have hats, primes, and now there comes the tilde! We shall simplify this notation by identifying the basic elements $\tilde{\sigma}$ with σ. Upon this identification, we get the identification of σ' with $-\sigma$. The *simplicial boundary operator* $\partial_k : C_k(\mathcal{S}) \to C_{k-1}(\mathcal{S})$ is defined on any basic element $[v_0, v_1, \ldots, v_n]$ by the formula

$$\partial_k [v_0, v_1, \ldots, v_n] = \sum_{i=0}^{n} (-1)^i [v_0, v_1, \ldots, v_{i-1}, v_{i+1}, \ldots, v_n]. \quad (11.2)$$

There is a bit of work involved in showing that this formula actually defines a boundary map: First, one needs to show that the formula is correct, namely that it does not depend on the choice of a representative of the equivalence class $[v_0, v_1, \ldots, v_n]$. Second, one needs to show that $\partial_{k-1} \partial_k = 0$. The reader may consult [69, Chapter 1] for the proofs.

Thus $\mathcal{C}(\mathcal{S}) := \{C_k(\mathcal{S}), \partial_k\}_{k \in \mathbf{Z}}$ has the structure of a chain complex as defined in Section 2.6. The homology of that chain complex is the sequence of abelian groups

$$H_*(\mathcal{S}) = \{H_n(\mathcal{S})\} = \{\ker \partial_n / \operatorname{im} \partial_{n+1}\}.$$

Exercises

11.1 Prove Proposition 11.5.

11.2 Define the chain complex $\mathcal{C}(T; \mathbf{Z}_2)$ for the triangulation discussed in Example 11.10 and use the Homology program to compute $H_*(T; \mathbf{Z}_2)$.

11.3 Label vertices, edges, and triangles of the triangulation of the torus in Example 11.10 displayed in Figure 11.2. Define the chain complex $\mathcal{C}(T)$. Use the Homology program to compute $H_*(T)$.

11.4 Let K be a polyhedron constructed as T in Example 11.10 but with one pair of sides twisted before gluing so that the directed edge $[a, d]$ is identified with $[c, b]$. The remaining pair of edges is glued as before, $[b, c]$ with $[a, d]$. Compute $H_*(K)$. What happens if we try to use the Homology program for computing $H_*(K; \mathbf{Z}_2)$?
This K is called the *Klein bottle*. Note that K cannot be visualized in $\mathbf{R}^3$; we need an extra dimension in order to glue two circles limiting a cylinder with twisting and without cutting the side surface of the cylinder.

11.5 Let P be a polyhedron constructed as T in Example 11.10 but with sides twisted before gluing so that the directed edge $[a, d]$ is identified with $[c, b]$ and $[b, c]$ with $[d, a]$. Compute $H_*(P)$. What happens if we try to use the Homology program for computing $H_*(P; \mathbf{Z}_2)$?
This P is called the *projective plane*. Note that P cannot be visualized in $\mathbf{R}^3$.

11.2 Comparison of Cubical and Simplicial Complexes

Keeping the brief introduction to simplicial homology provided in the previous section in mind, we now turn to a comparison of the use of simplicies versus cubes. Cubical complexes have several nice properties that simplicial complexes do not share:

1. As we have emphasized on several occasions, images and numerical computations naturally lead to cubical sets. Subdividing these cubes to obtain a triangulation is at this point artificial and increases the size of data significantly. For example, it requires $n!$ simplices to triangulate a single n-dimensional cube.
2. Because cubical complexes are so rigid, they can be stored with a minimal amount of information. For example, an elementary cube can be described using one vertex. Specifying a simplex typically requires knowledge of all the vertices. Similarly, for each elementary cube there is a fixed upper bound on the number of adjacent elementary cubes. For example, a vertex in a cubical set X in $\mathbf{R}^2$ is shared by at most four edges, in $\mathbf{R}^3$ by at most six edges. Similarly, an edge in $\mathbf{R}^2$ is shared by at most two squares, while in $\mathbf{R}^3$ there are at most four squares. As is indicated in Figure 11.4, this is not the case for simplicial complexes.
3. A product of elementary cubes is an elementary cube, but a product of simplices is not a simplex. For example, the product of two edges is a square, not a simplex. Thus, in the simplicial theory, one needs to perform subdivisions of products of simplices. This observation also implies that there is no natural projection from a higher-dimensional simplicial complex to a lower-dimensional complex, thus precluding the construction of simple chain maps such as that of Example 4.8.
4. In Chapter 6 we use rescaling to better approximate continuous functions. The rescaling formula is extremely simple and allows for uniform improvements of the approximation. The standard approximation procedure for simplicial theory makes use of barycentric subdivisions (see [69, Chapter II]). This process is much more complex both as a concept and as a numerical tool.
5. As indicated in the previous section, to define chain complexes based on simplices requires the nontrivial concept of orientation. The reader may wonder how we avoided this problem in our development of cubical homology? The answer is, we did not. However, because our cubical complexes are based on elementary cubes, which in turn are defined in terms of an integral lattice in $\mathbf{R}^d$, the orientation is completely natural. More precisely, we began by writing an elementary interval as $[l, l+1]$ and not $[l+1, l]$. In other words, a linear order of real numbers imposes a choice of an orientation on each coordinate axis in $\mathbf{R}^d$. Furthermore, we have always written a product of intervals as $I_1 \times I_2 \times \cdots \times I_d$. This implicitly chooses an ordering of the canonical basis for $\mathbf{R}^d$.

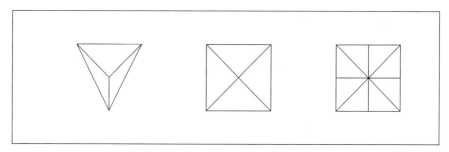

Fig. 11.4. There is no limit on the number of edges that can share a vertex in a simplicial complex.

Listing the strengths of the cubes is not meant to imply that cubical complexes are ideal for every computational problem. Computer graphics and visualization are based on the rendering of triangulated surfaces. Thus the basic building blocks are simplices and while every cubical set can be triangulated, it is not the case that every polyhedron can be expressed as a cubical set. There is a price to be paid for the rigidity of elementary cubes.

At this point we need to make an important observation. The fact that a polyhedron need not be a cubical set does not imply that we cannot use the cubical theory to define the homology of the polyhedron. We have seen this phenomenon before on an algebraic level. The CCR reduction algorithm discussed in Section 4.4 makes fundamental use of abstract chain complexes, which might not be realizable as cubical complexes. Thus we use the algebraic structure of chain complexes to circumvent the geometric rigidity inherent in the cubical complex. To define the homology of polyhedra, we will use topology to relate it to cubical sets. We start with a definition and observation.

Definition 11.16 Given any $d \geq 0$, the *standard d-simplex* Δ^d is given by $\Delta^d := \text{conv}\{0, \mathbf{e}_1, \mathbf{e}_2, \ldots, \mathbf{e}_d\}$, where 0 is the origin of coordinates in R^d and $\{\mathbf{e}_1, \mathbf{e}_2, \ldots, \mathbf{e}_d\}$ is the canonical basis for $\mathbf{R}^d$.

Since the set $\{\mathbf{e}_1, \mathbf{e}_2, \ldots, \mathbf{e}_d\}$ is linearly independent, it easily follows that $\{0, \mathbf{e}_1, \mathbf{e}_2, \ldots, \mathbf{e}_d\}$ is geometrically independent and hence that Δ^d is a d-simplex. The verification is left as an exercise. Note that the barycentric coordinates of a point $x = (x_1, x_2, \ldots, x_d)$ in Δ^d are $\lambda_i = x_i$ for $i = 1, 2, \ldots, d$ and
$$\lambda_0 = 1 - \sum_{i=1}^{d} x_i.$$

Theorem 11.17 ([8, Part III, Theorem 1.1]) *Every polyhedron P is homeomorphic to a cubical set. Moreover, given any triangulation $\mathcal{S}$ of P, there exists a homeomorphism $h : P \to X$, where X is a cubical subset of $[0,1]^d$ and $d+1$ is the number of vertices of $\mathcal{S}$, such that the restriction of h to any simplex of $\mathcal{S}$ is a homeomorphism of that simplex onto a cubical subset of X.*

11.2 Comparison of Cubical and Simplicial Complexes

Proof. In order to keep the idea transparent, we skip several technical verifications and refer the reader to [8, Part III] for more details. The construction of h is done in two steps.

Step 1. We construct a homeomorphic embedding of P into a standard simplex in a sufficiently high-dimensional space.

Indeed, let $\mathcal{S}$ be a triangulation of P and let $\mathcal{V} = \{v_0, v_1, v_2, \ldots, v_d\}$ be the set of all vertices of $\mathcal{S}$. Let Δ^d be the standard d-simplex in $\mathbf{R}^d$ described in Definition 11.16. Consider the bijection f_0 of $\mathcal{V}$ onto the set $\{0, \mathbf{e}_1, \mathbf{e}_2, \ldots, \mathbf{e}_d\}$ given by $f_0(v_0) = 0$ and $f_0(v_i) = \mathbf{e}_i$ for $i = 1, 2, \ldots, d$. For simplicity of notation we put $\mathbf{e}_0 := 0$. This way we avoid distinguishing the case $i = 0$ in the formula for f_0. Given any n-simplex $S = \text{conv}\{v_{p_0}, v_{p_1}, \ldots, v_{p_n}\}$ of $\mathcal{S}$, f_0 extends to a map $f_S : S \to \mathbf{R}^d$ by the formula

$$f_S(\sum \lambda_i v_{p_i}) = \sum \lambda_i \mathbf{e}_{p_i},$$

where λ_i are barycentric coordinates of a point in S. It follows that $f_S(S)$ is an n-simplex and f_S is a homeomorphism of S onto $f_S(S)$. If S and T are any two simplices of $\mathcal{S}$, $S \cap T$ is either empty or their common face. Hence, if $x \in S \cap T$, then

$$f_S(x) = f_{S \cap T}(x) = f_T(x).$$

Thus the maps f_S and f_T match on intersections of simplices. Since simplices are closed and there are finitely many of them, the maps f_S can be extended to a map $f : P \to \tilde{P} := f(P)$. Using the geometric independence of $\{0, \mathbf{e}_1, \mathbf{e}_2, \ldots, \mathbf{e}_d\}$, one shows that $\tilde{P}$ is a polyhedron triangulated by $\{f(S)\}$ and f is a homeomorphism. Moreover, by its construction, f maps simplices to simplices.

Step 2. We construct a homeomorphism g of Δ^d onto $[0,1]^d$ such that g restricted to any face of Δ^d is a homeomorphism onto a cubical subset of $[0,1]^d$. Once we do that, it will be sufficient to take $X := g(\tilde{P})$ and define the homeomorphism h as the composition of f and g.

The idea is to keep the vertex $\mathbf{e}_0 = 0$ constant and extend line segments in Δ^d emanating from it so that Δ^d is extended to the whole cube $[0,1]^d$. This leads to the formula for $g : \Delta^d \to [0,1]^d$,

$$g(x) := \begin{cases} 0 & \text{if } x = 0, \\ \lambda(x)x & \text{if } x \neq 0, \end{cases}$$

for $x \in \Delta^d$, where $\lambda(x)$ is a scalar function given by

$$\lambda(x) := \frac{x_1 + x_2 + \cdots + x_d}{\max\{x_1, x_2, \ldots, x_d\}}.$$

It is left as an exercise to verify that this map is a homeomorphism of Δ^d onto $[0,1]^d$ with the inverse given by

$$g^{-1}(y) := \begin{cases} 0 & \text{if } y = 0, \\ \frac{1}{\lambda(y)} y & \text{if } y \neq 0. \end{cases}$$

The property that g maps faces of $\mathcal{S}$ onto unions of cubical faces of $[0,1]^d$ can be proved by induction on d (see [8, Part III, Lemma 1.2]). □

Example 11.18 Let $P = \operatorname{conv}\{e_1, e_2, e_3\} \subset [0,1]^3$. Then P is a two-dimensional face of Δ^3. The map g leaves the vertices constant. The images of edges are

$$g([e_1, e_2]) = ([1] \times [0,1] \times [0]) \cup ([0,1] \times [1] \times [0]),$$

$$g([e_2, e_3]) = ([0] \times [1] \times [0,1]) \cup ([0] \times [0,1] \times [1]),$$

$$g([e_1, e_3]) = ([1] \times [0] \times [0,1]) \cup ([0,1] \times [0] \times [1]).$$

The image of P is

$$Y = ([1] \times [0,1] \times [0,1]) \cup ([0,1] \times [1] \times [0,1]) \cup ([0,1] \times [0,1] \times [1]).$$

Exercises

11.6 Prove that any cubical set can be triangulated.

11.7 Prove that the map constructed in the proof of Theorem 11.17 is a homeomorphism.

11.8 Download the file qprojpln.cub in the folder Examples of Homology program. The symbolic image at the beginning of the file provides a hint on how to construct a triangulation of the projective plane P with 6 vertices which leads to the presentation of a cubical subset of $[0,1]^5$ homeomorphic to P. Fulfill the details by presenting the image of each simplex of the triangulation.

11.9 Use Theorem 11.17 to create files analogous to qprojpln.cub from the previous exercise presenting cubical sets homeomorphic to

(a) the torus discussed in Example 11.10,
(b) the Klein bottle discussed in Exercise 11.4.

11.3 Homology Functor

As indicated in the introduction, we finish this chapter by extending the cubical theory to the class of polyhedra. The first step is done in Theorem 11.17 where it is demonstrated that every polyhedron is homeomorphic to a cubical set. Corollary 6.59 shows that if two cubical sets are homeomorphic, then their homology groups are isomorphic. With this in mind, let P_0 and P_1 be polyhedra and let $h_i : P_i \to X_i$ be homeomorphisms to cubical sets. This suggests declaring the homology groups of P_i to be isomorphic to $H_*(X_i)$. Of course, it remains to be shown that this is a well-defined concept. In particular, it needs to be shown that given a continuous map $f : P_0 \to P_1$, this approach leads to a well-defined homomorphism $f_* : H_*(P_0) \to H_*(P_1)$.

The most efficient method of proving this is through the use of category theory, which is the standard technique for comparing mathematical structures. For the reader who is not familiar with category theory, we begin by introducing a few basic definitions.

11.3.1 Category of Cubical Sets

We begin with the definition of a category. Keep in mind that category theory is used to relate and compare general mathematical structures.

Definition 11.19 A category Cat is a pair (Obj, Mor) consisting of a collection of sets, called *objects* and denoted by Obj, and a collection of *morphisms* between these objects, denoted by Mor.

Given A, B in Obj, the collection of morphisms from A to B is denoted by $\text{Cat}(A, B)$. For each A and B in Obj, it is assumed that $\text{Cat}(A, B)$ contains a morphism. Furthermore, $\text{Cat}(A, A)$ contains the identity morphism id $_A$, and the composition of a morphism in $\text{Cat}(A, B)$ with a morphism in $\text{Cat}(B, C)$ is a morphism in $\text{Cat}(A, C)$. Furthermore, this composition of morphisms is associative.

We have made use of a variety of categories in this book.

- The category Set has all sets as objects and all functions as morphisms.
- The category Top has all topological spaces as objects and all continuous maps as morphisms.
- The category Hom has all topological spaces as objects and all homotopy classes of maps as morphisms.
- The category of cubical sets Cub is defined as follows. The objects in Cub are cubical sets. If X, Y are cubical sets, then

$$\text{Cub}(X, Y) := \{f : X \to Y \mid f \text{ is continuous}\}.$$

- The category Ab has all abelian groups as objects and all group homomorphisms as morphisms.
- The category Ab_*, called the *category of graded abelian groups*, has all sequences $A_* = \{A_n\}_{n \in \mathbf{Z}}$ of abelian groups as objects and all sequences $\{f_n\}_{n \in \mathbf{Z}}$ of group homomorphisms $f_n : A_n \to B_n$ as morphisms.

As mentioned earlier, category theory is used to discuss the relationship between mathematical structures. Two categories that we have used throughout this book are Cub and Ab_*; Cub when we discuss cubical sets and continuous maps between them, and Ab_* when we discuss homology groups and homology maps. Furthermore, we have defined a series of algorithms by which we could move from the category Cub to the category Ab_*. More precisely, given a cubical set X (an object of Cub), $H_*(X)$ is a graded abelian group (an object of Ab$_*$). Furthermore, given a continuous function $f : X \to Y$ [a morphism in Cub(X,Y)], $f_* : H_*(X) \to H_*(Y)$ is a graded group homomorphism [a morphism in Ab$_*(X,Y)$]. Category theory formalizes going from one category to another via the following concept.

Definition 11.20 Let Cat and Cat$'$ be categories. A *functor* $\mathbf{F}$: Cat $\to$ Cat$'$ sends an object X of Cat to an object $\mathbf{F}(X)$ of Cat$'$ and a morphism f of Cat(X,Y) to a morphism $\mathbf{F}(f)$ of Cat$(\mathbf{F}(X), \mathbf{F}(Y))$. Furthermore, the functor $\mathbf{F}$ is *covariant* if
$$\mathbf{F}(g \circ f) = \mathbf{F}(g) \circ \mathbf{F}(f).$$

The contents of Proposition 6.40 and Theorem 6.58 may be reformulated as the following theorem.

Theorem 11.21 H_* : Cub $\to$ Ab$_*$ *is a covariant functor.*

We now introduce one more category, which contains the spaces to which we will extend our homology theory.

A compact metric space K is a *topological polyhedron* or *representable space* if there exists a cubical set X and a homeomorphism $s : K \to X$. The collection of all topological polyhedra with all continuous maps forms a category that we shall denote by Pol. In particular, every geometric polyhedron is an object of Pol by Theorem 11.17. The triple (K, s, X) is called a *representation* of X.

Definition 11.22 The category of *representations of topological polyhedra*, denoted by Repr, is defined as follows. Its objects are all triples (K, s, X), where K is a compact space, X a cubical set, and $s : K \to X$ a homeomorphism. If (K, s, X) and (L, t, Y) are two objects in Repr, then the morphisms from (K, s, X) to (L, t, Y) are all continuous maps from X to Y.

11.3.2 Connected Simple Systems

The careful reader will have noticed that in our discussion of categories we do not use the standard language and notation of set theory. This is dictated by issues of mathematical logic. Consider the category Set. As mentioned before, it has all sets as objects. This implies that the objects of Set do not form a set. To see why, recall that one of the principles of set theory is that if S is a set, then the statement $S \in S$ is inadmissible. However, if S is the set of all sets, then, by definition, $S \in S$, clearly a contradiction. This leads to the following definition. A category is *small* if the objects form a set.

We now introduce two new categories. Their purpose is to allow us to treat all representations of cubical sets simultaneously.

11.3 Homology Functor

Definition 11.23 Fix a category Cat. A *connected simple system* (CSS) in Cat is a small category $\mathcal{E}$ with objects and morphisms from Cat satisfying the property that for any two objects $E_1, E_2 \in \mathcal{E}$ there exists exactly one morphism in $\mathcal{E}(E_1, E_2)$, which is denoted by $\mathcal{E}_{E_1, E_2}$.

Since $\mathcal{E}_{E_2, E_1} \circ \mathcal{E}_{E_1, E_2} = \mathcal{E}_{E_1, E_1}$ and the identity must be a morphism, it follows that all morphisms in $\mathcal{E}$ are isomorphisms.

A simple example of a connected simple system in Hom is the following. Let the objects of $\mathcal{E}$ be the set of elementary cubes. Since all elementary cubes are homotopic to a single point, $\mathcal{E}(Q, P)$ consists of the homotopy class of the constant map, for any $Q, P \in \mathcal{K}$.

The following connected simple system will be used shortly. For every topological polyhedron K we define the connected simple system Rep(K) in Repr as follows. The objects in Rep(K) are all representations of K. If (K, s, X) and (K, t, Y) are two representations of K, then the unique morphism in Rep$(K)((K, X, s), (K, Y, t))$ is the map ts^{-1}.

Definition 11.24 Fix a category Cat. The *category of connected simple systems* over Cat is denoted by CSS(Cat) and defined as follows. Every connected simple system in Cat is an object in CSS(Cat). If $\mathcal{E}$ and $\mathcal{F}$ are two connected simple systems in Cat, then a morphism from $\mathcal{E}$ to $\mathcal{F}$ is any collection of morphisms in Cat

$$\varphi := \{\varphi_{FE} \in \mathrm{Cat}(E, F) \mid E \in \mathcal{E}, F \in \mathcal{F}\}$$

that satisfy

$$\varphi_{F'E'} = \mathcal{E}_{F'F} \varphi_{FE} \mathcal{E}_{EE'}$$

for any $E, E' \in \mathcal{E}, F, F' \in \mathcal{F}$. The elements of φ are called *representants* of φ.

Theorem 11.25 CSS(Cat) *is a category.*

Proof. If $\psi := \{\psi_{GF} \in \mathrm{Cat}(F, G) \mid F \in \mathcal{F}, G \in \mathcal{G}\}$ is a morphism from $\mathcal{F}$ to $\mathcal{G} \in$ CSS(Cat), then it is straightforward to verify that, for given objects $E \in \mathcal{E}, G \in \mathcal{G}$, the composition $\psi_{GF} \varphi_{FE}$ does not depend on the choice of an object $F \in \mathcal{F}$. Thus the morphism $(\psi\varphi)_{FE} := \psi_{GF}\varphi_{FE}$ is well defined and we can set

$$\psi\varphi := \{(\psi\varphi)_{GE} \mid E \in \mathcal{E}, G \in \mathcal{G}\}.$$

The commutativity of the diagram

$$\begin{array}{ccccc} E & \xrightarrow{\varphi_{FE}} & F & \xrightarrow{\psi_{GF}} & G \\ \downarrow & & \downarrow & & \downarrow \\ E' & \xrightarrow{\varphi_{F'E'}} & F' & \xrightarrow{\psi_{G'F'}} & G' \end{array}$$

implies that $\psi\phi$ is a well-defined composition of morphisms. It is now straightforward to verify that $\{\mathcal{E}_{E'E} \mid E, E' \in \mathrm{Obj}(\mathcal{E})\}$ is the identity morphism in $\mathcal{E}$. □

What may be unclear at this point is the purpose of CSS (Cat). With this in mind consider CSS (Ab$_*$). An object in this category is a connected simple system whose objects are isomorphic graded abelian groups. Consider now a topological polyhedron. Associated to it are many representations, that is many cubical sets for which we can compute the graded homology groups. However, if we can show that there is an appropriate functor from Repr to CSS (Ab$_*$), then we will be able to conclude that all graded homology groups are isomorphic and furthermore the homology maps commute. This will be made more explicit shortly.

The following proposition is straightforward.

Proposition 11.26 *Assume* $\mathbf{F} : \text{Cat} \to \text{Cat}'$ *is a functor that maps distinct objects in* Cat *into distinct objects in* Cat$'$ *and* $\mathcal{E}$ *is a connected simple system in* Cat. *Then* $\mathbf{F}(\mathcal{E})$ *is a connected simple system in* Cat$'$. *Moreover, if* $\varphi : \mathcal{E}_1 \to \mathcal{E}_2$ *is a morphism in* CSS (Cat), *then* $\mathbf{F}(\varphi) : \mathbf{F}(\mathcal{E}_1) \to \mathbf{F}(\mathcal{E}_2)$ *given by*

$$\mathbf{F}(\varphi) := \{\mathbf{F}(\varphi_{E_1, E_2}) \mid E_1 \in \mathcal{E}_1, E_2 \in \mathcal{E}_2\}$$

is a morphism in CSS (Cat$'$).

The above proposition lets us extend the functor $\mathbf{F} : \text{Cat} \to \text{Cat}'$ to a functor $\mathbf{F} : \text{CSS (Cat)} \to \text{CSS (Cat}')$.

Using the language of category theory, extending our homology theory from cubical sets to topological polyhedra is done by extending the functor $H_* : \text{Cub} \to \text{Ab}_*$ to $H_* : \text{Pol} \to \text{CSS (Ab}_*)$.

We begin by extending it to $H_* : \text{Repr} \to \text{Ab}_*$ by defining

$$H_*(K, s, X) := H_*(X) \times \{(K, s, X)\}.$$

By Proposition 11.26, $H_* : \text{Repr} \to \text{Ab}_*$ extends to the functor $H_* : \text{CSS (Repr)} \to \text{CSS (Ab}_*)$.

If K, L are two topological polyhedra and $f : K \to L$ is a continuous map, then we define $\text{Rep}(f) : \text{Rep}(K) \to \text{Rep}(L)$ as the collection of maps

$$\{tfs^{-1} \mid (X, s) \in \text{Rep}(K), (Y, t) \in \text{Rep}(L)\}.$$

One easily verifies that the above collection is a morphism in CSS (Repr).

We now define the homology functor $H_* : \text{Pol} \to \text{CSS (Ab}_*)$ by

$$H_*(K) := H_*(\text{Rep}(K))$$

and

$$H_*(f) := H_*(\text{Rep}(f)).$$

One can verify that $H_* : \text{Pol} \to \text{CSS (Ab}_*)$ defined above is indeed a functor. It extends the homology theory of cubical sets to the class of topological polyhedra. It is a lengthy but not very difficult task to show that most of the features of homology presented in this book carry over to the homology of topological polyhedra. We leave the details to the interested reader.

11.4 Bibliographical Remarks

The material presented in Section 11.1 can be found in most classical textbooks in algebraic topology. A reader interested in learning more about the simplicial theory is referred to [51, 69, 74, 36, 78]. For an insight in the relation between the simplicial theory and convex sets, one may consult [21].

Theorem 11.17 is due to Blass and Holsztyński [8]. In that series of papers the authors study properties of *cellular cubical complexes*.

Category theory providing the language for Section 11.3 has been introduced by Eilenberg and Mac Lane (see [50] and references therein) for the purposes of algebraic topology. The reader interested in learning more about categories is referred to [41, Basic Algebra II, Chapter 1].

Part III

Tools from Topology and Algebra

12
Topology

In this chapter we provide a brief presentation of topological preliminaries. The matters discussed here can be found in most standard topology textbooks or in topology chapters in analysis textbooks. Among recommended complementary references are Munkres [68] and Rudin [76].

12.1 Norms and Metrics in $\mathbf{R}^d$

In order to study geometric properties of subsets of the real d-dimensional vector space $\mathbf{R}^d$ and the behavior of maps from one set to another, we need to know how to measure how far two points, $x, y \in \mathbf{R}^d$, are from each other. In $\mathbf{R}$, the distance between two points is expressed by means of the absolute value of their difference $|x - y|$. In spaces of higher dimensions there are several natural ways of measuring the distance, each of them having some particular advantages. The most standard way is by means of the expression called *Euclidean norm*, which is given for each $x = (x_1, x_2, \ldots, x_d) \in \mathbf{R}^d$ by

$$||x||_2 := \sqrt{x_1^2 + x_2^2 + \cdots + x_d^2}. \tag{12.1}$$

The number $||x||_2$ represents the length of the vector beginning at 0 and ending at x, which we shall identify with the point x. This norm permits measuring the distance from x to y by the quantity

$$\mathrm{dist}_2(x, y) := ||x - y||_2. \tag{12.2}$$

Equation (12.1), however, becomes quite unpleasant in calculations since taking the square root is not a simple arithmetic operation. Thus, in numerical estimations, we often prefer to use the *supremum norm* of $x \in \mathbf{R}^d$ given by

$$||x||_0 := \sup_{1 \leq i \leq d} |x_i| = \max\{|x_1|, |x_2|, \ldots, |x_d|\}. \tag{12.3}$$

The related distance between points $x, y \in \mathbf{R}^d$ is

$$\text{dist}_0(x, y) := ||x - y||_0. \tag{12.4}$$

Yet another convenient norm is the *sum norm* given by

$$||x||_1 := \sum_{i=1}^{d} |x_i|. \tag{12.5}$$

with the related distance

$$\text{dist}_1(x, y) := ||x - y||_1. \tag{12.6}$$

Note that all three definitions of norms and distances coincide when the dimension is $d = 1$. In this case, all three distances introduced above reduce to the distance defined at the beginning of this chapter by means of the absolute value. We will refer to it as the standard distance in $\mathbf{R}$.

Example 12.1 Consider points $x = (0, 0)$ and $y = (a, a)$ in $\mathbf{R}^2$ where $a > 0$. Then

$$\text{dist}_2(x, y) = a\sqrt{2}, \; \text{dist}_0(x, y) = a, \; \text{dist}_1(x, y) = 2a.$$

The notions that grasp all those cases are described in the following definitions.

Definition 12.2 A function $||\cdot|| : \mathbf{R}^d \to [0, \infty]$ is called a *norm* if it has the following properties:

(a) $||x|| = 0 \Leftrightarrow x = 0$ for all $x \in \mathbf{R}^d$ (*normalization*);
(b) $||cx|| = |c| \cdot ||x||$ for all $x \in \mathbf{R}^d$ and $c \in \mathbf{R}$ (*homogeneity*);
(c) $||x + y|| \leq ||x|| + ||y||$ for all $x, y \in \mathbf{R}^d$ (*triangle inequality*).

Definition 12.3 Let X be any set. A function $\text{dist} : X \times X \to [0, \infty]$ is called a *metric* on X if it has the following properties:

(a) $\text{dist}(x, y) = 0 \Leftrightarrow x = y$ for all $x, y \in X$ (*normalization*);
(b) $\text{dist}(y, x) = \text{dist}(x, y)$ for all $x, y \in X$ (*symmetry*);
(c) $\text{dist}(x, z) \leq \text{dist}(x, y) + \text{dist}(y, z)$ for all $x, y, z \in X$ (*triangle inequality*).

A set X with a given metric dist is called a *metric space*. More formally, a metric space is a pair (X, dist) that consists of a set and a metric on it.

Given any norm $||\cdot||$ in X, the formula

$$\text{dist}(x, y) = ||x - y||$$

defines a metric in X. Equations (12.1), (12.3), and (12.5) define norms in $\mathbf{R}^d$, and Eqs. (12.2), (12.4), and (12.6) define metrics in $\mathbf{R}^d$. We leave the proofs of those observations as exercises.

Since we will always work with subsets of $\mathbf{R}^d$, why do we bother with the definition of a metric and not content ourselves just with the definition of a norm? The main reason is that norms are defined in linear spaces and subsets of $\mathbf{R}^d$ are not necessarily closed under addition and multiplication by scalars. As the reader can check, this fact does not affect the definition of a metric:

Proposition 12.4 *Let X be a metric space with a metric* dist *and let Y be any subset of X. Then the restriction of* dist *to $Y \times Y$ is a metric on Y.*

Example 12.5 Let $X = S_2^1 = \{x = (x_1, x_2) \in \mathbf{R}^2 \mid x_1^2 + x_2^2 = 1\}$. S_2^1 is called a *unit circle* in $\mathbf{R}^2$. The restriction of the Euclidean distance to S_2^1 defines a metric on S_2^1 even if $x - y$ is not necessarily in S_2^1 for $x, y \in S_2^1$. The distance between points is the length of the line segment joining them in the outer space $\mathbf{R}^2$.

Given a point x in a metric space (X, dist), we define the *ball* of radius r centered at x as
$$B(x, r) := \{y \in X \mid \text{dist}(x, y) < r\}.$$
The set
$$S(x, r) := \{y \in X \mid \text{dist}(x, y) = r\}$$
bounding the ball is called the *sphere* of radius r centered at x. The balls in $\mathbf{R}^d$ for the metrics dist_0, dist_1, dist_2 are denoted, respectively, by $B_0(x, r)$, $B_1(x, r)$, $B_2(x, r)$, and the spheres by $S_0(x, r)$, $S_1(x, r)$, and $S_2(x, r)$. Figure 12.1 shows the unit spheres $S_2 := S_2(0, 1)$ and $S_0 := S_0(0, 1)$ in $\mathbf{R}^2$. Drawing the sphere $S_1 := S_1(0, 1)$ is left as an exercise. When the dimension of a sphere or a ball is important but not clear from the context, we will add it as a superscript (e.g., B_0^d is a unit ball with respect to the supremum norm in $\mathbf{R}^d$ and S_0^{d-1} is the corresponding unit sphere in $\mathbf{R}^d$).

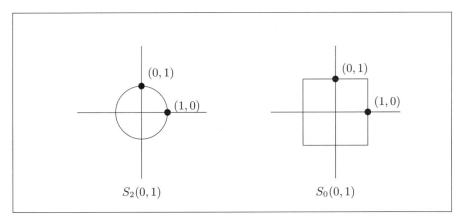

Fig. 12.1. The unit spheres with respect to the Euclidean and supremum norm.

The central notion in topology is the notion of the open set:

Definition 12.6 Let (X, dist) be a metric space. A set $U \subset X$ is *open* if and only if for every point $x \in U$ there exists an $\epsilon > 0$ such that $B(x, \epsilon) \subset U$.

First observe that the empty set is open, because there is nothing to verify in this case.

Example 12.7 The interval $(-1, 2) \subset \mathbf{R}$ is an open set with respect to the standard distance in $\mathbf{R}$. To prove this, let $x \in (-1, 2)$. This is equivalent to the conditions $-1 < x$ and $x < 2$. Choose $r_0 = (x+1)/2$ and $r_1 = (2-x)/2$. Then both $r_0 > 0$ and $r_1 > 0$. Let $\epsilon = \min\{r_0, r_1\}$. One easily verifies that $B_2(x, \epsilon) \subset (-1, 2)$. Since this is true for any $x \in (-1, 2)$, the conclusion follows.

Generalizing this argument leads to the following result. Its proof is left as an exercise.

Proposition 12.8 *Any interval of the form (a, b), (a, ∞), or $(-\infty, b)$ is open in $\mathbf{R}$.*

From Definition 12.6, it follows that the arbitrary union of intervals is open [e.g., $(a, b) \cup (c, d)$ is an open set].

Example 12.9 The unit ball B_2 is an open set with respect to the metric dist_2. Observe that if $x \in B_2$, then $||x||_2 < 1$. Therefore, $0 < 1 - ||x||_2$. Let $r = (1 - ||x||_2)/2$. Then $B_2(x, r) \subset B_2$.

Example 12.10 Of course, not every set is open. As an example consider $(0, 1] \subset \mathbf{R}$. By definition, $1 \in (0, 1]$ but, given any $\epsilon > 0$, $B_2(1, \epsilon) \not\subset (0, 1]$. Therefore, $(0, 1]$ is not open in $\mathbf{R}$. The same argument shows that any interval of the form $(a, b]$, $[a, b)$, or $[a, b]$ is not open in $\mathbf{R}$.

The following observation motivates the notion of topology introduced in the next section.

Proposition 12.11 *Let $U \subset \mathbf{R}^d$. The following conditions are equivalent.*

(a) U is open with respect to the metric dist_2.
(b) U is open with respect to the metric dist_0.
(c) U is open with respect to the metric dist_1.

Proof. Let $U \subset \mathbf{R}^d$. Suppose that U is open with respect to dist_0, and let $x \in U$. Then there exists $\epsilon > 0$ such that $B_0(x, \epsilon) \subset U$. It is left as an exercise to check that
$$B_1(x, \epsilon) \subset B_2(x, \epsilon) \subset B_0(x, \epsilon) \subset U.$$
Since this is true for all $x \in U$, it follows that U is open with respect to dist_2; hence (b) implies (a). By the same arguments, (a) implies (c).

It is also easy to check that $B_0(x, \epsilon/2) \subset B_1(x, \epsilon)$. Hence, by the same argument as above, (c) implies (b). Consequently, all three conditions are equivalent. □

12.1 Norms and Metrics in $\mathbf{R}^d$

In topological considerations we care only about open sets and not about the exact metric. In the light of the above proposition, we may choose any of the three discussed metrics. The metric most convenient for us is dist $_0$. For simplicity of notation, when we abandon the subscript in $||x||$ or in dist (x, y), we mean respectively $||x||_0$ and dist $_0(x, y)$.

Exercises

12.1 Show that the functions defined by Eqs. (12.3) and (12.5) are norms in the sense of Definition 12.2.

12.2 Show that the function defined by Eq. (12.1) is a norm in the sense of Definition 12.2.
Hint: For proving the triangle inequality, find, in linear algebra textbooks, the proof of the *Cauchy–Schwartz inequality*:

$$|x \cdot y| \leq ||x||_2 \cdot ||y||_2,$$

where

$$x \cdot y := \sum_{i=1}^{d} x_i y_i.$$

Observe that $||x||_2^2 = x \cdot x$. Deduce the triangle inequality from the Cauchy–Schwartz inequality.

12.3 Let $|| \cdot ||$ be a norm in $\mathbf{R}^d$ and $X \subset \mathbf{R}^d$. Show that the formula dist $(x, y) := ||x - y||$ for $x, y \in X$ defines a metric in X.

12.4 Sketch the set given by the equation $||x||_1 = 1$ in $\mathbf{R}^2$ as it is done for $||x||_2 = 1$ and $||x||_0 = 1$ in Figure 12.1.

12.5 Prove Proposition 12.8.

12.6 (a) Complete the proof of Proposition 12.11 by verifying that

$$B_1(x, \epsilon) \subset B_2(x, \epsilon) \subset B_0(x, \epsilon) \subset U.$$

(b) Show that

$$B_0\left(x, \frac{\epsilon}{\sqrt{d}}\right) \subset B_2(x, \epsilon)$$

for any $\epsilon > 0$, and use this fact to give a direct proof of the implication $(a) \Rightarrow (b)$ in Proposition 12.11.

12.2 Topology

As we saw in the previous section, different metrics may give the same class of open sets. Since many concepts such as the continuity of functions may be expressed entirely in terms of open sets, it is natural to search for a definition of an open set that would not refer to a particular metric. This is done by listing axioms that the collection of all open sets should satisfy.

Definition 12.12 A *topology* on a set X is a collection $\mathcal{T}$ of subsets of X with the following properties:

1. $\emptyset$ and X are in $\mathcal{T}$.
2. Any union of elements of $\mathcal{T}$ is in $\mathcal{T}$.
3. Any *finite* intersection of elements of $\mathcal{T}$ is in $\mathcal{T}$.

The elements of the topology $\mathcal{T}$ are called *open sets*. A set X for which a topology $\mathcal{T}$ has been specified is called a *topological space*.

This is a fairly abstract definition—fortunately, we don't need to work at this level of generality. In fact, in everything we do we will always assume that the set X is a subset of $\mathbf{R}^d$, so, by Proposition 12.4, it inherits the chosen metric d_0 from $\mathbf{R}^d$. Hence it is enough to know that the definition of an open set with respect to a given metric is consistent with the definition of a topology:

Theorem 12.13 *Let (X, d) be a metric space. Then the set of all subsets of X that are open in the sense of Definition 12.6 forms a topology on X.*

We leave the proof as an exercise. The topology, which is given by means of a metric and Definition 12.6 is called a *metric topology*. One can show that not every topology needs to be metric, but in this book we are not interested in such topologies. Therefore we assume that every topology considered in the rest of this chapter is metric.

In Proposition 12.11 we have proved that all three discussed metrics induce the same topology on $\mathbf{R}^d$. It will be called the *standard topology*. Unless it is explicitly stated otherwise, $\mathcal{T}$ will always be the standard topology.

As important as an open set is the notion of a closed set.

Definition 12.14 A subset K of a topological space X is *closed* if its complement
$$X \setminus K := \{x \in X \mid x \notin K\}$$
is open.

Example 12.15 The interval $[a, b]$ is a closed subset of $\mathbf{R}$. This is straightforward to see since its complement $\mathbf{R} \setminus [a, b] = (-\infty, a) \cup (b, \infty)$ is open. Similarly, $[a, \infty)$ and $(-\infty, b] \subset \mathbf{R}$ are closed.

Example 12.16 The set $\bar{B}_0 := \{x \in \mathbf{R}^d \mid ||x|| \leq 1\}$ is closed. This is equivalent to claiming that $\mathbf{R}^d \setminus \bar{B}_0$ is open, that is, that $\{x \in \mathbf{R}^d \mid ||x|| > 1\}$ is open. Observe that $||x|| > 1$ is equivalent to $\max_{i=1,\ldots,d}\{|x_i|\} > 1$. Thus there exists at least one coordinate, say the jth coordinate, such that $|x_j| > 1$. Then

$$B_0(x, \frac{|x_j| - 1}{2}) \subset \mathbf{R}^d \setminus \bar{B}_0.$$

What we have proved justifies calling $\bar{B}_0$ the *closed unit ball* (with respect to the supremum norm). By the same arguments, the set

$$\bar{B}_0(x, r) := \{y \in \mathbf{R}^d \mid ||y - x|| \leq r\}$$

is closed for every $x \in \mathbf{R}^d$ and every $r \geq 0$ so the name *closed ball* of radius r centered at x is again justified.

Remark 12.17 The reader should take care not to get lulled into the idea that a set is either open or closed. Many sets are *neither*. For example, the interval $(0, 1] \subset \mathbf{R}$ is neither open nor closed. As observed in Example 12.10, it is not open. Similarly, it is not closed since its complement is $(-\infty, 0] \cup (1, \infty)$, which is not an open set.

Theorem 12.18 *Let X be a topological space. Then the following statements are true.*

1. *$\emptyset$ and X are closed sets.*
2. *Arbitrary intersections of closed sets are closed.*
3. *Finite unions of closed sets are closed.*

Proof. (1) $\emptyset = X \setminus X$ and $X = X \setminus \emptyset$.
(2) Let $\{K_\alpha\}_{\alpha \in \mathcal{A}}$ be an arbitrary collection of closed sets. Then

$$X \setminus \bigcap_{\alpha \in \mathcal{A}} K_\alpha = \bigcup_{\alpha \in \mathcal{A}} (X \setminus K_\alpha).$$

Since, by definition, $X \setminus K_\alpha$ is open for each $\alpha \in \mathcal{A}$ and the arbitrary union of open sets is open, $X \setminus \bigcap_{\alpha \in \mathcal{A}} K_\alpha$ is open. Therefore, $\bigcap_{\alpha \in \mathcal{A}} K_\alpha$ is closed.
(3) The proof of this property is similar to (2) and is left as an exercise. □

Definition 12.19 Let X be a topological space and let $A \subset X$. The *closure* of A in X is the intersection of all closed sets in X containing A. The closure of A is denoted by $\operatorname{cl} A$ (the notation $\bar{A}$ is also used in literature).

By Theorem 12.18 the arbitrary intersection of closed sets is closed; therefore, the closure of an arbitrary set is a closed set.

Example 12.20 Consider $[0,1) \subset \mathbf{R}$. Then $\operatorname{cl}[0,1) = [0,1]$. This is easy to show. First one needs to check that $[0,1)$ is not closed. This follows from the fact that $[1,\infty)$ is not open. Then one shows that $[0,1]$ is closed by showing that $(-\infty, 0) \cup (1, \infty)$ is an open set in $\mathbf{R}$. Finally, one observes that any closed set that contains $[0,1)$ must contain $[0,1]$.

A similar argument shows that

$$\operatorname{cl}(0,1) = \operatorname{cl}[0,1) = \operatorname{cl}(0,1] = \operatorname{cl}[0,1] = [0,1].$$

We leave as an exercise the proof of the following.

Proposition 12.21 *Let $r > 0$. The closed ball $\bar{B}_0(x, r)$ defined in Example 12.16 is the closure of the ball $B_0(x, r)$. Explicitly,*

$$\bar{B}_0(x, r) = \operatorname{cl} B_0(x, r).$$

Up to this point, the only topological spaces that have been considered are those of $\mathbf{R}^d$ for different values of d. Proposition 12.4 shows that metrics can be restricted to subsets. Here is the analogy of that for topologies:

Definition 12.22 Let Z be a topological space with topology $\mathcal{T}$. Let $X \subset Z$. The *subspace topology* on X is the collection of sets

$$\mathcal{T}_X := \{X \cap U \mid U \in \mathcal{T}\}.$$

Before this definition can be accepted, the following proposition needs to be proved.

Proposition 12.23 *$\mathcal{T}_X$ defines a topology on X.*

Proof. The three conditions of Definition 12.12 need to be checked. First, observe that $\emptyset \in \mathcal{T}_X$ since $\emptyset = X \cap \emptyset$. Similarly, $X \in \mathcal{T}_X$ since $X = X \cap Z$.

The intersection and union properties follow from the following equalities:

$$\bigcap_{i=1}^{n}(X \cap U_i) = X \cap \left(\bigcap_{i=1}^{n} U_i\right),$$

$$\bigcup_{i \in I}(X \cap U_i) = X \cap \left(\bigcup_{i \in I} U_i\right). \qquad \square$$

Using this definition of the subspace topology, we can treat any set $X \subset \mathbf{R}^d$ as a topological space.

The following proposition is straightforward.

Proposition 12.24 *If A is open in Z, then $A \cap X$ is open in X. If A is closed in Z, then $A \cap X$ is closed in X.*

12.2 Topology

It is important to notice that while open sets in the subspace topology are defined in terms of open sets in the ambient space, the sets themselves may "look" different.

Example 12.25 Consider the interval $[-1,1] \subset \mathbf{R}$ with the subspace topology induced by the standard topology on $\mathbf{R}$. $(0,2)$ is an open set in $\mathbf{R}$, hence

$$(0,1] = (0,2) \cap [-1,1]$$

is an open set in $[-1,1]$. We leave it to the reader to check that any interval of the form $[-1,a)$ and $(a,1]$, where $-1 < a < 1$, is an open set in $[-1,1]$.

Example 12.26 Let $X = [-1,0) \cup (0,1]$. Observe that $[-1,0) = (-2,0) \cap X$ and $(0,1] = (0,2) \cap X$; thus both are open sets. However, $[-1,0) = X \setminus (0,1]$ and $(0,1] = X \setminus [-1,0)$, so both are also closed sets. This shows that for general topological spaces one can have nontrivial sets that are both open and closed.

Given a pair $A \subset X$, we want to consider the set of points that separate A from its complement $X \setminus A$.

Definition 12.27 Let X be a topological space and $A \subset X$. The set

$$\operatorname{bd}_X A := \operatorname{cl} A \cap \operatorname{cl}(X \setminus A)$$

is called the *topological boundary* or, for short, the *boundary* of A in X. If the space X is clear from the context, an abbreviated notation $\operatorname{bd} A$ may be used.

Example 12.28 Let $A := [-1,1] \times \{0\}$ and $X = \mathbf{R} \times \{0\}$. Then

$$\operatorname{bd}_X A = \{(-1,0), (1,0)\},$$

which corresponds to the geometric intuition that a boundary of an interval should be the pair of its endpoints. However,

$$\operatorname{bd}_{\mathbf{R}^2} A = A$$

because $\operatorname{cl}(\mathbf{R}^2 \setminus A) = \mathbf{R}^2$.

The above example shows that the concept of the boundary of A is relative to the space in which A is sitting. Note also that

$$\operatorname{bd}_X X = \emptyset$$

for any space X because $X \setminus X = \emptyset$.

The proof of the following is left as an exercise.

Proposition 12.29 $\operatorname{bd} \bar{B}_0(x,r) = S_0(x,r)$.

Exercises

12.7 Prove that any set consisting of a single point is closed in $\mathbf{R}^d$.

12.8 Prove Theorem 12.13.

12.9 Prove that the intersection of finitely many closed sets is closed.

12.10 (a) Prove Proposition 12.21.
(b) Prove Proposition 12.29.

12.11 Let $Q = [k_1, k_1 + 1] \times [k_2, k_2 + 1] \times [k_3, k_3 + 1] \subset \mathbf{R}^3$, where $k_i \in \mathbf{Z}$ for $i = 1, 2, 3$. Prove that Q is a closed set.

12.12 Let $\overset{\circ}{Q} = \{k_1\} \times (k_2, k_2 + 1) \times (k_3, k_3 + 1) \subset \mathbf{R}^3$, where $k_i \in \mathbf{Z}$ for $i = 1, 2, 3$.

(a) Prove that $\overset{\circ}{Q}$ is neither open nor closed in $\mathbf{R}^3$.
(b) Show that $\overset{\circ}{Q}$ is open in the hyperplane $X = \{x \in \mathbf{R}^3 : x_1 = k_1\}$.

12.13 Let Q be the *unit d-cube* in $\mathbf{R}^d$:
$$Q := \{x \in \mathbf{R}^d \mid 0 \leq x_i \leq 1\}.$$

Let
$$\overset{\circ}{Q} := \{x \in \mathbf{R}^d \mid 0 < x_i < 1\}$$

and
$$\Gamma^{d-1} := \operatorname{bd}_{\mathbf{R}^d} Q.$$

Prove the following:

1. Q is closed in $\mathbf{R}^d$;
2. $\overset{\circ}{Q}$ is open in $\mathbf{R}^d$;
3. Γ^{d-1} is closed in $\mathbf{R}^d$;
4. $\Gamma^{d-1} = \{x \in Q \mid x_i \in \{0, 1\} \text{ for some } i = 1, \ldots, d\}$;
5. Γ^{d-1} is the union of proper faces of Q (see Definition 2.7);
6. Γ^{d-1} is the union of free faces of Q (see Definition 2.60).

12.14 Let X be a cubical set in $\mathbf{R}^d$ (see Definition 2.9). Let X_1 be the union of its maximal faces (see Definition 2.60) of dimension d, X_2 the union of its maximal faces of X of dimension less than d, and $\texttt{Freeface}(X_1)$ the union of all free faces of X_1. Show that

$$\operatorname{bd}_{\mathbf{R}^d} X = \texttt{Freeface}(X_1) \cup X_2.$$

You may want to do Exercise 2.16 first.

12.3 Continuous Maps

As mentioned in the introduction, a frequently asked question is whether or not two given sets in $\mathbf{R}^d$ have a similar "shape." Such comparisons can be done by means of continuous maps.

Example 12.30 The square $X := [0,1] \times [0,1] \subset \mathbf{R}^2$ and a portion of the closed unit disk $Y := \{x \in \mathbf{R}^2 \mid ||x||_2 \leq 1,\ x_1 \geq 0,\ x_2 \geq 0\} \subset \mathbf{R}^2$ are clearly different from the geometric point of view: The first one is a polyhedron, the second one is not. However, we would like to think of them as being "equivalent" in a topological sense, since they can be transformed from one to the other and back by simply stretching or contracting the spaces.

To be more precise, observe that any element of Y has the form $y = (r\cos\theta, r\sin\theta)$, where $0 \leq r \leq 1$ and $0 \leq \theta \leq \pi/2$. Define $f : Y \to X$ by

$$f(r\cos\theta, r\sin\theta) := \begin{cases} (r, r\tan\theta) & \text{if } 0 \leq \theta \leq \pi/4, \\ (r\cot\theta, r) & \text{if } \pi/4 \leq \theta \leq \pi/2. \end{cases} \quad (12.7)$$

It can be verified that f is well defined. Moreover, one can check that f is continuous in the sense of any standard calculus textbook. Observe that this map just expands Y by moving points out along the rays emanating from the origin. One can also write down a map $g : X \to Y$ that shrinks X onto Y along the same rays (see Exercise 12.15).

Finally, one can construct a map from the square $\bar{B}_0^2 = [-1,1]^2$ onto the disk $\bar{B}_2^2 = \{x \in \mathbf{R}^2 \mid x_1^2 + x_2^2 \leq 1\}$ by repeating the same construction for each quadrant of the plane. Note that either set represents a closed unit ball: the first one with respect to the supremum norm and the second one with respect to the Euclidean norm (see Figure 12.1).

Maps such as the one given by (12.7) are typical examples of continuous functions presented in calculus textbooks. Since we introduced the notion of topology on an abstract level, we also want to define continuous functions in an equally abstract way.

Recall that a topological space consists of two objects, the set X and the topology $\mathcal{T}$.

Definition 12.31 Let X and Y be topological spaces with topologies $\mathcal{T}_X$ and $\mathcal{T}_Y$, respectively. A function $f : X \to Y$ is *continuous* if and only if for every open set $V \in \mathcal{T}_Y$ its preimage under f is open in X, that is,

$$f^{-1}(V) \in \mathcal{T}_X.$$

Another useful way to characterize continuous functions is as follows.

Proposition 12.32 *A function $f : X \to Y$ is continuous if and only if, for every closed set $K \subset Y$, $f^{-1}(K)$ is a closed subset of X.*

Proof. ($\Rightarrow$) Let $K \subset Y$ be a closed set. Then $Y \setminus K$ is an open set. Since f is continuous, $f^{-1}(Y \setminus K)$ is an open subset of X. Hence $X \setminus f^{-1}(Y \setminus K)$ is closed in X. Thus it only needs to be shown that $X \setminus f^{-1}(Y \setminus K) = f^{-1}(K)$. Let $x \in X \setminus f^{-1}(Y \setminus K)$. Then $f(x) \in Y$ and $f(x) \notin Y \setminus K$. Therefore, $f(x) \in K$ or equivalently $x \in f^{-1}(K)$. Thus $X \setminus f^{-1}(Y \setminus K) \subset f^{-1}(K)$. Now assume $x \in f^{-1}(K)$. Then, $x \notin f^{-1}(Y \setminus K)$ and hence $x \in X \setminus f^{-1}(Y \setminus K)$.

($\Leftarrow$) Let $U \subset Y$ be an open set. Then $Y \setminus U$ is a closed subset. By hypothesis, $f^{-1}(Y \setminus U)$ is closed. Thus $X \setminus f^{-1}(Y \setminus U)$ is open. But $X \setminus f^{-1}(Y \setminus U) = f^{-1}(U)$. □

Even in this very general setting we can check that some maps are continuous.

Proposition 12.33 *Let X and Y be topological spaces.*

(i) The identity map $\mathrm{id}_X : X \to X$ is continuous.
(ii) Let $y_0 \in Y$. The constant map $f : X \to Y$ given by $f(x) = y_0$ is continuous.

Proof. (i) Since id_X is the identity map, $\mathrm{id}_X^{-1}(U) = U$ for every set $U \subset X$. Thus, if U is open, its preimage under id_X is open.

(ii) Let $V \subset Y$ be an open set. If $y_0 \in V$, then $f^{-1}(V) = X$, which is open. If $y_0 \notin V$, then $f^{-1}(V) = \emptyset$, which is also open. □

Proposition 12.34 *If $f : X \to Y$ and $g : Y \to Z$ are continuous maps, then $g \circ f : X \to Z$ is continuous.*

Proof. Let W be an open set in Z. To show that $g \circ f$ is continuous, we need to show that $(g \circ f)^{-1}(W)$ is an open set. However, $(g \circ f)^{-1}(W) = f^{-1}(g^{-1}(W))$. Since g is continuous, $g^{-1}(W)$ is open, and since f is continuous, $f^{-1}(g^{-1}(W))$ is open. □

As before, we have introduced the notion of continuous functions on a level of generality greater than we need. The following result indicates that this abstract definition matches that learned in calculus. For its proof, see [76, Chapter 4, Theorem 4.8].

Theorem 12.35 *Let (X, dist_X) and (Y, dist_Y) be metric spaces and let $f : X \to Y$. Then f is continuous if and only if for every $x \in X$ and any $\epsilon > 0$, there exists a $\delta > 0$ such that if $\mathrm{dist}_X(x, y) < \delta$, then $\mathrm{dist}_Y(f(x), f(y)) < \epsilon$.*

For a function $f : X \to \mathbf{R}$, the continuity can be characterized by means of lower and upper semicontinuity.

Definition 12.36 A function $f : X \to \mathbf{R}$ is *upper semicontinuous* if the set $\{x \in X \mid f(x) < \alpha\}$ is open for any real α. It is *lower semicontinuous* if the set $\{x \in X \mid f(x) > \alpha\}$ is open for any real α.

Proposition 12.37 *A function $f : X \to \mathbf{R}$ is continuous if and only if it is both upper and lower semicontinuous.*

Proof. Since $\{x \in X \mid f(x) < \alpha\} = f^{-1}((-\infty, \alpha))$, it is clear that any continuous function is upper semicontinuous. The same argument shows that it also is lower semicontinuous. Let f be both lower and upper semicontinuous. Then the inverse image of any open interval (α, β) by f is

$$\{x \in X \mid f(x) > \alpha\} \cap \{x \in X \mid f(x) < \beta\},$$

so it is open. From that it follows that the inverse image of any open set is open. □

We now return to the question of comparing topological spaces. To say that two topological spaces are equivalent, it seems natural to require that both objects, the sets and the topologies, be equivalent. On the level of set theory the equivalence of sets is usually taken to be the existence of a bijection. To be more precise, let X and Y be sets. A function $f : X \to Y$ is an *injection* if for any two points $x, z \in X$, $f(x) = f(z)$ implies that $x = z$. The function f is a *surjection* if for any $y \in Y$ there exists $x \in X$ such that $f(x) = y$. If f is both an injection and a surjection, then it is a *bijection*. If f is a bijection, then one can define an inverse map $f^{-1} : Y \to X$ by $f^{-1}(y) = x$ if and only if $f(x) = y$.

Definition 12.38 Let X and Y be topological spaces with topologies $\mathcal{T}_X$ and $\mathcal{T}_Y$, respectively. A bijection $f : X \to Y$ is a *homeomorphism* if and only if both f and f^{-1} are continuous. We say that X is *homeomorphic* to Y and we write $X \cong Y$ if and only if there exists a homeomorphism $f : X \to Y$.

Proposition 12.39 *The relation $\cong$ is an equivalence relation on topological spaces.*

Proof. We need to show that the relation $X \cong Y$ is reflexive, symmetric, and transitive.

To see that it is reflexive, observe that, given any topological space X, the identity map $\mathrm{id}_X : X \to X$ is a homeomorphism from X to X.

Assume that X is homeomorphic to Y. By definition this implies that there exists a homeomorphism $f : X \to Y$. Observe that $f^{-1} : Y \to X$ is also a homeomorphism and hence Y is homeomorphic to X. Thus homeomorphism is a symmetric relation.

Finally, Proposition 12.34 shows that homeomorphism is a transitive relation, that is if X is homeomorphic to Y and Y is homeomorphic to Z, then X is homeomorphic to Z. □

Using Theorem 12.35 we can easily show that a variety of simple topological spaces are homeomorphic.

Proposition 12.40 *The following topological spaces are homeomorphic:*

(i) $\mathbf{R}$;
(ii) (a, ∞) *for any* $a \in \mathbf{R}$;
(iii) $(-\infty, a)$ *for any* $a \in \mathbf{R}$;

(iv) (a,b) for any $-\infty < a < b < \infty$.

Proof. We begin by proving that $\mathbf{R}$ and (a, ∞) are homeomorphic. Let $f : \mathbf{R} \to (a, \infty)$ be defined by
$$f(x) = a + e^x.$$
This is clearly continuous. Furthermore, $f^{-1}(x) = \ln(x-a)$ is also continuous.

Observe that $f : (a, \infty) \to (-\infty, -a)$ given by $f(x) = -x$ is a homeomorphism. Thus any interval of the form $(-\infty, b)$ is homeomorphic to $(-b, \infty)$ and hence to $\mathbf{R}$.

Finally, to see that (a, b) is homeomorphic to $\mathbf{R}$, observe that $f : (a, b) \to \mathbf{R}$ given by
$$f(x) = \ln\left(\frac{x-a}{b-x}\right)$$
is continuous and has a continuous inverse given by $f^{-1}(x) = (be^y + a)/(1+e^y)$.
□

Inverses of the most elementary functions are rarely given by explicit elementary formulas, so it would be nice to know whether or not an inverse of a given continuous bijective map is continuous without knowing it explicitly. In a vector calculus course we learn the inverse function theorem, which can often be applied for that purpose since any differentiable map is continuous. That theorem, however, discusses maps on open subsets of $\mathbf{R}^d$ and the local invertibility, but we often want to compare closed subsets. Here is a topological "partner" of the inverse function theorem.

Theorem 12.41 *Let X, Y be closed and bounded subsets of $\mathbf{R}^d$. If $f : X \to Y$ is continuous and bijective, then its inverse $f^{-1} : Y \to X$ also is continuous.*

The proof of this theorem requires the notion of *compactness*, and we postpone it to the last section, Exercise 12.32.

Proving that two spaces are homeomorphic is not always so elementary as in the previous examples.

Example 12.42 In Example 12.30 we show that the closed unit balls $\bar{B}_0^2$ and $\bar{B}_2^2$ in $\mathbf{R}^2$ are homeomorphic. Without going into details we shall indicate how to construct a homeomorphism in arbitrary dimension.

The boundary of $\bar{B}_0 = \bar{B}_0^d$ is the sphere
$$S_0^{d-1} = \{x \in [-1, 1]^d \mid x_i = -1 \text{ or } x_i = 1 \text{ for some } i\},$$
and the boundary of $\bar{B}_2 = \bar{B}_2^d$ is the sphere
$$S_2^{d-1} = \{x \in \mathbf{R}^d \mid \sum_{i=1}^{d} x_i^2 = 1\}.$$

Any $x \neq 0$ in $\bar{B}_0$ can be uniquely written as $x = t(x)y(x)$, where $t(x) \in (0, 1]$ and $y(x)$ is a vector in the intersection of S_0^{d-1} with the half-line emanating from the origin and passing through x. It can be proved that the map $f : \bar{B}_0 \to \bar{B}_2$ given by

$$f(x) := \begin{cases} t(x) \frac{x}{\|x\|} & \text{if } x \neq 0, \\ 0 & \text{otherwise,} \end{cases}$$

is a homeomorphism. Moreover, the restriction of this map to S_0^{d-1} is a homeomorphism of that set onto the sphere S_2^{d-1}.

Exercises

12.15 Referring to Example 12.30:

(a) Write down the inverse function for f.
(b) Prove that f is a continuous function.

12.16 Prove Theorem 12.35.

12.17 Construct a homeomorphism from any interval $[a,b]$ onto $[-1,1]$.

12.18 Show that the boundary Γ^{d-1} of the unit square defined in Exercise 12.13 is homeomorphic to the unit sphere S_0^{d-1} (with respect to the supremum norm) defined in Example 12.42.

12.19 Construct a homeomorphism from the triangle

$$T := \{(s,t) \in \mathbf{R}^2 \mid s + t \leq 1, s \geq 0, t \geq 0\}$$

onto the square $[0,1]^2$.

12.20 We say that a topological space X has the *fixed-point property* if every continuous self-map $f : X \to X$ has a fixed point, namely a point $x \in X$ such that $f(x) = x$.

(a) Show that the fixed-point property is a topological property, namely that it is invariant under a homeomorphism.
(b) Show that any closed bounded interval $[a,b]$ has the fixed-point property.

12.21 Complete the proof given in Example 12.42 for $d = 3$.

12.4 Connectedness

One of the most fundamental global properties of a topological space is whether or not it can be broken into two distinct open subsets. The following definition makes this precise.

Definition 12.43 Let X be a topological space. X is *connected* if the only subsets of X that are both open and closed are $\emptyset$ and X. If X is not connected, then it is *disconnected*.

Example 12.44 Let $X = [-1, 0) \cup (0, 1] \subset \mathbf{R}$. Then X is a disconnected space since by Example 12.26 $[-1, 0)$ and $(0, 1]$ are both open and closed in the subspace topology.

While it is easy to produce examples of disconnected spaces, proving that a space is connected is more difficult. Even the following intuitively obvious result is fairly difficult to prove.

Theorem 12.45 *Any interval in $\mathbf{R}$ is connected.*

Hints how to prove this theorem can be found in Exercise 12.22. The reader can also consult [68] or [76].

A very useful theorem is the following.

Theorem 12.46 *Let $f : X \to Y$ be a continuous function. If X is connected, then so is $f(X) \subset Y$.*

Proof. Let $Z = f(X)$. Suppose that Z is disconnected. Then there exists a set $A \subset Z$, such that $A \neq \emptyset$, $A \neq Z$, and A is both open and closed. Since f is continuous, $f^{-1}(A)$ is both open and closed. Moreover, $f^{-1}(A) \neq \emptyset$ and $f^{-1}(A) \neq X$, which contradicts the assumption that X is connected. □

We can now prove one of the fundamental theorems of topology.

Theorem 12.47 Intermediate value (Darboux theorem) *If $f : [a, b] \to \mathbf{R}$ is a continuous function and $f(a)f(b) < 0$, then there exists $c \in [a, b]$ such that $f(c) = 0$.*

Proof. The proof is by contradiction. Assume that there is no $c \in [a, b]$ such that $f(c) = 0$. Then

$$f([a, b]) \subset (-\infty, 0) \cup (0, \infty).$$

Let $U = (-\infty, 0) \cap f([a, b])$ and $V = (0, \infty) \cap f([a, b])$. Then U and V are open in $f([a, b])$ and $f([a, b]) = U \cup V$. Since $f(a)f(b) < 0$, U and V are not trivial. Therefore, $f([a, b])$ is disconnected, contradicting Theorems 12.45 and 12.46. □

The definition of connectedness appears difficult to most beginners in topology since the notion of a subset that is both open and closed violates their intuition. We now present a related concept that is closer to intuition.

Definition 12.48 A topological space X is *path connected* if for every two points $x, y \in X$ there exists a continuous map $f : [a, b] \to X$ such that $f(a) = x$ and $f(b) = y$. Such a map is called a *path joining x to y*.

Any interval obviously is path-connected. A special case of a path connected set is a convex set (see Definition 11.1).

Example 12.49 The unit circle S_2^1 in $\mathbf{R}^2$ is not convex, but it is path connected. Indeed, let $x, y \in S_2^1$. In polar coordinates, $x = (\cos \alpha, \sin \alpha)$ and $y = (\cos \beta, \sin \beta)$. Without a loss of generality, $\alpha \leq \beta$. Then $\sigma : [\alpha, \beta] \to S_2^1$ given by $\sigma(\theta) = (\cos \theta, \sin \theta)$ is a path joining x to y.

Theorem 12.50 *Any path-connected space is connected.*

Proof. We argue by contradiction. Suppose that X is disconnected, and let $B \subset X$ be nonempty, different from X, both open and closed. Then there exist two points $x \in B$ and $y \in X \setminus B$. Let $\sigma : [a, b] \to X$ be a path joining x to y and let $A := \sigma^{-1}(B)$. Since σ is continuous, A is both open and closed in $[a, b]$. Since $\sigma(a) = x$, $a \in A$ so $A \neq \emptyset$. It cannot be $b \in A$, because otherwise $y = \sigma(b) \in B$. Thus $A \neq [a, b]$, which contradicts Theorem 12.45. □

A connected space does not need to be path connected, but such spaces are quite "pathological" (see Exercise 12.27). For a large class of spaces such as geometric polyhedra, the two definitions are equivalent: We discuss this topic when studying graphs and cubical sets.

Proving that two spaces are not homeomorphic is often hard, because showing that there exists no homeomorphism from one space onto another is difficult. Knowing *topological properties*—properties of spaces that are preserved by homeomorphisms—is helpful. By Theorem 12.46, connectedness is a topological property. This gives a simple criterion for distinguishing between two spaces.

Example 12.51 The half-closed interval $(0, 1]$ is not homeomorphic to the open interval $(0, 1)$. We will argue by contradiction. Suppose that $f : (0, 1] \to (0, 1)$ is a homeomorphism, and let $t := f(1)$. Then the restriction of f to $(0, 1)$ is a homeomorphism of $(0, 1)$ onto the set $(0, t) \cup (t, 1)$. That is impossible since the first set is connected and the second is not.

Definition 12.52 A subset A of a topological space X is called *connected* if A is connected as a topological space with subset topology.

Theorem 12.53 *Assume $\{X_\iota\}_{\iota \in J}$ is a family of connected subsets of a topological space Z. If $\bigcap_{\iota \in J} X_\iota \neq \emptyset$, then $\bigcup_{\iota \in J} X_\iota$ is connected.*

Proof. Let $X := \bigcup_{\iota \in J} X_\iota$ and let A be a nonempty subset of X that is both open and closed. Let $x \in A$. Then $x \in X_{\iota_0}$ for some $\iota_0 \in J$. It follows that $A \cap X_{\iota_0}$ is nonempty and both open and closed in X_{ι_0}. Since X_{ι_0} is connected, it must be $A \cap X_{\iota_0} = X_{\iota_0}$ (i.e., $X_{\iota_0} \subset A$). Recalling that $\bigcap_{\iota \in J} X_\iota$ is nonempty, we get $A \cap X_\iota \neq \emptyset$ for every $\iota \in J$. This means, by the same argument as in the case of X_{ι_0}, that $X_\iota \subset A$ for every $\iota \in J$. Hence $X = A$, that is, X is connected. □

Theorem 12.54 *Every rectangle in $\mathbf{R}^d$, namely a product of closed, bounded intervals, is connected.*

Proof. Let $X := [a_1, b_1] \times [a_2, b_2] \times \cdots \times [a_d, b_d]$. Given any points $x, y \in X$, the function $f : [0, 1] \to \mathbf{R}^d$ given by $f(t) = (1-t)x + ty$ is a continuous path connecting x to y. It has values in X because $a_i \leq (1-t)x_i + ty_i \leq b_i$ for all $t \in [0, 1]$, $x, y \in X$ and $i = 1, 2, \ldots, d$. Hence X is path connected. □

It is easy to see that having exactly n connected components is a topological property: If X has k connected components, then the same is true about any space homeomorphic to X.

Exercises

12.22 Let A and B be two disjoint nonempty open sets in $I = [0, 1]$. The following arguments will show that $I \neq A \cup B$ and therefore that I is a connected set.

Let $a \in A$ and $b \in B$. Then either $a < b$ or $a > b$. Assume without loss of generality that $a < b$.

(a) Show that the interval $[a, b] \subset I$.

Let $A_0 := A \cap [a, b]$ and $B_0 := B \cap [a, b]$.

(b) Show that A_0 and B_0 are open in $[a, b]$ under the subspace topology.

Let c be the least upper bound for A_0, that is,

$$c := \inf\{x \in \mathbf{R} \mid x > y \text{ for all } y \in A_0\}.$$

(c) Show that $c \in [a, b]$.
(d) Show that $c \notin B_0$. Use the facts that c is the least upper bound for A_0 and that B_0 is open.
(e) Show that $c \notin A_0$. Again use the facts that c is the least upper bound for A_0 and that A_0 is open.

Finally, observe that $c \in I$, but $c \notin A_0 \cup B_0$ and, therefore, that $I \neq A_0 \cup B_0$.

12.23 Prove the converse of Theorem 12.45: If $X \subset \mathbf{R}$ is connected, then X is an interval (by interval we mean any type of bounded interval, half-line, or $\mathbf{R}$).

12.24 Use Theorem 12.45 to show that S_2^1 is connected.

12.25 Show that the unit circle $S_2^1 = \{x \in \mathbf{R}^2 \mid \|x\|_2 = 1\}$ is not homeomorphic to an interval (whether it is closed, open, or neither).
Hint: Use arguments similar to that in Example 12.51.

12.26 A *simple closed curve* in $\mathbf{R}^d$ is the image of an interval $[a, b]$ under a continuous map $\sigma : [a, b] \to \mathbf{R}^d$ (called a *path*) such that $\sigma(s) = \sigma(t)$ for any $s < t, s, t \in [a, b]$ if and only if $s = a$ and $t = b$. Prove that any simple closed curve is homeomorphic to the unit circle S_2^1.
Hint: Recall the path in Example 12.49 and use Theorem 12.41.

12.27 Let $X \subset \mathbf{R}^d$ be the union of $\{0\} \times [-1, 1]$ and the set Σ given by

$$\Sigma := \{(s, t) \in \mathbf{R}^2 \mid t = \sin\frac{1}{s}, 0 < s \le \frac{1}{2\pi}\}.$$

Draw X. Show that X is connected but not path connected.

12.5 Limits and Compactness

Limits of sequences play a central role in any textbook on analysis, numerical analysis, and metric space topology. So far we have purposely avoided speaking about limits, in order to emphasize our combinatorial approach to topology. This notion, however, becomes hard to avoid when discussing compactness and properties of continuous functions on compact sets. Due to space limitations we can only provide a brief presentation of the topic. We assume that the reader is familiar with the calculus of limits of sequences of real numbers, and we refer to [76] for the omitted proofs.

Definition 12.55 Let (X, dist) be a metric space. A sequence

$$(x_n)_{n \in N} = (x_1, x_2, x_3, \ldots)$$

[or for short (x_n)] of points of X *converges* to $x \in X$ if and only if for every $\epsilon > 0$ there exists n_0 such that, for any $n > n_0$, $\text{dist}(x_n, x) < \epsilon$. The point x is called the *limit* of (x_n), and we write either $(x_n) \to x$ or $\lim_{n \to \infty} x_n = x$.

If a sequence does not converge, we say that it *diverges*.

Example 12.56 Let $X = \mathbf{R}$.

1. $\lim_{n \to \infty}(1/n) = 0$.
2. The sequence $(n)_{n \in N} = (1, 2, 3, \ldots)$ diverges.
3. The sequence $(x_n) = ((-1)^n) = (-1, 1, -1, 1, \ldots)$ diverges. However, one can extract from it two convergent subsequences $(x_{2n}) = (1) \to 1$ and $(x_{2n-1}) = (-1) \to -1$.

Proposition 12.57 *Let $(x(n))_{n \in N}$ be a sequence of points*

$$x(n) = (x_1(n), x_2(n), \ldots, x_d(n))$$

in $\mathbf{R}^d$. The sequence $(x(n)_{n \in N})$ converges in $\mathbf{R}^d$ if and only if $(x_i(n))_{n \in N}$ converges in $\mathbf{R}$ for every $i = 1, 2, \ldots, d$.

Proof. We have chosen dist_0 as the standard metric in $\mathbf{R}^d$. Thus the condition $\text{dist}_0(x(n), x) < \epsilon$ is equivalent to $|x_i(n) - x_i| < \epsilon$ for all $i = 1, 2, \ldots, d$. Suppose that $(x(n)) \to x$, take any $\epsilon > 0$, and let n_0 be as in the definition of convergence. Then $|x_i(n) - x_i| < \epsilon$ for all $n > n_0$ and all i so $x_i(n) \to x_i$ for all i. Conversely, if, given $\epsilon > 0$, n_i is such that $\text{dist}(x_i(n), x) < \epsilon$ for any

$n > n_i$, it is enough to take $n_0 := \min\{n_1, n_2, \ldots, n_d\}$ and the condition for the convergence of $(x(n))$ follows. □

Many topological notions discussed in the previous sections can be characterized in terms of convergent sequences. Here are some most useful facts:

Proposition 12.58 *Let (X, dist) be a metric space and $A \subset X$. Then A is closed if and only if for every sequence (a_n) of points in A, $(a_n) \to x \in X$ implies $x \in A$.*

Proof. We argue by contradiction. Suppose that A is closed, and let $(a_n) \to x$, where $a_n \in A$ but $x \notin A$. Since $X \setminus A$ is open, there exists $\epsilon > 0$ such that $B(x, \epsilon) \cap A = \emptyset$. That contradicts $\text{dist}(a_n, x) < \epsilon$ for all sufficiently big n.

Conversely, suppose all convergent sequences in A have limits in A. We show that $U := X \setminus A$ is open. If not, then there exists $x \in U$ such that every ball $B(x, r)$ intersects A. For any $n \in \mathbf{N}$ choose $a_n \in B(x, 1/n) \cap A$. Then $(a_n) \to x \notin A$, a contradiction. □

Proposition 12.59 *Let (X, dist_X) and (Y, dist_Y) be metric spaces and $f : X \to Y$. Then f is continuous if and only if for every $x \in X$ and every sequence $(x_n) \to x$ we have $(f(x_n)) \to f(x)$.*

Proof. Suppose that f is continuous, let $(x_n) \to x$, and take any $\epsilon > 0$. Let δ be as in Theorem 12.35 and let n_0 be such that $\text{dist}_X(x_n, x) < \delta$ for all $n > n_0$. Then $\text{dist}_Y(f(x_n), f(x)) < \epsilon$ for all those n so $(f(x_n)) \to f(x)$.

Conversely, suppose the second condition is satisfied but f is not continuous. Let x and ϵ be such that the conclusion of Theorem 12.35 fails. Then for any $n \in \mathbf{N}$ there exists x_n such that $\text{dist}_X(x_n, x) < 1/n$, but $\text{dist}_Y(f(x_n), f(x)) > \epsilon$. Then $(x_n) \to x$, but $f(x_n) \not\to f(x)$. □

The following theorem displays the fundamental property of the set $\mathbf{R}$ of real numbers that distinguishes it, for example, from the set $\mathbf{Q}$ of rational numbers.

Theorem 12.60 (Bolzano–Weierstrass theorem [76, Chapter 2, Theorem 2.42]) *Every bounded sequence in $\mathbf{R}^d$ contains a convergent subsequence.*

This prompts the following definition.

Definition 12.61 A metric space X is called *compact* if every sequence in X contains a convergent subsequence.

The above definition is not the most general one but the most elementary one. The general definition for topological spaces uses the concept of covering, which will be presented at the end. The Bolzano–Weierstrass theorem can now be reformulated as follows.

Theorem 12.62 *A subset X of $\mathbf{R}^d$ is compact if and only if it is closed and bounded. In particular, every cubical set is compact.*

Proof. That every compact subset of $\mathbf{R}^d$ is closed and bounded is proved by contradiction. Suppose that X is not closed in $\mathbf{R}^d$. Then there exists a sequence (x_n) in X convergent to a point $y \in \mathbf{R}^d \setminus X$. Since every subsequence of a convergent sequence tends to the same limit, it follows that (x_n) has no subsequence convergent to a limit in X. The hypothesis that X is not bounded implies that there is a sequence (x_n) in X with $\|x_n\| \to \infty$, and this again contradicts compactness of X.

Suppose now that X is closed and bounded. Then every sequence in X is bounded, and a limit of any sequence in X convergent in $\mathbf{R}^d$ is in X, so the conclusion instantly follows from Theorem 12.60. □

We now present some useful properties of compact sets.

Definition 12.63 Let (X, dist_X) and (Y, dist_Y) be metric spaces. A function $f : X \to Y$ is called *uniformly continuous* if, for any $\epsilon > 0$, there exists a $\delta > 0$ such that, for any $x, y \in X$, if $\text{dist}_X(x, y) < \delta$, then $\text{dist}_Y(f(x), f(y)) < \epsilon$. In other words, f is uniformly continuous if δ in Theorem 12.35 does not depend on the choice of x.

Theorem 12.64 Let (X, dist_X) and (Y, dist_Y) be metric spaces. If X is compact, than every continuous function $f : X \to Y$ is uniformly continuous.

Proof. We argue by contradiction. Suppose that there exists $\epsilon > 0$ such that for any $\delta > 0$ there is a pair of points $x, y \in X$ such that $\text{dist}_X(x, y) < \delta$ but $\text{dist}_Y(f(x), f(y)) \geq \epsilon$. By taking $\delta = 1/n$ we get a pair of sequences (x_n) and (y_n) such that $\text{dist}_X(x_n, y_n) < 1/n$ but $\text{dist}_Y(f(x_n), f(y_n)) \geq \epsilon$. By compactness, we obtain subsequences $(x_{n_k}) \to x$ and $(y_{n_k}) \to y$. But

$$\text{dist}_X(x, y) \leq \text{dist}_X(x, x_{n_k}) + \text{dist}_X(x_{n_k}, y_{n_k}) + \text{dist}_X(y_{n_k}, y) \to 0,$$

hence $\text{dist}_X(x, y) = 0$, so $x = y$. By Proposition 12.59,

$$\epsilon \leq \text{dist}_Y(f(x_{n_k}), f(y_{n_k})) \leq \text{dist}_Y(f(x_{n_k}), f(x)) + \text{dist}_Y(f(y_{n_k}), f(x)) \to 0,$$

so $\epsilon = 0$, which contradicts that $\epsilon > 0$. □

Proposition 12.65 Let $f : X \to Y$ be a continuous function. If X is compact, then $f(X)$ is also compact.

Proof. Let (y_n) be a sequence in $f(X)$ and let $x_n \in f^{-1}(y_n)$. Then (x_n) contains a convergent subsequence $(x_{n_k}) \to x$. By Proposition 12.59, $(y_{n_k}) = (f(x_{n_k})) \to f(x)$. □

The proof of the following is left as an exercise.

Theorem 12.66 Let X be a compact metric space. Any continuous function $f : X \to \mathbf{R}$ assumes its minimum and maximum in X, namely there are points $x_0, x_1 \in X$ such that $f(x_0) \leq f(x)$ for all $x \in X$ and $f(x_1) \geq f(x)$ for all $x \in X$.

Definition 12.67 Let X be a topological space. A family $\{U_\iota\}_{\iota \in J}$ of subsets of X is called a *covering* of X if its union is all X. It is called an *open covering* if all U_ι are open. A *subcovering* of $\{U_\iota\}_{\iota \in J}$ is a subfamily of $\{U_\iota\}_{\iota \in J}$ that is a covering of X.

The statement of the next theorem is actually used as a definition of a compact space in courses of general topology. The proof is too long to present here, so the reader is referred to [76, Chapter 2, Theorem 2.41].

Theorem 12.68 *A set $X \subset \mathbf{R}^d$ is compact if and only if every open covering of X contains a finite open subcovering.*

Exercises

12.28 Let X be a metric space and consider $f : X \to \mathbf{R}^d$,
$$f(x) = (f_1(x), f_2(x), \ldots, f_d(x)).$$
Show that f is continuous if and only if the coordinate functions $f_i(x)$ are continuous for all i.

12.29 Let A be a subset of a metric space X. Show that $\operatorname{cl} A$ is the set of all limit points of A, that is, $x \in \operatorname{cl} A$ if and only if there exists a sequence of points of A converging to x.

12.30 Let be $\overset{\circ}{Q}$ an elementary cell (see Definition 2.13). Show that $\operatorname{cl} \overset{\circ}{Q} = Q$.

12.31 Prove Theorem 12.66.

12.32 Let X, Y be metric spaces and $f : X \to Y$ a continuous bijection. Prove that if X is compact, then f^{-1} is continuous.

12.33 Let $S_1 \supset S_2 \supset S_3 \supset \cdots$ be a decreasing sequence of nonempty compact sets. Prove that $S := \bigcap_{i=1}^{\infty} S_i$ is a nonempty compact set.

13
Algebra

We provide in this chapter a brief presentation of algebraic preliminaries. The matters discussed here can be found in most of standard textbooks on either abstract algebra or linear algebra. The book by S. Lang [47] is a recommended complementary reference.

13.1 Abelian Groups

13.1.1 Algebraic Operations

Algebra is devoted to the study of the properties of operations. The operations we use in everyday life are addition, subtraction, multiplication, and division of numbers. Formally, a *(binary) operation* on a set G is a mapping $q : G \times G \to G$. Rather than writing the operation in this functional form [e.g., $q(a,b)$], one typically selects a symbol, for instance a diamond, and uses a notation such as $a \diamond b$. Frequently the symbol is borrowed from the operations on numbers, especially when this emphasizes some similarities. Since we want to study similarities to addition of numbers, we will use the symbol $+$ in the sequel.

The fundamental property of the addition of numbers is that for any numbers a, b, c

$$a + (b + c) = (a + b) + c, \tag{13.1}$$

that is, if we have to add three numbers then it does not matter in which order we perform the addition. This is more or less obvious for addition (and also for multiplication) of numbers but needn't be true for other operations. If an operation on a set G satisfies (13.1) for any $a, b, c \in G$, then we say that it is *associative*. An example of an operation that is not associative is subtraction of numbers.

Another property, which is obvious for addition of numbers, is that it is commutative. We say that an operation on a set G is *commutative* if, for any $a, b \in G$,

$$a + b = b + a. \tag{13.2}$$

We say that $e \in G$ is the *identity element* if, for all $a \in G$,

$$a + e = e + a = a. \tag{13.3}$$

It is easy to see that if such an element exists, then it is unique. In the case of addition of numbers, the identity element is zero. For this reason 0 is traditionally used to denote the identity element, and we do so in the sequel.

We say that $a' \in G$ is an *inverse* of $a \in G$ if

$$a + a' = a' + a = 0. \tag{13.4}$$

Of course, this makes sense only if the identity element exists. If the operation is associative, then the inverse element is unique, because if a'' is another inverse of a, then

$$a'' = 0 + a'' = (a' + a) + a'' = a' + (a + a'') = a' + 0 = a'.$$

For addition of numbers, it is obvious that the inverse of a number a is just $-a$. For this reason $-a$ is traditionally used to denote the inverse element of a, and we do so in the sequel.

So far we have spoken of addition of numbers without stating precisely what set of numbers we consider. Not every set is acceptable if we want to consider addition as an operation. For instance, the interval $[0, 1]$ is not good, because the fractions $0.6, 0.7 \in [0, 1]$ but $1.3 = 0.6 + 0.7 \notin [0, 1]$. However, there are several sets of numbers where addition is an operation, for instance the set of all real numbers $\mathbf{R}$, the set of rational numbers $\mathbf{Q}$, the set of integers $\mathbf{Z}$, the set of natural numbers $\mathbf{N}$, and the set of nonnegative integers $\mathbf{Z}^+$. Notice, however, that for addition in $\mathbf{N}$ there is no identity element, because $0 \notin \mathbf{N}$. Also, although $0 \in \mathbf{Z}^+$, the addition in $\mathbf{Z}^+$ does not admit the existence of inverse elements. For instance, $2 \in \mathbf{Z}^+$ but $-2 \notin \mathbf{Z}^+$.

13.1.2 Groups

As we will see in the sequel, several operations on sets other than numbers have the same properties as the four discussed properties of the addition of numbers: associativity, commutativity, the existence of the identity element, and the existence of inverse elements. Whatever we are able to prove for the addition of numbers utilizing only these four properties will also be true for the other operations as long as they have the same four properties. Therefore, it seems useful to do some abstraction and give a name to a set with an operation satisfying those four properties.

For this reason we give the following definition.

Definition 13.1 An *abelian group* is a pair $(G, +)$ consisting of a set G and a binary operation $+$ defined on G and satisfying the following four axioms:

1. the operation is associative;
2. the operation is commutative;
3. the operation admits the identity element;
4. the operation admits the inverse for every element of G.

In practice the operation is often clear from the context. In such a case we often simply say that G is a group. Also, when not explicitly stated otherwise, we denote the operation by $+$.

The word "abelian" refers to the commutativity property. If G satisfies all other axioms but not necessarily commutativity, it is simply called a *group*. Since throughout this text we shall only consider abelian groups, we shall often say "group" when we mean "abelian group."

Example 13.2 The sets $\mathbf{Z}, \mathbf{Q}, \mathbf{R}$, and $\mathbf{C}$ of , respectively, integer numbers, rational numbers, real numbers, and complex numbers with the standard addition operation are all abelian groups. The sets $\mathbf{Z}^+$ and $\mathbf{N}$ of, respectively, non-negative integers and natural numbers (i.e., positive integers) are not groups.

The preceding example may tempt us to say that a group must contain an infinite number of elements. That this needn't be true shows the following example.

Example 13.3 Let $G = \{0, 1, 2\}$ and let the operation $+$ in G be defined by the following table.

$+$	0	1	2
0	0	1	2
1	1	2	0
2	2	0	1

It is straightforward to see that the operation is associative and commutative, 0 is the identity element, the inverse of 0 is 0, the inverse of 1 is 2, and the inverse of 2 is 1. Therefore, we have an example of an abelian group.

Actually, the last example is a special case of the following more general example.

Example 13.4 Given a positive integer n, put $\mathbf{Z}_n := \{0, 1, 2, \ldots, n-1\}$ and define an operation $+_n : \mathbf{Z}_n \times \mathbf{Z}_n \to \mathbf{Z}_n$ by

$$a +_n b := (a + b) \bmod n,$$

where $(a + b) \bmod n$ is the remainder of $a + b \in \mathbf{Z}$ after division by n, that is, the smallest integer $c \geq 0$ such that $a + b - c$ is divisible by n. One can check that $(\mathbf{Z}_n, +_n)$ is a group. Its identity element is just 0. The inverse of $a \in \mathbf{Z}_n$ is $n - a$. Notice that the previous example is just $(\mathbf{Z}_3, +_3)$. In the sequel we shall abandon the subscript n in $+_n$ when it is clear from the context that we mean the addition in $\mathbf{Z}_n$ and not in $\mathbf{Z}$.

The smallest possible group consists of just one element, which must be the identity element of this group (i.e., 0). This group is called the *trivial* group and is denoted by {0} or simply 0. Given n abelian groups $G_1, G_2, \ldots, G_n$,

$$G_1 \times G_2 \times \cdots \times G_n \tag{13.5}$$

is an abelian group with the operation being coordinatewise addition:

$$(a_1, a_2, \ldots, a_n) + (b_1, b_2, \ldots, b_n) := (a_1 + b_1, a_2 + b_2, \ldots, a_n + b_n).$$

One can easily check that the identity element in this group is $(0, 0, \ldots, 0)$, and the inverse of $(a_1, a_2, \ldots, a_n)$ is $(-a_1, -a_2, \ldots, -a_n)$.

Example 13.5 $\mathbf{R}^n$ and $\mathbf{Z}^n$ with the addition operation defined above are examples of abelian groups.

The elements of $\mathbf{R}^n$ are called *vectors* or *number vectors* to distinguish them from the abstract vectors we will introduce later. For $\mathbf{x} = (x_1, x_2, \ldots, x_n) \in \mathbf{R}^n$, the number x_i will be called the ith coordinate of $\mathbf{x}$. Number vectors will be used frequently in this book. Throughout the book we adopt the convention that whenever we use a lowercase bold letter, it denotes a number vector whose coordinates are denoted using the same but boldless letter.

$$\mathbf{x} = (x_1, x_2, \ldots, x_n).$$

If all the numbers x_i are real numbers, then we call $\mathbf{x}$ a *real vector* or just a vector. If all x_i are integers, we call it an *integer vector*. Obviously, integer vectors are elements of $\mathbf{Z}^n$.

Definition 13.6 Let G be an abelian group with the binary operation $+$. A nonempty subset $H \subset G$ is a *subgroup* of G if

1. $0 \in H$,
2. for every $a \in H$, its inverse $-a \in H$,
3. H is *closed* under $+$, namely, given $a, b \in H$, $a + b \in H$.

Example 13.7 $(\mathbf{Z}, +)$ is a subgroup of $(\mathbf{Q}, +)$, which in turn is a subgroup of $(\mathbf{R}, +)$. $(\mathbf{Z}^d, +)$ is a subgroup of $(\mathbf{R}^d, +)$.

Proposition 13.8 *Let H be a nonempty subset of G with the property that if $a, b \in H$, then $a - b \in H$. Then H is a subgroup of G.*

13.1.3 Cyclic Groups and Torsion Subgroup

Let G be a group and let $g \in G$. Given $m \in \mathbf{Z}$, we use the notation

$$mg := \underbrace{g + g + \cdots + g}_{m \text{ terms}} \tag{13.6}$$

to denote the sum of g with itself m times. If m is a negative integer, then this should be interpreted as the m-fold sum of $-g$.

For $a \in G$ put
$$\langle a \rangle := \{ na \in G \mid n \in \mathbf{Z} \}.$$
It is straightforward to verify that $\langle a \rangle$ is a subgroup of G. The group G is called *cyclic* if there exists $a \in G$ such that $G = \langle a \rangle$. In particular, if $a \in G$, then $\langle a \rangle$ is a cyclic subgroup of G.

The *order* of G, denoted by $|G|$, is the number of elements of G. Thus $|\mathbf{Z}| = \infty$ and $|\mathbf{Z}_n| = n$. The *order of an element* $a \in G$, denoted by $o(a)$, is the smallest positive integer n such that $na = 0$, if it exists, and ∞ if not. Observe that $|\langle a \rangle| = o(a)$.

Proposition 13.9 *The set of all elements in G of finite order is a subgroup of G.*

The proof of this proposition is left as an exercise.

Definition 13.10 The subgroup of G of all elements of finite order is called the *torsion subgroup* of G. This subgroup is denoted by $T(G)$.

Example 13.11 The addition table for $\mathbf{Z}_6$ is as follows:

+	0	1	2	3	4	5
0	0	1	2	3	4	5
1	1	2	3	4	5	0
2	2	3	4	5	0	1
3	3	4	5	0	1	2
4	4	5	0	1	2	3
5	5	0	1	2	3	4

Using the table, it is easy to check that 0 has order 1; 1 and 5 have order 6, thus each of them generates the whole group; 2 has order 3; and 3 has order 2. Note the relation between the divisors of 6 and orders of elements of $\mathbf{Z}_6$.

Example 13.12 In the group $\mathbf{Z}_2^2 = \mathbf{Z}_2 \times \mathbf{Z}_2$ of order 4, all three nonzero elements $(0,1)$, $(1,0)$, and $(1,1)$ have order 2. Thus this is not a cyclic group. Consider the group $\mathbf{Z}_2 \times \mathbf{Z}_3$. Here are the orders of its elements:

$$o(0) = 1, \ o((1,0)) = 2, \ o((0,1)) = o((0,2)) = 3, \ o((1,1)) = o((1,2)) = 6.$$

Thus $\mathbf{Z}_2 \times \mathbf{Z}_3$ is cyclic of order 6, generated by $(1,1)$ and by $(1,2)$. The notion of isomorphism introduced in the next section will permit us to identify this group with $\mathbf{Z}_6$. The same consideration applies to $\mathbf{Z}_n \times \mathbf{Z}_m$, where n and m are relatively prime (see the exercises).

We end this section with the following observation.

Lemma 13.13 *Any subgroup of a cyclic group is cyclic.*

Proof. Let G be a cyclic group and let H be a subgroup of G. We may assume that $H \neq 0$. Let $a \in G$ be such that $G = \langle a \rangle$. Choose a nonzero element of H. It has the form ka for some nonzero integer $k \in \mathbf{Z}$. Let k_0 be the smallest positive integer such that $k_0 a \in H$. Then, obviously, $nk_0 a \in H$ for all integers n. We need to show that all elements of H are of this form. Indeed, if not, then there exists $h \in H$ of the form $h = (nk_0 + r)a$, where $0 < r < k_0$. Since $nk_0 a \in H$, we get $ra \in H$, which contradicts the minimality of k_0. □

13.1.4 Quotient Groups

Let H be a subgroup of G and let $a \in G$. The set

$$a + H := \{a + h \ : \ h \in H\}$$

is called a *coset* of H in G. The element a is called its *representative*. Typically, a coset will have many different representatives. For example, let $h_0 \in H$, $a \in G$, and $b = a + h_0$; then a and b are representatives for the same coset. The following proposition makes this precise.

Proposition 13.14 *Let H be a subgroup of G and $a, b \in G$. Then*

(a) The cosets $a + H$ and $b + H$ are either equal or disjoint.
(b) $a + H = b + H$ if and only if $b - a \in H$.

Proof. (a) Suppose that $(a + H) \cap (b + H) \neq \emptyset$. Then there exist $h_1, h_2 \in H$ such that $a + h_1 = b + h_2$. Hence, for any $h \in H$, $b + h = a + h_1 - h_2 + h \in a + H$ so $b + H \subset a + H$. The reverse inclusion holds by the symmetric argument.

(b) Let $a + H = b + H$ and let h_1, h_2 be as in (a). Then $b - a = h_1 - h_2 \in H$. To prove the converse assume that $b - a \in H$. If $b = a$, the conclusion is obvious. Thus assume $b \neq a$. Then at least one of a, b is different from 0. Therefore, without a loss of generality we may assume that $b \neq 0$. Since $b = b + 0 = a + (b - a) \in (b + H) \cap (a + H)$, the conclusion follows from (a). □

Writing cosets in the form of $a + H$ is a bit cumbersome, so we shorten it to $[a] := a + H$. Notice that to use this notation it is essential that we know the subgroup H that is being used to form the cosets.

We can define a binary operation on the set of cosets by setting

$$[a] + [b] := [a + b]. \tag{13.7}$$

Observe that $[0] + [a] = [0 + a] = [a]$, so $[0]$ acts like an identity element. Furthermore, $[a] + [-a] = [a + -a] = [0]$, so there are inverse elements. It is also easy to check that this operation is associative and commutative. The only serious issue is whether this new operation is well defined; in other words, does it depend on which representative we use?

Proposition 13.15 *Equation (13.7) does not depend on the choice of coset representative used and therefore defines a group structure on $\{a + H\}_{a \in G}$.*

Proof. If $a' + H = a + H$ and $b' + H = b + H$, then, by Proposition 13.14, $a' - a \in H$, $b' - b \in H$, and so $(a' + b') - (a + b) = (a' - a) + (b' - b) \in H$. Hence $a' + b' + H = a + b + H$. □

Definition 13.16 The group of cosets described by Proposition 13.15 is called the *quotient group* of G by H and denoted by G/H.

An alternative way of introducing the quotient group is in terms of an equivalence relation. Define the relation $a \sim b$ if and only if $b - a \in H$. Note that this is an *equivalence relation* in G, that is,

i) $a \sim a$, for all $a \in G$;
ii) $a \sim b \Leftrightarrow b \sim a$, for all $a, b \in G$;
iii) $a \sim b$ and $b \sim c \Rightarrow a \sim c$, for all $a, b, c \in G$.

The *equivalence class* of $a \in G$ is the set of all $b \in G$ such that $b \sim a$. Thus, by Proposition 13.14, the group of cosets is exactly the group of equivalence classes of $a \in G$.

Proposition 13.17 *Let G be a finite group and H its subgroup. Then each coset $a + H$ has the same number of elements. Consequently,*

$$|G| = |G/H| \cdot |H|.$$

Proof. The first conclusion is an obvious consequence of the cancellation law for the group addition: $a + h_1 = a + h_2 \Leftrightarrow h_1 = h_2$. The second conclusion is an immediate consequence of the first one and Proposition 13.14(a). □

Example 13.18 Let $G = \mathbf{Z}$ and $H = k\mathbf{Z}$ for some $k \in \mathbf{Z} \setminus \{0\}$. The group $G/H = \mathbf{Z}/k\mathbf{Z}$ has k elements $[0], [1], \ldots, [k-1]$. Since in this case the coset $[a]$ is also represented by the remainder of the division of a by k, this group may be identified with Z_k discussed in the previous section. What "identification" means will become clear later, when we talk about isomorphisms.

Example 13.19 Let $G = \mathbf{Z}^2$ and $H = \{(-n, n) : n \in \mathbf{Z}\}$. Since any element $(m, n) \in \mathbf{Z}^2$ can be written as $(m+n)(1, 0) + n(-1, 1) \in (m+n)(1, 0) + H$, we have $G/H = \{[m(1,0)]\}_{m \in \mathbf{Z}}$. It is easily seen that $[k(1,0)] \neq [m(1,0)]$ whenever $k \neq m$, thus there is a bijection between G/H and $\mathbf{Z}$.

Example 13.20 Consider $\mathbf{Z}$ as a subgroup of $\mathbf{R}$ and the quotient $\mathbf{R}/\mathbf{Z}$. Since any real number is an integer translation of a number in the interval $[0, 1)$, $\mathbf{R}/\mathbf{Z}$ is represented by the points of that interval. Moreover, there is a bijection between $\mathbf{R}/\mathbf{Z}$ and $[0, 1)$, since no two numbers in that interval may differ by an integer. For any $\alpha, \beta \in [0, 1)$, the coset $[\alpha + \beta]$ is represented in $[0, 1)$ by the fractional part of $\alpha + \beta$. Since $1 \sim 0$, $\mathbf{R}/\mathbf{Z}$ may be visualized as a circle obtained from the interval $[0, 1]$ by gluing 1 to 0.

A very similar example explaining the concept of polar coordinates is the quotient group $\mathbf{R}/2\pi\mathbf{Z}$. The equivalence relation is now $\alpha \sim \beta \Leftrightarrow \beta - \alpha =$

$2n\pi$ for some $n \in \mathbf{Z}$. The representatives may be searched, for example, in the interval $[0, 2\pi)$. Thus the elements of $\mathbf{R}/2\pi\mathbf{Z}$ may be identified with the points on the circle $x^2 + y^2 = 1$ in the plane, via the polar coordinate θ in $(x, y) = (\cos\theta, \sin\theta)$.

13.1.5 Direct Sums

Let A and B be subgroups of G. We define their *sum* by

$$A + B := \{c \in G : c = a + b \text{ for some } a \in A, b \in B\}. \tag{13.8}$$

We say that G is a *direct sum* of A and B and write

$$G := A \oplus B$$

if $G = A + B$ and for any $c \in G$ the decomposition $c = a + b$, where $a \in A$ and $b \in B$, is unique.

We have the following simple criterion for a direct sum.

Proposition 13.21 *Let G be the sum of its subgroups A and B. Then $G = A \oplus B$ if and only if $A \cap B = \{0\}$.*

Proof. Suppose that $A \cap B = \{0\}$ and that $c = a_1 + b_1 = a_2 + b_2$ are two decompositions of $c \in G$ for some $a_1, a_2 \in A$ and $b_1, b_2 \in B$. Then $a_1 - a_2 = b_2 - b_1 \in A \cap B = \{0\}$, which implies that $a_1 = a_2$ and $b_1 = b_2$. Hence the decomposition is unique. Conversely, let $A \cap B \neq \{0\}$ and let $c \in A \cap B$, $c \neq 0$. Then c can be decomposed as $c = a + b$ in at least two ways: by setting $a := c$, $b := 0$ or $a := 0$, $b := c$. □

In a similar way one defines the sum and direct sum of any family $G_1, G_2, \ldots, G_n$ of subgroups of a given abelian group G. The group G is the *direct sum* of $G_1, G_2, \ldots, G_n$ if every $g \in G$ can be uniquely written as $a = \sum_{i=1}^{n} g_i$, where $g_i \in G$ for all $i = 1, 2, \ldots, n$. In this case we write

$$G = \bigoplus_{i=1}^{n} G_i = G_1 \oplus G_2 \oplus \cdots \oplus G_n. \tag{13.9}$$

There is a close relation between direct products and direct sums. Let $G = G_1 \times G_2 \times \cdots \times G_n$. We may identify each G_i with the subgroup

$$G'_i := \{0\} \times \cdots \times \{0\} \times \underbrace{G_i}_{i\text{th place}} \times \{0\} \times \cdots \times \{0\}.$$

Then

$$G = G'_1 \oplus G'_2 \oplus \cdots \oplus G'_n$$

and, for the simplicity of notation, we may write

$$G = G_1 \oplus G_2 \oplus \cdots \oplus G_n.$$

This identification of direct products and sums will become more formal when we talk about isomorphisms of groups in the next section.

When infinite families of groups are considered, their direct sum may only be identified with a subgroup of the direct product consisting of sequences that have zeros in all but finitely many places.

Exercises

13.1 Let G be an abelian group. Prove that if there exists an identity element 0 of an operation, then it is unique.

13.2 Prove Proposition 13.8.

13.3 Determine the orders of all elements of $\mathbf{Z}_5, \mathbf{Z}_6, \mathbf{Z}_8$.

13.4 Prove Proposition 13.9.

13.5 (a) Let m, n be relatively prime. Show that $\mathbf{Z}_m \oplus \mathbf{Z}_n$ is cyclic of order mn.
(b) Let $G = \mathbf{Z}_{12} \oplus \mathbf{Z}_{36}$. Express G as a direct sum of cyclic groups whose orders are powers of primes.

13.6 (a) Prove that a group of prime order has no proper subgroup.
(b) Prove that if G is a cyclic group and p is a prime dividing $|G|$, then G contains an element of order p.

13.2 Fields and Vector Spaces

13.2.1 Fields

Is the multiplication of numbers an operation that can be used to build a group? The multiplication of numbers is associative and commutative. Since for any number a
$$1 \cdot a = a \cdot 1 = a,$$
the identity element for multiplication of numbers is number 1. Also for every nonzero number a the inverse of a under multiplication is $a^{-1} = 1/a$. Therefore, we see that $(\mathbf{R} \setminus \{0\}, \cdot)$ is an abelian group. Another example is $(\mathbf{R}^+ \setminus \{0\}, \cdot)$. However, neither $(\mathbf{Z} \setminus \{0\}, \cdot)$ nor $(\mathbf{N} \setminus \{0\}, \cdot)$ is a group, because for instance $1/2 \notin \mathbf{Z}$. As we will see in the sequel, there are examples of groups when it is more natural to denote the group operation by $\cdot$. In that case we denote the identity element by 1 and the inverse element of a by a^{-1}. We then speak of *multiplicative groups* as being in contrast to additive groups, when the group operation is denoted by $+$, the identity element by 0, and the inverse of a by $-a$. However, the reader should remember that there is no

difference between the additive and the multiplicative groups other than just notation.

In a multiplicative group we typically simplify the expression of multiplication and write ab instead of $a \cdot b$.

In case of numbers, there is a useful property that involves both addition and multiplication. Namely, for any numbers a, b, c,

$$a \cdot (b + c) = a \cdot b + a \cdot c. \tag{13.10}$$

Given a set F and operations $+ : F \times F \to F$ and $\cdot : F \times F \to F$, we say that the operation $\cdot$ *distributes* over the operation $+$ if for any $a, b, c \in F$ property (13.10) is satisfied.

We can summarize the properties of addition and multiplication of real numbers by saying that $(\mathbf{R}, +)$ and $(\mathbf{R} \setminus \{0\}, \cdot)$ are abelian groups and the multiplication distributes over addition.

Definition 13.22 A *field* is a triple $(F, +, \cdot)$, where $+ : F \times F \to F$ and $\cdot : F \times F \to F$ are operations in F such that $(F, +)$ and $(F \setminus \{0\}, \cdot)$ are abelian groups and $\cdot$ is distributive over $+$.

As in the case of groups, if the operations are clear from the context, we simply say that F is a field.

Example 13.23 The set of complex numbers $\mathbf{C}$ and the set of rational numbers $\mathbf{Q}$ are fields.

Example 13.24 The integers $\mathbf{Z}$ do not form a field. In particular, $2 \in \mathbf{Z}$, but $2^{-1} = 1/2 \notin \mathbf{Z}$.

Example 13.25 A very useful field is $\mathbf{Z}_2$, the set of integers modulo 2. The rules for addition and multiplication are as follows:

+	0	1
0	0	1
1	1	0

$\cdot$	0	1
0	0	0
1	0	1

We leave it to the reader to check that $(\mathbf{Z}_2, +, \cdot)$ is a field.

Example 13.26 Another field is $\mathbf{Z}_3$, the set of integers modulo 3. The rules for addition and multiplication are as follows:

+	0	1	2
0	0	1	2
1	1	2	0
2	2	0	1

$\cdot$	0	1	2
0	0	0	0
1	0	1	2
2	0	2	1

Again, we leave it to the reader to check that $(\mathbf{Z}_3, +, \cdot)$ is a field. However, we note that the inverse of 1 under addition in $\mathbf{Z}^3$ is 2 and the inverse of 2 under multiplication in $\mathbf{Z}_3$ is 2.

Example 13.27 $\mathbf{Z}_4$, the set of integers modulo 4, is not a field. The rules for addition and multiplication are as follows:

+	0	1	2	3
0	0	1	2	3
1	1	2	3	0
2	2	3	0	1
3	3	0	1	2

·	0	1	2	3
0	0	0	0	0
1	0	1	2	3
2	0	2	0	2
3	0	3	2	1

Observe that the element 2 does not have an inverse with respect to multiplication in $\mathbf{Z}_4$.

13.2.2 Vector Spaces

Let $(G, +)$ be a group. We have an operation $\mathbf{Z} \times G \to G$, which assigns mg to $m \in \mathbf{Z}$ and $g \in G$. Such an operation is an example of an *external operation*, because the first operand is not from the group G. In the case of real vectors, such an external operation makes sense even if the first operand is an arbitrary real number. More precisely, given a real vector $\mathbf{x} := (x_1, x_2, \ldots, x_d) \in \mathbf{R}^n$ and a real number $a \in \mathbf{R}$, we define

$$a\mathbf{x} := (ax_1, ax_2, \ldots, ax_n). \tag{13.11}$$

In this case, or more generally when the first operand is from a field, the external operation is referred to as a *scalar multiplication* and the first operand is called a *scalar*. One easily verifies that when $a \in \mathbf{Z}$ the definition (13.11) coincides with (13.6).

The group $(\mathbf{R}^n, +)$ considered together with the scalar multiplication $\mathbf{R} \times \mathbf{R}^n \to \mathbf{R}^n$ given by (13.11) is an example of a *vector space* over the field $\mathbf{R}$. An example we are most familiar with is $\mathbf{R}^3$, because it is used as coordinate space to describe our everyday world.

In the general definition below the field may be arbitrary.

Definition 13.28 A *vector space* over a field F is a set V with two operations, vector addition $+ : V \times V \to V$ and scalar multiplication $\cdot : F \times V \to V$ such that $(V, +)$ is an abelian group and the scalar multiplication satisfies the following rules:

1. For every $v \in V$, 1 times v equals v, where $1 \in F$ is the unique element one in the field.
2. For every $v \in V$ and $\alpha, \beta \in F$,

$$\alpha(\beta v) = (\alpha\beta)v.$$

3. For every $\alpha \in F$ and all $u, v \in V$,

$$\alpha(u + v) = \alpha u + \alpha v.$$

4. For all $\alpha, \beta \in F$ and every $v \in V$,
$$(\alpha + \beta)v = \alpha v + \beta v.$$

Example 13.29 Let F be a field and $n > 0$ an integer. For $(x_1, x_2, \ldots, x_n)$, $(y_1, y_2, \ldots, y_n) \in F^n$ and $\alpha \in F$, define

$$(x_1, x_2, \ldots, x_n) + (y_1, y_2, \ldots, y_n) := (x_1 + y_1, x_2 + y_2, \ldots, x_n + y_n) \quad (13.12)$$

$$\alpha(x_1, x_2, \ldots, x_n) := (\alpha x_1, \alpha x_2, \ldots, \alpha x_n). \quad (13.13)$$

It is straightforward to verify that F^n with these two operations is a vector space over the field F.

From the viewpoint of this book, the important case is the case of vector spaces $\mathbf{Z}_2^n$ over the finite field $\mathbf{Z}_2$. Such a vector space is finite itself. For instance,

$$\mathbf{Z}_2^3 = \{(0,0,0), (0,0,1), (0,1,0), (0,1,1), (1,0,0), (1,0,1), (1,1,0), (1,1,1)\}$$

consists of eight vectors.

We already know that the operation given by (13.12) may also be considered in $\mathbf{Z}^n$, and $\mathbf{Z}^n$ with this operation is an abelian group. We also know that the multiplication of an integer vector by an integer makes sense and it is just shorthand for adding a number of copies of a vector. However, we should be aware that $\mathbf{Z}^n$ is not a vector space, because $\mathbf{Z}$ is not a field. Nevertheless, there are some similarities of the group $\mathbf{Z}^n$ and a vector space. Since the group $\mathbf{Z}^n$ is very important from the point of view of applications to homology, in the sequel we will study these similarities as well as differences.

13.2.3 Linear Combinations and Bases

Let V be a vector space over a field F. Given k vectors $w_1, w_2, \ldots, w_k \in V$ and k scalars $\alpha_1, \alpha_2, \ldots, \alpha_k \in F$, we can form the expression

$$\alpha_1 w_1 + \alpha_2 w_2 + \cdots + \alpha_k w_k,$$

called the *linear combination* of vectors $w_1, w_2, \ldots, w_k$ with *coefficients* $\alpha_1, \alpha_2, \ldots, \alpha_k$.

Slightly more general is the *linear combination of a family of vectors* $\{w_j\}_{j \in J} \subset V$ with coefficients $\alpha_1, \alpha_2, \ldots, \alpha_k \in F$, which has the form

$$\sum_{j \in J} \alpha_j w_j. \quad (13.14)$$

Of course, in order to guarantee that such a linear combination makes sense, we need to assume that for all but a finite number of $j \in J$ the coefficients $\alpha_j = 0$. In the sequel, whenever we write a linear combination of the form (13.14) we assume that all but a finite number of coefficients are zero.

The following concept is fundamental in the study of vector spaces.

Definition 13.30 A family $\{w_j\}_{j\in J} \subset V$ of vectors in a vector space V is a *basis* of V if for every vector $v \in V$ there exists a unique family of scalars $\{\alpha_j\}_{j\in J} \subset F$ such that all but a finite number of them are zero and

$$v = \sum_{j\in J} \alpha_j w_j.$$

One can prove that in every vector space there exists a basis and any two bases have the same cardinality, although the proof is not easy (see [41, II, Chapter III, Theorem 3.12]). A vector space V is called *finitely dimensional* if there exists a basis that is finite. In this case the cardinality of a basis is simply the number of its elements. This number is called the *dimension* of the vector space.

Assume V is a finitely dimensional vector space over a field F and

$$\{w_1, w_2, \ldots, w_n\} \subset V$$

is a basis. Then for every $v \in V$ there exist unique scalars $\alpha_1, \alpha_2, \ldots, \alpha_n$ such that

$$v = \alpha_1 w_1 + \alpha_2 w_2 + \ldots + \alpha_n w_n.$$

They are called the *coordinates* of v in the basis $\{w_1, w_2, \ldots, w_n\}$. These coordinates form a vector $(\alpha_1, \alpha_2, \ldots, \alpha_n) \in F^n$, called the *coordinate vector* of v.

In an arbitrary vector space there is in general no natural way to select a basis. There is, however, an important special case where there is a natural choice of basis, namely the vector space F^n. The *canonical basis* of F^n consists of vectors $\{\mathbf{e}_1^n, \mathbf{e}_2^n, \ldots, \mathbf{e}_n^n\}$, where

$$\mathbf{e}_i := \mathbf{e}_i^n := (0, 0, \ldots, 0, 1, 0, \ldots, 0)$$

has all coordinates zero except the ith coordinate, which is 1. To see that this is a basis, observe that for every $\mathbf{x} = (x_1, x_2, \ldots, x_n) \in F^n$,

$$\mathbf{x} = \sum_{i=1}^n x_i \mathbf{e}_i^n$$

and the uniqueness of this decomposition is obvious.

This basis will be used throughout the book. We will usually drop the superscripts n in $\mathbf{e}_i^n$ if it is clear from the context what n is.

To see why the canonical basis is a natural choice, observe that coordinate vector of the element $\mathbf{x} = (x_1, x_2, \ldots, x_n) \in F^n$ in this basis is $(x_1, x_2, \ldots, x_n)$, that is, it is indistinguishable from the element $\mathbf{x}$ itself.

There are two other important concepts of linear algebra based on linear combinations: linear independence and spanning.

Definition 13.31 Let V be a vector space over a field F. A set of vectors $\{v_j\}_{j \in J} \subset V$ is *linearly independent* if the only way of writing 0 as a linear combination of vectors $\{v_j\}_{j \in J}$ is by taking all coefficients equal to zero, that is, if for any $\{\alpha_j\}_{j \in J} \subset F$,

$$\sum_{j \in J} \alpha_j v_j = 0 \Rightarrow \alpha_j = 0 \quad \text{for all } j \in J.$$

The set of vectors $\{v_j\}_{j \in J}$ *spans* V if every element $w \in V$ can be written (not necessarily in a unique way) as a linear combination of vectors in $\{v_j\}_{j \in J}$, that is, if

$$w = \sum_{j \in J} \alpha_j v_j$$

for some $\{\alpha_j\}_{j \in J} \subset F$ such that all but a finite number of them are zero.

The following theorem is a convenient criterion for a set to be a basis.

Theorem 13.32 *A set of vectors* $\{v_j\}_{j \in J} \subset V$ *is a basis if and only if it is linearly independent and spans* V.

Proof. Let $\{v_j\}_{j \in V} \subset V$ be a basis in a vector space V. Then obviously every vector is a linear combination of elements in $\{v_j\}_{j \in J}$. Therefore, the set $\{v_j\}_{j \in J}$ spans V. Its linear independence follows from the uniqueness of the decomposition of $0 \in V$.

Conversely, if $\{v_j\}_{j \in J}$ is linearly independent and spans V, then every element $w \in V$ equals $\sum_{j \in J} \alpha_j v_j$ for some coefficients $\{\alpha_j\}_{j \in J}$. Assume that also $v = \sum_{j \in J} \beta_j v_j$. Then

$$0 = \sum_{j \in J} (\alpha_j - \beta_j) v_j$$

and from the linear independence of $\{v_j\}_{j \in J}$ it follows that $\alpha_j - \beta_j = 0$, that is $\alpha_j = \beta_j$ for all $j \in J$. We conclude that the representation $w = \sum_{j \in J} \alpha_j v_j$ is unique. □

The following theorem is a very convenient characterization of bases in vector spaces.

Theorem 13.33 *Let V be a vector space. A set $\{w_j\}_{j \in J} \subset V$ is a basis of V if and only if it is a maximal linearly independent set. Also $\{w_j\}_{j \in J} \subset V$ is a basis of V if and only if it is a minimal set spanning V.*

Exercises _____

13.7 (a) Write down the tables of addition and multiplication for $\mathbf{Z}_5, \mathbf{Z}_6, \mathbf{Z}_8$.
(b) If $\mathbf{Z}'_n := \mathbf{Z}_n \setminus 0$, show that $\mathbf{Z}'_5$ is a multiplicative group but $\mathbf{Z}'_6, \mathbf{Z}'_8$ are not.

(c) Now let $\mathbf{Z}_n^* := \{k \in \mathbf{Z}_n : k \text{ and } n \text{ are relatively prime}\}$. Show that $\mathbf{Z}_n^*$ is a multiplicative group for any positive integer n.

13.8 Determine the orders of all elements of $\mathbf{Z}_5^*, \mathbf{Z}_6^*, \mathbf{Z}_8^*$, where $\mathbf{Z}_n^*$ is defined in the preceding exercise.

13.9 Prove that the set of rational numbers $\mathbf{Q}$ is a field.

13.10 Let $\mathbf{Z}_n$ denote the set of integers modulo n. For which n is $\mathbf{Z}_n$ a field?

13.3 Homomorphisms

If we wish to compare two groups or two vector spaces, then we need to be able to talk about functions between them. Of course, these functions need to preserve the operations.

13.3.1 Homomorphisms of Groups

In the case of groups this leads to the following definition.

Definition 13.34 *Let G and G' be two abelian groups. A map $f : G \to G'$ is called a* homomorphism *if*

$$f(g_1 + g_2) = f(g_1) + f(g_2)$$

for all $g_1, g_2 \in G$.

There are some immediate consequences of this definition. For example, as the following argument shows, homomorphisms map the identity element to the identity element.

$$f(0) = f(0+0) = f(0) + f(0),$$
$$f(0) - f(0) = f(0),$$
$$0 = f(0).$$

A trivial induction argument shows that

$$f(mg) = mf(g)$$

for all $m \in \mathbf{Z}$ and $g \in G$. As a consequence, we obtain

$$f(m_1 g_1 + m_2 g_2) = m_1 f(g_1) + m_2 f(g_2)$$

for any $m_1, m_2 \in \mathbf{Z}$ and $g_1, g_2 \in G$.

Proposition 13.35 *Let $f : G \to G'$ be a homomorphism of groups. Then*

(a) for any subgroup H of G, its image $f(H)$ is a subgroup of G';

(b) *for any subgroup H' of G', its inverse image $f^{-1}(H')$ is a subgroup of G;*
(c) *if f is bijective (i.e., one-to-one and onto), then its inverse $f^{-1} : G' \to G$ also is a bijective homomorphism.*

Proof. (a) Let $f(h_1), f(h_2)$ be two elements of $f(H)$. Then

$$f(h_1) - f(h_2) = f(h_1 - h_2) \in f(H).$$

Therefore, by Proposition 13.8, $f(H)$ is a subgroup of G'.
 (b) Let $g_1, g_2 \in f^{-1}(H')$. Then

$$f(g_1 - g_2) = f(g_1) - f(g_2) \in H'.$$

Therefore, $g_1 - g_2 \in f^{-1}(H')$ and, again by Proposition 13.8, $f^{-1}(H')$ is a subgroup of G.
 (c) The last assertion follows from similar types of arguments and is left to the reader. □

Definition 13.36 The set $\operatorname{im} f := f(G)$ is called the *image* or *range* of f in G'. By the previous proposition it is a subgroup of G'. The set

$$\ker f := f^{-1}(0) = \{a \in G \mid f(a) = 0\}$$

is called the *kernel* of f. Again by the previous proposition it is a subgroup of G.

Definition 13.37 A homomorphism $f : G \to G'$ is called an *epimorphism* if it is surjective (or onto) (i.e., $\operatorname{im} f = G'$) and a *monomorphism* if it is injective (or 1–1), [i.e., for any $a \neq b$ in G, $f(a) \neq f(b)$]. This latter condition obviously is equivalent to the condition $\ker f = 0$. Finally, f is called an *isomorphism* if it is both a monomorphism and an epimorphism.

The last definition requires some discussion since the word "isomorphism" takes different meanings in different branches of mathematics. Let X and Y be any sets and $f : X \to Y$ any map. Then f is called *invertible* if there exists a map $g : Y \to X$, called the *inverse* of f, with the property

$$gf = \operatorname{id}_X \text{ and } fg = \operatorname{id}_Y, \tag{13.15}$$

where id_X and id_Y denote the identity maps on X and Y, respectively. It is easy to show that f is invertible if and only if it is bijective. If this is the case, g is uniquely determined and denoted by f^{-1}. When we speak about a particular class of maps, by an invertible map or an isomorphism we mean a map that has an inverse in the same class of maps. For example, if continuous maps are of concern, an isomorphism would be a continuous map that has a continuous inverse. The continuity of a bijective map does not guarantee, in general, the continuity of its inverse. Proposition 13.35(c) guarantees that this problem does not occur in the class of homomorphisms. Thus a homomorphism is an isomorphism if and only if it is invertible in the class of homomorphisms.

When $G = G'$, a homomorphism $f : G \to G$ is also called an *endomorphism* and an isomorphism $f : G \to G$ is called an *automorphism*.

Groups G and G' are called *isomorphic* if there exists an isomorphism $f : G \to G'$. We then write $f : G \xrightarrow{\sim} G'$ or just $G \cong G'$ if f is irrelevant. It is easy to see that $G \cong G'$ is an equivalence relation. We shall often permit ourselves to identify isomorphic groups.

Example 13.38 Let G be a cyclic group of infinite order generated by a. Then $f : \mathbf{Z} \to G$ defined by $f(n) = na$ is an isomorphism with the inverse defined by $f^{-1}(na) = n$. By the same argument, any cyclic group of order k is isomorphic to $\mathbf{Z}_k$.

Example 13.39 Let H be a subgroup of G and define $q : G \to G/H$ by the formula $q(a) := a + H$. It is easy to see that q is an epimorphism and its kernel is precisely H. This map is called the *canonical quotient homomorphism*.

Example 13.40 Let A, B be subgroups of G such that $G = A \oplus B$. Then the map $f : A \times B \to G$ defined by $f(a, b) = a + b$ is an isomorphism with the inverse defined by $f^{-1}(c) = (a, b)$, where $c = a + b$ is the unique decomposition of $c \in G$ with $a \in A$ and $b \in B$. This can be generalized to direct sums and products of any finite number of groups.

Example 13.41 Let A, B, and G be as in Example 13.40. The inclusion map $i : A \to G$ is a monomorphism and the projection map $p : G \to A$ defined by $p(c) = a$, where $c = a + b$ with $a \in A$ and $b \in B$, is an epimorphism. Note that $pi = \mathrm{id}_A$, hence p may be called a *left inverse* of i and i a *right inverse* of p. Note that a left inverse is not necessarily unique. Indeed, take subgroups $A = \mathbf{Z}\mathbf{e}_1, B = \mathbf{Z}\mathbf{e}_2$ of $\mathbf{Z}^2$. Another choice of a left inverse of i is $p'(n\mathbf{e}_1 + m\mathbf{e}_2) = (n+m)\mathbf{e}_1$ (a "slanted" projection).

We leave as an exercise the following useful result.

Proposition 13.42 *If $f : G \to G$ is a group homomorphism, then $f(T(G)) \subset T(G)$, where $T(G)$ is the torsion subgroup of G.*

Now let $f : G \to G'$ be a homomorphism and $H = \ker f$. Then, for any $a \in G$ and $h \in H$, we have $f(a + h) = f(a)$. Hence the image of any coset $a + H$ under f is
$$f(a + H) = \{f(a)\}.$$
Moreover, that image is independent of the choice of a representative of a coset $a + H$. Indeed, if $a + H = b + H$, then $b - a \in H$; thus $f(b) = f(a)$. We may now state the following.

Theorem 13.43 *Let $f : G \to G'$ be a homomorphism and $H = \ker f$. Then the map*

$$\bar{f} : G/H \to \operatorname{im} f$$

defined by $\bar{f}(a + H) = f(a)$ is an isomorphism, called the quotient isomorphism.

Proof. By the preceding discussion, the formula for $\bar{f}$ is independent of the choice of coset representatives, thus $\bar{f}$ is well defined. Since

$$\bar{f}((a+H) + (b+H)) = \bar{f}(a+b+H) = f(a+b) = f(a) + f(b),$$

it is a homomorphism. $\bar{f}$ is a monomorphism since $\bar{f}(a + H) = 0$ implies $f(a) = 0$, that is, $a \in \ker f = H$ or $a + H = H$.

Finally, $\bar{f}$ is also an epimorphism since $\operatorname{im} \bar{f} = \operatorname{im} f$. □

As a corollary we obtain the following special case known as the fundamental epimorphism theorem in group theory.

Corollary 13.44 *If $f : G \to G'$ is an epimorphism, then*

$$\bar{f} : G/\ker f \to G'$$

is an isomorphism.

Example 13.45 Let $q : G \to G/H$ be the canonical homomorphism from Example 13.39. Then $\bar{q} = \operatorname{id}_{G/H}$, so this is the trivial case of Theorem 3.1.

Example 13.46 Let $f : \mathbf{Z} \to \mathbf{Z}_n$ be given by $f(a) = a \bmod n$. Then f is a well-defined epimorphism with $\ker f = n\mathbf{Z}$. Thus $\bar{f} : \mathbf{Z}/n\mathbf{Z} \xrightarrow{\cong} \mathbf{Z}_n$.

Example 13.47 Let's go back to p' in Example 13.41: $\operatorname{im} p' = \mathbf{Z}\mathbf{e}_1 = A$ and $\ker p' = \mathbf{Z}(\mathbf{e}_2 - \mathbf{e}_1)$. Thus $\bar{f} : \mathbf{Z}^2/\mathbf{Z}(\mathbf{e}_2 - \mathbf{e}_1) \xrightarrow{\cong} \mathbf{Z}\mathbf{e}_1$. Note that $\mathbf{Z}^2 = \mathbf{Z}\mathbf{e}_1 \oplus \mathbf{Z}(\mathbf{e}_2 - \mathbf{e}_1) = \operatorname{im} p' \oplus \ker p'$. This observation will be generalized later.

Example 13.48 Consider Example 13.20 in terms of the quotient isomorphism. Let S^1 be the unit circle in the complex plane, namely the set defined by $|z| = 1$, $z = x + iy \in \mathbf{C}$, i the primitive square root of -1. Then S^1 is a multiplicative group with the complex number multiplication and the unity $1 = 1 + i0$. We define $f : \mathbf{R} \to S^1$ by $f(\theta) = e^{i\theta} = \cos\theta + i\sin\theta$. Then f is a homomorphism from the additive group of $\mathbf{R}$ to the multiplicative group S^1. It is an epimorphism with the kernel $\ker f = 2\pi\mathbf{Z}$. Thus $\bar{f} : \mathbf{R}/2\pi\mathbf{Z} \xrightarrow{\cong} S^1$.

Theorem 13.49 *Assume $G_1, G_2, \ldots, G_n$ and $H_1, H_2, \ldots, H_n$ are groups and $H_i \subset G_i$ are subgroups. Then*

$$G_1 \oplus G_2 \oplus \cdots \oplus G_n / H_1 \oplus H_2 \oplus \cdots \oplus H_n \cong G_1/H_1 \oplus G_2/H_2 \oplus \cdots \oplus G_n/H_n.$$

Proof. We have an epimorphism $\pi : G_1 \oplus G_2 \oplus \cdots \oplus G_n \to G_1/H_1 \oplus G_2/H_2 \oplus \cdots \oplus G_n/H_n$ defined by

$$\pi(g_1 + g_2 + \cdots + g_n) := [g_1] + [g_2] + \cdots + [g_n].$$

It is straightforward to verify that $\ker \pi = H_1 \oplus H_2 \oplus \cdots \oplus H_n$. Therefore, the conclusion follows from Corollary 13.44. □

13.3.2 Linear Maps

In the case of a vector space V over a field F, we have two operations: vector addition $+$, which makes $(V, +)$ an abelian group, and scalar multiplication. The definition of a map preserving the structure of a vector space must take care of both of them. Therefore, we have the following definition.

Definition 13.50 *Assume V, V' are two vector spaces over the same field F. A map $f : V \to V'$ is called a linear map if*

$$f(\alpha_1 v_1 + \alpha_2 v_2) = \alpha_1 f(v_1) + \alpha_2 f(v_2)$$

for all $\alpha_1, \alpha_2 \in F$ and $v_1, v_2 \in V$.

The terminology "linear map" comes from the fact that the graph of a linear map of $f : \mathbf{R}^1 \to \mathbf{R}^1$ is a line.

We will study only linear maps of finitely dimensional vector spaces.

Let V and V' be two finitely dimensional vector spaces with bases, respectively, $W := \{w_1, w_2, \ldots, w_n\}$ and $W' := \{w'_1, w'_2, \ldots, w'_m\}$. Let $f : V \to V'$ be a linear map and let $v = \sum_{j=1}^n x_j w_j$ be an element of V. Then

$$f(v) = f\left(\sum_{j=1}^n x_j w_j\right) = \sum_{j=1}^n x_j f(w_j). \tag{13.16}$$

This means that f is uniquely determined by its values on the elements of a basis in V. In particular, in order to define a linear map it is enough to define its values on elements of a selected basis of V.

On the other hand, the values $f(w_j)$ may be written as linear combinations of the elements of the basis in V'

$$f(w_j) = \sum_{i=1}^m a_{ij} w'_i. \tag{13.17}$$

From (13.16) and (13.17) we obtain

$$f(v) = \sum_{i=1}^m \left(\sum_{j=1}^n a_{ij} x_j\right) w'_i. \tag{13.18}$$

Therefore, the coordinates $x'_1, x'_2, \ldots, x'_m$ of $f(v)$ in the basis W' are described by the following system of linear equations:

$$\begin{aligned}
x'_1 &= a_{11} x_1 + a_{12} x_2 + \ldots + a_{1n} x_n, \\
x'_2 &= a_{21} x_1 + a_{22} x_2 + \ldots + a_{2n} x_n, \\
&\ldots \quad \ldots \quad \ldots \quad \ldots \quad \ldots \\
x'_m &= a_{m1} x_1 + a_{m2} x_2 + \ldots + a_{mn} x_n.
\end{aligned} \tag{13.19}$$

13.3.3 Matrix Algebra

In the study of systems of linear equations such as (13.19), it is convenient to use the language of matrices. Recall that an *m by n matrix* with *entries* a_{ij} in the field F is an array of elements of F of the form

$$\begin{bmatrix} a_{11} & a_{12} & \cdots & a_{1n} \\ a_{21} & a_{22} & \cdots & a_{2n} \\ \vdots & & & \\ a_{m1} & a_{m2} & \cdots & a_{mn} \end{bmatrix}. \tag{13.20}$$

The number m is referred to as the number of rows and n as the number of columns of the matrix. When these numbers are clear from the context, we often use a compact notation $[a_{ij}]$. Even more compact is the convention of denoting the whole matrix by one capital letter. Usually the entries are denoted by lowercase letters and the whole matrix by the corresponding uppercase letter, so in the case of (13.20) we write $A = [a_{ij}]$. We denote the set of matrices with m rows and n columns by $\mathrm{M}_{m,n}(F)$, where F is the field from which the entries of the matrix are taken.

We define the multiplication of matrices as follows. If $B = [b_{jk}] \in \mathrm{M}_{n,p}(F)$ is another matrix whose number of columns is the same as the number of rows of A, then the *product* AB is the matrix $C = [c_{ik}] \in \mathrm{M}_{m,p}(F)$ whose entries are defined by

$$c_{ik} = \sum_{j=1}^{m} a_{ij} b_{jk}.$$

It should be emphasized that matrix multiplication is in general not commutative.

Using matrix notation, Eq. (13.19) may be written as

$$\begin{bmatrix} x'_1 \\ x'_2 \\ \vdots \\ x'_m \end{bmatrix} = \begin{bmatrix} a_{11} & a_{12} & \cdots & a_{1n} \\ a_{21} & a_{22} & \cdots & a_{2n} \\ \vdots & & & \\ a_{m1} & a_{m2} & \cdots & a_{mn} \end{bmatrix} \begin{bmatrix} x_1 \\ x_2 \\ \vdots \\ x_n \end{bmatrix}. \tag{13.21}$$

The *identity matrix*

$$I_{n \times n} := \begin{bmatrix} 1 & 0 & \cdots & 0 \\ 0 & 1 & \cdots & 0 \\ \vdots & & & \\ 0 & 0 & \cdots & 1 \end{bmatrix}$$

with n rows and columns has all entries zero except the diagonal entries, which are 1. It is straightforward to verify that the identity matrix is an identity element for matrix multiplication, that is,

$$A I_{n \times n} = A \quad \text{and} \quad I_{n \times n} B = B$$

for $A \in M_{m,n}(F)$ and $B \in M_{n,p}(F)$.

Matrices whose number of rows is the same as the number of columns are called *square matrices*. A square matrix $A \in M_{n,n}(F)$ is *invertible* or *nonsingular* if there exists a matrix $B \in M_{n,n}(F)$ such that

$$AB = BA = I_{n \times n}.$$

One easily verifies that if such a matrix B exists, then it is unique. It is called the *inverse of matrix* A and it is denoted by A^{-1}.

If $A_1, A_2, \ldots, A_k$ are matrices with the same number of rows, then the matrix built of all columns of A_1 followed by all columns of A_2 and so on up to the last matrix will be denoted by

$$[\, A_1 \; A_2 \; \ldots \; A_k \,].$$

Sometimes, to emphasize the process in which the matrix was obtained, we will put vertical bars between the matrices A_i, namely we will write

$$[\, A_1 \mid A_2 \mid \ldots \mid A_k \,].$$

The *transpose* of a matrix $A = [a_{ij}] \in M_{m,n}(F)$ is the matrix

$$A^T := [a_{ji}] = \begin{bmatrix} a_{11} & a_{21} & \cdots & a_{m1} \\ a_{12} & a_{22} & \cdots & a_{m2} \\ \vdots & & & \\ a_{1n} & a_{2n} & \cdots & a_{mn} \end{bmatrix}.$$

We will identify an element $\mathbf{x} = (x_1, x_2, \ldots, x_n) \in F^n$ with the row matrix $[x_1, x_2, \ldots, x_n]$.

Using the introduced notation, Eq. (13.19) becomes

$$(\mathbf{x}')^T = A\mathbf{x}^T.$$

In situations when this does not lead to confusion, we will allow ourselves the freedom of identifying the vector $\mathbf{x}$ with its transpose $\mathbf{x}^T$ to reduce the number of transpose signs. With this convention, Eq. (13.19) becomes just

$$\mathbf{x}' = A\mathbf{x}. \tag{13.22}$$

Going back to the linear map $f : V \to V'$, let $\mathbf{x} := (x_1, x_2, \ldots, x_n)$ be the vector of coordinates of $v \in V$ in the basis W and let $\mathbf{x}' := (x'_1, x'_2, \ldots, x'_n)$ be the vector of coordinates of $f(v) \in V'$ in the basis W'. Then the equation (13.19) describing the relation between the coordinates of v and $f(v)$ may be written in the compact form as (13.22), where the matrix $A = [a_{ij}]$ consists of coefficients in (13.18). This matrix is referred to as the *matrix of f in the bases W and W'*.

Conversely, if $A = [a_{ij}]$ is an $m \times n$ matrix, then Eq. (13.18) defines a unique linear map $f : V \to V'$.

If $f : F^n \to F^m$ is a linear map, then its matrix in the canonical bases is simply called *the matrix of f*. Since this matrix is frequently needed, we will use A_f as the notation for it.

Note that in the case of a homomorphism $f : F^n \to F^m$, the equation

$$\mathbf{y} = f(\mathbf{x}) \quad \text{for } \mathbf{x} \in F^n, \mathbf{y} \in F^m$$

may be rewritten using matrix notation as

$$\mathbf{y} = A_f \mathbf{x},$$

because the elements $\mathbf{x}, \mathbf{y} \in F^n$ are indistinguishable from their coordinates in the canonical bases. Nevertheless, we want to have a clear distinction between the matrix A, which is an array of numbers, and the linear map defined by this matrix. However, given a matrix $A \in M_{m,n}(F)$, it is customary to identify it in some situations with the linear map $f : F^m \to F^n$ given by

$$f(x) := Ax.$$

This is done especially when speaking about the kernel $\ker A$ and the image $\operatorname{im} A$ of the matrix A.

An elementary calculation shows that if $f' : F^m \to F^k$ is another linear map, then

$$A_{f'f} = A_{f'} A_f. \tag{13.23}$$

Actually, this fact is the motivation for the definition of the product of two matrices.

Let $\operatorname{id}_{F^n} : F^n \to F^n$ be the identity map. It is straightforward to verify that id_{F^n} is an isomorphism and its matrix is $I_{n \times n}$.

Therefore, it follows from (13.23) that if $f : F^n \to F^n$ is an isomorphism, then the matrix A_f is nonsingular and

$$A_f^{-1} = A_{f^{-1}}.$$

Exercises

13.11 Show that $\mathbf{Z}_6 \cong \mathbf{Z}_2 \times \mathbf{Z}_3$.

13.12 If m and n are relatively prime, show that $\mathbf{Z}_m \oplus \mathbf{Z}_n \simeq \mathbf{Z}_{mn}$ (see Exercise 13.7).

13.13 Prove Proposition 13.42.

38. M. W. Hirsch and Stephen Smale, *Differential Equations, Dynamical Systems, and Linear Algebra*, Pure and Appl. Math., Vol. 60. Academic Press, New York-London, 1974.
39. J. M. Hyde, M. K. Miller, M. G. Hetherington, A. Cerezo, G. D. W. Smith, and C. M. Elliott, Spinodal decomposition in Fe-Cr alloys: Experimental study at the atomic level and comparison with computer models, II, III, Acta Metallurgica et Materialia, **43** (1995), 3403–3413; 3415–3426.
40. C. S. Illiopoulos, Worst case complexity bounds on algorithms for computing the canonical structure of finite abelian groups and Hermite and Smith normal form of an integer matrix, SIAM J. Computing 18 (1989), 658–669.
41. N. Jacobson, *Basic Algebra I and II*, W.H. Freedman & Co., San Francisco, 1974 (I), 1980 (II).
42. T. Kaczynski, Recursive coboundary formula for cycles in acyclic chain complexes, Top. Meth. Nonlin. Anal., 18(2) 2002, 351–372.
43. T. Kaczynski. K. Mischaikow, and M. Mrozek, Computing homology, Homotopy, Homology and Applications, 5(2) 2003, 233-256.
44. T. Kaczynski and M. Mrozek, Conley index for discrete multivalued dynamical systems, Topology & Its Appl., 65 (1995), 83–96.
45. T. Kaczynski, M. Mrozek, and M. Ślusarek, Homology computation by reduction of chain complexes, Computers and Math. Appl., 35 (1998), 59–70.
46. W. Kalies, K. Mischaikow, and G. Watson, Cubical approximation and computation of homology, *Conley Index Theory* (eds. K. Mischaikow, M. Mrozek, and P. Zgliczyński), Banach Center Publications 47, Warsaw (1999), 115–131.
47. S. Lang, *Algebra* (3d ed.), GTM Vol. 211, Springer-Verlag, New York, 2002.
48. S. Lefschetz, Continuous transformations on manifolds, Proc. NAS USA, 9 (1923), 90–93.
49. N. G. Lloyd, *Degree Theory*, Cambridge Tracts in Math. 73, Cambridge Univ. Press, 1978.
50. S. Mac Lane, *Homology*, Springer-Verlag, New York, 1963.
51. W. S. Massey, *A Basic Course in Algebraic Topology*, Springer-Verlag, New York, 1991.
52. M. Mazur and J. Szybowski, Algorytm Allili-Kaczynskiego Konstrukcji Homomorfizmu Łańcuchowego Generowanego Przez Odwzorowanie Wielowartościowe Reprezentowalne, M.Sc. Diss. (in polish), Comp. Sci. Inst., Jagellonian Univ. Krakow, 1999.
53. M. K. Miller, J. M. Hyde, M. G. Hetherington, A. Cerezo, G. D. W. Smith, and C. M. Elliott, Spinodal decomposition in Fe-Cr alloys: Experimental study at the atomic level and comparison with computer models — I. Introduction and methodology, Acta Metallurgica et Materialia, **43** (1995), 3385–3401.
54. J. Milnor, *Morse Theory*, Princeton University Press, Princeton, NJ, 1963.
55. K. Mischaikow, Topological techniques for efficient rigorous computation in dynamics, Acta Numerica, **11** Cambridge University Press, 2002, 435–478.
56. K. Mischaikow and M. Mrozek, Chaos in Lorenz equations: A computer assisted proof, Bull. Amer. Math. Soc. (N.S.), 33 (1995), 66–72.
57. K. Mischaikow and M. Mrozek, *Conley Index*, Handbook of Dynamical Systems (ed. B. Fiedler), North-Holland, 2002 393–460.
58. K. Mischaikow, M. Mrozek, and Paweł Pilarczyk, Graph approach to the computation of the homology of continuous maps, to appear.
59. Ramon E, Moore, *Interval Analysis*, Prentice-Hall, Inc., Englewood Cliffs, NJ, 1966.

60. M. Mrozek, Index pairs and the fixed point index for semidynamical systems with discrete time, Fundamenta Mathematicae, 133 (1989), 179–194.
61. M. Mrozek, Leray functor and the cohomological Conley index for discrete time dynamical systems, Trans. Amer. Math. Soc., 318(1), (1990), 149–178.
62. M. Mrozek, Open index pairs, the fixed point index and rationality of zeta functions, Erg. Th. and Dyn. Syst., 10 (1990), 555–564.
63. M. Mrozek, Shape index and other indices of Conley type for continuous maps in locally compact metric spaces, Fundamenta Mathematicae, 145 (1994), 15–37.
64. M. Mrozek, Topological invariants, multivalued maps and computer assisted proofs, Computers & Mathematics, 32 (1996), 83–104.
65. M. Mrozek, An algorithmic approach to the Conley index theory, J. Dynam. Diff. Equ. 11(4) (1999), 711–734.
66. M. Mrozek, Index pairs algorithms, to appear.
67. M. Mrozek, Čech Type Approach to Computing Homology of Maps, to appear.
68. J. R. Munkres, *Topology: A first course*, Prentice-Hall, Inc. Englewood Cliffs, NJ, 1975.
69. J. R. Munkres, *Elements of Algebraic Topology*, Addison-Wesley, Reading, MA, 1984.
70. J. D. Murray, *Mathematical Biology*, Springer, New York, 1993.
71. Y. Nishiura, *Far-from Equilibrium Dynamics*, Translation of Mathematical Monographs **209**, AMS, Providence, RI, 2002.
72. P. Pilarczyk, Algorithmic homology computation over a Euclidean domain, preprint Comp. Sci. Inst. Jagellonian Univ., Kraków, 2000.
73. C. Robinson, *Dynamical Systems: Stability, Symbolic Dynamics, and Chaos*, CRC Press, Boca Raton, FL, 1995.
74. J. J. Rotman, *An Introduction to Algebraic Topology*, G.T.M. Vol. 119, Springer-Verlag, New York, 1988.
75. J. Robbin and D. Salamon, *Dynamical systems, shape theory, and the Conley index*, Ergodic Theory and Dynamical Systems, 8* (1988), 375–393.
76. W. Rudin, *Principles of Mathematical Analysis*, 3d ed., McGraw Hill Book Co., New York, 1976.
77. H.W. Sieberg and G. Skordev, Fixed point index and chain approximations, Pacific J. Math., 102 (1982), 455–485.
78. E. Spanier, Algebraic Topology, McGraw-Hill, New York, 1966.
79. G. Strang, *Linear Algebra and Its Applications*, Second ed., Academic Press, New York 1980.
80. A. Szymczak, The Conley index for discrete semidynamical systems, Topology and Its Applications 66 (1995), 215–240.
81. A. Szymczak, A combinatorial procedure for finding isolating neighbourhoods and index pairs, Proc. Roy. Soc. Edinburgh Sect. A 127 (5) (1997), 1075–1088.
82. A. Szymczak, The Conley index for decompositions of isolated invariant sets, Fundamenta Mathematicae 148 (1995), 71–90.
83. A. Szymczak, *Index pairs: From dynamics to combinatorics and back*, Ph.D. thesis, Georgia Inst. Tech., Atlanta, 1999.
84. C. Uras and A. Verri, On the recognition of the alphabet of the sign language through size functions. In *Proc. 12th IAPR Intl. Conf. on Pattern Recognition*, Jerusalem, Vol II Conf. B. 334–338, IEEE Comp. Soc. Press, Los Alamitos, 1994.

85. T. Ważewski, Une méthode topologique de l'examen du phénomène asymptotique relativement aux équations différentielles ordinaires, Rend. Accad. Nazionale dei Lincei, Cl. Sci. fisiche, mat. e naturali, Ser. VIII, III (1947), 210–215.

Symbol Index

Ab, 389
A_f, 449
$\alpha = (\alpha_1, \ldots, \alpha_d)$, 222
$\alpha(\gamma_x, f)$, 311
$\alpha(x, \varphi)$, 319
$a + H$, 424
$A[k : l, k' : l']$, 104
A^T, 100

B_0, B_1, B_2, 399
B_0^d, S_0^{d-1}, 399
B_k, 85
$B_k(X, A)$, 281
$\langle a \rangle$, 423
$[a]$, 424
$B(x, r)$, 399
$\bar{B}_0$, 403

$\langle c_1, c_2 \rangle$, 49
C, 421
$\mathcal{C}$, 85
card (S), 45
Cat, 389
$cc_X(x)$, 66
$(\mathcal{C}, \partial)$, 162
ceil (x), 44
$\mathcal{C}^f$, 169
ch (A), 202
Chest $_k(c)$, 247
$G \circ F$, 215
C_k, 48, 85
$C_k(X)$, 53
cl A, 403
$C_n(\mathcal{S}; \mathbf{Z}_2)$, 381

col$(\mathcal{X})$, 344
conv A, 378
CSS, 391
Cub, 389
$\mathcal{C}(X)$, 59
$C_k(X, A)$, 280

D_k, 151
$\deg(f)$, 333
$\deg(f, U)$, 337
∂, 54, 85, 381
Δ^d, 386
$Q(\Delta, k)$, 79
∂_k^X, 59
$\partial^{(X,A)}$, 281
diam A, 220
dim Q, 41
dist, 398
dist $_0$, dist $_1$, dist $_2$, 398
$x \cdot t, x \cdot [0, t]$, 320
$\dot{x}$, 318
$\widehat{P} \diamond \widehat{Q}, c_1 \diamond c_2$, 51

E^α, 273
$E_{i,j}, E_i, E_{i,j,q}$, 100
emb Q, 41
$\epsilon : C_0(X) \to \mathbf{Z}$, 70
$E(X)$, 331

$F(A)$, 207
f^α, 224
$\bar{f}_*$, 287
F : Cat $\to$ Cat$'$, 390
$\lfloor \mathcal{F} \rfloor, \lceil \mathcal{F} \rceil$, 206

Symbol Index

$\bar{f}(Q)$, 237
F^{-1}, 207
$F: X \rightrightarrows Y$, 179
$\mathcal{F}: \mathcal{K}_{\max}(X) \rightrightarrows \mathcal{K}(Y)$, 178, 206
floor (x), 44
$\mathcal{F}^{-1}$, 348
f_{P*}, 355
F_*, 214
f_*, 227
F^{*-1}, 207
F_{*k}, 214
f_{*k}, 227

Γ^1, 19
Γ^{d-1}, 406
Γ_Q, 347
γ_x, 308
G/H, 126, 425
$(G, +)$, 420
graph(F), 209

$\hat{Q}$, 48
$\widehat{s_i}$, 445
$H_k(G; \mathbf{Z}_2)$, 29
$H_k(X), H_*(X)$, 60
$H_k(X, A)$, 281
Hom, 389

id, 408
$I_i(Q)$, 41
im f, 434
Inv (N, f), 315, 346
Inv $(\mathcal{N}, \mathcal{F})$, 348
Inv (N, φ), 319
ι_*, 354
$\cong$, 75
i_*, 287

$\mathcal{K}, \mathcal{K}^d$, 40
$\mathcal{K}(X), \mathcal{K}_k(X), \mathcal{K}_k^d(X)$, 43
$\mathcal{K}_k^d$, 41
ker f, 434
$\mathcal{K}_{\max}$, 71, 343
(K, s, X), 390

Λ^α, 222
$P \prec Q$, 208
$L(f)$, 326

M_f, 216

$M(j)$, 169
$\mathrm{M}_{m,n}(F)$, 438
Mor, 389

N, 346
$n(\alpha, \beta)$, 273
$\mathbf{N}$, 421
$\mathcal{N}^i$, 364
$\| x \|$, 398
$\| x \|_0, \| x \|_1, \| x \|_2$, 397
$\mathcal{N}$, 348, 350

oh (A), 202
$\overset{\circ}{I}, \overset{\circ}{Q}$, 44
$\overset{\circ}{Q}$, 406
Obj, 389
Ω_X^α, 223
$\omega(x, f)$, 311
$\omega(x, \varphi)$, 319
$A \oplus B$, 426

$P = (P_1, P_0)$, 346
(X, A), 280
φ, 144
φ_*, 145
$\varphi(t, x)$, 318
π, π_q, 159
Π_A, 372
Pol, 390
Pre (z), 244

$\mathbf{Q}$, 421

$\mathbf{R}$, 397, 421, 429
$\mathbf{R}^d$, 397
$\mathcal{R}^d$, 199
Rep (K), 391
Repr, 390
Rib (V_1), 245
$\mathbf{R}^n$, 429

S_0, S_1, S_2, 399
Set, 389
Σ_n, 309
$\sigma: \Sigma_n \to \Sigma_n$, 309
Σ_n^+, 310
Σ_B, 373
$X \sim Y$, 231
$b \sim a$, 425

$f \sim g$, 231
$x \stackrel{\beta}{\sim} y$, 273
$\sim_X$, 67
$\mathcal{S}$, 379

$T(G)$, 423
Θ, 373
$\tilde{H}_k(X)$, 89
$\tilde{\sigma}$, 384
$\tilde{C}_k(X)$, 89
Top, 389
tr A, 326
$\mathcal{T}$, 402

$\overrightarrow{[k,l]}$, 245

W, W^0, W^-, 320
W_k, 305
wrap($\mathcal{X}$), wrap(A), 344

X^α, 222
$\xi_V(v_j)$, 94
$(x_n)_{n \in N}$, 415
$\mathcal{X}$, 206, 257
$|\mathcal{X}|$, 206, 257, 344

Z, 421
Z_k, 85
$Z_k(X)$, 60
$Z_k(X, A)$, 281
$\mathrm{M}_{m,n}(\mathbf{Z})$, 447
$\mathbf{Z}^+$, 421

Subject Index

Abelian group, 420
  free, 441
  generated by, 445
abs, 453
`acyclityTest`, 240
Algorithm, 451
  CCR, 165, 170
  cubical homology, 138
  homology of a map, 243
  index pair, 350
  linear equation solver, 124
  preboundary
    algebraic, 140
    cubical, 252
  quotient group finder, 128
  reduction of a pair, 170
  Smith normal form, 122
and, 454
Array, 455
array, 455
Automorphism, 435
average, 451

Ball, 399
  closed, 403
Barycentric
  coordinates, 379
  subdivision, 385
Basis, 431
  canonical, 431, 445, 449
  of a group, 441
`basis`, 242
Betti number, 93

Bijection, 409
Bolzano–Weierstrass theorem, 416
`bool`, 454
Borsuk theorem, 324
Boundary
  f-boundary, 346
  group of boundaries, 85
  map, 54
    simplicial, 381
  of a chain, 23, 60
  operator, 21, 85
    cubical, 54
    simplicial, 384
  relative, 281
  topological, 405
  weak, 305, 328
`boundaryMap`, 134, 169, 241
`boundaryOperator`, 137
`boundaryOperatorMatrix`, 137
Branching, 459
break, 460
Brouwer fixed-point theorem, 324

Cahn–Hilliard equation, 10, 259
`canonicalCoordinates`, 134
Category, 389
  of cubical sets, 389
  of representations of cubical
    polyhedra, 390
  small, 390
Ceiling, 44
Cell
  elementary, 44
Chain, 27

elementary, 47
equivalence, 153
group, 48, 53
homotopy, 151
isomorphism, 144
map, 144
relative, 280
selector, 185, 212
selector theorem, 210
subcomplex, 87
chain, 134, 169, 252
Chain complex
 abstract, 85
 augmented, 89
 cubical, 59
 finitely generated free, 86
 free, 85
 relative, 281
chainComplex, 169
chainFromCanonicalCoordinates, 135
chainMap, 241
chainSelector, 241
Chaotic dynamics, 15
checkForDivisibility, 120
chest, 252
Collapse
 elementary, 71
Collar, 344
columnAdd, 104
columnAddOperation, 106
columnEchelon, 117
columnExchange, 104
columnExchangeOperation, 106
columnMultiply, 104
columnMultiplyOperation, 106
combinatorialEnclosure, 345
Commutative diagram, 145
Component
 nondegenerate, 41
Conjugacy
 semiconjugacy, 358
 topological, 358
Conley index, 357
Connected
 simple system, 391
 component, 66
Contractible, 232
Convex
 hull, 378

set, 378
Coordinate
 isomorphism, 94
 vector, 431
coordinates, 242
Coset, 424
 of a subgroup, 424
Covering, 418
Cube
 elementary, 40, 154
 dual, 48
 unitary, 45
cube, 134
Cubical
 acyclic set, 79
 carrier, 216
 pair, 280
 product, 51
 set, 43
 wrap, 344
cubicalChainComplex, 137
cubicalChainGroups, 136
cubicalSet, 134, 345
cubicalWrap, 345
cuPreBoundary, 252
cutFirst, 454
Cycle, 22, 60
 group of cycles, 85
 reduced, 89
 relative, 281

Darboux theorem, 412
Data
 structure, 452
 type, 452
Decomposition
 disjoint, 364
 normal, 131
defined, 459
Deformation
 retract, 76
 retraction, 76, 232
degenerateDimensions, 237, 240
Degree
 of a map, 333
 topological, 337
dim, 140
Dimension
 of a cube, 41

of a vector space, 431
Direct sum, 426
divides, 454
Dynamical system, 308
  invertible, noninvertible, 308

Echelon form
  column, 108
  row, 107
Edge, 17, 43
  connected, 67
  path, 67
Elementary
  cell, 179
`elementaryCubes`, 238
Embedding number, 41
Enclosure
  combinatorial, 344, 345
Endomorphism, 435
`endpoint`, 134, 453
Epimorphism, 290, 434
Equivalence
  class, 425
  relation, 425
Euler number, 331
`evaluate`, 236, 241
`evaluateInverse`, 350
Excision isomorphism theorem, 287
Existence and uniqueness theorem, 318

Face
  free, 71
  maximal, 71
Field, 428, 429
firstIndex, 455
FitzHugh-Nagumo equations, 267
Floor, 44
floor, 453
Flow, 318
Full
  cubical set, 344
  solution, 308, 347
Function
  between pairs, 286
  continuous, 407
  lower semicontinuous, 408
  measuring, 271
  overloading, 461
  rational, 174

  size, 271, 273, 277, 282
  uniformly continuous, 417
  upper semicontinuous, 408
Functor, 390
  homology, 390, 392
Fundamental decomposition theorem, 94, 130

Gaussian elimination, 107
`generator`, 169
`generatorsOfHomology`, 138
Graph, 17
  combinatorial, 18
    equivalent, 46
  connected, 18
  cubical transition, 365
  of a cubical map, 209
Grey-scale, 6
Group
  cyclic, 423
  of weak boundaries, 305, 328

Hash, 458
  element, 458
  key, 458
  value, 458
hash, 458
Hénon map, 362
Homeomorphism, 409
Homology
  complex, 86
  cubical, 60
    group, 60
  functor, 390, 392
  map, 214
  of a chain complex, 86
  reduced, 89
  relative, 281
  simplicial, 384
  with $\mathbf{Z}_2$ coefficients, 29, 382
`homology`, 138
`homologyGroupOfChainComplex`, 133
`homologyOfMap`, 243
Homomorphism, 433
  connecting, 293
Homotopy
  class, 234
  invariance theorem, 233
  of maps, 231

478 Subject Index

of spaces, 231
on pairs, 338
type, 231
Hopf trace theorem, 328
Hull
  closed, 202
  open, 202

identityMatrix, 457
Image, 434
  binary, 8
  grey-scale, 8
in, 457
Index
  map, 355
  pair, 346
invariantPart, 350
Injection, 409
int, 453
Interface, 262
Intermediate value theorem, 412
intersection, 457
Interval
  elementary, 40
    degenerate, 40
    nondegenerate, 40
interval, 134
Interval arithmetic, 174, 176
Invariant
  isolated set, 346
  maximal set, 315, 319, 348
  subset, 309, 319
invariantPart, 349
Isolating neighborhood, 346, 350
Isomorphism
  coordinate, 94
  of groups, 434

join, 457

Kernel, 434
kernelImage, 116
keys, 458
Keyword, 452
Klein bottle, 384

lastIndex, 455
Lefschetz
  number, 326

Lefschetz fixed-point theorem, 327
  relative, 363
left, 453
Length, 454
length, 454
Level set, 273
Limit
  of a sequence, 415
Limit set
  omega, alpha, 311, 319
List, 454
list, 454
Literal, 452

Map
  antipodal, 335
  bilinear, 50
  boundary
    cubical, 54
  chain, 144
  cubical, 179, 206
    acyclic-valued, 210
    combinatorial, 178, 206, 343
  Hénon, 362
  inclusion, 145
  index, 355
  induced in homology, 214
  linear, 437
  logistic, 15, 358
  multivalued, 178, 206
    lower semicontinuous, 207
    upper semicontinuous, 207
  projection, 147
  shift, 309
Matrix, 438
  Z-invertible, 448
  canonical, 449
  change of coordinates, 95
  elementary, 100
  integer, 447
  nonsingular, 439
  of f in bases, 448
  transition, 315, 360, 373
  transpose, 100, 439
matrix
  submatrix, 103
matrix, 103, 456
Mayer–Vietoris sequence, 299
Metallurgy, 9

Metric, 398
    space, 398
minNonzero, 120
minRectangle, 252
Modulo 2, 21
Monomorphism, 290, 434
Morphism, 389
Morse theory, 277
moveMinNonzero, 120
MRI, 259
multivaluedMap, 241
multivaluedRepresentation, 239, 244

Neighbor, 237
neighbors, 237
next, 460
Nilpotent, 355
Norm, 398
    Euclidean, 397
    sum, 398
    supremum, 397
not, 454

Object, 389
of, 457
or, 454
Orbit
    backward, 308
    forward, 308
    heteroclinic, 310
    of a flow, 318
    of a point, 308
    periodic, 14, 311
Order
    of a group, 423
    of an element, 423
Orientation
    of vertices, 382
Outer approximation, 180

partColumnReduce, 111
partRowReduce, 110
partSmithForm, 121
Path, 17
Period
    minimal, 14
Phase space, 308
Pivot position, 107
Pixel, 6

Poincaré–Brouwer theorem, 335
Point
    equilibrium, 319
    fixed, 310
    heteroclinic, 310
    homoclinic, 310
    periodic, 311, 319
Polyhedron, 379
    geometric, 413
    topological, 390
polynomialMap, 236, 239
Polytope, 379
Preboundary, 140, 244
preBoundary, 140
Preimage, 207, 407
    weak, 207
primaryFaces, 135
Projection, 149
projection, 252
Projective plane, 384

Quotient
    canonical homomorphism, 435
    group, 126, 425
    isomorphism, 436
quotientGroup, 128

Range, 434
Rank, 93, 126, 443
rationalMap, 236
Rectangle, 83, 413
rectangle, 236, 252
rectangleIntersection, 238
rectangularCover, 238
reduce, 170
reduceChainComplex, 170
Reduction
    elementary, 155
    pair, 159
Reference point, 271
remove, 458
Representable
    set, 199
    space, 390
Representant, 391
Representation, 180
    acyclic-valued, 219
    cubical, 215
    minimal, 216

triple, 390
Rescaling, 182, 222
restrictedMap, 350
Retract, 77
Retraction, 77, 232
return, 453
right, 453
Root, 336
rowAdd, 104
rowAddOperation, 106
rowEchelon, 114
rowExchange, 104
rowExchangeOperation, 105
rowMultiply, 104
rowMultiplyOperation, 105
rowPrepare, 112
rowReduce, 113

Scalar product
  of chains, 49
Scaling, 181, 222
Sea of Tranquillity, 6
selectVertex, 241
Sequence
  exact, 289
    pair, 297
  Mayer–Vietoris
    relative, 303
  of points, 415
  short exact, 290
    chain complexes, 292
    pair, 291
    splits, 291
set, 457
setminus, 457
Shift
  equivalent, 356, 357
  map, 309
  subshift, 315, 373
Sign laguage
  alphabet, 271
Simplex
  geometric, 379
  ordered, 382
  standard, 386
Simplicial complex, 379
Size function, 282
smallestNonzero, 112
Smith normal form, 123, 124

smithForm, 122
Solve, 124
Space
  compact, 416
  connected, 411
  path-connected, 412
  topological, 402
Sphere, 399
Spinodal Decomposition, 262
Spiral waves, 267
Star-shaped, 234
Star-shaped set, 83, 220
String, 454
Submap, 214
subset, 457
subsets, 457
Surjection, 409
Symbol space, 309

Tetral, 12
Thresholding, 8
Topological invariance theorem, 230
Topology, 402
  standard, 402
  subspace, 404
Torsion
  coefficients, 93
  subgroup, 304, 423
Torus, 380
Trace, 326
transpose, 457
Tree, 18
Triangulation, 379
typedef, 456

union, 457

Variable, 451
vector, 456
Vector space, 429
Vertex, 17, 43
  free, 19
Voxel, 12, 259

Ważewski
  principle, 321
  set, 320

Zig-zag lemma, 292
$\mathbf{Z}$-invertible, 97

Applied Mathematical Sciences

(continued from page ii)

60. *Ghil/Childress:* Topics in Geophysical Dynamics: Atmospheric Dynamics, Dynamo Theory and Climate Dynamics.
61. *Sattinger/Weaver:* Lie Groups and Algebras with Applications to Physics, Geometry, and Mechanics.
62. *LaSalle:* The Stability and Control of Discrete Processes.
63. *Grasman:* Asymptotic Methods of Relaxation Oscillations and Applications.
64. *Hsu:* Cell-to-Cell Mapping: A Method of Global Analysis for Nonlinear Systems.
65. *Rand/Armbruster:* Perturbation Methods, Bifurcation Theory and Computer Algebra.
66. *Hlaváček/Haslinger/Necasl/Lovísek:* Solution of Variational Inequalities in Mechanics.
67. *Cercignani:* The Boltzmann Equation and Its Applications.
68. *Temam:* Infinite-Dimensional Dynamical Systems in Mechanics and Physics, 2nd ed.
69. *Golubitsky/Stewart/Schaeffer:* Singularities and Groups in Bifurcation Theory, Vol. II.
70. *Constantin/Foias/Nicolaenko/Temam:* Integral Manifolds and Inertial Manifolds for Dissipative Partial Differential Equations.
71. *Catlin:* Estimation, Control, and the Discrete Kalman Filter.
72. *Lochak/Meunier:* Multiphase Averaging for Classical Systems.
73. *Wiggins:* Global Bifurcations and Chaos.
74. *Mawhin/Willem:* Critical Point Theory and Hamiltonian Systems.
75. *Abraham/Marsden/Ratiu:* Manifolds, Tensor Analysis, and Applications, 2nd ed.
76. *Lagerstrom:* Matched Asymptotic Expansions: Ideas and Techniques.
77. *Aldous:* Probability Approximations via the Poisson Clumping Heuristic.
78. *Dacorogna:* Direct Methods in the Calculus of Variations.
79. *Hernández-Lerma:* Adaptive Markov Processes.
80. *Lawden:* Elliptic Functions and Applications.
81. *Bluman/Kumei:* Symmetries and Differential Equations.
82. *Kress:* Linear Integral Equations, 2nd ed.
83. *Bebernes/Eberly:* Mathematical Problems from Combustion Theory.
84. *Joseph:* Fluid Dynamics of Viscoelastic Fluids.
85. *Yang:* Wave Packets and Their Bifurcations in Geophysical Fluid Dynamics.
86. *Dendrinos/Sonis:* Chaos and Socio-Spatial Dynamics.
87. *Weder:* Spectral and Scattering Theory for Wave Propagation in Perturbed Stratified Media.
88. *Bogaevski/Povzner:* Algebraic Methods in Nonlinear Perturbation Theory.
89. *O'Malley:* Singular Perturbation Methods for Ordinary Differential Equations.
90. *Meyer/Hall:* Introduction to Hamiltonian Dynamical Systems and the N-body Problem.
91. *Straughan:* The Energy Method, Stability, and Nonlinear Convection, 2nd ed.
92. *Naber:* The Geometry of Minkowski Spacetime.
93. *Colton/Kress:* Inverse Acoustic and Electromagnetic Scattering Theory, 2nd ed.
94. *Hoppensteadt:* Analysis and Simulation of Chaotic Systems, 2nd ed.
95. *Hackbusch:* Iterative Solution of Large Sparse Systems of Equations.
96. *Marchioro/Pulvirenti:* Mathematical Theory of Incompressible Nonviscous Fluids.
97. *Lasota/Mackey:* Chaos, Fractals, and Noise: Stochastic Aspects of Dynamics, 2nd ed.
98. *de Boor/Höllig/Riemenschneider:* Box Splines.
99. *Hale/Lunel:* Introduction to Functional Differential Equations.
100. *Sirovich (ed):* Trends and Perspectives in Applied Mathematics.
101. *Nusse/Yorke:* Dynamics: Numerical Explorations, 2nd ed.
102. *Chossat/Iooss:* The Couette-Taylor Problem.
103. *Chorin:* Vorticity and Turbulence.
104. *Farkas:* Periodic Motions.
105. *Wiggins:* Normally Hyperbolic Invariant Manifolds in Dynamical Systems.
106. *Cercignani/Illner/Pulvirenti:* The Mathematical Theory of Dilute Gases.
107. *Antman:* Nonlinear Problems of Elasticity.
108. *Zeidler:* Applied Functional Analysis: Applications to Mathematical Physics.
109. *Zeidler:* Applied Functional Analysis: Main Principles and Their Applications.
110. *Diekmann/van Gils/Verduyn Lunel/Walther:* Delay Equations: Functional-, Complex-, and Nonlinear Analysis.
111. *Visintin:* Differential Models of Hysteresis.
112. *Kuznetsov:* Elements of Applied Bifurcation Theory, 2nd ed.
113. *Hislop/Sigal:* Introduction to Spectral Theory: With Applications to Schrödinger Operators.
114. *Kevorkian/Cole:* Multiple Scale and Singular Perturbation Methods.
115. *Taylor:* Partial Differential Equations I, Basic Theory.
116. *Taylor:* Partial Differential Equations II, Qualitative Studies of Linear Equations.

(continued on next page)

Applied Mathematical Sciences

(continued from previous page)

117. *Taylor:* Partial Differential Equations III, Nonlinear Equations.
118. *Godlewski/Raviart:* Numerical Approximation of Hyperbolic Systems of Conservation Laws.
119. *Wu:* Theory and Applications of Partial Functional Differential Equations.
120. *Kirsch:* An Introduction to the Mathematical Theory of Inverse Problems.
121. *Brokate/Sprekels:* Hysteresis and Phase Transitions.
122. *Gliklikh:* Global Analysis in Mathematical Physics: Geometric and Stochastic Methods.
123. *Le/Schmitt:* Global Bifurcation in Variational Inequalities: Applications to Obstacle and Unilateral Problems.
124. *Polak:* Optimization: Algorithms and Consistent Approximations.
125. *Arnold/Khesin:* Topological Methods in Hydrodynamics.
126. *Hoppensteadt/Izhikevich:* Weakly Connected Neural Networks.
127. *Isakov:* Inverse Problems for Partial Differential Equations.
128. *Li/Wiggins:* Invariant Manifolds and Fibrations for Perturbed Nonlinear Schrödinger Equations.
129. *Müller:* Analysis of Spherical Symmetries in Euclidean Spaces.
130. *Feintuch:* Robust Control Theory in Hilbert Space.
131. *Ericksen:* Introduction to the Thermodynamics of Solids, Revised ed.
132. *Ihlenburg:* Finite Element Analysis of Acoustic Scattering.
133. *Vorovich:* Nonlinear Theory of Shallow Shells.
134. *Vein/Dale:* Determinants and Their Applications in Mathematical Physics.
135. *Drew/Passman:* Theory of Multicomponent Fluids.
136. *Cioranescu/Saint Jean Paulin:* Homogenization of Reticulated Structures.
137. *Gurtin:* Configurational Forces as Basic Concepts of Continuum Physics.
138. *Haller:* Chaos Near Resonance.
139. *Sulem/Sulem:* The Nonlinear Schrödinger Equation: Self-Focusing and Wave Collapse.
140. *Cherkaev:* Variational Methods for Structural Optimization.
141. *Naber:* Topology, Geometry, and Gauge Fields: Interactions.
142. *Schmid/Henningson:* Stability and Transition in Shear Flows.
143. *Sell/You:* Dynamics of Evolutionary Equations.
144. *Nédélec:* Acoustic and Electromagnetic Equations: Integral Representations for Harmonic Problems.
145. *Newton:* The N-Vortex Problem: Analytical Techniques.
146. *Allaire:* Shape Optimization by the Homogenization Method.
147. *Aubert/Kornprobst:* Mathematical Problems in Image Processing: Partial Differential Equations and the Calculus of Variations.
148. *Peyret:* Spectral Methods for Incompressible Viscous Flow.
149. *Ikeda/Murota:* Imperfect Bifurcation in Structures and Materials: Engineering Use of Group-Theoretic Bifucation Theory.
150. *Skorokhod/Hoppensteadt/Salehi:* Random Perturbation Methods with Applications in Science and Engineering.
151. *Bensoussan/Frehse:* Regularity Results for Nonlinear Elliptic Systems and Applications.
152. *Holden/Risebro:* Front Tracking for Hyperbolic Conservation Laws.
153. *Osher/Fedkiw:* Level Set Methods and Dynamic Implicit Surfaces.
154. *Bluman/Anco:* Symmetry and Integration Methods for Differential Equations.
155. *Chalmond:* Modeling and Inverse Problems in Image Analysis.
156. *Kielhöfer:* Bifurcation Theory: An Introduction with Applications to PDEs.
157. *Kaczynski/Mischaikow/Mrozek:* Computational Homology.
158. *Oertel:* Prandtl-Essentials of Fluid Mechanics, 10th ed.

THE UNI'

VNDERGRAD